DC Distribution Systems and Microgrids

Edited by
Tomislav Dragičević, Pat Wheeler
and Frede Blaabjerg

The Institution of Engineering and Technology

Published by The Institution of Engineering and Technology, London, United Kingdom

The Institution of Engineering and Technology is registered as a Charity in England & Wales (no. 211014) and Scotland (no. SC038698).

First published 2018

The Institution of Engineering and Technology
Michael Faraday House
Six Hills Way, Stevenage
Herts, SG1 2AY, United Kingdom

www.theiet.org

British Library Cataloguing in Publication Data
A catalogue record for this product is available from the British Library

ISBN 978-1-78561-382-1 (hardback)
ISBN 978-1-78561-383-8 (PDF)

Typeset in India by MPS Limited

Contents

About the editors

Tomislav Dragičević is an Associate Professor at Aalborg University, Denmark. His research interests are in modelling, control and design of reliable power electronic based systems. In 2018, he was a visiting professor at University of Nottingham, UK. He serves as an Associate Editor for IEEE Transactions on Industrial Electronics and Journal of Power Electronics. He has been involved in numerous industrial projects involving terrestrial, shipboard and aircraft DC microgrids and has published widely in this area.

Pat Wheeler is a Professor in Power Electronic Systems and Head of the Department of Electrical and Electronic Engineering at the University of Nottingham, UK. In 1996 he became a lecturer in the Power Electronics, Machines and Control Group at the University of Nottingham, becoming a Senior Lecturer in 2003, and a full Professor in January 2008. He is also the Li Dak Sum Chair Professor in Electrical and Aerospace Engineering at the University of Nottingham, Ningbo, China. His research focuses on various aspects of power converter topologies and their applications.

Frede Blaabjerg is Professor and Villum Investigator at Department of Energy Technology of Aalborg University, Denmark. His research encompasses control, power electronics, energy savings, reliability, power quality, solar power and wind power. He has been Visiting Professor at several Chinese institutions and has been a member of many international scientific societies and boards such as the ERC advanced grant panel 2010-2017 and Board of Danish National Advanced Technology Foundation, DK from 2007-2014. He has published widely in the field.

Preface

The current revolution in power electronics has enabled direct-current (DC) distribution architectures. These architectures have a potential for higher efficiency and reliability, better current carrying capacity, and faster response when compared to conventional alternating-current (AC) architectures. DC distribution also has advantages in terms of inexistence of reactive power flows, power quality issues and frequency regulation, resulting in a simpler control architecture and more robust stability when compared to the AC coupled systems. For these reasons, DC is more and more used in the modern society at different voltage and power levels. The most recent emergence of DC is happening at low voltage levels, where practically all modern appliances (e.g., laptops, cellphones, LED lights, displays, etc.), air, land and sea vehicles, residential and commercial facilities, as well as renewable energy plants increasingly rely on the DC power distribution technology. All these emerging applications can be categorized under the same label, i.e., the DC "Microgrid". However, as the utility power system will very likely continue to be operated using the AC architecture, development and accommodation of DC microgrids in the overall power system holds many integration challenges. On the other hand, the adoption of DC microgrids in standalone and transport applications has already begun, but lack of existing standards and protection methodologies creates a specific set of challenges in this area as well. The aim of this book is to overview the technologies that have been developed over the last 10–15 years in the area of DC microgrids.

The book is divided into three parts – control circuits, power architectures, and real-world applications. The first part comprises seven chapters, which provides a comprehensive overview of different control approaches focused both on local control functions of individual converters as discussed in Chapters 1, 3, and 7 (e.g., voltage/current control, DC-bus stabilization, point of load converters support), and on functions that allow their effective coordinated performance as presented in Chapters 2, 4, and 5 (e.g., distributed power sharing/voltage regulation and energy management systems) and safe operation (coordinated protection systems). Except for Chapter 6 that focuses specifically on the control of solid-state transformers, this portion of the book is rather generic and is thereby applicable to any type of the power converter topology and DC microgrid architecture. This part will thus provide essential reading for anyone involved in research in design and implementation of control algorithms.

Building upon the set of control oriented chapters, the three chapters in the second part overview several prominent DC microgrid architectures. Architectures that are different and more complex compared to the most conventional single bus structure were intentionally selected for a detailed overview here, as they can offer additional degrees of flexibility and redundancy. Moreover, due to the rapid development of power electronics technology that has resulted in increased reliability of power electronic devices, additional hardware required to establish such architectures is not considered as a drawback any more. In particular, two architectures overviewed are relatively generic (i.e., the solid-state transformers in Chapter 9, and bipolar-type DC microgrids in Chapter 10), while the architecture detailed in Chapter 8 for electrical vehicle charging infrastructure is more tailored toward that specific application.

The final part, comprised of seven chapters, presents an analysis of a number of applications in which the control methods and architectures discussed in the first two parts can be applied in the specific context of a given application. The selected applications for this book include aircraft and shipboard systems, electrical vehicles, data centers, residential buildings, and photovoltaic powered systems, respectively. The book ends with an overview of some of the DC microgrid demonstration sites around the world.

The Editors would like to thank all the contributors for their excellent work and cooperation in the preparation of this book. Furthermore, we would like to acknowledge the excellent and timely assistance of the editorial and production staff at the Institution of Engineering and Technology (IET).

The Editors
Tomislav Dragičević, Aalborg University, Denmark
Patrick Wheeler, University of Nottingham, UK and China
Frede Blaabjerg, Aalborg University, Denmark

Chapter 1

DC microgrid control principles – hierarchical control diagram

Linglin Chen[1], Tao Yang[1], Fei Gao[1], Serhiy Bozhko[1], and Patrick Wheeler[1]

1.1 Introduction

Different electrical components can be conveniently connected to a DC grid as there are no issues of reactive power, power factor, different frequencies, etc. as in AC grids. Control is one of the key disciplines in achieving the objective of smart grid initiatives. The control of DC microgrids (MGs) covers a wide range of control objectives and responsive dynamics. Therefore, to better understand the concept and implement the MG control, division of the MG control into relatively independent sublevels is essential. The utility AC power system has a history of over a hundred year of development. Due to its unique physical characteristic, MGs have more feasibility of controlling individual distributed generation (DG) subsystems due to their power electronics interfaces. In order to fully exploit the potential of power electronic converters in the MG, a hierarchical control for AC MGs was developed in literature [1].

In the hierarchical structure, MG controls are classified into three groups according to their different functionalities. Generally, in the primary control level, controllers are responsible for the active and reactive power sharing among DGs. In the secondary control level, frequency variation and AC voltage amplitude offset are compensated. In the tertiary control level, power management algorithm is conducted. The recent development of solid-state circuit breakers for the DC network has made the DC MG an alternative option to the AC MGs. Compared with AC networks, the DC network has no reactive power issues and less power cables, which can potentially reduce weight and cost of the power grid. A similar hierarchical control structure can also be implemented in DC MGs as shown in Figure 1.1. The details of each control level will be explained in the following section.

[1]Department of Electrical and Electronics Engineering, The University of Nottingham, United Kingdom

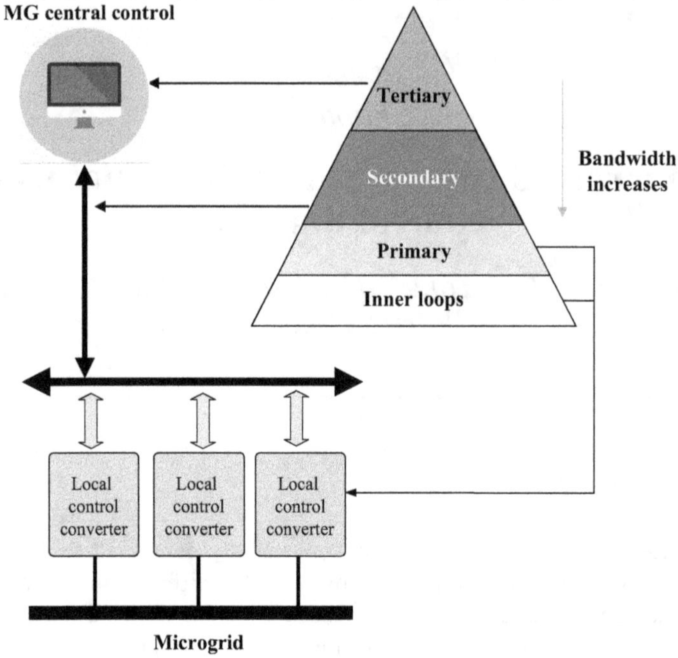

Figure 1.1 Hierarchical control of an MG

1.2 The hierarchical control for DC MGs

In the past few years, different concepts for primary, secondary and tertiary control have been developed for DC MGs applications [1–4]. The explanation of each level can be given as follows:

- The primary control is implemented strictly local within the DG subsystems. In a DC MG, installation of more than one power source is common. Therefore, how to manage the DGs to share the load proportionally to their rated power is an issue to consider. Conventionally, droop control [5] is a preferred option in this control level as it provides autonomous control features to the DGs in the presence of communication failure. It effectively deals with the load sharing among DGs connected on the same DC bus and provides references to sub-control loops (voltage or current loops). Alternatively, the droop control can also be perceived as virtual resistance, connected in series with converter on the DC bus side.
- The secondary control, predominately, sets in an MG central controller which is usually, together with tertiary control level, remotely located to DGs, thus communications are required. It generates voltage shifting signals to primary control level in aim to eliminate the DC voltage deviation caused by primary droop controls. Moreover, it is also possible to implement the secondary control level locally inside DGs.

- The tertiary control takes responsibility of power management between MG and upper grids and is also commonly referred to as the energy management system. In addition, tertiary control level has to prepare dispatch schedules inside MG to achieve the optimum operation [1].

The hierarchical control essentially groups the MG control co-ordinately from the bottom level power sharing among electronics converters to top level power management of DC MGs. The response dynamic ranges from milliseconds to a few days. The primary control guarantees reliable operation of DC MGs in the presence of communication failure, and the tertiary level provides flexibilities and intelligence to power governance through the intermedium secondary control.

1.3 Primary control

The functionality of this control level is to deal with the power sharing among different DG units. Generally, the load sharing can be achieved in both centralized and decentralized ways. In the centralized way [6], usually the DG with the largest power capacity is chosen to be the dominant power source, and it is responsible for the regulation of the DC bus voltage (voltage support mode). Other DGs in the DC MG work in power mode (current mode). They receive power (current) reference signals from a central controller. In this case, a reliable operation of power sharing largely depends on the effective communication. As the control bandwidth of the primary control is the highest among the three control levels (in terms of kHz), communication within this level between DGs also increases infrastructure investment. Thus, for large-scale MGs with remotely located DGs, communication links are not practical in the primary control level, and only local current and voltage information is available. Thus, as a fully decentralized approach, the droop control has shown favourable features in large-scale MGs for load sharing. With the droop control, all measurements take place locally within DGs. The droop control in DC MGs can be intuitively appreciated as 'virtual resistance' [2]. This allows paralleled DGs to share DC bus currents proportionally in line with their respective power rating. Due to its modularity and reliability, the droop control has been widely used in DC MGs in the primary control level [7–9].

1.3.1 Basics of droop control

The original idea of droop control is from AC generators in legacy grid, where the power generation is controlled by mechanical outputs of prime movers. A speed governor measures the variation of a synchronous generator (SG) speed (frequency) and provides the power regulation according to generator's droop characteristic [10]. With such characteristic, SGs on a common AC bus can adjust to load changes according to the variation of their terminal frequencies and voltages. This allows power sharing of different SGs without extra communication links. A similar idea was initially used in paralleled inverters for load sharing [11]. If neglecting the line resistance, the active power (P) delivered from each DG unit is only dependent on frequency (f), and the reactive power (Q) is solely related to voltage (V). Therefore, by adjusting the frequency f and the terminal V, the active power and reactive power from each DG unit can be regulated. These results in the famous droop regulation in an AC grid as

$$f - f_0 = -k_p(P - P_0) \tag{1.1}$$

$$V - V_0 = -k_q(Q - Q_0) \tag{1.2}$$

with measured terminal voltage V and frequency f, the required P and Q can be derived from (1.1) and (1.2). The derived P and Q are then sent to the inverter inner controllers (control level 0), and the DG unit can operate in desired conditions. A similar concept can also be used to parallel DC converters for power sharing. Within DC MGs, there is no reactive power, and the active power is only controlled by terminal DC voltage. Hence, a linear droop characteristic line between V and I can be implemented. As shown in Figure 1.2, the DC-link voltage v_{DC} and the DC current i_{DC} is in a linear relation. With the DC current increasing, the DC-link voltage will be reduced. The relation between DC-link voltage v_{DC} and the DC current i_{DC} can be expressed as

$$v_{DC} = V_0 - R_D i_{DC} \tag{1.3}$$

where i_{DC} is the power converter output DC current; V_0 is the voltage at no load ($i_{DC} = 0$ A); v_{DC} is the DG terminal voltage; the slope R_D is the droop gain and can also be viewed as a virtual resistor, which will be explained later.

To explain the power sharing using DC V–I droop control concept, a simple DC MG system is shown in Figure 1.3. The droop curves for the two DGs are illustrated as in Figure 1.4. The control structure details are depicted in Figure 1.5. DC currents and voltages of DG1 and DG2 are measured locally and fed into their own primary controllers. Based on the measured current i_{DC1} and i_{DC2}, terminal DC voltage references can be calculated accordingly with the V–I droop curves. Eventually at steady state, the DC voltage equals to V_L. Because the droop gains R_{D1} and R_{D2} are designed based on the rated power of the DG, this will be discussed later. As shown in Figure 1.4, DG1 and DG2 spend different 'efforts' to share the load current, with i_{DC1} and i_{DC2}, respectively.

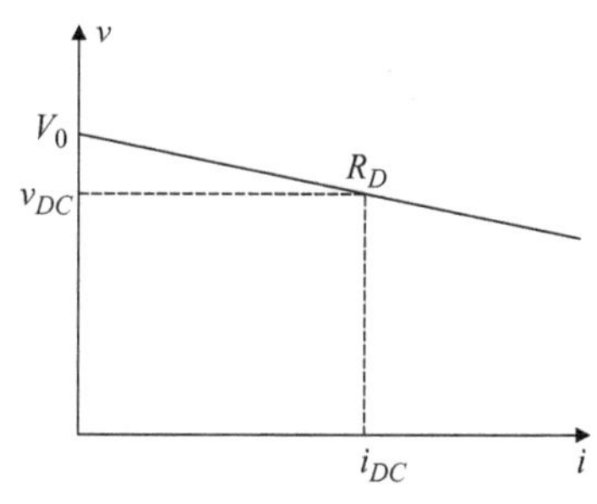

Figure 1.2 V–I droop characteristic

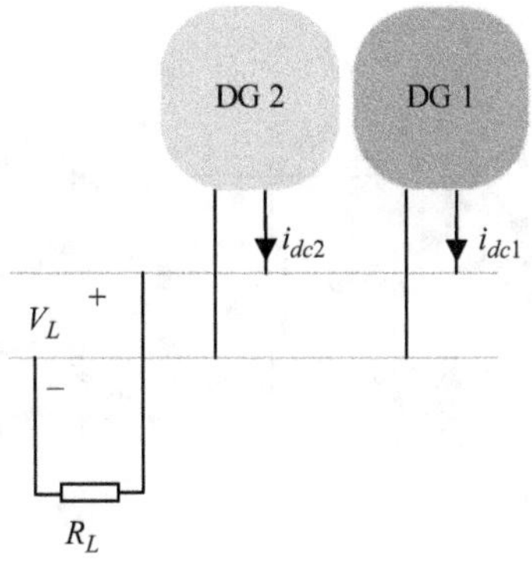

Figure 1.3 A simple MG with two DGs and a resistive load

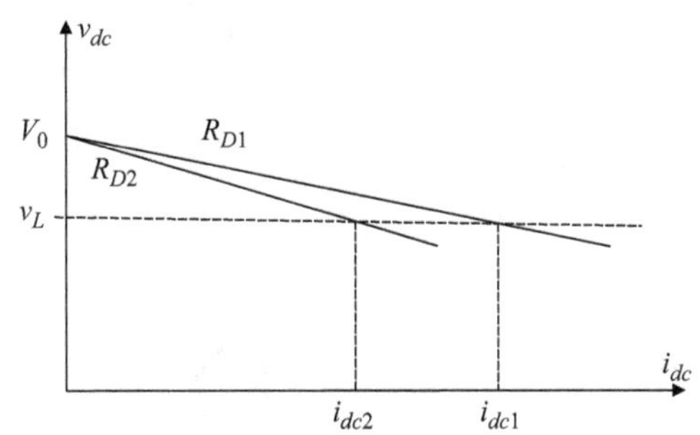

Figure 1.4 Droop curves for DG1 and DG2

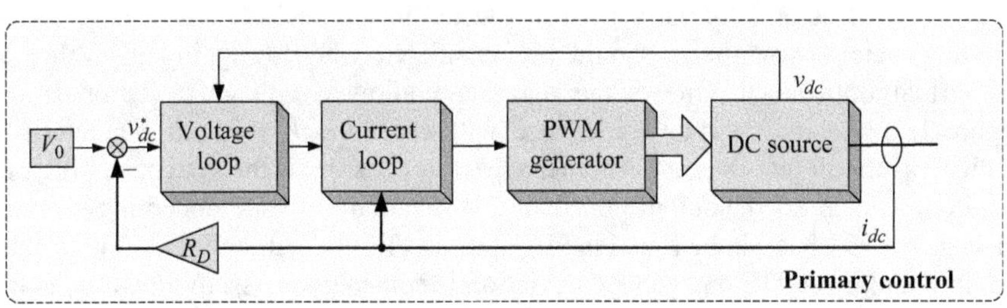

Figure 1.5 Practical primary control implementation with V–I droop method [2]

Usually, the inner voltage and current loops have fast dynamic response. The output impedance of the converter is predominately dependant on the droop gain. Therefore, the *V–I* droop gain is termed also as virtual resistance R_D [5]. Figure 1.6 illustrates a simplified DC system diagram, where V_{o1} and V_{o2} are DC voltages at no load from *V–I* droop controllers for DG1 and DG2, respectively; R_{D1} and R_{D2} are virtual resistance for DG1 and DG2, respectively. If there is no virtual resistance, a small difference between V_{o1} and V_{o2} will result in a large circulating

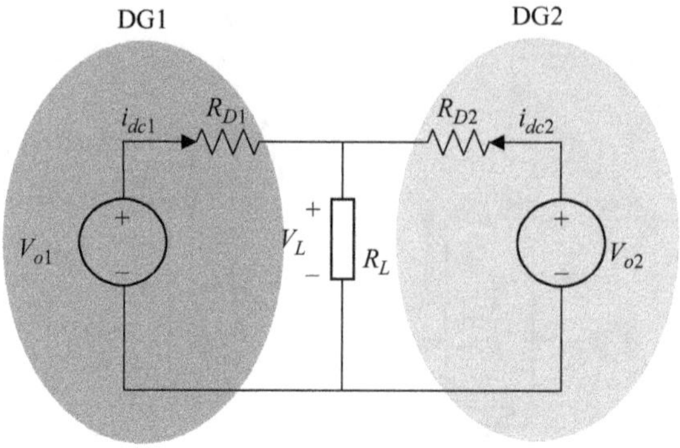

Figure 1.6 The equivalent circuit of two DC power supplies

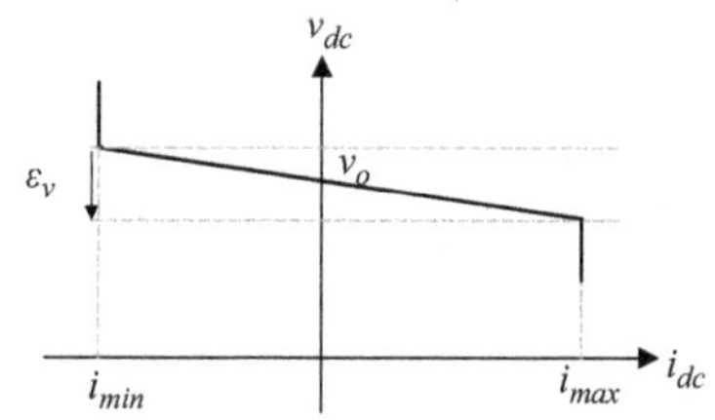

Figure 1.7 An illustration of V–I droop with maximum allowed deviation

current between DG1 and DG2. This is because without virtual resistance R_{D1} and R_{D2}, the circulating current between DG1 and DG2 is only subject to cable resistance which is essentially very small (neglected in Figure 1.6).

In general, two aspects should be considered when designing the value of the virtual impedance. One is the maximum allowed voltage deviation (ε_v in Figure 1.7). A larger virtual impedance will make the *V–I* output characteristic 'softer' (as the slope is larger). Another aspect to consider is the system stability, as the droop gain alters the output impedance. A detailed analysis concerning stability analysis in DC MGs can be found in [5]. Thus, in this chapter, only constraint of the maximum allowed DC bus voltage deviation is considered. Assuming that ε_v is the maximum allowed voltage deviation, the allowed maximum R_D is given as

$$R_D = \frac{\varepsilon_v}{2i_{max}} \tag{1.4}$$

where i_{max} is the maximum output current. The relation of ε_v and i_{max} is illustrated in Figure 1.7.

1.3.2 Power sharing errors

The droop *V–I* droop control provides autonomous local control to DGs at the price of 'soft' *V–I* characteristic (here the term 'soft' is used compared with an ideal stiff voltage source). Therefore, voltage deviation will be expected when different loads are applied to the DC system [7]. Another issue for using the conventional *V–I* droop is that some power sharing error would occur during operations. Two main reasons for this error are discussed below:

1. Error in DC voltage control. In practical situations, due to the measurement error from voltage and current sensors, the DC bus voltage at each DG unit might be different even if the references are set the same. If we consider no load conditions, this control error gives different V_0 for each DG unit (V_{o1} and V_{o2}), as shown in Figure 1.8. The voltage error within DGs will potentially cause large power sharing discrepancies in MG. As can be seen from Figure 1.8, if two DGs are with the same power rating and virtual impedances (i.e. R_D), the current error between two DGs is smaller (I_1-I_2) with a larger droop gain. On the other hand, although a smaller droop gain produces larger current sharing errors ($I_{1o}-I_{2o}$), it means a stiffer voltage source and a smaller voltage deviation due to load changes. This results in a dilemma in designing the droop gain for primary control.
2. Cable resistance: In large-scale DC MGs, DG terminal voltages might be a variant with load power in presence of cable resistance. A simple DC MG which takes accounts of cable resistance is shown in Figure 1.9. The cable resistance between the converter DC terminal and the common DC bus is considered as R_{line1} and R_{line2}.

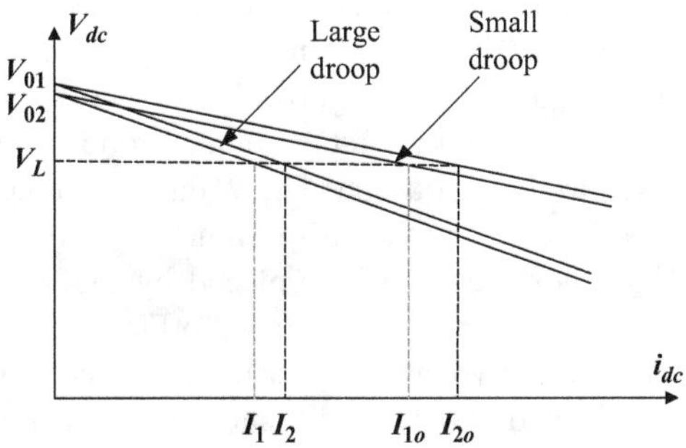

Figure 1.8 Unequal load sharing due to load distribution in DC MG

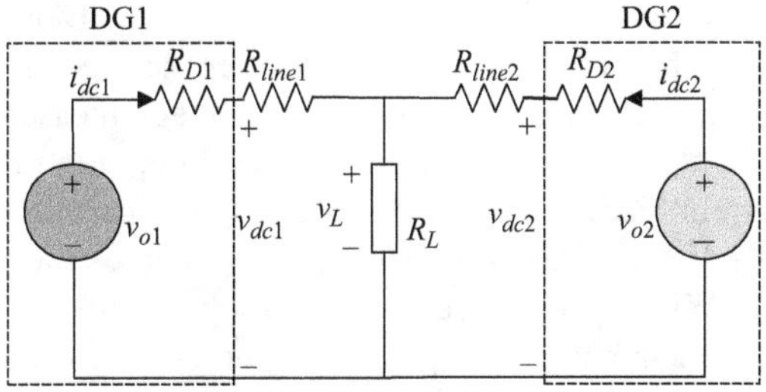

Figure 1.9 A simplified DC MG with cable resistance

The DG1 node voltage v_{DC1} can be derived as in the following equation:

$$v_{DC1} = V_0 - R_{D1}i_{DC1} \tag{1.5}$$

where v_{DC1} is the voltage reference produced by *V–I* droop control in MG1; i_{DC1} is the output current from DG1; R_{D1} is the virtual output resistance (droop gain) for DG1; and V_0 is the output voltage at no load. If the line impedance of R_{line1} and R_{line2} is considered, (1.5) can be written as

$$v_L = V_{01} - R_{D1}i_{DC1} - R_{line1}i_{DC1} \tag{1.6}$$

$$v_L = V_{02} - R_{D2}i_{DC2} - R_{line2}i_{DC2} \tag{1.7}$$

Combining (1.6) and (1.7), and assuming V_{o1} equals to V_{o2}, yields

$$i_{DC1} : i_{DC2} = \frac{1}{R_{D1} + R_{line1}} : \frac{1}{R_{D2} + R_{line2}} \tag{1.8}$$

From (1.8), we can conclude that the current sharing between DGs is not only dependent on the droop gain but also the line impedance. In large-scale DC systems where the line impedance is not negligible, the current sharing error cannot be avoided with conventional droop control.

1.3.3 Droop strategies

Different droop control techniques for power electronics converters in DC MGs have been developed [5,12]. These MGs implement the basic concept of droop control in different ways and can be categorized into two types, i.e. voltage-mode droop and current-mode droop. In the voltage-mode droop control, the output current is measured and fed to the *V–I* droop. The voltage reference is calculated and fed into inner voltage controller, as shown in Figure 1.10(a). In the voltage-mode droop control, the *V–I* droop can also be replaced by the *V–P* droop as the DC current is essential reflecting the power, as shown in Figure 1.10(b). In the current-mode droop control, the output voltage is measured and fed to the *I–V* droop. The current reference is calculated and fed into inner current controller, as shown in Figure 1.10(c). Similarly, the *I–V* droop can be replaced by the *P–V* droop as shown in Figure 1.10(d) [13].

Depending on the DC control strategy, the converter can be operated either in the current mode or voltage mode [14]. The current-mode droop control scheme is shown in Figure 1.11(b), with the current reference derived from the *I–V* droop characteristic in Figure 1.11(a), based on the DC voltage measurement.

Although the voltage-mode and current-mode droop controls seem to be different, the DC current sharing in steady states for both methods is the same in presence of cable resistance [13]. The difference for them is the implication on the system stability [5]. When using current-mode droop control, the current output of the converter is dependent on the droop gain. An increased DC current control loop bandwidth is helpful for stability enhancement of the system. However, the upper boundary of DC current control bandwidth is limited by the right half plane (RHP)

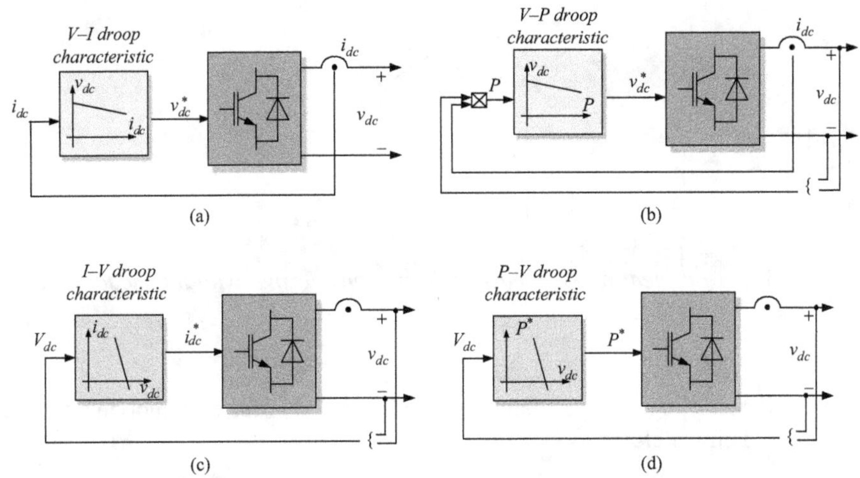

Figure 1.10 Droop characteristic employed in VSCs: (a) V–I droop, (b) V–P droop, (c) I–V droop and (d) P–V droop

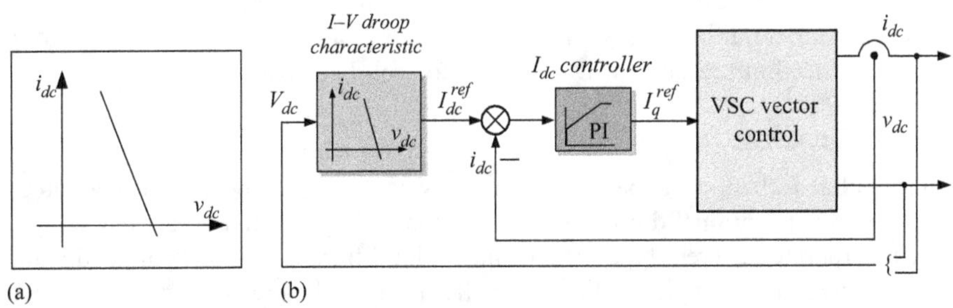

Figure 1.11 I–V droop and its corresponding current-mode control scheme: (a) I–V droop characteristic and (b) current-mode control scheme

zero in the transfer function. For the voltage-mode droop-controlled system, the DC voltage dynamics are affected by the DC voltage control bandwidth and the droop gain. Increasing the droop gain will reduce the voltage loop bandwidth. In contrast to the current-mode approach, the voltage-mode droop control regulates the terminal voltage based on current measurements. The RHP zero causes high gain instability within the v_{DC} controller, i.e. the system will easily go unstable when the v_{DC} controller has a high gain value. Furthermore, when utilizing voltage-mode droop control, the voltage loop bandwidth is mainly determined by the voltage controller rather than the droop gain. Greater details on stability analysis can be found in literature [5].

The discussed, droop control techniques so far have assumed linear relations between the DC voltage and DC current. As opposed to linear droop regulations mentioned above, a nonlinear droop curve is proposed in [12]. The droop gain can

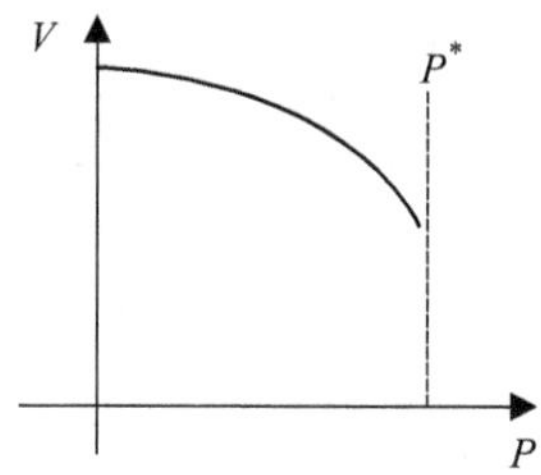

Figure 1.12 Nonlinear droop control scheme considering available headroom for each DGs

be adapted according to the available headroom available for each individual DGs. The droop gain is defined as

$$K = K_0\left(\frac{P^* - |P|}{P^*}\right)^{\lambda} \tag{1.9}$$

where K_0 is the rated droop gain, K is the adapted coefficient value, P^* is the rated power and P is the real-time output power. Essentially, the droop gain K gets larger in heavy load conditions, as shown in Figure 1.12. Therefore, the power sharing error is reduced in presence of DC voltage control discrepancy.

1.3.4 Dynamic power sharing

Under sudden change of load, a desired power sharing may not be guaranteed solely with conventional droop controls. This is mainly due to the fact that an MG may incorporate different types of DG units with different dynamic bandwidth and performance. For example, fuel cells usually have much slower dynamics than PV panels due to the limitation of chemical reactions. Although the rated power of fuel cell could be larger than other renewables and storage sources, in presence of a sudden load change, extra power required by a load with faster dynamics could only be provided by faster DGs (PVs, batteries, super-capacitors, etc.). In this section, power sharing in the transient condition is addressed. Therefore, to ensure a reasonable power sharing in the transition process, dynamic power sharing is inevitable and essential for MGs.

The basic idea of the dynamic power sharing is to compensate the required power from different DGs according to the load and DG dynamic spectrums. Dynamic power sharing can be achieved by configuring a dynamic compensator in series with the droop control loop. The approach is referred as frequency-coordinating virtual impedance [15]. It separates the dynamic spectrum into different frequency ranges and implements the compensation by inserting a filter in series with the *I–V* droop admittance as shown in Figure 1.13. The error between the reference voltage V_o at no load and measured output voltage v_{DC} is sent to the *P–V* droop block. In series with the droop block is a high-pass filter. The output of the filter is the power reference P^*, and the power reference P^* is then converted

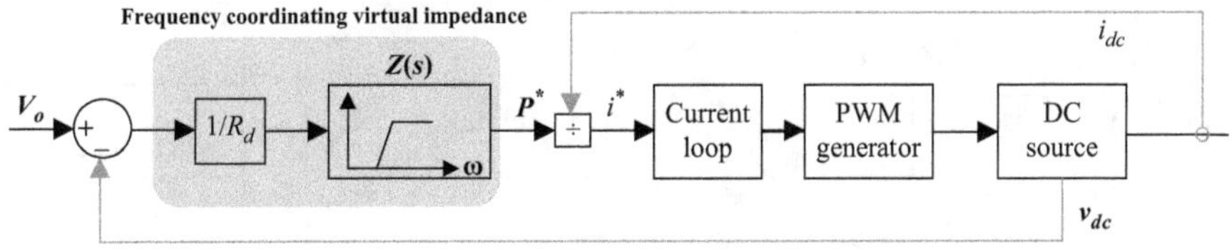

Figure 1.13 Frequency coordinating virtual impedance with I–V droop [15]

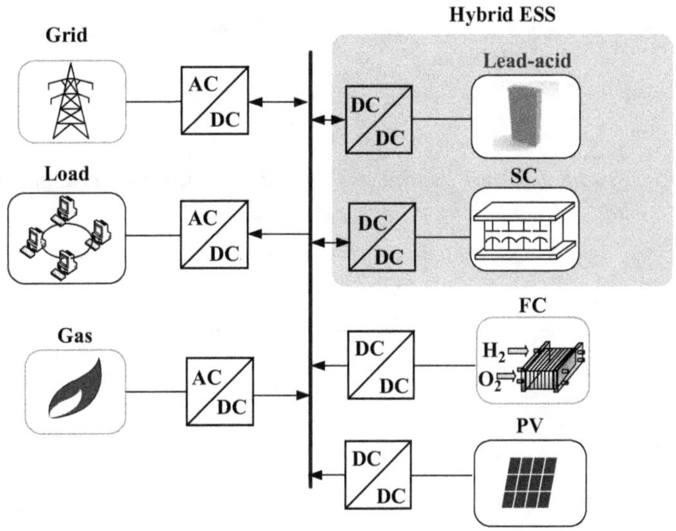

Figure 1.14 Composite of a DC MG with hybrid energy storages

into a current reference i^*. Consequently, the high-pass filter alters the virtual impedance, and the equivalent output impedance becomes

$$Z_{droop}(s) = \frac{Z(s)}{R_d} \tag{1.10}$$

As an example, an MG with both lead acid and super-capacitor energy storages is shown in Figure 1.14. The lead-acid battery has slower dynamics than the super-capacitor. Thus, the slower load dynamic spectrum is loaded to battery converter, while the faster dynamic load is assigned to super-capacitor converter as depicted in Figure 1.15. A low-pass filter is inserted in series with the droop admittance in battery control loop to shape the equivalent virtual impedance as illustrated in Figure 1.16. This allows the output impedance of the battery converter increases with frequency. On the contrary, a high-pass filter is implemented in the control loop of super capacitor, and the output impedance of the super capacitor converter shows low impedance in the medium frequency range. Dynamic range greater than control bandwidth is mainly managed by passive DC bus filtering capacitors.

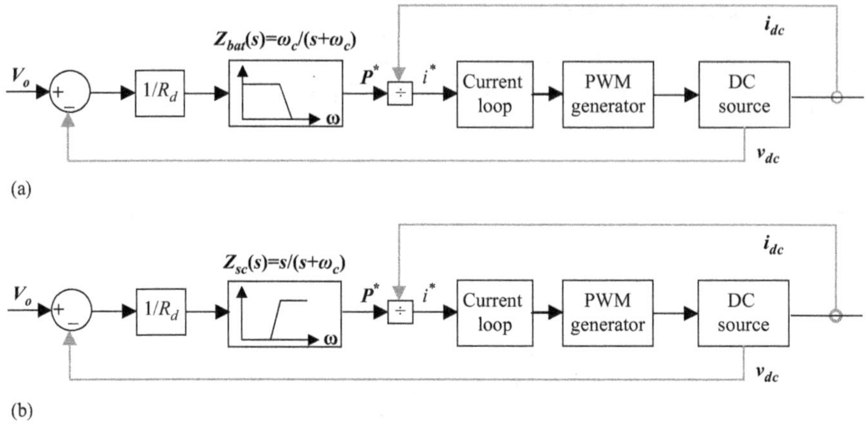

Figure 1.15 Frequency coordination for hybrid ESS: (a) control loop for battery and (b) control loop for super-capacitor [15]

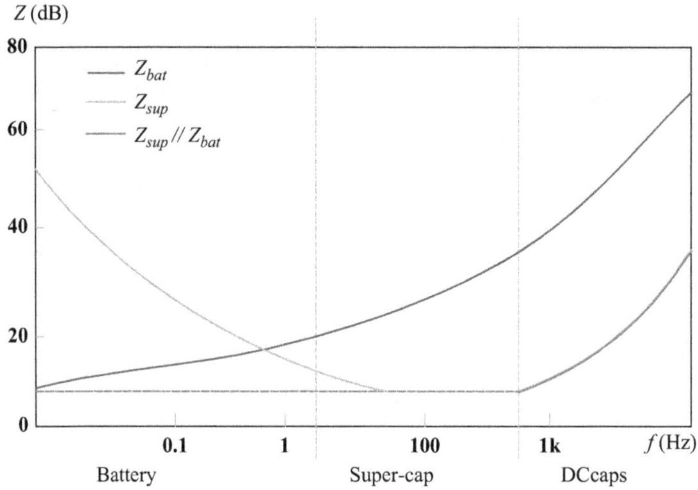

Figure 1.16 Frequency-domain impedance comparison of converters [15]

1.3.5 Interfaces to upper levels

Primary control level needs interfaces to interact with upper levels. According to above discussions on the primary control, interfaces are no load voltage reference value (V_0), droop gain (R_D) and saturation limits (P^- and P^+). These variables can be manipulated by secondary and tertiary control levels to ensure power quality and accommodate flexible power management. They are illustrated in Figure 1.17. Implications of changing interface variables on droop curve are shown in Figure 1.18. Typical saturation limits P^- and P^+ are normally controlled according

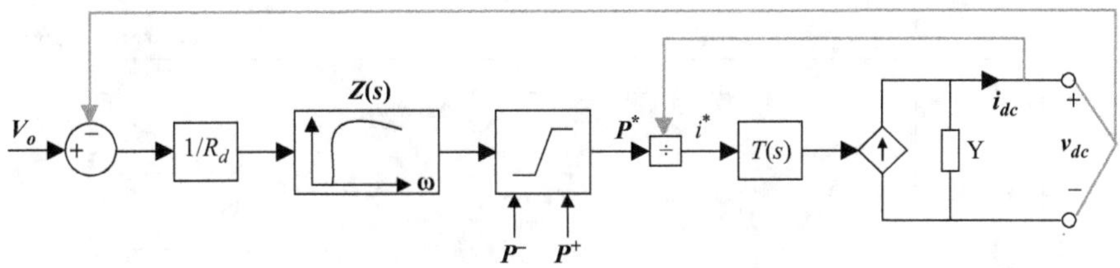

Figure 1.17 Interfaces of primary level to upper levels

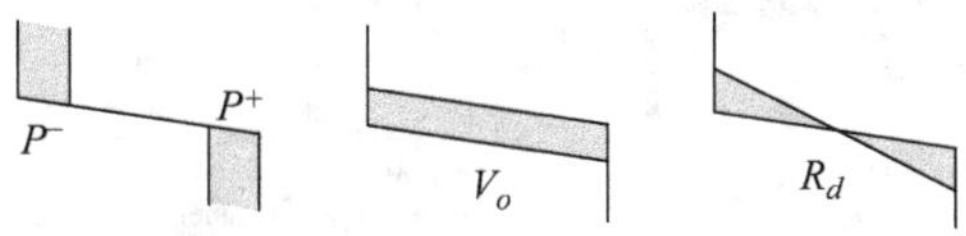

Figure 1.18 Implications of changing interface variables to the droop curve

to the requirement of energy management [16]. The voltage reference value V_0 is adjusted to restore the DC voltage offset caused by the primary droop control. The droop gain R_D is modified to reschedule the power sharing between DGs.

1.4 Secondary control

The secondary control aims to restore the DC bus voltage deviation caused by primary droop control. When MG is connected to an upper grid, the secondary controller receives command from upper tertiary controllers. This allows voltages at common coupling points to track the upper grid voltage. The secondary control can be implemented locally inside DGs. It can also be integrated within the tertiary control in an MG central controller, as shown in Figure 1.1. When the secondary control is implemented locally inside DGs, this level of control will be decentralized and is referred to as distributed control. On the other hand, if the secondary control is integrated into the upper level controller (controller in the tertiary level), it will be centralized (thus referred as centralized approach).

1.4.1 Centralized approach

As mentioned before, the MG secondary controller is used to recover the DC bus voltage deviation caused by primary control. In order to achieve this functionality, the DC bus voltage is measured and sent to a central controller. A low bandwidth communication channel can be used for this application [2]. The central controller will collect the data from local sensors, process the data and send out commands to local primary controllers. As the signals are always transmitted in one direction, the communication is normally very reliable.

As shown in Figure 1.19, the DC bus voltage is measured remotely and sent to a central controller. The error between the measured DC bus voltage and its reference value $V_{MG\text{-}ref}$ is proceed in the controller in the secondary control level.

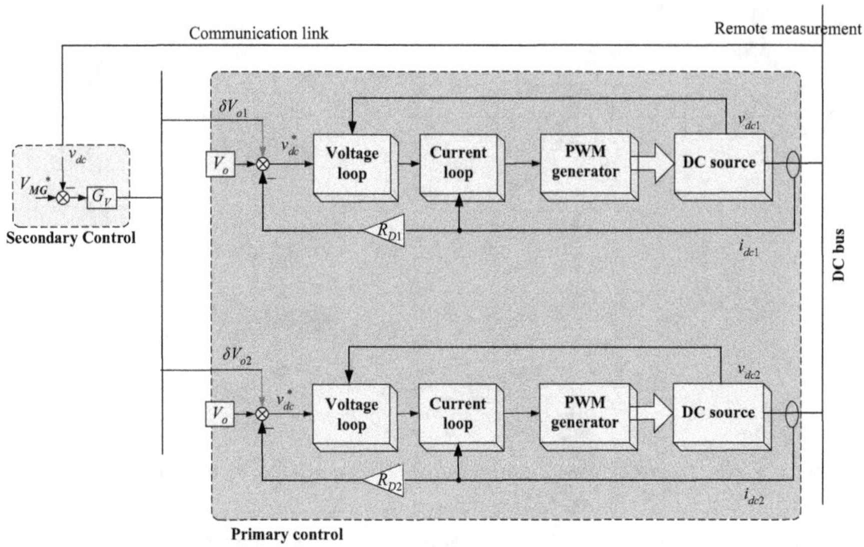

Figure 1.19 A conceptual diagram of centralized primary and secondary controls of a DC MG with power and information flow

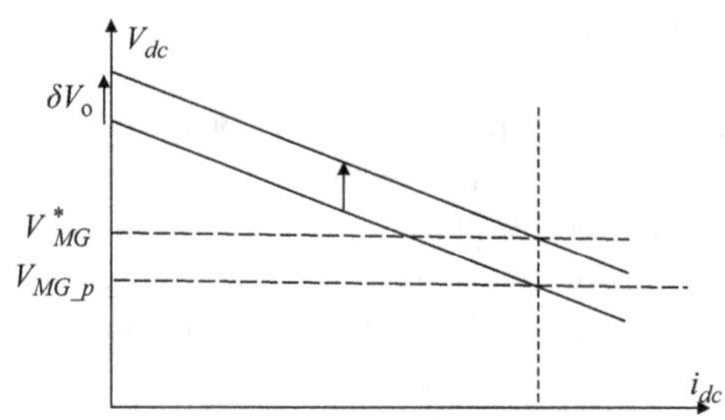

Figure 1.20 Droop curve lifted up by the secondary controller

A droop shifting value δv_o is generated from the secondary controller and sent to each DGs through communication links. To adjust power sharing dynamically, different shifting values can be sent to different DGs. When embedding the droop shifting into the DGs, the *V–I* droop (1.1) becomes

$$v_{DC} = v_0 + \delta v_0 - R_D i_{DC} \tag{1.11}$$

Another figure showing the curve shifting is shown in Figure 1.20. As can be seen, a droop shifting value δV_0 from the secondary controller is used to adjust the no-load voltage V_0 in the primary control level. After this compensation, the droop curve is lifted up, and DC voltage is increased from V_{MG_P} to V^*_{MG}.

As can be seen from Figures 1.19 and 1.20, a reliable operation of secondary control depends on effective communication between the two levels and a reliable MG central controller. Any fault that occurs in the communication channel or the central controller may jeopardize the functionality of the secondary control. Under these faulty conditions, the deviation of the DC bus voltage is totally subjected to load change.

1.4.2 Distributed approach

To improve system reliabilities, the distributed secondary control have been proposed [7,17–19]. In this approach, the DC bus voltage regulation is entirely realized locally within DGs. This is the most significant difference compared with the centralized approach. The communication between DGs can be achieved either through dedicated communication channels or through the power line.

1.4.2.1 Communication through dedicated communication channels

A general diagram is shown in Figure 1.21. In this diagram, the average current sharing (ACS) [7] technique is implemented in the secondary control level. This technique is incorporated within each DG. The communication links can be either analogue or digital.

The operation principle of an analogue communication bus can be explained in Figure 1.22. The voltage V_{Nn} can be calculated as

$$V_{Nn} = \frac{\sum_{i=1}^{n} V_i/R_i}{\sum_{i=1}^{n} 1/R_i} \tag{1.12}$$

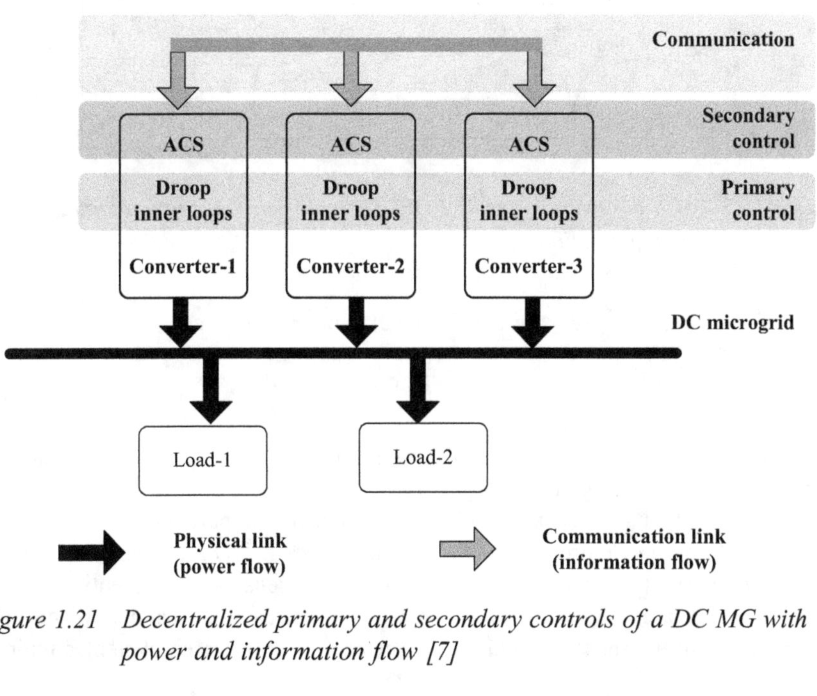

Figure 1.21 Decentralized primary and secondary controls of a DC MG with power and information flow [7]

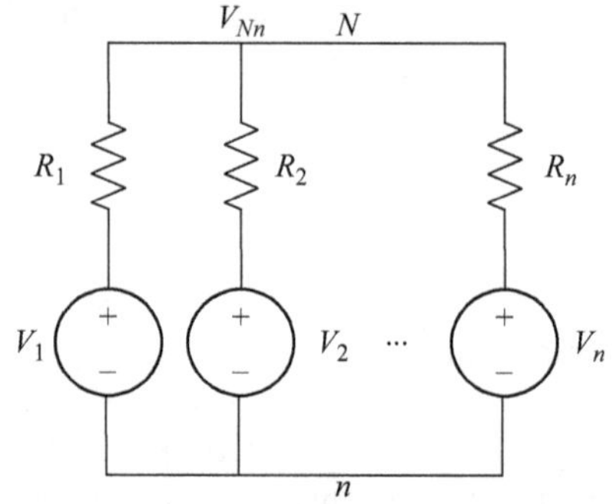

Figure 1.22 Equivalent circuit for the average current signal bus

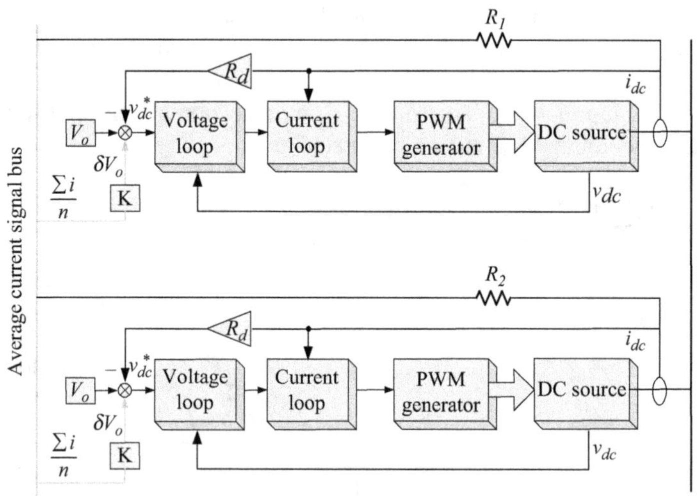

Figure 1.23 ACS control for parallel DC–DC converters using analogue communication

If resistance $R_1, R_2, \ldots, R_n$ are equal, (1.12) can be written as

$$V_{Nn} = \frac{\sum_{i=1}^{n} V_i}{n} \tag{1.13}$$

From (1.13), we can conclude that if the resistance $R_1, R_2, \ldots, R_n$ are equal, the voltage V_{Nn} is the average of $V_1, V_2, \ldots, V_n$.

This concept has been adopted in ACS control for paralleled DC–DC converters as shown in Figure 1.23 [7]. The currents of DC/DC converters are measured with current sensors. The outputs of these sensors are essentially some voltages which are proportionate to the value of measured currents. When connecting the current sensor output to a resistor (R_1, R_2), an equivalent circuit in

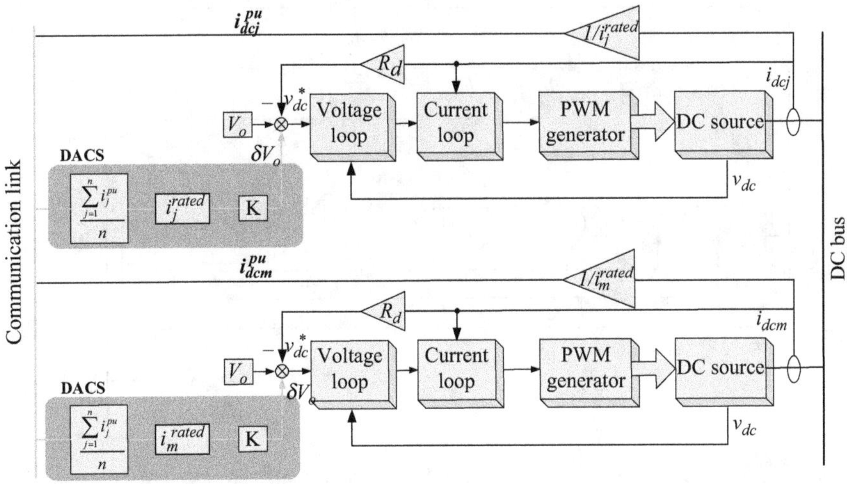

Figure 1.24 ACS-based distributed control using CAN bus as the communication channel

Figure 1.22 will be created. The voltage on the average current signal bus is, hence, reflecting the average value of DC currents within different DC/DC converters. As analysed in the primary control section, with conventional droop control, the voltage variation is subjected to load changes (current changes). If the load increases, according to the droop curve, the DC voltage will decrease. Thus, the average current measured from average current signal bus will also increase. Therefore, the droop shifting value is increased, and the droop curve is shifted up, as shown in Figure 1.20. The voltage deviation caused by primary droop control is compensated.

Although ACS method is an effective approach to restore DC bus voltage, this method is only effective when DGs are configured close to each other. When DGs are widely spread with long distance in between, since the average current signal bus has to be deployed along with power lines, this makes average current bus susceptible to noise. This will degrade the system performance when using ACS method for the secondary control.

To overcome the above-mentioned shortcomings of ACS, ACS-based distributed method using controller area network (CAN bus) has been proposed, as shown Figure 1.24 [7]. The nature of CAN is to provide reliable digital communications between controllers and devices without a host. This makes CAN bus rather suitable for the secondary control application. Since all the current information shared on the CAN bus are normalized values, this makes the current sharing algorithm simpler and MG easier to expand. To distinct from analogue-based ACS, the CAN bus-based implementation is termed as digital ACS (DACS). Normalized averaged current is calculated in DACS and transformed into a droop shifting signal by multiplying its current normal base i_j^{rated} and shifting gain k_j.

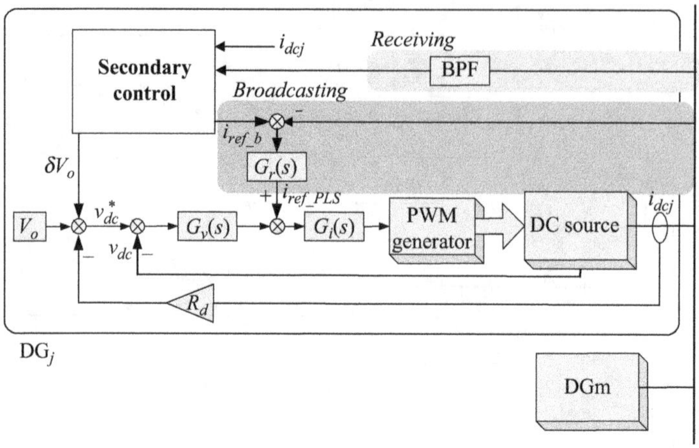

Figure 1.25 Block diagram of the PLS-based decentralized secondary control scheme [19]

Afterwards, the droop shifting signal Δv_j^0 (the same purpose as δv_o in Figure 1.20) is sent to the primary controller to restore the bus voltage.

1.4.2.2 Communication through power lines

The utilization of power distribution line as a communication channel to carry and transmit both power and information is nothing new. The investigation on Power Line Signalling (PLS) has been started not long after electrical power supply became widespread. The idea of PLS has been introduced to DC MG in [19] as shown in Figure 1.25. Sinusoidal signal is injected into the power line by using a resonant controller in DG$_j$. The sinusoidal reference is generated from the local secondary controller. A band-pass filter is employed to extract the frequency component signal embedded in the DC bus voltage. By using such communication, the current information can be shared among DGs, and DC voltage compensation can be achieved in the same way as discussed in above sub-section.

1.5 Tertiary control

Whenever there is a requirement to connect MG to an upper stiff grid, a DC bus voltage synchronization process has to be initiated. Distinctively from AC MGs, only DC bus voltage needs to be synchronized in the DC MGs as shown in Figure 1.26. A bypass switch is required to remain off to avoid inrush current before the secondary control tracks down V_{DC}. Once the deviation between V_{MG} and V_{DC} is controlled under a predefined threshold, a mode transfer signal will be sent from tertiary controller to turn on the bypass switch. The tertiary controller at this point will also alter the secondary reference (V_{MG}^*) from V_{DC} to V_{MG}^T. By doing this, the whole MG will be seen by the stiff DC grid as a current source, since there is a

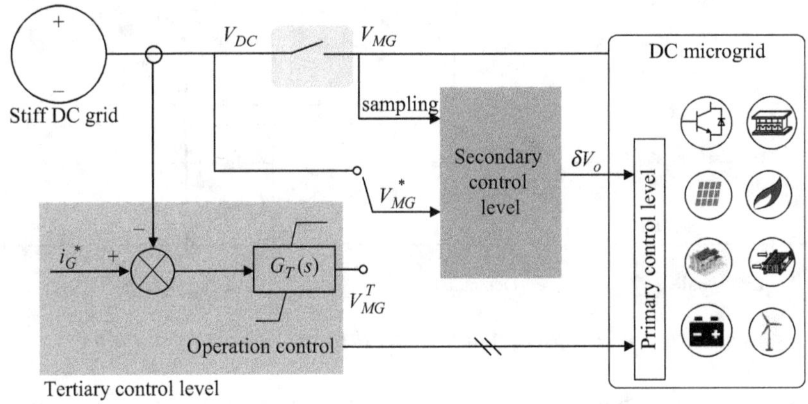

Figure 1.26 Tertiary control to manage power exchange between MG and upper grid with synchronization loop of a DC MG [2]

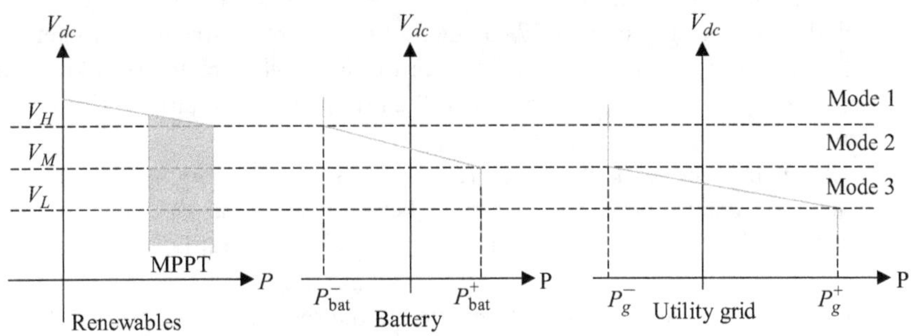

Figure 1.27 Adaptive autonomous mode structure

current (or power) control loop in tertiary control level and the current reference is set by the command i_G^*.

Once the MG is connected to the upper grid, it communicates with the distribution system operator (DSO) or transmission system operator (TSO) [1] and the secondary control level. System power management can be implemented based on DC Bus Signalling (DBS). The concept of DBS technique has been introduced in literature [15,16]. In this method, the DC bus voltage itself is used as a communication signal between DGs. Agreements have been made beforehand among DGs. DGs will take different operation actions under different DC bus voltage ranges. Sources in the DC MGs can be classified into three main categories: renewables, storages and utility grid as shown in Figure 1.27. When the bus voltage is higher than V_H (Mode 1), DC bus voltage is regulated by renewables and battery/utility grid receives power generated from renewables. When bus voltage falls below V_H but stays above V_M (Mode 2), DC bus voltage is supported by battery. In Mode 2, renewables are working in Maximum Power Point Tracking and power

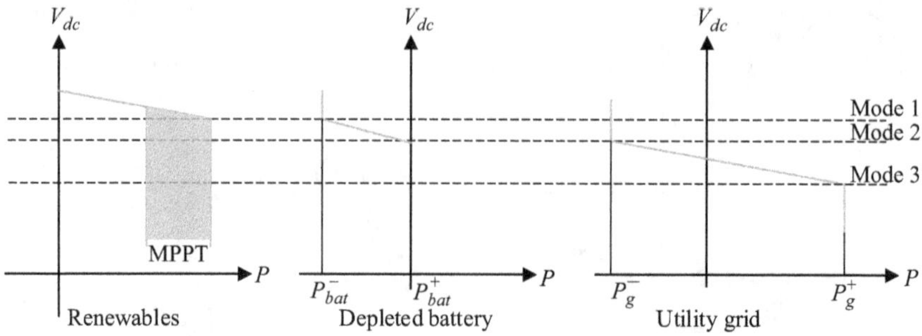

Figure 1.28 Adaptive autonomous droops with depleted battery

is feeding into the utility grid. When the bus voltage drops below V_M (Mode 3), grid converters adjust the DC bus voltage. Both renewables and battery work as current sources. Due to the existence of the secondary control level, the final steady state bus voltage is always restored to the reference. All the curves shown in Figure 1.27 can be shifted up or down as an entity by the secondary controller.

Each droop curve in Figure 1.27 can be modified in real time online by tertiary control according to system requirements. For instance, when battery has low State of Charge (SOC), P^+_{bat} is forced to zero and the droop curve for utility grid is lifted up accordingly, as shown in Figure 1.28.

Based on the above mode structure, DSO/TSO gives command to tertiary control level to determine the power flow between MG and the upper grid by changing P^+_g and P^-_g. Based on information gathered from both DSO/TSO and the MG, the tertiary controller prepares the source and storage dispatch schedule to optimize the operation which is communicated to the secondary and primary levels. From this point, tertiary control level shares similarity for both AC and DC MGs. Therefore, the existing solutions on energy management for AC MGs can be adapted for DC MG applications. Greater details concerning energy management will be discussed in other chapters.

1.6 Summary

In this chapter, the hierarchical control of DC MG is introduced. The definitions for each control levels have been discussed. Primary control is responsible for DGs load sharing and is predominately implemented using the droop control. The droop control can be perceived as a virtual resistance, and its value can affect system stability and maximum DC bus voltage deviation. Two inherent issues with conventional droop control are discussed. Both terminal DC voltage control accuracy and cable resistance have impacts on the power sharing among DGs. Droop control can be mainly divided into two groups: current-based droop and voltage-based droop. The difference is the implication on system stability. In addition, another nonlinear droop curve is also mentioned with adaptive droop gain. It can reduce power sharing error in heavy load. Furthermore, dynamic power sharing during load transition is also discussed in this section. Different load dynamic spectrums are assigned to different DGs according to their dynamic response. Finally, to receive commands from upper control levels, interfaces are pointed out without changing the basic primary control loop structure.

The purpose of the secondary control is to restore the DC bus voltage deviation caused by conventional droop control from primary control level. It can be implemented remotely in an MG central controller (Centralized approach) or locally inside each DG (distributed approach). Both dedicated analogue and digital communication links can be used in a decentralized secondary control to transmit current signals. To enhance the reliability of the control, a communication link based on power lines is also possible.

The tertiary control level is responsible for the connecting process of MG to the upper grid. A basic mode structure based on DBS is introduced to accommodate energy-management algorithms.

References

[1] T. L. Vandoorn, J. C. Vasquez, J. De Kooning, J. M. Guerrero, and L. Vandevelde, "Microgrids: Hierarchical Control and an Overview of the Control and Reserve Management Strategies," *IEEE Ind. Electron. Mag.*, vol. 7, no. 4, pp. 42–55, Dec. 2013.

[2] J. M. Guerrero, J. C. Vasquez, and R. Teodorescu, "Hierarchical control of droop-controlled DC and AC microgrids—a general approach towards standardization," in *2009 35th Annual Conference of IEEE Industrial Electronics*, 2009, pp. 4305–4310.

[3] C. N. Papadimitriou, E. I. Zountouridou, and N. D. Hatziargyriou, "Review of Hierarchical Control in DC Microgrids," *Electr. Power Syst. Res.*, vol. 122, pp. 159–167, May 2015.

[4] C. Jin, P. Wang, J. Xiao, Y. Tang, and F. H. Choo, "Implementation of Hierarchical Control in DC Microgrids," *IEEE Trans. Ind. Electron.*, vol. 61, no. 8, pp. 4032–4042, Aug. 2014.

[5] F. Gao, S. Bozhko, A. Costabeber, *et al.*, "Comparative Stability Analysis of Droop Control Approaches in Voltage-Source-Converter-Based DC Micro-grids," *IEEE Trans. Power Electron.*, vol. 32, no. 3, pp. 2395–2415, Mar. 2017.

[6] D. Xu, H. Li, Y. Zhu, K. Shi, and C. Hu, "High-surety Microgrid: Super Uninterruptable Power Supply with Multiple Renewable Energy Sources," *Electr. Power Compon. Syst.*, vol. 43, no. 8–10, pp. 839–853, Jun. 2015.

[7] S. Anand, B. G. Fernandes, and J. Guerrero, "Distributed Control to Ensure Proportional Load Sharing and Improve Voltage Regulation in Low-Voltage DC Microgrids," *IEEE Trans. Power Electron.*, vol. 28, no. 4, pp. 1900–1913, Apr. 2013.

[8] Y. A.-R. I. Mohamed, "Robust Droop and DC-Bus Voltage Control for Effective Stabilization and Power Sharing in VSC Multiterminal DC Grids," *IEEE Trans. Power Electron.*, vol. 33, no. 5, pp. 4373–4395, May 2018.

[9] P. Karlsson and J. Svensson, "DC Bus Voltage Control for a Distributed Power System," *IEEE Trans. Power Electron.*, vol. 18, no. 6, pp. 1405–1412, Nov. 2003.
[10] J. Jiang, "Design of an Optimal Robust Governor for Hydraulic Turbine Generating Units," *IEEE Trans. Energy Convers.*, vol. 10, no. 1, pp. 188–194, Mar. 1995.
[11] K. De Brabandere, B. Bolsens, J. Van den Keybus, A. Woyte, J. Driesen, and R. Belmans, "A Voltage and Frequency Droop Control Method for Parallel Inverters," *IEEE Trans. Power Electron.*, vol. 22, no. 4, pp. 1107–1115, Jul. 2007.
[12] N. R. Chaudhuri and B. Chaudhuri, "Adaptive Droop Control for Effective Power Sharing in Multi-Terminal DC (MTDC) Grids," *IEEE Trans. Power Syst.*, vol. 28, no. 1, pp. 21–29, Feb. 2013.
[13] J. Beerten and R. Belmans, "Analysis of Power Sharing and Voltage Deviations in Droop-Controlled DC Grids," *IEEE Trans. Power Syst.*, vol. 28, no. 4, pp. 4588–4597, Nov. 2013.
[14] F. Gao, Y. Gu, S. Bozhko, G. Asher, and P. Wheeler, "Analysis of droop control methods in DC microgrid," in *2014 16th European Conference on Power Electronics and Applications*, 2014, pp. 1–9.
[15] Y. Gu, W. Li, and X. He, "Frequency-Coordinating Virtual Impedance for Autonomous Power Management of DC Microgrid," *IEEE Trans. Power Electron.*, vol. 30, no. 4, pp. 2328–2337, Apr. 2015.
[16] Y. Gu, X. Xiang, W. Li, and X. He, "Mode-Adaptive Decentralized Control for Renewable DC Microgrid with Enhanced Reliability and Flexibility," *IEEE Trans. Power Electron.*, vol. 29, no. 9, pp. 5072–5080, Sep. 2014.
[17] K. Sun, L. Zhang, Y. Xing, and J. M. Guerrero, "A Distributed Control Strategy Based on DC Bus Signaling for Modular Photovoltaic Generation Systems With Battery Energy Storage," *IEEE Trans. Power Electron.*, vol. 26, no. 10, pp. 3032–3045, Oct. 2011.
[18] L. Zhang, T. Wu, Y. Xing, K. Sun, and J. M. Gurrero, "Power control of DC microgrid using DC bus signaling," in *2011 Twenty-Sixth Annual IEEE Applied Power Electronics Conference and Exposition (APEC)*, 2011, pp. 1926–1932.
[19] T. Dragicevic, J. M. Guerrero, and J. C. Vasquez, "A Distributed Control Strategy for Coordination of an Autonomous LVDC Microgrid Based on Power-Line Signaling," *IEEE Trans. Ind. Electron.*, vol. 61, no. 7, pp. 3313–3326, Jul. 2014.

Chapter 2

Distributed and decentralized control of dc microgrids

Saeed Peyghami[1], *Hossein Mokhtari*[2], *and Frede Blaabjerg*[1]

2.1 Introduction

Power sharing control of dc power sources in dc microgrids is an important issue in order to obtain a stable and reliable operation [1–7]. Power sharing approaches in three levels of the hierarchal control system are described in the last chapter. Two major objectives of this method of control are proportional power sharing and appropriate voltage regulation in dc microgrids which can be carried out using centralized and decentralized/distributed approaches [2,3,6,7]. Centralized approaches deal with lower reliability and resiliency. In order to enhance the resiliency of the power sharing methods, distributed and decentralized controls have been introduced. Distributed approaches employ sparse communication systems instead of point-to-point communication control among different converters (and maybe a central control unit). On the contrary, decentralized approaches use no communication (or even physical communication link) in the corresponding control system.

Decentralized approaches can be categorized as

1. Mode adaptive (autonomous) droop control
2. Nonlinear droop control
3. Frequency droop control

The voltage droop method is the simplest decentralized approach employed in dc microgrids. This method can perfectly control the load sharing among power sources in the case of short distances where the line resistances can be neglected. In this case, mode adaptive or autonomous droop approach can be used to control the output power of different (dispatchable and nondispatchable) sources in the dc microgrid. Considering the line resistances of long distances among converters, the conventional droop method cannot properly carry out the power sharing objectives,

[1]Department of Energy Technology, Aalborg University, Denmark
[2]Department of Electrical Engineering, Sharif University of Technology, Iran

and hence, nonlinear droop approach and frequency droop methods have been introduced in order to reach the power management objectives.

Furthermore, some distributed approaches have been presented which can be classified as

1. Fully communicated control
2. Sparse communicated (consensus-based) control
3. Sparse communicated control using current information.

In the distributed approaches, each converter is equipped with a secondary controller cascaded by the primary controller. A communication system is employed to share the voltage and current information between all of the converters. In a simple distributed secondary control, it is required that all of the converters have access to the information each other. This approach may be complex and has a lower resiliency; however, consensus strategy with a sparse communication network has been presented and can use the neighboring converter information to reach an acceptable operating condition. Furthermore, another sparse communicated control approach can reach the power management objectives by only employing the current information of the converters. In this chapter, decentralized and distributed control concepts are discussed.

2.2 Decentralized approaches

In this section, the decentralized power sharing approaches are conceptually explained. Decentralized methods do not require any communication link between the converters and hence introduce higher reliability and stability. The voltage droop control is the simplest decentralized method employed in dc microgrids. This method can perfectly control the load sharing among the power sources in the case of short distances where the line resistances can be neglected. In this case, mode adaptive or autonomous droop approach can be used to control the output power of different (dispatchable and nondispatchable) sources in the dc microgrid. Considering the line resistances of long distances among converters, the conventional droop method cannot properly control the power sharing, and hence, nonlinear droop approach and frequency droop methods have been presented in order to appropriately manage the load power among converters. The decentralized approaches are discussed in the following section.

2.2.1 Mode-adaptive (autonomous) droop control

dc Microgrids, in practice, contain dispatchable, nondispatchable and storage units. A voltage droop-based approach can be employed to control the power and energy flow in the microgrid by defining suitable droop characteristics in order for different kinds of energy units to operate in an autonomous mode. A simplified dc microgrid is shown in Figure 2.1. In the following section, possible droop schemes for different kinds of energy units are explained.

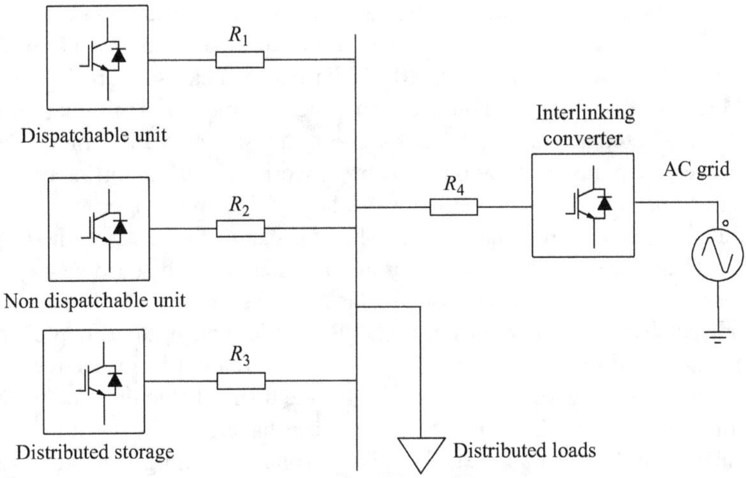

Figure 2.1 Typical dc microgrid with dispatchable and nondispatchable units, distributed storage, interlinking converter and distributed loads—DG. DG, distributed generation

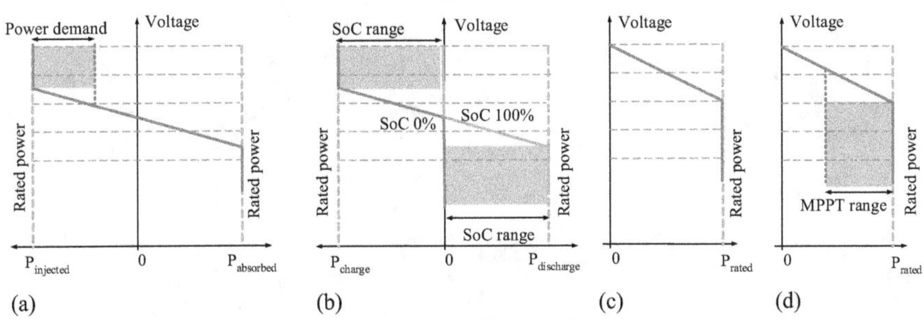

Figure 2.2 Droop characteristics for different energy units in dc microgrid with constant droop slopes; (a) IC, (b) storage converter, (c) dispatchable units and (d) nondispatchable units

Typical droop characteristics for an interlinking converter (IC), a distributed storage unit, and dispatchable and nondispatchable units are shown in Figure 2.2 [8]. For an IC, a bidirectional droop can be considered, such as the one shown in Figure 2.2(a). The maximum injected power into the grid must be determined by the transmission/distribution system operator (TSO/DSO). However, the required power can be supported by the grid which can be determined by the droop characteristic shown in Figure 2.2(a). For the storage droop control shown in Figure 2.2(b), such as a battery, a regenerative fuel cell, or a flywheel, the maximum and minimum limits can be determined by the energy level or state of charge (SoC) level of the battery. Dispatchable energy units such as microturbines and fuel cells can be controlled with a droop curve as shown in Figure 2.2(c). Finally, as shown in Figure 2.2(d), nondispatchable units such as photovoltaic arrays and wind turbines can operate in a droop mode from zero to an Maximum Power Point Tracking (MPPT) power. Therefore, MPPT-based units can support the microgrid under MPPT power. With this approach, only the maximum and minimum limits of the droop characteristics are updated, based on the SoC or MPPT limits. Furthermore, in the droop mode operation, converters work as grid supporting converters, where at a constant power mode, they work as grid feeding converters.

Adjusting power limits may cause voltage instability because of the lack of a voltage forming or supporting converter at some loading and sourcing conditions. For example, if the power demand of an IC is zero, the storage is fully charged, and the nondispatchable unit operates at an MPPT mode; hence, the droop characteristics must be modified as shown in Figure 2.3(a), (b) and (d), respectively. If the load is equal to the sum of the rated powers of the dispatchable unit and the MPPT power of a nondispatchable unit, the voltage can have any value within the gray region in Figure 2.3. In this area, there is no voltage forming converter, and the voltage cannot converge to a steady-state value.

To avoid all converters operating at constant power mode at the same time, which may occur at some loading and sourcing conditions, in addition to adapting the power limits, the slope of the droop controls should be adjusted, and this approach is illustrated in Figure 2.4. However, a control system needs to be

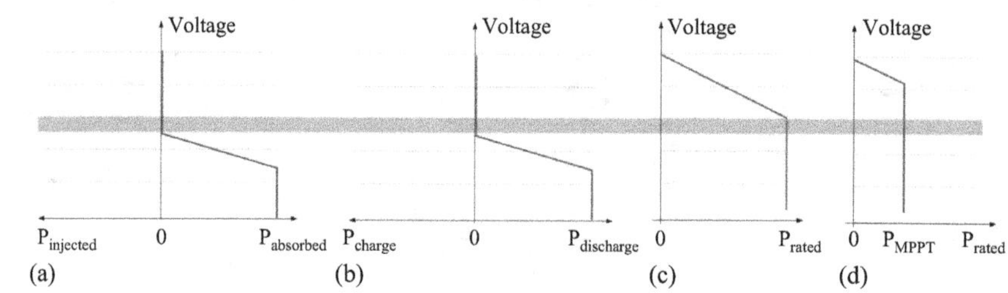

Figure 2.3 An unstable region in droop characteristics of different energy units in a droop controlled dc microgrid with constant droop slopes; (a) IC, (b) storage converter, (c) dispatchable units and (d) nondispatchable units

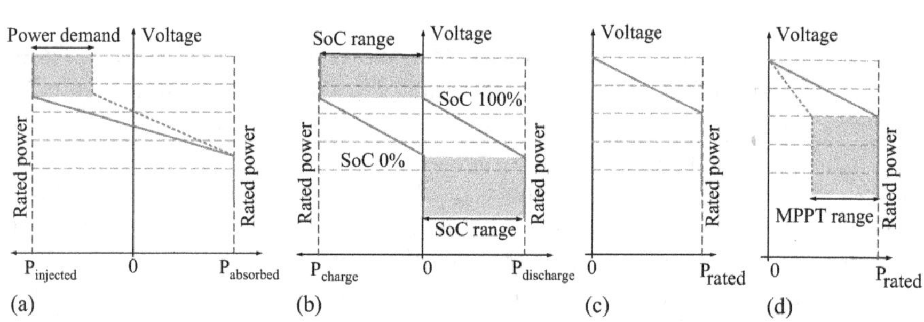

Figure 2.4 Droop characteristics for different energy units with adjustable droop slopes; (a) IC, (b) storage converter, (c) dispatchable units and (d) nondispatchable units

designed to guarantee the system's stability at different loading and sourcing conditions, since the droop gains affect the system's stability.

The droop characteristics shown in Figures 2.2 and 2.4 guarantee parallel operation of the ICs in the dc microgrid. However, in the case of islanded hybrid ac–dc microgrids, as well as the absence of the tertiary control DSO/TSO, it would be better to support the microgrid demand by its internal sources. Hence, the excess power or required power can be supported by the ICs. This approach is capable of controlling the energy flow between ac–dc microgrids [9]. The droop characteristics of two parallel ICs are shown in Figure 2.5(b) and (c). If the voltage lies between V_H and V_L, the power of the ICs is zero. Otherwise, the output power of the ICs can be determined by the droop characteristics as shown in Figure 2.5(b). Furthermore, the power sharing between ICs can be properly managed by this droop control approach [9].

Different types of droop characteristics may be defined for various energy sources in the dc microgrid. To operate the microgrid in an autonomous mode, it is important to determine a voltage range for the droop gain of each unit in order to

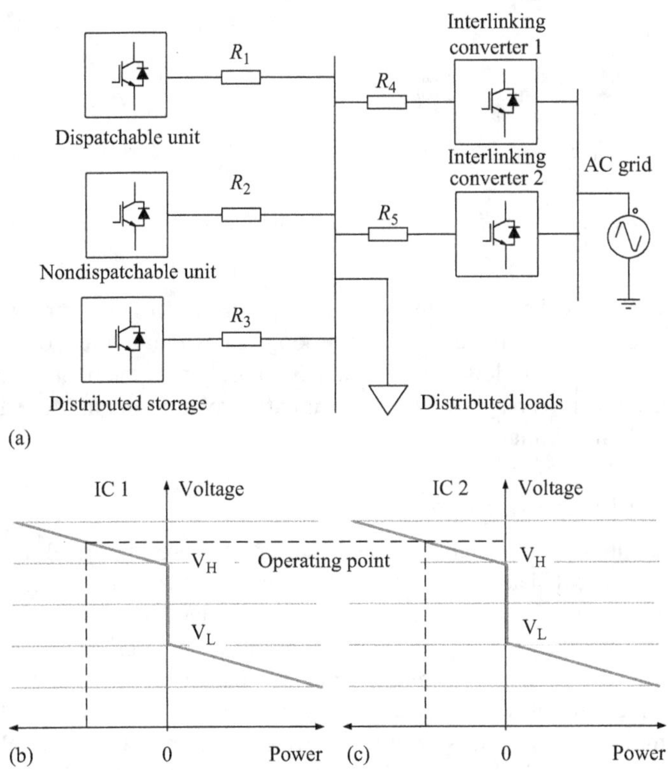

Figure 2.5 Typical ac–dc microgrid with two ICs; (a) the single line diagram of the power system and (b), (c) the droop characteristics of the interlinking converters (ICs)

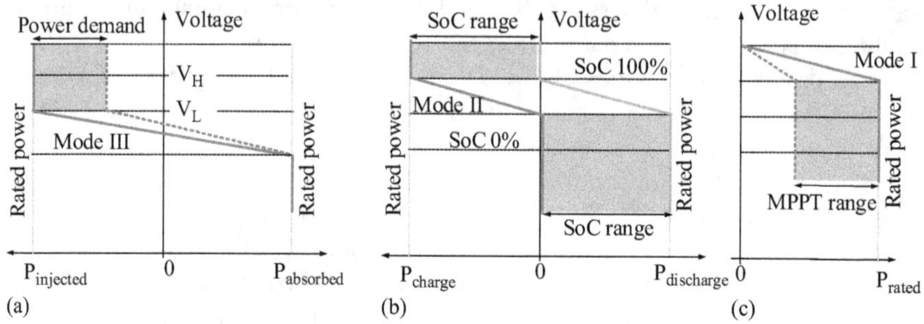

Figure 2.6 An autonomous droop-based power sharing in dc microgrid; (a) IC, (b) battery converter, (c) photovoltaic array

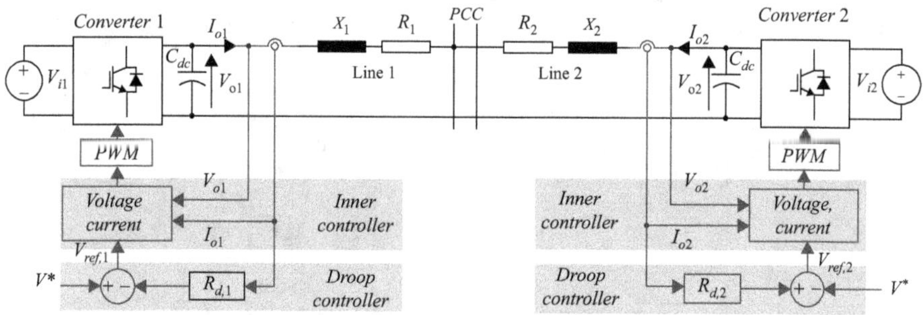

Figure 2.7 Simplified dc microgrid with primary droop control

have at least one voltage forming converter in the microgrid. For example, an autonomous control system based on the droop is shown in Figure 2.6 [10]. As can be seen in Figure 2.6, in Mode I, the voltage is controlled by the photovoltaic array. In Mode II, a battery is responsible for forming the voltage, and in Mode III, the IC controls the dc-link voltage.

2.2.2 Nonlinear droop control

In the conventional droop control, considering the higher virtual resistor $R_{d,i}$ improves the current sharing accuracy but deteriorates the voltage regulation. The relationship between the current sharing error and voltage regulation by increasing the virtual resistors $R_{d,1} = R_{d,2}$ for the simplified dc microgrid is shown in Figure 2.7 and is illustrated in Figure 2.8. In heavy-load conditions, unlike light load conditions, the current sharing accuracy is very important, since overloading a converter may affect its corresponding lifecycle. Therefore, large virtual resistors need to be employed in heavy-load condition. However, larger virtual resistors create a large voltage drop in the microgrid; hence, an adaptive technique is required to compensate for the voltage drop. This is a secondary controller.

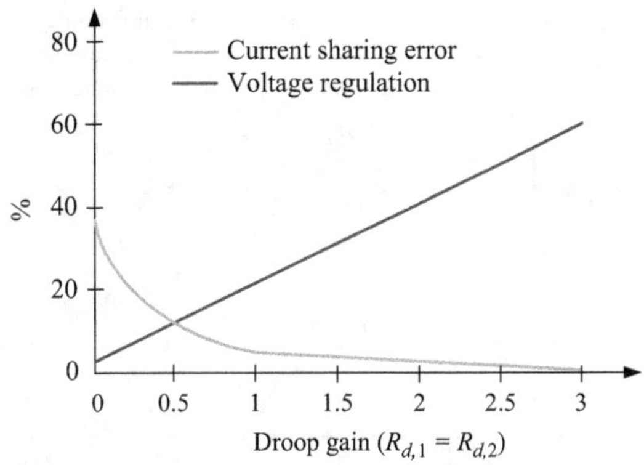

Figure 2.8 Performance of the conventional droop control for the simplified dc microgrid, current sharing error and voltage regulation

Secondary control approaches are explained in the first chapter. The secondary controller requires communication infrastructures to share the voltage and current information of converters to regulate the dc-link voltage. In order to avoid utilizing a communication system, an adaptive nonlinear approach has been used, presented in [4]. In this approach, the reference voltage $V_{ref,i}$ of the ith converter can be defined as (2.1), where m_i and α are the droop curve constant and coefficients, respectively [4]. m_i can be determined by (2.2), where V_{min} is the minimum allowable voltage and I_{max} are the maximum current of ith converter.

$$V_{ref,i} = V^* - m_i I_i^\alpha \tag{2.1}$$

$$m_i = \frac{V^* - V_{min}}{I_{max,i}^\alpha} \tag{2.2}$$

The nonlinear droop curve is graphically shown in Figure 2.9. As can be seen, the slope of the droop curve slope is larger in the higher currents, which reflects accurate current sharing in heavy-loading condition. Furthermore, the dc-link voltage remains in the acceptable interval. Considering the effective droop curves (the tangent line of the nonlinear curve) shown in Figure 2.9, the droop parameters including the slope and voltage references can automatically be adapted by varying the load current. Actually, with the conventional droop approach, the droop curve needs to be shifted by a secondary controller to regulate the dc voltage at an acceptable interval. However, by utilizing the nonlinear droop method, the effective voltage reference can be determined by the droop curve [4], where the effective voltage shifting can be calculated as follows:

$$\Delta V_i = \frac{(\alpha - 1)(V^* - V_{min})}{I_{max,i}^\alpha} I_i^\alpha \tag{2.3}$$

The current sharing error and voltage regulation performance of the conventional and nonlinear droop methods are illustrated in Figure 2.10. As is shown, employing

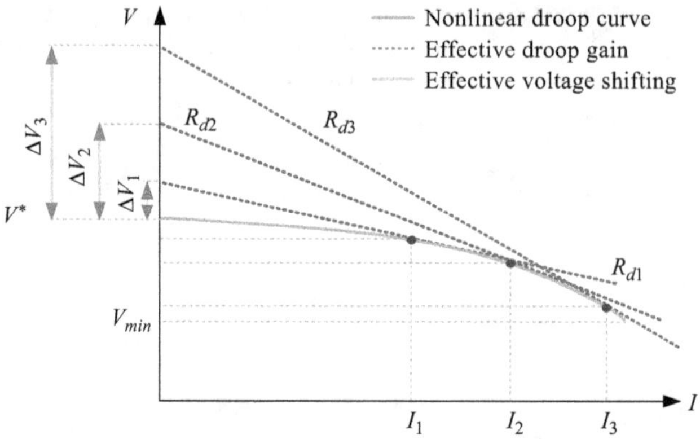

Figure 2.9 Characteristics of the nonlinear droop control method

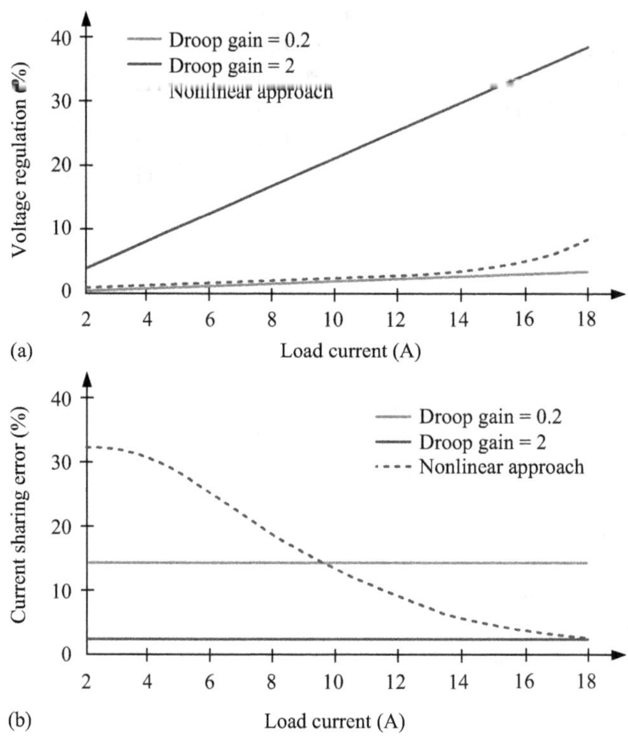

Figure 2.10 Performance comparison of the nonlinear method with the conventional droop control; (a) voltage regulation and (b) current sharing error

small droop gains causes acceptable voltage regulation but leads to high current sharing error. Although selecting higher droop gains causes appropriate current sharing, it leads to a large voltage drop in the higher load currents. However, employing the nonlinear curve introduces an acceptable current sharing in a heavy-load condition and appropriates voltage regulation in different load currents as can be seen from Figure 2.10.

2.2.3 Frequency droop control

A typical dc microgrid with distributed loads is shown in Figure 2.11(a). The dc source can be a dispatchable unit such as a fuel cell module or a hybrid battery/nondispatchable unit such as a photovoltaic array, which can control the dc-link voltage as a voltage source converter. The proposed control approach for the *k*th unit is shown in Figure 2.11(b). Each converter modulates a small ac voltage

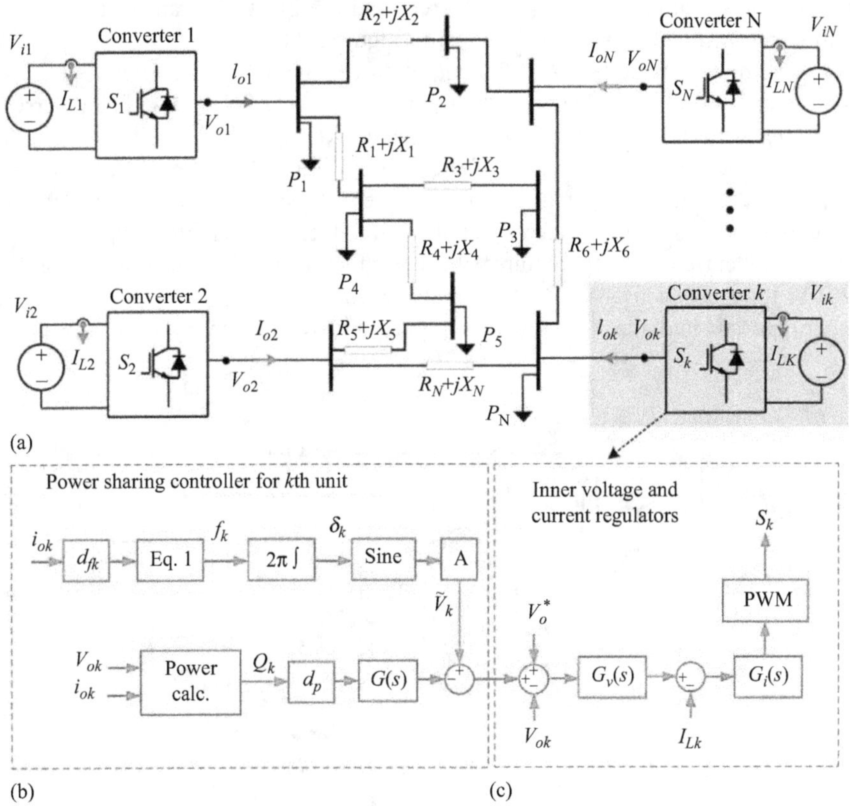

Figure 2.11 Block diagram of (a) a typical dc MG with distributed loads, (b) proposed control structure for the kth converter, (c) inner voltage and current control loops—$G_v(s)$ and $G_i(s)$ are PI-based inner voltage and current controllers. PI, proportional–integrator

superimposed onto the dc voltage, where the frequency of the ac voltage is proportional to the output dc current of the converter. The rated frequency should be selected to be smaller than the bandwidth of the inner voltage controller, and hence, to be regulated by a proportional–integrator (PI)-based voltage regulator. Therefore, the inner voltage ($G_v(s)$) and current ($G_i(s)$) controllers in Figure 2.11(c) can modulate the reference voltage including dc voltage and superimposed ac voltage. From the ac voltage point of view, the converters are working like a synchronous generator; hence, they can be coordinated together with the common frequency [2,11,12]. From the power system dynamics and control theory, for analyzing the dynamic behavior of a Synchronous Generator (SG) in an ac power system, it can be modeled as two SGs; one being the specified SG and the other modeling the entire power system. Moreover, the two SGs can be simplified as a single-machine-infinite-bus, where the infinite bus is considered as a stiff ac source [13]. Therefore, since the proposed approach is based on the SG principles, without losing the generality, a simplified dc MG, with two converters connected to a load at a point of common coupling (PCC), is considered, and the block diagram of the system with the corresponding signals is shown in Figure 2.11.

According to Figure 2.12, if the output dc voltage of the converters (V_{o1}, V_{o2}) is settled at a reference value (V_o^*), their output dc current (I_{o1}, I_{o2}) will be inversely proportional to the corresponding line resistances (i.e., $I_{o1}/I_{o2} = R_2/R_1$), where R_1 and R_2 denote the line resistance of the first and second converter, respectively. Adjusting the output dc voltage of the converters is the only option for controlling the corresponding output currents at a desired value, for example, proportional to their rated current, which requires the coordination of the converters. For the coordination between converters, a small ac voltage, i.e., $\tilde{v}_k = A \cdot \sin(2\pi f_k t)$, is superimposed onto the dc voltage reference and modulated by each converter.

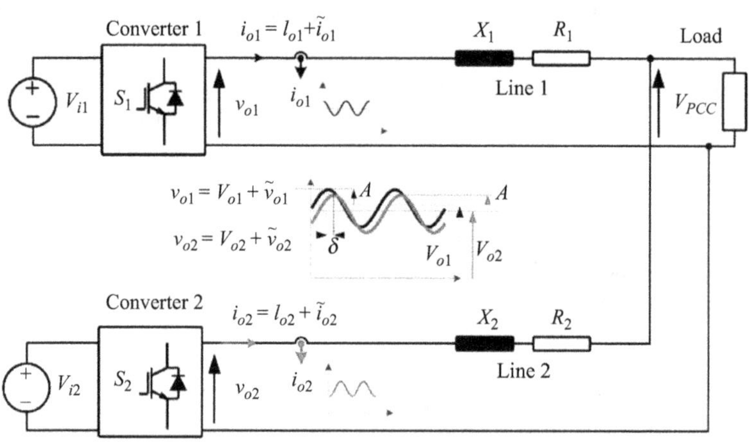

Figure 2.12 Conceptual illustration of the proposed strategy showing the injected ac voltages and corresponding currents in a simplified dc MG based on two dc–dc converters

The amplitude of the superimposed voltage (denoted as A) is considered as a small constant value with a small ripple factor; however, it should be detectable by measurement. Furthermore, the corresponding frequency should be proportional to the output current of the converter. It can be defined as [2,11,12]

$$f_k = f^* - d_{fk} i_{ok} \tag{2.4}$$

where f^* (50 Hz) is the nominal frequency, i_{ok} being the output current, and d_{fk} is the frequency droop gain of the kth converter determined by the following equation:

$$d_{fk} = \frac{f_{max} - f_{min}}{I_{n,k}}; \quad k = 1, 2 \tag{2.5}$$

with f_{max}/f_{min} being the maximum/minimum frequency for tuning the droop gain and $I_{n,k}$ is the nominal current of kth converter. At steady-state condition, the frequency of the units has the same value; hence, the output current of the units has to be shared in proportion to their rated current as shown in (2.6), where $\langle\cdot\rangle$ denotes the average value and ξ is the ratio of the rated current of the converters.

$$\frac{\langle i_{o1}\rangle}{\langle i_{o2}\rangle} = \frac{I_{o1}}{I_{o2}} = \frac{I_{n,1}}{I_{n,2}} = \frac{d_{f2}}{d_{f1}} = \xi \tag{2.6}$$

Frequency droop can be used to coordinate the converters by a common injected frequency; therefore, the dc currents need to be regulated by the frequency to control the power sharing. However, the dc currents are determined by the dc voltages as

$$I_{o1} = \frac{V_{o1} - V_{PCC}}{R_1}; \quad I_{o2} = \frac{V_{o2} - V_{PCC}}{R_2} \tag{2.7}$$

with R_1 and R_2 being the line resistances. Therefore, the dc voltages should be adjusted by an ac injected variable related to the frequency. According to Figure 2.11 and (2.6), the phase angle of each unit (δ_1, δ_2) can be found as

$$\delta_1 = \frac{2\pi}{S}\left(f^* - d_{f1} I_{o1}\right), \quad \delta_2 = \frac{2\pi}{S}\left(f^* - d_{f2} I_{o2}\right) \tag{2.8}$$

where S is the Laplace operator. The relative phase (δ) between the converters, thus, is equal to

$$\delta = \delta_1 - \delta_2 = \frac{2\pi}{S}\left(d_{f2} I_{o2} - d_{f1} I_{o1}\right) \tag{2.9}$$

If the output currents are not proportional to the rated currents, the phase angle will not be zero; hence, this phase difference causes a small ac power flow. As the load impedance is higher than the line impedances, the small ac power will only flow between the converters. According to the ac power flow theory, ac power is proportional to the ac currents ($\tilde{i}_1$,$\tilde{i}_2$). Furthermore, the ac currents are proportional to the line impedances. Thereby, the ac power contains the information of the line impedances. On the other hand, in LV systems with low X/R ratio, the reactive power can accurately be controlled by the frequency [14]. Therefore, employing the injected reactive power (Q) of the converters to adjust the dc voltage reference (V_o^*) causes proper current sharing. Applying the proposed control algorithm, the output dc voltage of the converters can be written as

$$V_{o1} = V_o^* - d_p Q_1 G(s), \quad V_{o2} = V_o^* - d_p Q_2 G(s) \tag{2.10}$$

where d_p is the coupling gain between dc voltage and reactive power, and $G(s) = \omega_c/(s+\omega_c)$ is a low pass filter to eliminate the high-frequency component of the calculated reactive power. Therefore, the frequency droop can be used to coordinate the converters, and the small ac power can be employed to adjust the dc voltage and, consequently, the dc currents. Each converter can be controlled by the local measured values, and hence, like SGs, there is no need for any communication network [2,11,12]. Furthermore, the injected ac voltage by the converters must be synchronized with the ac component of the grid voltage at the startup time. The phase of the connection bus voltage can be extracted using a phase locked loop (PLL) block. On the other hand, in ac systems, synchronization methods are employed to make the converter voltage close to the grid voltage in order to limit the inrush current at the start time, which may damage the converter switches for the high currents. However, the injected ac voltage and consequently the ac currents are very small in the proposed approach, and hence, the converters can be connected without utilizing a PLL. Hence, they can be synchronized based on the droop control functionality as the grid supporting voltage source converters in ac grids [15].

2.3 Distributed approaches

In the distributed approaches, each converter is equipped with a secondary controller cascaded by the primary controller. A communication system is employed to share the voltage and current information of converters between all of them. In a simple distributed secondary control, it is a requirement that all of the converters have access to the information of all the others. This approach may be complex and have a lower resiliency. However, consensus strategy with a sparse communication network has been presented which can use the neighboring converter information to reach an acceptable operating condition, i.e., suitable load sharing and appropriate voltage regulation. Furthermore, another sparsely communicated control approach can fulfill the power management objectives by only employing the current information of the converters. The distributed approaches are explained in the following section.

2.3.1 Fully communicated control

In this approach, secondary controllers can locally regulate the voltage of the microgrid. A general distributed secondary controller is shown in Figure 2.13.

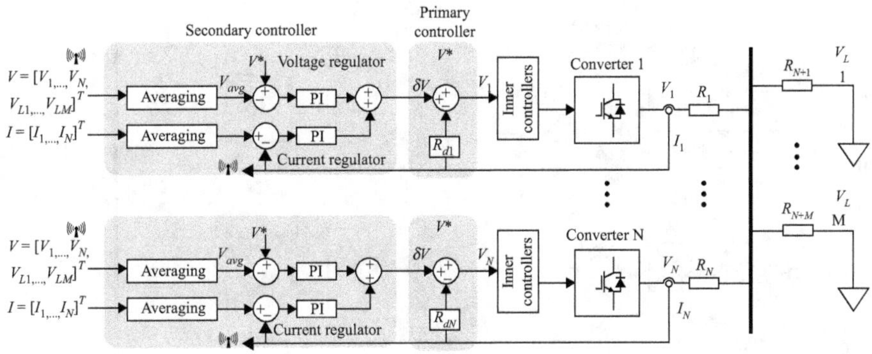

Figure 2.13 Distributed secondary controller

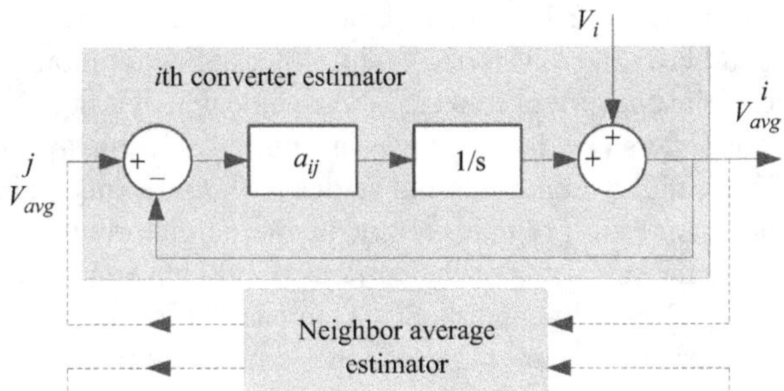

Figure 2.14 Consensus global average voltage estimator [19]

A communication network collects the information of the voltage of the desired buses, $\boldsymbol{V} = [V_1, V_2, \ldots, V_M]^T$, and the output current of the converters, $\boldsymbol{I} = [I_1, I_2, \ldots, I_N]^T$. The secondary controller of each converter calculates the average voltage (V_{avg}) and weighted average current of the converters and, then, regulates its output voltage and current accordingly. Implementing the secondary controller in the distributed policy improves the reliability of the system, since the central controller is replaced by distributed regulators.

2.3.2 Sparse communicated (consensus-based) control

However, regulating the average of the voltages and currents requires more data transmission through the microgrid. To overcome this problem, consensus algorithms are employed to regulate the global variables in distributed systems [3,16]. In consensus algorithms, each converter needs to communicate with the neighboring converters. An updating protocol as shown in Figure 2.14, also called *dynamic consensus protocol*, carries out the estimated global average voltage of the neighboring converter and the local voltage to estimate the global average voltage [3,17–20]. The average voltage of the *j*th converter, $V^j{}_{avg}$, can be used to calculate the average voltage of the *i*th converter, $V^i{}_{avg}$, with the dynamic consensus

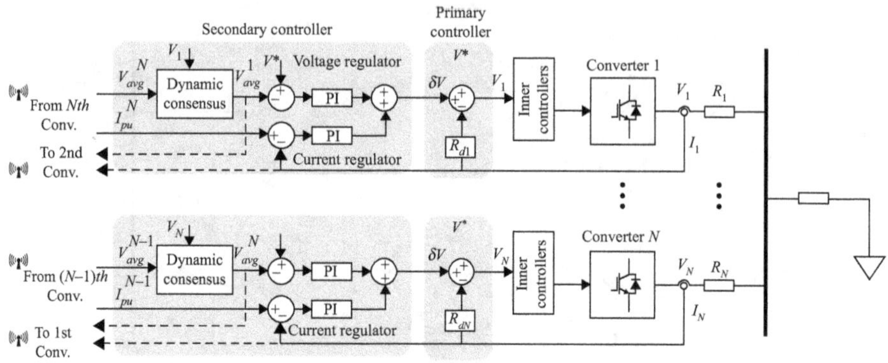

Figure 2.15 Distributed secondary controller

estimator shown in Figure 2.14, where the coefficient a_{ij} is the weight of information exchanged between converters i and j. After some iterations, the estimated average voltage of the converters converges to a value which is equal to the average voltage of the converters (V_{avg}). As shown in Figure 2.15, this average voltage is regulated by the secondary controller and settles at the reference value.

On the other hand, the per-unit current of the neighboring converter can be used to compensate the mismatch of the currents of all converters. This approach is analogous to the circular-chain-control (CCCs) in parallel inverters in ac microgrids, which is used to improve the current mismatches among the inverters [14,21–24].

In this approach, each converter shares the average voltage and per-unit current with its neighboring converter. Therefore, sparse communication among the converters with low volume of transmitted information improves the reliability and stability of the system [3]. However, in consensus algorithms, the average voltage of the grid supporting buses is only regulated. In practice, the loads may not be connected to the grid supporting converters and might be distributed over the microgrid. Therefore, regulating the voltage of the grid supporting buses may not guarantee the voltage regulation at load buses.

2.3.3 Sparse communicated control using current information

Distributed secondary control using current information is shown in Figure 2.16 including two controllers for load sharing and voltage restoring in the microgrid [25]. A current regulator is used to increase the load sharing accuracy of the conventional droop controller, and a voltage regulator regulates the average voltage of the microgrid to the nominal values, which are explained in the following section.

2.3.3.1 Current regulator

Current sharing among converters is conventionally performed by a droop gain R_{dk} which can be defined as follows for the kth converter:

$$R_{dk} = \frac{V_{max} - V_{min}}{I_{nk}} \tag{2.11}$$

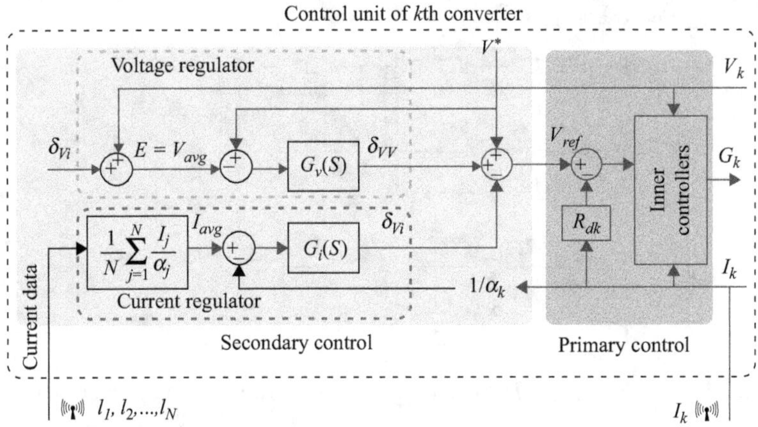

Figure 2.16 Proposed control approach; voltage regulator G_v, current regulator G_i, and low-bandwidth communication link with delay function of $G_d(s) = (1/(1+\tau s))$ [25]

where V_{max} and V_{min} are the maximum and minimum allowable voltage range, respectively, and I_{nk} is the rated current of the kth converter. Due to the differences in line resistances, the output current of converters cannot be proportionally dispatched among the converters. The current regulator calculates the weighted average currents of converters and regulates the corresponding output current proportional to the rated current of the converter. The average current (I_{avg}) can be calculated as [25]

$$I_{avg} = \frac{1}{N}\sum_{j=1}^{N}\frac{I_j}{\alpha_j} \quad j = 1:N, \tag{2.12}$$

where N is the number of converters, I_j is the measured current and α_j is the sharing coefficient of jth converter, respectively. Furthermore, as with the CCCs approach, the converters can only employ the neighboring converters information to reach proportional load sharing.

A simplified single line model of a dc microgrid with two converters is shown in Figure 2.17. In steady state, droop gain acts as a series resistor (R_{d1} and R_{d2}). The secondary current regulator behaves as a small positive/negative resistor (r_{d1} and r_{d2}) such that the total resistance of each line becomes proportional to the corresponding rated current. Hence, the relationship between rated current (I_{nj}) and sharing coefficient (α_j) and total line resistance between jth converter and PCC can be written as

$$\frac{I_{ni}}{I_{nj}} = \frac{\alpha_j}{\alpha_i} = \frac{R_{dj} + r_{dj} + r_j}{R_{di} + r_{di} + r_i}; \quad i,j = 1:N, \quad i \neq j \tag{2.13}$$

where r_j is the resistance of the line connected to jth converter.

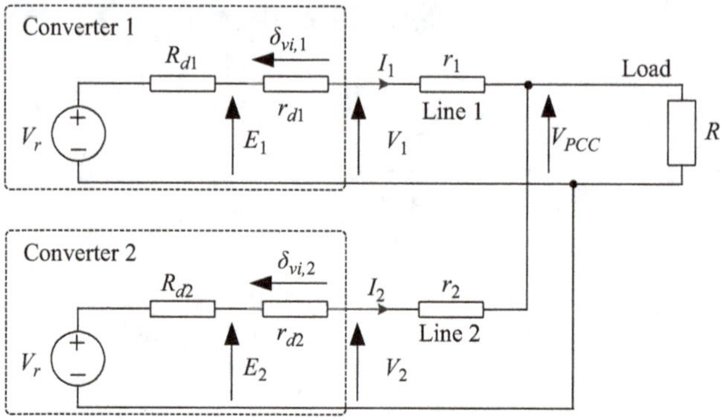

Figure 2.17 Simplified model of two converter-based dc microgrid

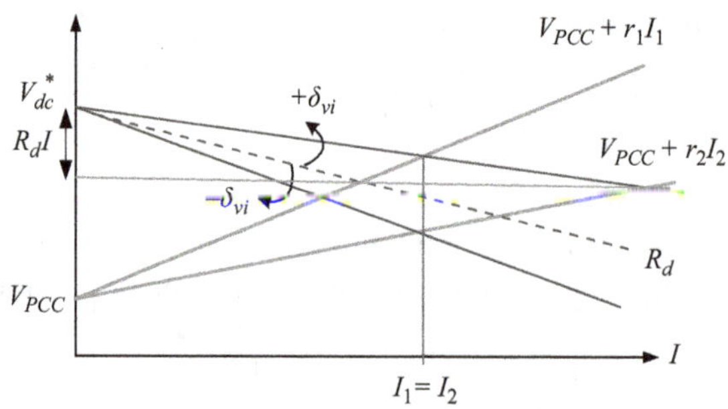

Figure 2.18 Proposed droop characteristic adjustment

The effect of a current regulator in a power sharing system is schematically described in Figure 2.18. The dashed graph shows the effect of conventional droop gain. The secondary current regulator alters the slope of this droop characteristics to reach the same current between the two converters. Sharing coefficients are assumed to be one for both converters. Therefore, the accurate current sharing is obtained with a droop controller and a current regulator. However, the dc voltage of the microgrid is dropped, and a distributed voltage regulator is required to restore the average voltage of the microgrid.

2.3.3.2 Voltage regulator

This regulator compensates for the voltage drop though the droop gains of the converters. From Figure 2.17, the output voltage of each converter (V_1 and V_2) can be calculated as

$$\begin{cases} V_1 = V_r - R_{d1}I_1 - \delta_{vi1} \\ V_2 = V_r - R_{d2}I_2 - \delta_{vi2} \end{cases} \tag{2.14}$$

where V_r is the reference value of the voltage loop. If sharing coefficients are assumed to be one, then the droop gains must be equal ($R_{d1} = R_{d2} = R_d$) and $I_1 = I_2$ at steady state. The output values of the current regulators can be obtained as

$$\begin{cases} \delta_{vi1} = \left(\dfrac{I_1 + I_2}{2} - I_1\right) G_d G_i(s) = \left(\dfrac{I_2 - I_1}{2}\right) G_d G_i(s) \\ \delta_{vi2} = \left(\dfrac{I_1 + I_2}{2} - I_2\right) G_d G_i(s) = \left(\dfrac{I_1 - I_2}{2}\right) G_d G_i(s) \end{cases} \tag{2.15}$$

where $G_i(s)$ is the PI controller and $G_d(s)$ is the delay of communication link. At steady state, $I_1 = I_2$, and hence $\delta_{vi1} + \delta_{vi2} = 0$. Therefore, the average voltage of the microgrid is

$$V_{avg} = \frac{1}{2}(V_1 + V_2) = V_r - R_d I \tag{2.16}$$

Applying the primary droop controller and secondary current regulator causes an average voltage drop equal to $R_d I$. From the single line model of the microgrid shown in Figure 2.17, internal voltages (i.e., E_1 and E_2) are equal to the average voltage calculated by (2.16). Therefore, the distributed voltage regulator can estimate the internal voltage and regulate it at the reference value. In fact, the correction term (δ_{vv}) shifts up the droop characteristics in Figure 2.18 to restore the average voltage of the microgrid which can be calculated as (2.17), where V^* is the rated voltage of the microgrid.

$$\begin{cases} \delta_{vv1} = (V^* - (V_1 + \delta_{vi1}))G_v(s) \\ \delta_{vv2} = (V^* - (V_2 + \delta_{vi2}))G_v(s) \end{cases} \tag{2.17}$$

The relationship between converter voltage and PCC voltage can be shown as

$$\begin{cases} V_1 - V_{PCC} = r_1 I_1 \\ V_2 - V_{PCC} = r_2 I_2 \end{cases} \tag{2.18}$$

where

$$V_{PCC} = R_L(I_1 + I_2) \tag{2.19}$$

The set point value for the primary controller is

$$\begin{cases} V_1^* = V_{ref1} - R_d I_1 \\ V_2^* = V_{ref2} - R_d I_2 \end{cases} \tag{2.20}$$

and the set point value for the inner voltage loop can be determined by the primary controller as

$$\begin{cases} V_1^* = V_{ref1} - R_d I_1 \\ V_2^* = V_{ref2} - R_d I_2 \end{cases} \tag{2.21}$$

Substituting (2.17), (2.18) and (2.20) in (2.21) results in

$$\begin{cases} V_1^* = V^* - \delta_{vi1} - \dfrac{R_d I_1}{1 + G_v(s)} \\ V_2^* = V^* - \delta_{vi2} - \dfrac{R_d I_2}{1 + G_v(s)} \end{cases} \tag{2.22}$$

This equation shows that the term of $R_d I$ which is related to the primary controller can be eliminated in low frequencies, i.e., in the secondary controller frequency bandwidth. Therefore, the primary controller shares the load current between converters based on droop gain, and the secondary controller reduces the mismatch in current sharing as well as decreasing the voltage drop of the droop gain. This approach uses the only current information to appropriately control the load sharing and voltage regulation in the dc microgrid.

2.4 Conclusion and future study

In this chapter, distributed and decentralized control approaches for dc microgrids are discussed. Distributed approaches employ a communication system among different converters in order to regulate the dc voltage and improve the load sharing accuracy. Some of the distributed methods utilize point-to-point communication links among converters; however, some of them use a sparse communication system based on consensus protocol. Sparse communication-based control approaches are more reliable and resilient than the fully communicated methods.

On the contrary, the decentralized methods use no communication (physical link) between converters to reach the power sharing objectives. In these approaches, the control system of each converter uses local voltage and current information to control the output power (current) of the corresponding converter. Since these converters do not need to communicate with other converters, the overall stability and reliability can be enhanced. The centralized control approaches can be categorized as mode-adaptive (autonomous) droop control, nonlinear droop control and frequency droop control, and these methods are conceptually discussed in this chapter.

For future studies, it is important to establish new power sharing controls with higher resiliency and reliability other than droop-based methods. These methods should be applicable for both dispatchable and nondispatchable resources. Furthermore, transient conditions such as connecting/disconnecting a converter, or fault conditions should be taken into account in the control system to reach stable and reliable operation. Moreover, stability issues such as constant power load problems should be analyzed with the newly suggested control algorithms to ensure the viability of control systems in real-operation conditions.

References

[1] Peyghami, S., Mokhtari, H., Davari, P., Loh, P.C., Blaabjerg, F.: 'On Secondary Control Approaches for Voltage Regulation in DC Microgrids' *IEEE Trans. Ind. Appl.*, 2017, **53**, (5), pp. 4855–4862.

[2] Peyghami, S., Davari, P., Mokhtari, H., Loh, P.C., Blaabjerg, F.: 'Synchronverter-Enabled DC Power Sharing Approach for LVDC Microgrids' *IEEE Trans. Power Electron.*, 2017, **32**, (10), pp. 8089–8099.

[3] Nasirian, V., Davoudi, A., Lewis, F.L., Guerrero, J.M.: 'Distributed Adaptive Droop Control for dc Distribution Systems' *IEEE Trans. Energy Convers.*, 2014, **29**, (4), pp. 944–956.

[4] Khorsandi, A., Ashourloo, M., Mokhtari, H., Iravani, R.: 'Automatic Droop Control for a Low Voltage DC Microgrid' *IET Gener. Transm. Distrib.*, 2016, **10**, (1), pp. 41–47.

[5] Khorsandi, A., Ashourloo, M., Mokhtari, H.: 'A Decentralized Control Method for a Low-Voltage DC Microgrid' *IEEE Trans. Energy Convers.*, 2014, **29**, (4), pp. 793–801.

[6] Guerrero, J.M., Vasquez, J.C., Matas, J., De Vicuña, L.G., Castilla, M.: 'Hierarchical Control of Droop-Controlled AC and DC Microgrids—A General Approach Toward Standardization' *IEEE Trans. Ind. Electron.*, 2011, **58**, (1), pp. 158–172.

[7] Peyghami, S., Mokhtari, H., Blaabjerg, F.: 'Hierarchical Power Sharing Control in DC Microgrids', in Magdi S Mahmoud (Ed.): 'Microgrid: advanced control methods and renewable energy system integration' (Oxford: Elsevier Science & Technology, 2017, first), pp. 63–100.

[8] Boroyevich, D., Cvetković, I., Dong, D., Burgos, R., Wang, F., Lee, F.: 'Future electronic power distribution systems—A contemplative view' *Proc. Int. Conf. Optim. Electr. Electron. Equipment, OPTIM*, 2010, pp. 1369–1380.

[9] Loh, P.C., Li, D., Chai, Y.K., Blaabjerg, F.: 'Autonomous Operation of ac–dc Microgrids with Minimised Interlinking Energy Flow' *IET Power Electron.*, 2013, **6**, (8), pp. 1650–1657.

[10] Gu, Y., Xiang, X., Li, W., He, X.: 'Mode-Adaptive Decentralized Control for Renewable DC Microgrid with Enhanced Reliability and Flexibility' *IEEE Trans. Power Electron.*, 2014, **29**, (9), pp. 5072–5080.

[11] Peyghami, S., Mokhtari, H., Loh, P.C., Davari, P., Blaabjerg, F.: 'Distributed Primary and Secondary Power Sharing in a Droop-Controlled LVDC Microgrid with Merged AC and DC Characteristics' *IEEE Trans. Smart Grid*, 2018, **9**, (3), pp. 2284–2294.

[12] Peyghami, S., Mokhtari, H., Blaabjerg, F.: 'Decentralized Load Sharing in a Low-Voltage Direct Current Microgrid with an Adaptive Droop Approach Based on a Superimposed Frequency' *IEEE J. Emerg. Sel. Top. Power Electron.*, 2017, **5**, (3), pp. 1205–1215.

[13] Kundur, P., Balu, N., Lauby, M.: *'Power system stability and control'* (New York: McGraw-Hill, 1994).

[14] Guerrero, J.M., Hang, L., Uceda, J.: 'Control of Distributed Uninterruptible Power Supply Systems' *IEEE Trans. Ind. Electron.*, 2008, **55**, (8), pp. 2845–2859.

[15] Rocabert, J., Luna, A., Blaabjerg, F., Rodriguez, P.: 'Control of Power Converters in AC Microgrids' *IEEE Trans. Power Electron.*, 2012, **27**, (11), pp. 4734–4749.

[16] Lu, X., Guerrero, J.M., Sun, K., Vasquez, J.C.: 'An Improved Droop Control Method for DC Microgrids Based on Low Bandwidth Communication with DC Bus Voltage Restoration and Enhanced Current Sharing Accuracy' *IEEE Trans. Power Electron.*, 2014, **29**, (4), pp. 1800–1812.

[17] Meng, L., Dragicevic, T., Vasquez, J.C., Guerrero, J.M.: 'Tertiary and Secondary Control Levels for Efficiency Optimization and System Damping in Droop Controlled DC–DC Converters' *IEEE Trans. Smart Grid*, 2015, **6**, (6), pp. 2615–2626.

[18] Meng, L., Dragicevic, T., Guerrero, J.M., Vasquez, J.C.: 'Dynamic consensus algorithm based distributed global efficiency optimization of a droop controlled DC microgrid' *ENERGYCON 2014—IEEE Int. Energy Conf.*, 2014, pp. 1276–1283.

[19] Shafiee, Q., Dragicevic, T., Vasquez, J.C., Guerrero, J.M.: 'Hierarchical Control for Multiple DC-Microgrids Clusters' *IEEE Trans. Energy Convers.*, 2014, **29**, (4), pp. 922–933.

[20] Moayedi, S., Davoudi, A.: 'Distributed Tertiary Control of DC Microgrid Clusters' *IEEE Trans. Power Electron.*, 2015, **31**, (2), pp. 1717–1733.

[21] Wu, T.-F., Chen, Y.-K., Huang, Y.-H.: '3C Strategy for Inverters in Parallel Operation Achieving an Equal Current Distribution' *IEEE Trans. Ind. Electron.*, 2000, **47**, (2), pp. 273–281.

[22] Ding, G., Gao, F., Zhang, S., Loh, P.C., Blaabjerg, F.: 'Control of Hybrid AC/DC Microgrid under Islanding Operational Conditions' *J. Mod. Power Syst. Clean Energy*, 2014, **2**, (3), pp. 223–232.

[23] Hatziargyriou, N., Asano, H., Iravani, R., Marnay, C.: 'Microgrids' *IEEE Power Energy Mag.*, 2007, **5**, (4), pp. 78–94.

[24] Lu, X., Guerrero, J.M., Sun, K., Vasquez, J.C., Teodorescu, R., Huang, L.: 'Hierarchical Control of Parallel AC–DC Converter Interfaces for Hybrid Microgrids' *IEEE Trans. Smart Grid*, 2014, **5**, (2), pp. 683–692.

[25] Peyghami-Akhuleh, S., Mokhtari, H., Loh, P.C., Blaabjerg, F.: 'Distributed secondary control in DC microgrids with low-bandwidth communication link' *2016 7th Power Electronics and Drive Systems Technologies Conference (PEDSTC)*, IEEE, 2016, pp. 641–645.

Chapter 3
Stability analysis and stabilization of DC microgrids

Alexis Kwasinski[1]

3.1 Dynamic characteristics of DC microgrids

Stability issues in DC microgrids can be characterized by dynamics associated to time constants ranging various time scales. At longer time scales, sometimes ranging from seconds to hours, stability concerns are related to the need of matching generated and consumed powers. That is, statically, generated power needs to match consumed power in loads and sources, whereas, dynamically, power generation sources response ramps should be able to follow load changes. Still, these stability concerns associated to generation and demand matching, both statically and dynamically, are fundamentally similar to same stability concerns observed in AC power systems and microgrids. These stability issues due to electric power generation and consumption mismatches tend to be more noticeable in microgrids than in power due to the lower combined inertia of microsources and because microsources maximum power ratings tend to be closer to individual loads power consumption. The common approach to address these stability concerns is to add energy storage [1] that can act as temporary sources by discharging when power demand is higher than supply or temporarily as loads by charging when power demand is lower than supply. Additionally, several studies have proposed control approaches that simulate the presence of inertia in microgrids [2,3]. In the case of DC microgrids, [4] discusses analogies between stored energy in electrical generators' rotors of conventional AC grids and the role of stored energy in capacitors of DC microgrids.

As indicated, stability issues associated to power generation and consumption mismatches are similar in AC and DC microgrids. However, stability issues at shorter time scales are more commonly observed in DC microgrids than in AC systems because these issues originate in the presence of loads interfaced by power electronic converters acting as instantaneous constant-power loads (CPLs), which tend to be more prevalent in DC microgrids than on AC ones because in DC microgrids almost all type of loads tend to be powered through a point-of-load

[1]Department of Electrical and Computer Engineering, University of Pittsburgh, USA

(POL) converter. Hence, within the context of this book, it is more relevant to focus the discussion of DC microgrids stability on the effect of CPLs and suitable approaches for mitigating the stability issues that these loads introduce.

As Figure 3.1(a) exemplifies, instantaneous CPLs are commonly realized by very efficient POL converters with fast controllers that regulate their output voltage tightly. Since the output voltage of the POL converter remains sufficiently constant, its output power with a resistive load will remain approximately constant, too. Hence, its input power will also remain constant despite input voltage changes because the POL converter is assumed to have a sufficiently high efficiency so that the input power equals to the output power. Figure 3.1(b) shows a common practical instantaneous CPL: a data center server. In the particular case of this figure, the servers are powered in part by photovoltaic arrays through a 380 V DC power distribution network.

Mathematically, ideal instantaneous CPLs are modeled by

$$i(t) = \frac{P_L}{v(t)} \tag{3.1}$$

where $v(t)$ is the voltage at the CPL, $i(t)$ is the current through the CPL and P_L is the power of the CPL. In practical applications, CPLs do not maintain indefinitely a same value for P_L, but although P_L may change over time, changes in P_L occur at time scales much longer than the time constants associated to the dynamics of the system under study. Additionally, real CPLs only show the behavior modeled by (3.1) above a given threshold voltage. Below such voltage, protections that are part of the CPL controller act in order to disconnect such load to prevent the current to reach excessively high values. Nevertheless, in order to provide a systematic analysis of DC microgrids stability characteristics and control that is not dependent on

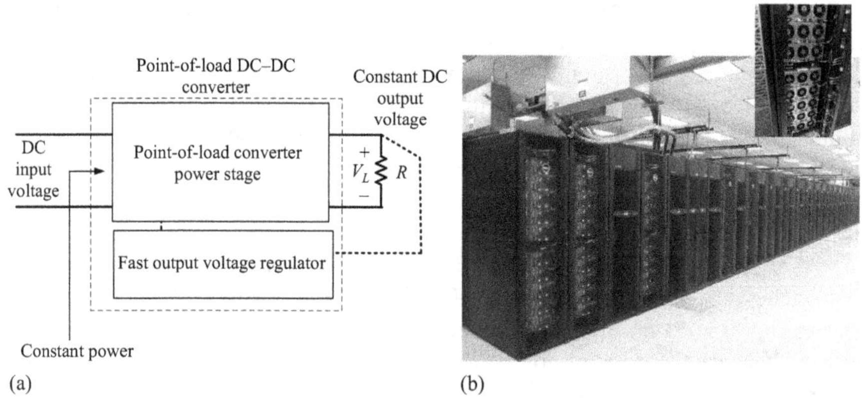

Figure 3.1 DC load systems. (a) A point-of-load converter acting as a CPLs. (b) An actual CPL: servers in a data center (Texas Advanced Computing Center). The detail on the upper-right corner show the servers power supplies acting as CPLs

specific design choices, such as selection of CPLs low voltage threshold, for the rest of this chapter, a voltage threshold is omitted and, thus, it is assumed that CPLs behave as described by (3.1) irrespective of the value of $v(t)$.

The next section of this chapter discusses the effect of CPLs on DC microgrids stability characteristics followed by two sections describing approaches for mitigating the destabilizing effect of CPLs. These mitigating approaches are described first considering passive methods followed by actively controlled strategies.

3.2 DC microgrids stability analysis

Stability challenges appear with CPLs when connecting at least two converters in a cascade configuration, as shown in Figure 3.2. In a DC microgrid, one of these converters would be a POL converter, whereas the other would be a converter interfacing an electric power source with the power distribution grid powering the POL converter. In Figure 3.2, the converter interfacing the power source is represented by a buck converter. However, the same general stability characteristics are observed when using other type of converters to interface a power source with the microgrid power distribution grid [5,6]. In order to initiate the discussion of DC microgrids stability characteristics with CPLs, consider that the buck converter interfacing a power source is controlled with a fixed duty cycle D and that it operates in continuous conduction mode. Hence, this buck converter can be mathematically represented by its average model given by

$$\begin{cases} L\dfrac{d\bar{i}_L}{dt} = DE - \bar{v}_C \\ C\dfrac{d\bar{v}_C}{dt} = \bar{i}_L - \dfrac{P_L}{\bar{v}_C} \\ \bar{v}_C > \varepsilon,\ \bar{i}_L \geq 0 \end{cases} \tag{3.2}$$

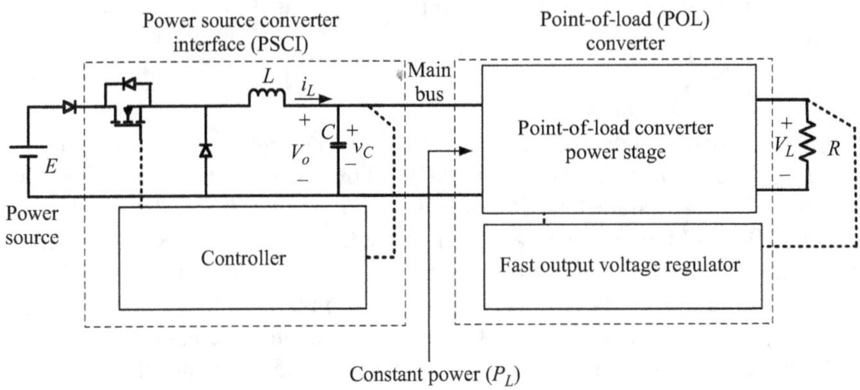

Figure 3.2 A POL converter acting as a CPL to a PSCI located upstream in a cascade configuration

where E is the power source voltage; $\overline{v}_C$ is the power source converter interface (PSCI) average capacitor voltage, which equals the DC bus average voltage; i_L is the PSCI average inductor current; P_L is the CPL power; L is the PSCI inductance; and C is the PSCI output capacitor capacitance. The equilibrium point for this converter is then

$$\begin{pmatrix} V_C \\ I_L \end{pmatrix} = \begin{pmatrix} DE \\ \dfrac{P_L}{DE} \end{pmatrix} = \begin{pmatrix} V_O \\ \dfrac{P_L}{V_O} \end{pmatrix} \tag{3.3}$$

It is possible to observe that this equilibrium point is not stable. In order to prove this characteristic, it is relatively simple to find that the characteristic equation for the small signal model of the buck PSCI with a CPL is given by

$$\lambda^2 - \frac{P_L}{CV_C^2}\lambda + \frac{1}{LC} = 0 \tag{3.4}$$

where λ are the unknown eigenvalues. Since the coefficient for the first-order term is negative, the characteristic polynomial associated to the buck PSCI with a CPL does not satisfy the Routh–Hurwitz criterion and, thus, the equilibrium point for a system represented by (3.1) is not stable.

Although small signal models allows to determine the stability characteristics of the equilibrium point in a relatively simple way, linearization analytic methods omit important behavior characteristics of converters with CPLs. A nonlinear analysis approach, such as the one in [5–7], shows that depending on the PSCI capacitor voltage and inductor current initial conditions, it is possible to observe that the PSCI shows two behaviors: for sufficiently large initial capacitor voltages, the voltage and current waveforms eventually settle into an oscillatory behavior, which in a state space plot represents a limit cycle behavior. Otherwise, the capacitor voltage drops and the inductor current increases to very high values. As [5–7] also show, it is possible to find that the curve—a separatrix—in the state space that acts as a boundary to both behaviors can be approximated to

$$i_L = \frac{P_L}{v_C} - \frac{Cv_C^2}{LP_L}(E - v_C) \tag{3.5}$$

A simulation based on the model represented by (3.1) can be used to exemplify the two possible behaviors of a PSCI with a CPL. For example, consider a buck PSCI with the following parameters: $E = 400$ V, $C = 3$ mF, $L = 0.15$ mH, $D = 0.5$ and $P_L = 50$ kW. As Figure 3.3 shows, different choices for initial conditions result in either the oscillatory behavior or the voltage collapse-high current conditions. Evidently, both of these behaviors are undesirable in a DC microgrid in which the PSCI output voltage is intended to be constant. Although the assumption for the analysis is that the converter operates in continuous conduction mode, in [7], it was shown that in most practical conditions, a more exact model, than (3.1) that considers operation in discontinuous conduction mode, does not deviate significantly from the already discussed observations. Additionally, as it was indicated above,

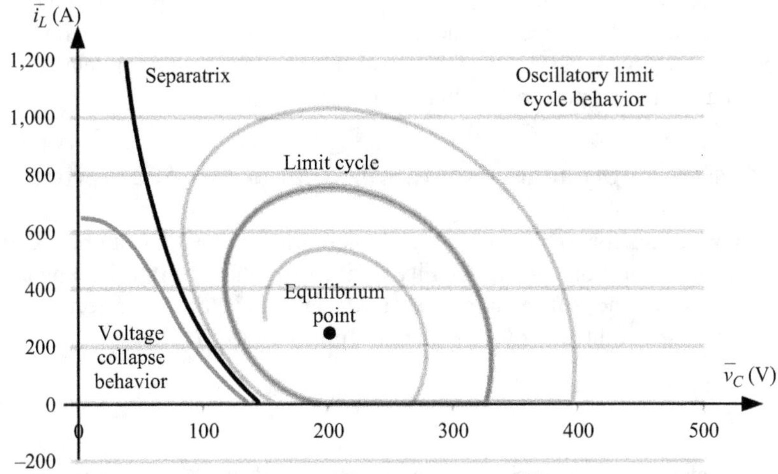

Figure 3.3 Various trajectories from different conditions in state space showing the two possible behaviors of a PSCI with a CPL shown in Figure 3.2

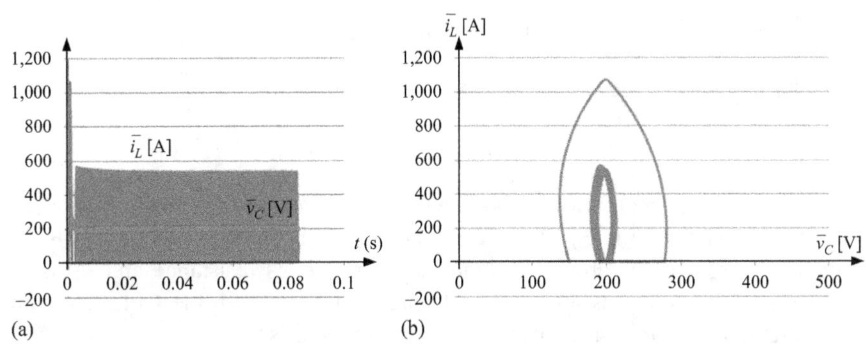

Figure 3.4 Time domain (a) and state-space response (b) of a buck PSCI attempting to regulate its output voltage that is supplied to a CPL

although the analysis uses a buck converter as an example for the discussion of the observations, as [5,6] shows, the same general behavior described for the buck converter is also observed in other DC–DC converter topologies.

One common initial thought to address the stability issues introduced by CPLs and to achieve a constant DC voltage on the microgrid main bus is to attempt to regulate the PSCI output voltage with a PI controller. However, as Figure 3.4 shows with the same PSCI used in Figure 3.3 but now with a PI controller with $i_L(t=0) = 1$ A, $v_C(t=0) = 150$ V, $k_i = 10$, $k_p = 0.1$ that attempts to regulate the output voltage at 200 V, the oscillatory behavior still persists exemplifying the fact that a common PI controller, that could achieve a stable equilibrium point with a PSCI powering a resistive load, may not achieve a stable equilibrium point when the load is an equivalent CPL. Hence, the next sections will discuss various strategies to mitigate the stabilizing effect of CPLs and to achieve a constant regulated DC voltage in a microgrid power distribution grid.

3.3 Passive approaches for stabilization of DC microgrids

Passive approaches for stabilization of DC microgrids with CPLs can be identified by a necessary condition for the stability derived from (3.5). As it is explained in [7], a necessary but not sufficient condition for stability is that the trajectory of any PSCI stays on the right side of the separatrix given by (3.5). That is,

$$i_L > \frac{P_L}{v_C} - \frac{Cv_C^2}{LP_L}(E - v_C) \tag{3.6}$$

Hence, the stability characteristics of a DC microgrid with CPLs improves when C increases, L decreases or PL decreases. Still, (3.6) only provides an initial approximation of suitable passive approaches for stabilizing DC microgrids.

In order to further determine passive stabilization approaches for DC microgrids, consider the expanded model of a buck PSCI

$$\begin{cases} L\dfrac{di_L}{dt} = q(t)(E - R_{SW}i_L) - (1 - q(t))(V_D + i_L R_D) - i_L R_L - v_C \\ C\dfrac{dv_C}{dt} = i_L - \dfrac{P_L}{v_C} - \dfrac{v_C}{R_O} \end{cases} \tag{3.7}$$

$$\textit{with } i_L \geq 0,\ v_C > \varepsilon$$

in which $q(t)$ is the main switch switching function, R_{SW} is the conduction resistance of the main switch, R_D is the conduction resistance of the diode, V_D is the diode's forward voltage drop, R_L is the inductor's series resistance and R_O is an output resistance acting as an additional load in parallel with the CPL. The new characteristic equation of the linearized average model is

$$\lambda^2 + \left(\frac{R_i}{L} - \frac{P_L}{CV_O^2} + \frac{1}{R_O C}\right)\lambda + \frac{R_i}{L}\left(\frac{1}{R_O C} - \frac{P_L}{CV_O^2}\right) + \frac{1}{LC} = 0 \tag{3.8}$$

where R_i is the sum of $R_{SW}D$, R_L and $R_D(1-D)$. Based on the Routh–Hurwitz criterion, the two necessary and sufficient conditions for a stable equilibrium point are

$$P_L < V_O^2\left(\frac{C}{L}R_i + \frac{1}{R_O}\right) \tag{3.9}$$

and

$$P_L < V_O^2\left(\frac{1}{R_i} + \frac{1}{R_O}\right) \tag{3.10}$$

Although both (3.9) and (3.10) need to be satisfied in order to ensure a stable equilibrium point, from a practical perspective in most applications (3.10) is satisfied if (3.9) is satisfied because R_i is usually less than 1 Ω and C is within the same order or an order of magnitude higher than L.

From (3.9), it is possible to identify several passive approaches for improving the stability characteristics of DC microgrids. One of such approaches is to reduce L. However, this is not a practical approach because the inductance is usually determined by other design needs and constraints. Another approach is to add an output resistance in parallel to the CPL. If it is assumed in (3.9) that $R_i = 0$, then the maximum output resistance that needs to be added in parallel with the CPL equals V_O^2/P_L. As Figure 3.5 exemplifies, adding an output resistance of 0.5 Ω to the PSCI simulated in Figure 3.3 achieves a stable equilibrium point, but if such resistor is added for the only purpose to achieve a stable equilibrium point, then the overall power efficiency would be 50% at best (power efficiency in the example in Figure 3.5 equals 0.38 because only 38% of the PSCI output power is dissipated in the CPL).

Another approach to achieve a stable equilibrium point suggested by (3.9) is to increase the PSCI's output capacitance. However, since R_i tends to be in the order of milli-Ohms or tens of milli-Ohms, the minimum capacitance needed in order to achieve stable equilibrium points are relatively high, leading to large and bulky designs. Such result is exemplified in Figure 3.6 in which the PSCI simulated in Figure 3.3 was augmented with an R_{SW} of 50 mΩ and with a new output capacitance of 50 mF. One practical alternative in DC microgrids to increase the PSCI's output capacitance is to add batteries directly connected to the microgrid power distribution network. However, such solution may lead to some control limitations, particularly if droop control is sought.

As it has been discussed, although passive approaches allow achieving a stable operating point, they tend to present important practical limitations. Thus, the next section discusses alternative strategies to achieve the same goal of a stable equilibrium point though controller action.

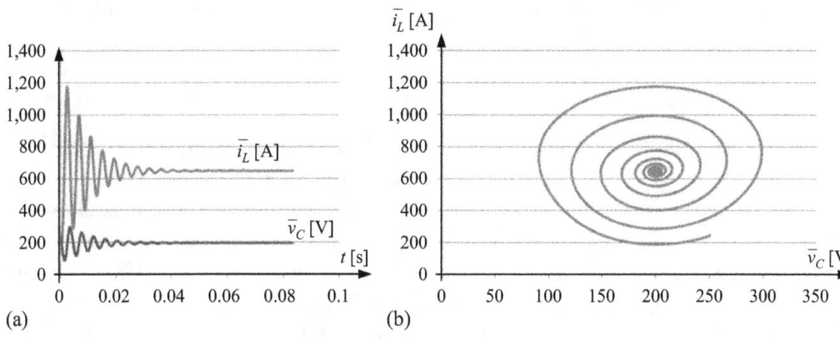

Figure 3.5 Time domain (a) and state-space behavior (b) of a buck PSCI powering a CPL and a resistive load

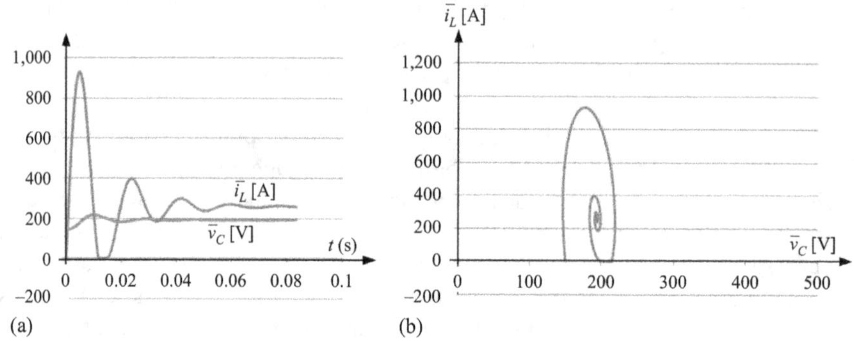

Figure 3.6 Time domain (a) and state-space behavior (b) of a buck PSCI with increased output capacitance and powering a CPL

3.4 Control strategies for stable DC microgrids operation

Control-based approaches for achieving stable operation points in DC microgrids with CPLs can be divided into time-based approaches and geometric control strategies. Arguably, the most commonly discussed time-based approach involves the insertion of virtual resistances to PSCIs through controller action. Since these inserted resistances are not real, their insertion do not affect the converter power efficiency. As it is shown in [5–7], such approach may lead in the case of buck PSCIs like that represented by (3.2) to a linear proportional-differential (PD) controller of the form

$$d = \frac{R_{i1}C}{E}\frac{de}{dt} + \frac{R_{i1}}{R_{i2}E}e + D \tag{3.11}$$

where e is the error signal resulting from the difference between the measured output voltage and the desired output voltage, R_{i1} is a virtual resistance in series with the PSCI's inductor and R_{i2} is a virtual resistance in parallel to the PSCI's output capacitor. As indicated in [5], asymptotic stability is observed only if

$$\frac{R_{i1}}{L} > \frac{P_L}{V_O^2 C} \tag{3.12}$$

Still, global asymptotic stability is not observed in practical applications due to the fact that the duty cycle is constrained to values between 0 and 1. An example of the result of using a PD controller to achieve a stable equilibrium point is shown in Figure 3.7. In this figure, a buck PSCI was simulated with the same parameters of the converter used in Figure 3.3 with initial conditions $i_L(t = 0) = 1$ A and $v_C(t = 0) = 150$ V, and a proportional gain of 0.5 and differential gain of $1 \cdot 10^{-3}$, implying a value of R_{i1} of 133.33 Ω and R_{i2} of 0.66 Ω.

The work in [5,6,8] shows that PD controllers also achieve a stable equilibrium point for other types of PSCIs, such as boost and buck-boost converters. However, although a properly designed PD controller may achieve a stable operation point,

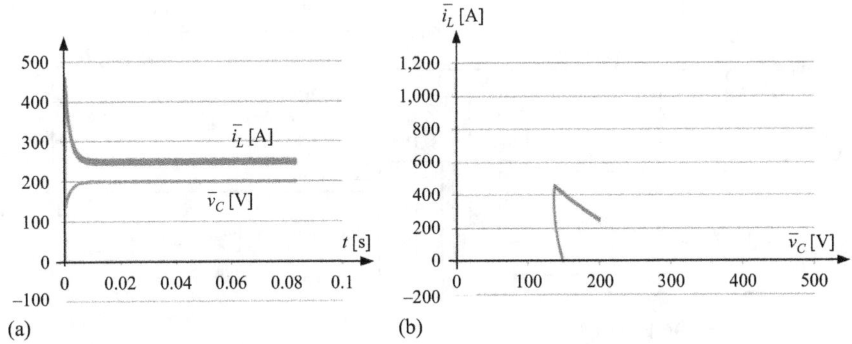

Figure 3.7 Time domain (a) and state-space behavior (b) of a buck PSCI powering a CPL and controlled with a PD controller

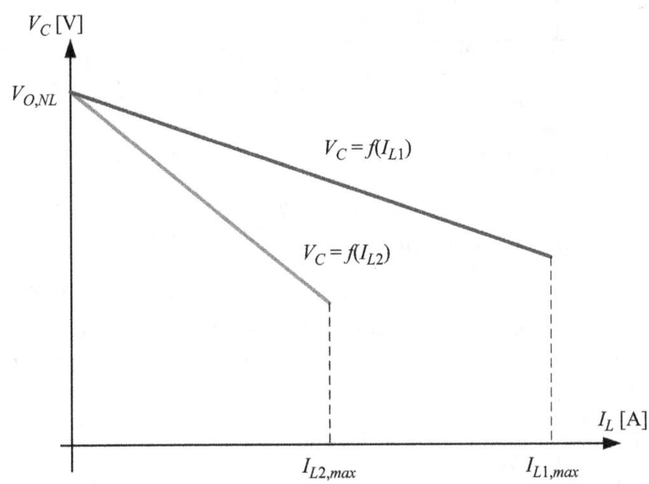

Figure 3.8 Droop lines for two parallel connected PSCIs

it does not provide output voltage regulation. Nevertheless, as shown in [5], a stable equilibrium point with output voltage regulation can be achieved by adding an integral term to the PD controller in (3.11) provided that the integral gain is low enough so that the dynamics of the integral term are slower than those associated to the PD terms.

Dynamics introduced by droop controllers create a control action somewhat analogous to that associated to the aforementioned PD controllers. The main concept of conventional droop controllers, at what it is commonly referred as the primary controller, is to insert a virtual resistance at the PSCI output via controller action with the goal of achieving autonomous control goals, such as load sharing among PSCIs connected in parallel. As a result, the output voltage vs. current relationship follows a line such as the one shown in Figure 3.8 for two buck PSCIs connected in parallel. Still, droop relationships show static operational relationships.

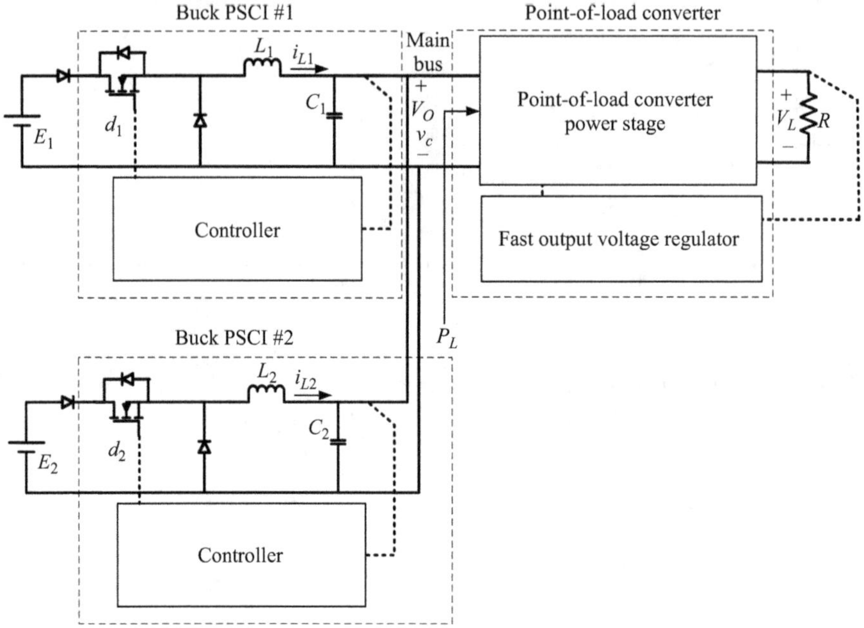

Figure 3.9 Two parallel-connected buck PSCIs sharing a CPL with a droop controller in each PSCI

In order to evaluate the dynamic effect of adding virtual droop resistances to two PSCIs with a CPL, consider the following dynamic average model of the resulting system [9]

$$\begin{cases} L_1 \dfrac{d\bar{i}_{L1}}{dt} = d_1 E_1 - \bar{v}_C \\ L_2 \dfrac{d\bar{i}_{L2}}{dt} = d_2 E_2 - \bar{v}_C \\ (C_1 + C_2) \dfrac{d\bar{v}_C}{dt} = \bar{i}_{L1} + \bar{i}_{L2} - \dfrac{P_L}{\bar{v}_C} \end{cases} \tag{3.13}$$

in which the corresponding variables and parameters are depicted in Figure 3.9. As also shown in [9], the duty cycles d_1 and d_2 are controlled so that if $V_{O,NL}$ is the no-load output voltage, then

$$d_1 = \frac{V_{0,NL} - R_{d,1}\bar{i}_{L1}}{E_1} \tag{3.14}$$

and

$$d_2 = \frac{V_{0,NL} - R_{d,2}\bar{i}_{L2}}{E_2} \tag{3.15}$$

imply the insertion of virtual resistances $R_{d,1}$ and $R_{d,2}$ in series with each of the inductors L_1 and L_2, respectively. Since the average current across these components equal the output current of the PSCIs, and the steady state average output voltage of a buck converter equals the product of the input voltage and the duty cycle, (3.14) and (3.15) imply a static linear droop relationship between the output voltage and output current. Dynamically, [9] shows that the addition of a droop resistance allows limit cycle oscillations to be damped provided that the following conditions obtained by studying the characteristic equation of the linearized system are met:

$$\frac{R_{d,1}}{L_1}+\frac{R_{d,2}}{L_2}>\frac{P_L}{(C_1+C_2)V_O^2} \tag{3.16}$$

$$\frac{L_1+L_2+R_{d,1}R_{d,2}(C_1+C_2)}{R_{d,2}L_1+R_{d,1}L_2}>\frac{P_L}{V_O^2} \tag{3.17}$$

$$\left(R_{d,1}||R_{d,2}\right)<\frac{V_O^2}{P_L} \tag{3.18}$$

$$\begin{aligned}&\left[\frac{P_L^2}{V_O^4(C_1+C_2)^2}+\frac{R_{d,1}R_{d,2}}{L_1L_2}\right]\left[\frac{R_{d,1}}{L_1}+\frac{R_{d,2}}{L_2}\right]+\frac{1}{C_1+C_2}\left[\frac{R_{d,1}}{L_1^2}+\frac{R_{d,2}}{L_2^2}\right]\\&-\left[\frac{P_L}{V_O^2(C_1+C_2)}\left[\frac{R_{d,1}}{L_1}+\frac{R_{d,2}}{L_2}\right]^2\right]-\left[\frac{P_L}{V_O^2(C_1+C_2)^2}\left[\frac{1}{L_1}+\frac{1}{L_2}\right]\right]>0\end{aligned} \tag{3.19}$$

where

$$V_O=\frac{V_{O,NL}+\sqrt{V_{O,NL}^2-4P_L\left(R_{d,1}||R_{d,2}\right)}}{2}$$

The effectiveness of the droop resistances insertion in order to damp the effect of the CPL was verified experimentally with a reduced scale microgrid in [9] by using a CPL of 100 W and two parallel buck PSCIs with $E_1=35$ V, $E_2=30$ V, $C_1=C_2=1{,}000$ μF, $L_1=640$ μH, $L_2=630$ μH and $V_{O,NL}=25.7$ V. Initially, the PSCIs were controlled in open loop with a constant duty cycle equal to 0.734 for the buck converter #1 and equal to 0.857. As Figure 3.10 shows, the result of such operation is limit cycle oscillations for all state variables. Then, a primary droop controller as the one represented by (3.15) and (3.14) with $R_{d1}=2.075$ Ω and $R_{d2}=1.709$ Ω is enabled causing, as Figure 3.10 shows, damping of the oscillations and verifying reaching a stable equilibrium point.

Although the insertion of virtual droop resistance through controller action damps the limit cycle oscillations, they also naturally cause voltage deviations, which in some cases may be undesirable. The conventional solution to this issue is to add a so-called secondary droop control that regulates the main bus voltage by shifting vertically the static droop lines in Figure 3.8. Thus, as indicated in [9],

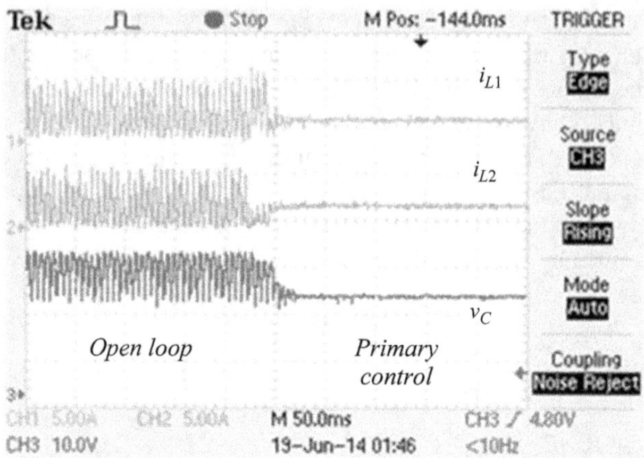

Figure 3.10 Experimental results showing the transition from open loop operation to primary droop control for two buck PSCIs with a CPL. Image courtesy of Mahesh Srinivasan

a secondary control regulating action can be achieved by adding an integral term with a gain k_i in (3.15) and (3.14). The resulting control action is then given by

$$d_1 = \frac{V_{0,NL} - R_{d,1}\bar{i}_{L1} + k_i \int \left(V_{O,NL} - \bar{v}_C\right)dt}{E_1} \tag{3.20}$$

and

$$d_2 = \frac{V_{0,NL} - R_{d,2}\bar{i}_{L2} + k_i \int \left(V_{O,NL} - \bar{v}_C\right)dt}{E_2} \tag{3.21}$$

with this controller, the new conditions for a stable equilibrium point become [9]

$$\frac{R_{d,1}}{L_1} + \frac{R_{d,2}}{L_2} > \frac{P_L}{(C_1 + C_2)V_{O,NL}^2} \tag{3.22}$$

$$\frac{1}{C_1 + C_2}\left(\frac{k_i}{L_2} + \frac{R_{d,1}}{L_1L_2} + \frac{k_i}{L_1} + \frac{R_{d,2}}{L_1L_2} - \frac{R_{d,1}R_{d,2}P_L}{L_1L_2V_{O,NL}^2}\right) > 0 \tag{3.23}$$

$$\frac{k_i(R_{d,1} + R_{d,2})}{L_1L_2(C_1 + C_2)} > 0 \tag{3.24}$$

$$\begin{aligned}&\left[\frac{P_L^2}{V_{O,NL}^4(C_1 + C_2)^2} + \frac{R_{d,1}R_{d,2}}{L_1L_2}\right]\left[\frac{R_{d,1}}{L_1} + \frac{R_{d,2}}{L_2}\right] + \frac{1}{C_1 + C_2}\left[\frac{R_{d,1}}{L_1^2} + \frac{R_{d,2}}{L_2^2}\right]\\&-\left[\frac{P_L}{V_{O,NL}^2(C_1 + C_2)}\left[\frac{R_{d,1}}{L_1} + \frac{R_{d,2}}{L_2}\right]^2\right] - \left[\frac{1}{(C_1 + C_2)}\left[\frac{1}{L_1} + \frac{1}{L_2}\right]\right] > 0\end{aligned} \tag{3.25}$$

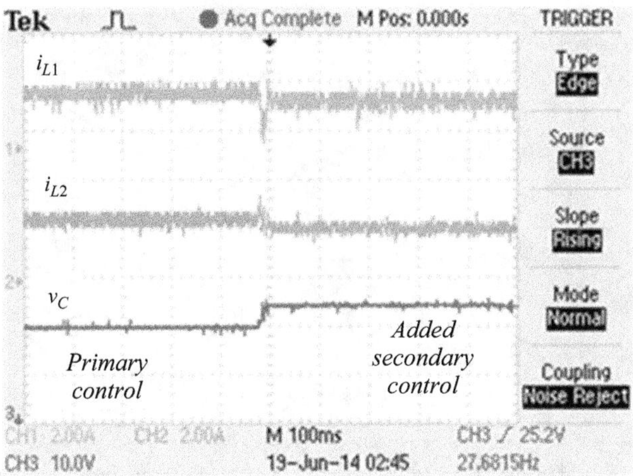

Figure 3.11 Experimental results showing the addition of regulation through a secondary droop controller for two buck PSCIs with a CPL. Image courtesy of Mahesh Srinivasan

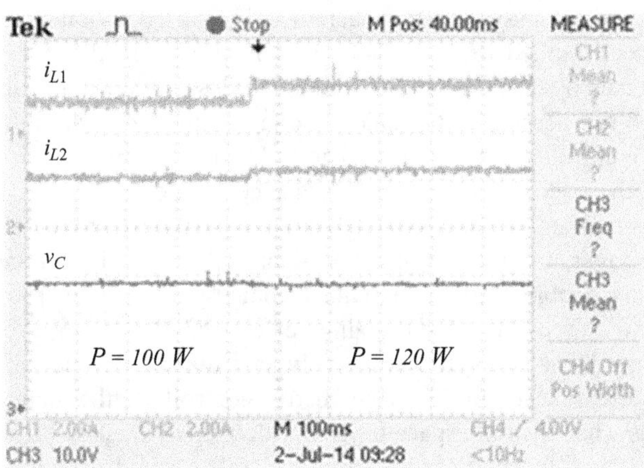

Figure 3.12 Experimental results showing droop control with load regulation for two buck PSCIs with a CPL. Image courtesy of Mahesh Srinivasan

Output voltage regulation while still achieving a stable operating point was also verified experimentally in [9]. As Figure 3.11 shows, the addition of integral regulators with $k_i = 4.5$ to the system used to produce Figure 3.10 compensate for the primary droop controller voltage deviations and restores the main bus voltage to its nominal value of 25.7 V. As it can be verified by comparing the current traces before and after the secondary controller is enabled, the currents share the load current maintaining a constant ratio of 1.22. As Figures 3.12 and 3.13 exemplify, the addition of the integral stage of the controller allows to regulate the main bus

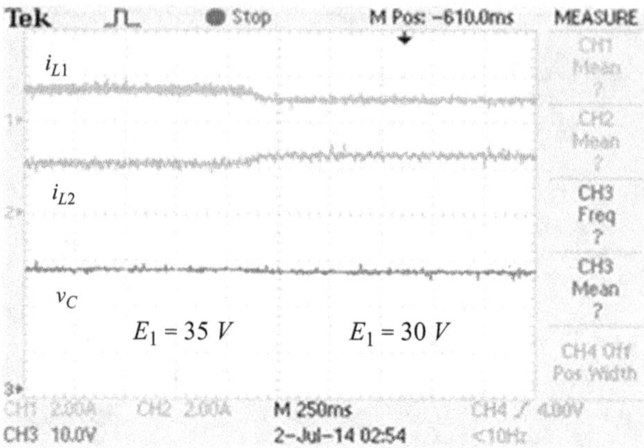

Figure 3.13 Experimental results showing droop control with line regulation for two buck PSCIs with a CPL. Image courtesy of Mahesh Srinivasan

voltage at a constant reference value even when the load's power or one or more input voltages change.

Although the discussion in this section still focused on buck PSCIs, virtual droop resistances applied to other DC–DC converters will damp the limit cycle oscillations caused by CPLs. In addition to the discussion in [9], the validity of this conclusion and the possibility to achieve voltage regulation with an integral secondary droop controller was also described in [10,11].

Although the addition of a virtual resistance through controller action makes the operating point stable, such controllers tend to have relatively slow dynamic response. Moreover, in the case of the PD controller, the differential term introduces noise susceptibility to the PSCIs. Geometric controllers provide another approach to mitigate the destabilizing effect of CPLs and to achieve a stable operating point but with a relatively fast dynamic response and without introducing noise susceptibility. In geometric controllers, the state of the main switch changes when the state space trajectory defined by the PSCI's state variables crosses a boundary. In practice, the boundary (also called switching surface) is replaced by a hysteresis band containing such boundary in order to avoid chattering when the trajectory tends to be confined along the switching surface. In general, it is possible to distinguish three behaviors when a trajectory intersects a switching surface [12]:

- Refractive behavior when the trajectories on one side of the boundary are incident, but they are exiting on the other side of the boundary.
- Rejective behavior when trajectories on either side of the boundary are exiting.
- Reflective behavior when trajectories on both sides of the boundary are incident. Once a trajectory reaches a switching surface in a reflective behavior region, it remains confined to the boundary. Hence, a reflective behavior could

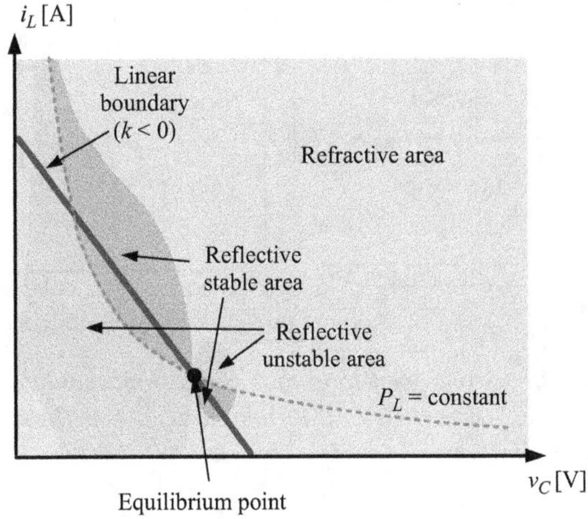

Figure 3.14 Boundary control behavior areas for a buck PSCI with a CPL

be associated to a stable equilibrium point when the confined trajectory tends along the switching surface to such an equilibrium point. However, a reflective behavior region could be associated to an unstable operating point when the confined trajectory moves away from such point.

As it was identified in [13] and it is shown with Figure 3.14, both refractive and reflective (stable and unstable) behaviors are observed when a linear boundary of the form

$$i_L = k(v_C - V_O) + \frac{P_L}{V_O} \tag{3.26}$$

is used for a buck converter powering a CPL. In (3.26), k is a design parameter that mathematically represents the slope of the line acting as the switching boundary. In order to achieve a stable operating point by having the trajectory intersecting the switching boundary at a stable reflective region instead of an unstable reflective region k needs to be negative [13]. Such need for a negative value of k in order to achieve a stable operating point is also observed for other converters powering CPLs, such as boost and buck-boost converters [6,14]. Also, [6,13,14] show that although for some of these converters, a linear boundary naturally achieves line or load regulation, it is also possible to simply enhance the controller so it can automatically displace the switching boundary to achieve line or load regulation. Figure 3.15 shows an example of the trajectories observed both in time domain and in the state space when a buck PSCI with $L = 500$ μH, $C = 1$ mF, $E = 22.2$ V, $V_O = 18$ V and a CPL of 108 W transitions from open loop operation to a linear boundary controller with $k = -1$.

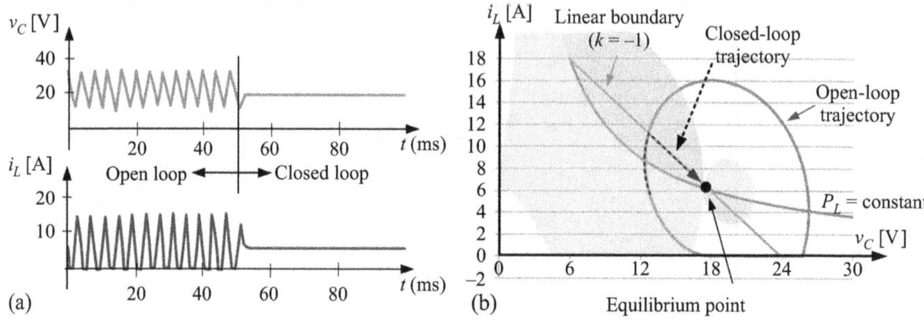

Figure 3.15 Transition from open-loop to a closed-loop boundary control for a buck PSCI with a CPL. (a) Time domain and (b) state-space representation

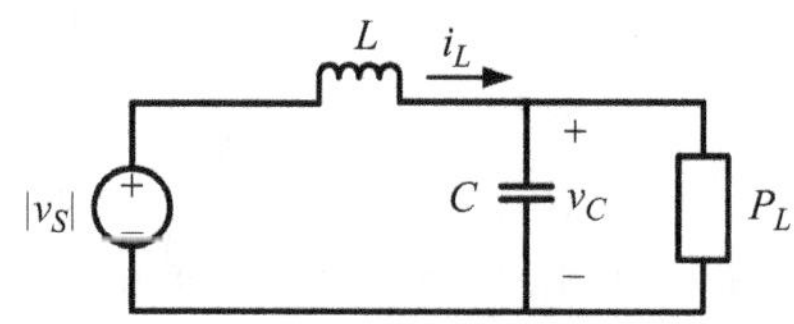

Figure 3.16 Simplified model of a rectifier with a CPL in which $|v_S|$ represents a rectified unfiltered sinusoidal voltage signal

3.5 Operation of rectifiers with instantaneous constant power loads

Although this chapter has focused primarily on the operation of DC–DC converters with CPLs, in DC microgrids, it is possible to have rectifiers directly interfacing a source and the microgrid main bus. Still, it is expected that in most applications, there would be a DC–DC converter at the output of the rectifier in order to improve regulation and filter design. However, the case of rectifiers directly interfacing an electric power source to a CPL was explored in [14]. In [14], it was shown that without an adequate mitigating approach, CPLs may cause either an oscillatory, limit cycle, behavior or a voltage collapse and high current condition similar to those discussed for a buck PSCI with a CPL. Since [14] explains that a rectifier, such as the one with a simplified model in Figure 3.16, with a second-order low-pass output filter behaves analogously to a buck converter with a CPL, then a necessary condition for the operating point to be an unstable focus that may show an oscillatory limit cycle behavior if

$$\frac{P_L}{V_{S,O}^2} < 2\sqrt{\frac{C}{L}} \tag{3.27}$$

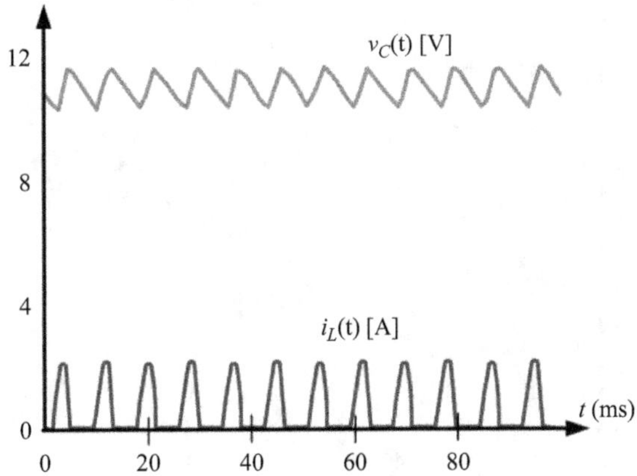

Figure 3.17 Example of limit cycle period-1 behavior of a rectifier with a CPL

where C and L are the rectifier output filter capacitor and inductor, respectively, and $V_{S,O}$ is an equivalent constant input voltage that equals $|v_s|$ when $|v_s| = v_C$ [14]. Condition (3.27) is necessary but not sufficient to observe a limit cycle behavior considering the rectifier output capacitor voltage and inductor current as state variables. In order to observe a limit cycle behavior, in addition to satisfy (3.27), the initial capacitor voltage needs to be high enough. An approximate value for such minimal initial capacitor voltage is given by [14]

$$v_C(0) > \sqrt{\frac{1}{2fV_{in}}\left(\frac{P_L^2 L}{4C}\right)^{1/4}\left[2fV_{in}\left(\frac{P_L^2 L}{4C}\right)^{1/4} + \frac{P_L}{C}\right]} \tag{3.28}$$

where V_{in} is the peak AC input voltage to the rectifier.

Both conditions (3.27) and (3.28) are necessary conditions for observing limit cycle oscillations at the rectifier output which are typically presented as the waveforms shown in Figure 3.17. Such behavior is called period-1 in [14]. However, since (3.27) and (3.28) are not sufficient conditions, it is still possible to observe the alternative rectifier behavior: voltage collapse with increasing inductor currents. As indicated in [14], closer to the limit between period-1 to voltage collapse behavior it is possible to observe another oscillatory behavior called period-2 in [14], and it is shown in Figure 3.18.

Although observing the limit cycle oscillatory behavior tends to be better than the voltage collapse behavior, the oscillations still cause output voltage variations that can be approximated to

$$\Delta v_C = V_{in} - \sqrt{V_{in}^2 - \frac{P_L}{Cf}} \tag{3.29}$$

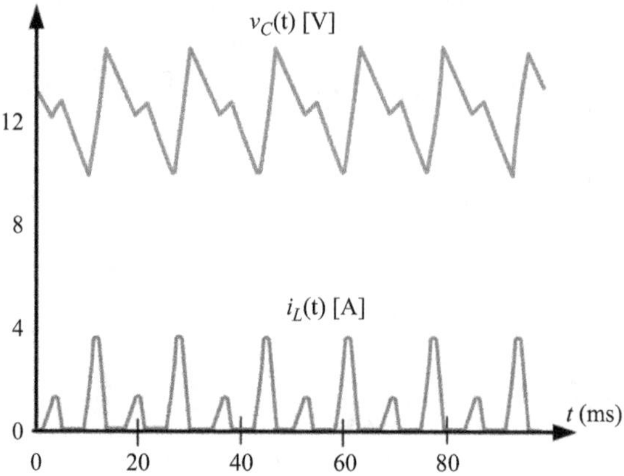

Figure 3.18 Example of limit cycle period-2 behavior of a rectifier with a CPL

This voltage variation is typically higher than conventional voltage ripple in rectifiers with an equivalent resistive load [14].

3.6 Summary

This chapter has discussed the effects of CPLs in DC-microgrids. Such loads are commonly observed in DC microgrids because of the extensive use of power electronic circuits to interface loads. As it is explained, due to their nonlinear nature, CPL creates a complex dynamic system that present two possible behaviors with an unstable equilibrium point unless PSCIs are properly controlled. In this chapter, both time domain and geometric control approaches that achieve stable equilibrium points in the presence of CPLs are explained. Passive approaches to compensate the destabilizing effect of CPLs are also discussed, but these approaches often lead to costly and bulky designs. Complex dynamics are also observed in rectifiers with CPLs, which often present excessively large output voltage variations.

References

[1] A. Toliyat and A. Kwasinski, "Energy Storage Sizing for Effective Primary and Secondary Control of Low-Inertia Microgrids," in Proc. 2015 IEEE Power Electronics for Distributed Generation Conference, pp. 1–7, Aachen, Germany, June 2015.

[2] N. Sony, S. Doolla and M. C. Chandorkar, "Improvement of Transient Response in Microgrids Using Virtual Inertia," IEEE Transactions on Power Delivery, vol. 28, no. 3, pp. 1830–1838, July 2013.

[3] X. Zhu, J. Cai, Q. Yan, J. Chen and X. Wang, "Virtual Inertia Control of Wind-Battery-Based Islanded DC Micro-Grid," in Proc. International Conference on Renewable Power Generation (RPG 2015), pp. 1–7, October 2015.

[4] T. L. Vandoorn, B. Meersman, J. D. M. De Kooning and L. Vandevelde, "Analogy Between Conventional Grid Control and Islanded Microgrid Control Based on a Global DC-Link Voltage Droop," IEEE Transactions on Power Delivery, vol. 27, no. 3, pp. 1405–1414, July 2012.

[5] A. Kwasinski, W. Weaver, and R. Balog, "Micro-Grids in Local Area Power and Energy Systems," Cambridge University Press, Cambridge, U.K., 2016.

[6] A. Kwasinski and C. N. Onwuchekwa, "Dynamic Behavior and Stabilization of DC Microgrids with Instantaneous Constant-Power Loads," IEEE Transactions on Power Electronics, vol. 26, no. 3, pp. 822–834, March 2011.

[7] A. Kwasinski and P. T. Krein, "Passivity-Based Control of Buck Converters with Constant-Power Loads," in Proc. Power Electronics Specialist Conference 2007 (PESC), pp. 259–265, Orlando, FL, USA, June 17–21, 2007.

[8] A. Kwasinski and P. T. Krein, "Stabilization of Constant Power Loads in DC–DC Converters Using Passivity-Based Control," in Proc. 2007 International Telecommunications Energy Conference (INTELEC), pp. 867–874, Rome, Italy, September 30–October 4, 2007.

[9] M. Srinivasan, "Hierarchical Control of DC Microgrids with Constant Power Loads," Ph.D. Dissertation, The University of Texas at Austin, Austin, Texas, USA, 2017.

[10] M. Srinivasan and A. Kwasinski, "Decentralized Control of a Vehicular Microgrid with Constant Power Loads," in Proc. 2014 IEEE International Electric Vehicle Conference, pp. 1–8, Florence, Italy, December 2014.

[11] M. Srinivasan and A. Kwasinski, "A Linear Damping Scheme for Higher Order dc–dc Converters Supplying Constant-Power Loads in a dc Microgrid," in Proc. 19th European Conference on Power Electronics and Applications (EPE) – Energy Conversion Congress and Exposition (ECCE) Europe, pp. 1–10, Warsaw, Poland, September 2017.

[12] P. T. Krein, "Elements of Power Electronics," Oxford University Press, New York, NY, 1998.

[13] C. N. Onwuchekwa and A. Kwasinski, "Analysis of Boundary Control for Buck Converters with Instantaneous Constant-Power Loads," IEEE Transactions on Power Electronics, vol. 25, no. 8, pp. 2018–2032, August 2010.

[14] C. N. Onwuchekwa and A. Kwasinski, "Dynamic Behavior of Single-Phase Full-Wave Uncontrolled Rectifiers with Instantaneous Constant-Power Loads," in Proc. 2011 IEEE ECCE, pp. 3472–3479, Phoenix, AZ, USA, September 17–22, 2011.

Chapter 4

Coordinated protection of DC microgrids

Jae-Do Park[1], *Md Habib Ullah*[1], *and Bhanu Babaiahgari*[1]

4.1 Introduction

In the 2000s, the development of renewable energy resources and the advance of power electronics technology, as well as the issues of fossil-fuel depletion and their impact on environment, have driven extensive research on distributed power systems. The energy policy of many governments in the world competitively increases the requirement of the penetration of renewable energy resources and distributed generation. DC power systems have received renewed attention as a solution for the distributed generation to replace or augment AC systems in various applications in transmission and distribution systems. The amount of equipment that operates on DC power is increasing, including photovoltaic (PV) panels, batteries, fuel cells, and even AC motors driven by adjustable speed drives. Improved power electronics converters allow DC power to be converted easily and efficiently. Unlike AC systems, multiple sources can readily operate attached to a single DC bus because they do not need to be synchronized. The system efficiency and reliability can be improved with DC power delivery, because the DC-to-AC power conversion layer can be eliminated from the back-to-back conversion scheme. Moreover, DC systems have an advantage in cable size because they do not suffer from skin effect, and there is no need for reactive power. Multiterminal DC (MTDC) systems have been investigated for high-voltage DC (HVDC) transmission systems and low-voltage DC (LVDC) distribution systems as well. DC power systems have been investigated for various microgrid systems, such as naval ships, traction systems, data centers, buildings, aircrafts, and microgrids.

However, compared to the AC power systems that have been in operation as the standard power system for the last two centuries after the "war of current" in the early 1800s, the DC power systems have much shorter operation history. Standards and accumulated experience are substantially insufficient. Moreover, it has been applied to the systems in confined areas in many cases. Lack of knowledge base is true as well when it comes to the system protection, especially for large systems.

[1]Department of Electrical Engineering, University of Colorado Denver, USA

Although the AC system protection technology is very mature, most of their devices cannot be directly applied in DC systems. Typically, DC switchgear is more expensive and may not always be available for certain systems.

The protection of electrical power systems is fundamentally essential for safe and stable operation; it prevents personnel injury and minimizes the service interruption and system damage from equipment failure, human error, and disastrous natural events. And this is one of the biggest challenges of DC power systems, because of the nonalternating nature and fast dynamics of DC power. Furthermore, although there are various devices that monitor the power system, detect abnormalities, and perform automatic actions to protect the system and sustain the operation, development of a well-coordinated protection scheme for DC power system is not a trivial task. As DC power system design is not quite a standardized procedure yet, its protection system is developed on the case-by-case basis and designer's empirical techniques and methods would play an important role in design.

Typically, a DC microgrid has multiple terminals connected to the bus, to which individual systems such as load, source, and bidirectional devices can interface through power converters, such as voltage-source converters (VSCs) and line-commutated converters (LCCs). Thyristors in LCCs have high-voltage and power capability, but lower control performance compared to IGBTs in VSCs. LCCs have been used for simpler PtP configurations. Recently, multimodular converter systems have been widely investigated for increased voltage and power capacity of DC bus systems. One of the challenges associated with the protection of VSC-based DC microgrid systems is that the fault current must be detected and extinguished quickly as the fault current withstand rating of typical VSCs is only twice the converter full-load current. Unlike AC systems, the typical sources in DC microgrid interfaced through VSCs do not have enough inertia to supply fault current while maintaining bus voltage. And the VSCs will trip and disconnect themselves from the bus when the bus voltage drops below a certain threshold. These make it difficult to locate the fault because the transient that contains fault circuit information might not be available to protection devices. Recently, semiconductor-based and hybrid circuit breakers (CBs) and fault current limiters have been investigated, and they have been used for fast DC current interruption and overcurrent limitation. Typically, CBs were proposed for converter and battery protection, while fuses and molded-case CBs (MCCBs) were used for feeder protection [1].

As well as the detection and interruption where traditional differential and overcurrent relaying techniques and main/back-up scheme can be applied, location and isolation of the fault in the microgrid are important tasks of a protection system for fast recovery and impact minimization, which requires cooperation between protective devices in the system. Although the line impedance method and traveling wave method have been adopted as industry standard for AC systems [2], it is difficult to directly apply to DC systems due to the inherent absence of frequency and phase parameters and much faster current dynamics.

In this chapter, nature of the faults in DC microgrid due to lack of inductance in the lines, typical DC power system configurations, coordinated protection schemes, and their applications for DC power systems will be reviewed.

4.2 Faults in DC power systems

4.2.1 Fault types and behavior

A *line-to-ground fault* is said to have occurred in a line segment when a path between either positive or negative pole and ground is created, and it is the most common type of fault. A *line-to-line fault* is defined as a path between positive and negative pole created in a line segment. A typical line-to-ground and line-to-line fault in a bipolar DC bus are shown in Figure 4.1.

If a short-circuit fault occurs in a DC bus, a large current flows in the line accompanied with terminal voltage drop. Relays will send trip signals to CBs and VSCs. Since the DC link voltage has dropped to a low level even to zero, the terminal capacitors will discharge through the fault and the fault current increases from AC main as well. This overcurrent condition might damage the reverse diodes and possibly the entire system, if the fault is not isolated quickly. The severity of a fault depends on the fault impedance and system configuration.

DC system grounding plays a critical role from the protection perspective. For a system where both AC and DC side are ungrounded, a line-to-ground fault does not create a current circulating in the ground, so the system still can operate; however, the fault has to be cleared before another line-to-ground fault occurs to generate a short circuit between lines. If the grounding points in a DC network include the neutral-ground link of the AC side or the DC-link midpoint, a line-to-ground fault in a transmission system shown in Figure 4.2 will form a ground loop with the grounding points and the AC main. In this case, the VSC cannot block the current flowing in from the AC side even if all the IGBTs are open, because the current bypasses IGBTs in VSC and flows through the antiparallel diodes as it does in a bridge rectifier. The fault current will flow from the bus capacitors that discharge through the fault path, making the fault current high-peak and fast-rising. In a system with positive or negative pole solidly grounded, a line-to-ground fault on

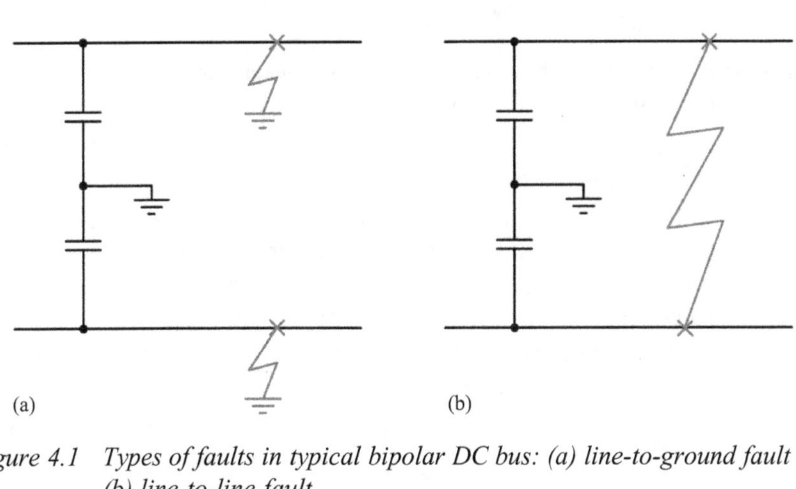

Figure 4.1 Types of faults in typical bipolar DC bus: (a) line-to-ground fault and (b) line-to-line fault

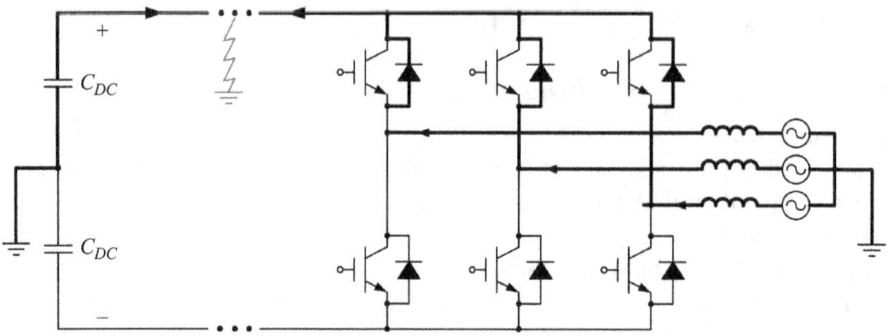

Figure 4.2 *Transmission line-to-ground fault fed by AC main and bus capacitance. The AC source feeds the fault current through antiparallel diodes regardless of the VSC switch status*

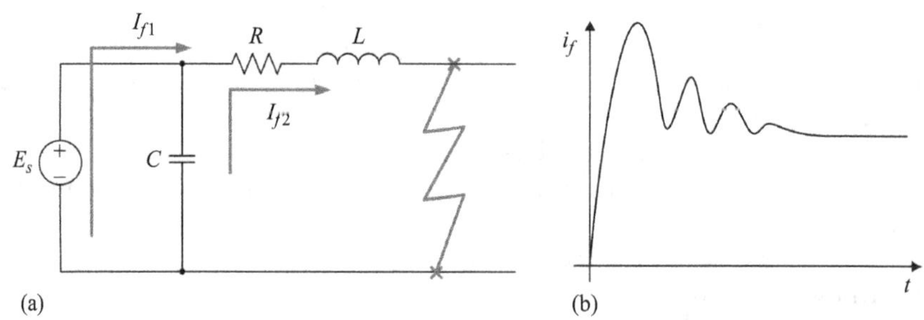

Figure 4.3 *(a) Line-to-line fault equivalent circuit model. The fault current consists of two current components from the source and bus capacitors, respectively. (b) Typical fault current waveform. The transient can be seen at the peak of the fault current*

the grounded pole will not have an impact on the system, but a fault on the other pole will cause a line-to-line short circuit through ground.

4.2.2 Fault current analysis

To explain further about the fault currents mathematically, a simple equivalent circuit model is considered and a line-to-line fault is assumed as shown in Figure 4.3(a). The current feeding the fault i_{fault} consists of two current components i_{f1} and i_{f2} from the source and bus capacitors, respectively. A mathematical expression for the fault current when the equivalent resistance in the fault circuit is small enough to cause an oscillation ($R < 2\sqrt{L/C}$) can be given as

$$I_{fault}(t) = \frac{E_s}{R}\left(1 - e^{-(R/L)t}\right) + \frac{E_s}{\omega_o\sqrt{1-\zeta^2}L}e^{-\alpha t}\sin\left(\omega_o\sqrt{1-\zeta^2}t\right) \tag{4.1}$$

where E_s is the line voltage, R, L, and C is the equivalent resistance, inductance, and capacitance in the fault path, respectively. The equivalent impedance of the fault path, including fault and line impedance, determines the natural frequency $\omega_o = 1/\sqrt{LC}$ and damping factor of the fault current $\zeta = \alpha/\omega_o$, where $\alpha = R/(2L)$ [3]. A typical waveform of fault current with this model is shown in Figure 4.3(b), and the transient can be seen at the peak of the fault current. However, in case the fault resistance or bus capacitance is large enough, such as the line-to-ground fault where the ground resistance is considerably high, the fault current can be modeled as a first-order damping thus having no oscillation.

It is very important to study short-circuit currents in a DC power system and how it can be calculated. It has to be noted here that the load current is not taken into consideration while calculating short-circuit current. A standard approximation function of short-circuit fault current proposed by International Electrotechnical Commission (IEC) [4] is shown in Figure 4.4, and the equations governing the function are given by

$$i_1(t) = i_p \frac{1 - e^{-t/\tau_1}}{1 - e^{-t_p/\tau_1}}, \quad 0 \le t \le t_p \tag{4.2}$$

where i_p is peak short-circuit current, t_p is the peak time, I_k is the quasi-steady-state short-circuit current, τ_1 is the rise time constant, τ_2 is the decay time constant, and T_k is the short-circuit duration.

$$i_2(t) = I_k + (i_p - I_k)e^{(-(t-t_p)/\tau_2)} \tag{4.3}$$

The two currents $i_1(t)$ and $i_2(t)$ determine the maximum short-circuit current that governs the rating of electrical equipment and the minimum short-circuit current that can be taken as the basis for fuse and protection device ratings, respectively. It is assumed that the quasi-steady-state current I_k is the value of the short-circuit current 1 s after the beginning of short circuit.

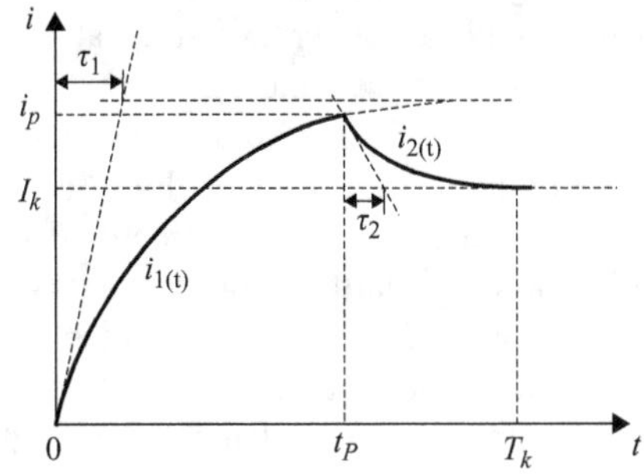

Figure 4.4 Standard approximation of a short-circuit fault current as a function of time [4], where i_p is peak short-circuit current, t_p is peak time, I_k is quasi-steady-state short-circuit current, τ_1 is rise time constant, τ_2 is decay time constant, and T_k is short-circuit duration

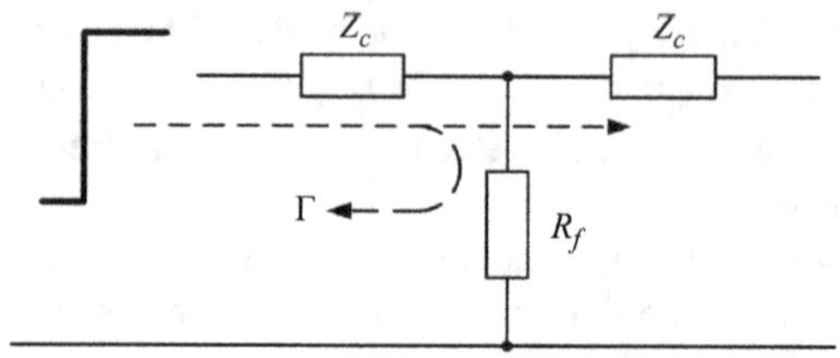

Figure 4.5 Reflection of voltage surge at the fault location. As the surge arrives at the fault location from VSC terminal, some parts of it are reflected according to the reflection coefficient Γ as a function of fault resistance R_f

For a HVDC system that has a long transmission line, a line-to-ground or line-to-line fault can cause a voltage reflection, especially when the fault resistance is small [5,6]. When a short-circuit fault happens, the voltage at the fault location decreases and the fault current rapidly increases. Immediately negative voltage surges start to flow from the fault location into the VSC terminals, where a part of the negative voltage surge is reflected back as a positive surge due to the termination of cable capacitance. This traveling wave that is reflected between the fault location and VSC terminal could increase to a very high magnitude. Figure 4.5 shows the fault resistance R_f, reflection of surges, reflection coefficient Γ, and characteristic impedance Z_c, which can be given as follows with the line impedance parameters.

$$\Gamma = \frac{1}{1 + 2(R_f/Z_c)}, \quad Z_c = \sqrt{\frac{R + j\omega L}{G + j\omega C}} \tag{4.4}$$

4.2.3 Faults in various bus configurations

4.2.3.1 Point-to-point

The *point-to-point* (PtP) configuration has conventionally been employed in HVDC power transmission systems with a converter station at each end, which can be classified into monopolar and bipolar schemes.

In the *monopolar* configuration (Figure 4.6(a) and (b)), there is one converter in each terminal connected through a transmission line. There are two variations: asymmetrical and symmetrical structure. In asymmetrical monopolar system, the converters are connected between an earthed pole and a positive pole. Earthed pole provides earthing through a low-voltage conductor which gives a return path with ground. A metallic return path can also be used. On the contrary, in symmetrical monopolar scheme, the converters are connected between high-voltage independent poles of opposite polarity: positive and negative. Symmetrical monopolar system can be grounded with high-resistance grounding or ungrounded; high-resistance grounding prevents larger fault current but causes higher ground potential.

Bipolar configuration (Figure 4.6(c)) is widely employed in HVDC systems [7]. It is the combination of two monopolar systems where two converters are connected

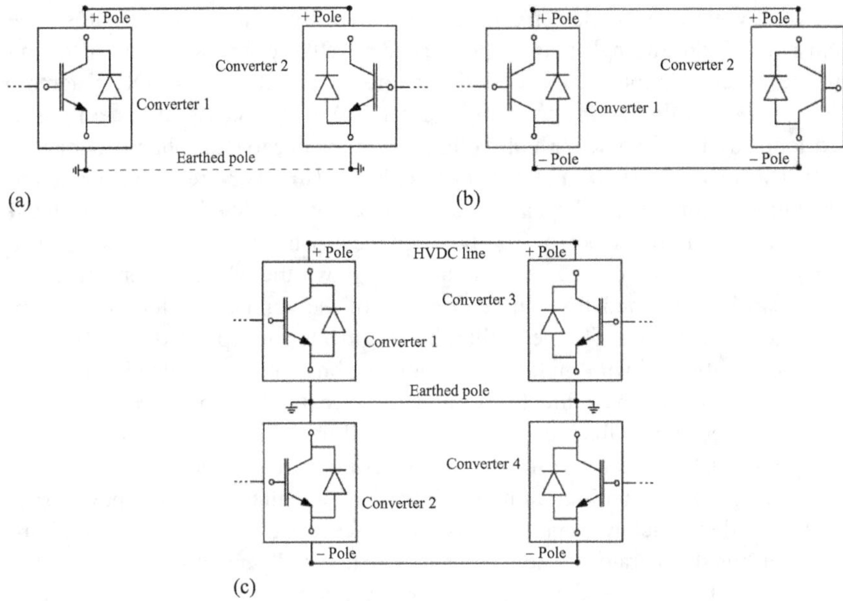

Figure 4.6 Point-to-point configurations: (a) asymmetrical monopole, (b) symmetrical monopole, and (c) bipole

in series between two poles at each terminal. The polarities of the poles are positive and negative similar to the symmetric monopolar system, but there is an earthing or a neutral point between the poles. The neutral points in the system can either be connected to or disconnected from each other. No current flows through these points in balanced operation. One of the poles can maintain the normal system operation while there is disturbance in the other pole, and the AC side protection can be used to ensure the converter protection, as well as the DC side protection schemes.

The possible faults in PtP configurations are line-to-ground fault, line-to-line fault, and AC side faults. The monopolar system is highly vulnerable to a line-to-ground fault as the power transmission gets fully interrupted unless a parallel path is available, but normal system operation can be maintained with other poles in bipolar configurations and it turns into an asymmetrical configuration; however, 50% reduced power capacity can cause an overload to the sending side converters. Although less likely to happen, the line-to-line fault causes very high current in the system. As well as the direct contact between lines, two line-to-ground faults in both poles can cause line-to-line fault if the system is ungrounded in symmetrical monopolar system. The unit protection such as distance, directional, and differential relaying can be effective solutions against faults in PtP configurations [8–10].

4.2.3.2 Multiterminal

In *multiterminal* configurations, a number of DC terminals are combined to form a power system that has improved efficiency, reliability, and flexibility, especially during fault conditions as power can be transmitted by using multiple paths. The system can withstand a single point of failure and effectively operate without affecting the whole system. Various network configurations are possible for MTDC systems.

In the *ring* configuration shown in Figure 4.7(a), a DC microgrid system uses a closed-loop bus. It is suitable for high-voltage operation due to less voltage fluctuation [11,12]. This type of system can be operated either in closed or open loop depending on operational conditions. Typically, converters are connected at each terminal through CBs to control the power flow between the bus and connected branches. The CBs can isolate faulted devices without shutting down the whole system. Because the load can be supported from the either side of the connection, the system can continue to operate even if some faulted bus segments are separated. The bus can typically be configured with multiple zones or sections with static switches or CBs; hence, the unit protection or directional protection techniques can be readily applicable for fault detection, which can be implemented by a number of intelligent CBs placed in the bus to protect the different zones during fault conditions.

Generally, *radial* configuration is utilized in low-voltage and low-power systems [13]. In this structure, separate distribution feeders from a single DC grid are fed by distributed generators. The feeders and distributed generators are connected in the grid with interface converters and protected by CBs or other protective devices. Figure 4.7(b) shows a radial DC power system. It is widely used in power distribution systems due to its major advantages such as low initial cost, simplicity,

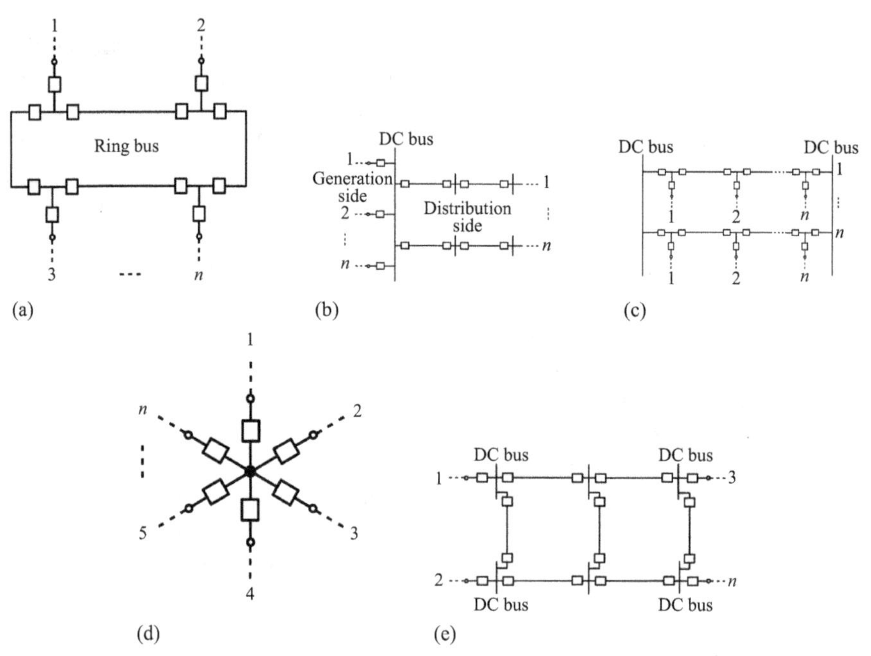

Figure 4.7 Multiterminal configurations: (a) ring, (b) radial, (c) ladder, (d) star, and (e) mesh. Rectangular boxes denote circuit breakers

and effective coordination of protective devices [14]. However, it is not as reliable since any fault on a feeder close to the main grid can hamper the downstream consumers. Also, disconnecting one transmission feeder would cause power interruption to the whole system.

The *ladder* configuration is a dual DC bus architecture that consists of distributed arrays of power generation, converters, and loads which are connected to the two buses as rungs in a ladder. Figure 4.7(c) shows a typical ladder configuration of DC power system. The rungs can be operated as an individual or as a cascaded power system. During the cascaded operation mode, each rung can share power with others using multiple paths through the DC buses. Since there is a redundancy in the arrays of primary power sources, this configuration can prevent the system from single point of failure, and it can withstand multiple failures without affecting the critical loads and power output of the system.

The *star* configuration is simple and secured topology that has been widely used in large-scale offshore HVDC wind farms [15,16]. In a star configuration, all the DC terminals are connected in a common star node through HVDC CBs as illustrated in Figure 4.7(d). The main drawback of this configuration is that any fault in the common node may cause voltage collapse and shut off the whole system [17]. Sophisticated protection schemes are required in order to protect the central node and enhance the system reliability.

The *mesh* configuration (Figure 4.7(e)) would be the most reliable configuration in electrical distribution system although the network can be complex. This configuration consists of multiple paths between two points so that power can reroute in fault conditions. In a meshed network, power can flow in any direction regardless of upstream or downstream concept. The redundancy of path further improves the degree of flexibility in power system planning. However, interface converters need to be placed at the key points of the network to control the power flow in the system [18]. The converters and transmission lines are protected by CBs from both ends. This structure can also withstand single or multiple points of failure. The drawback of this configuration is power losses in the redundant paths and the associated cable cost.

4.3 Coordinated protection techniques

Coordination between protective devices is important to improve the protection selectivity and operation continuity, as well as its capability to properly shut down or separate the system from the possible faults or abnormal conditions. Conventional concepts of AC system's protective coordination, such as main/back-up protection, time-interval/current settings, instantaneous trip (IT) devices, overcurrent/distance relay settings, zone overlapping, and automatic reclosing, can mostly be applied to the protection of DC power systems; however, the characteristics of DC power needs to be taken into consideration.

Typically, DC microgrids have generation sources and bidirectional devices, such as energy storage, as well as the loads. In case they are connected to the AC grid through PWM converters, the power can flow bidirectionally through the point of AC/DC coupling. Furthermore, the presence of energy storage or distributed generators should be carefully considered for the fault detection/location and restoration after the fault. Continuously flowing DC fault current is difficult to interrupt due to the arc, and its fast-rising dynamics can impact the instantaneous selectivity and time-interval/current settings. The location of the fault is challenging as well because the fault transient that contains information about the fault location happens at the peak of the fault current, which has to be interrupted as soon as possible due to the lack of capability to sustain the fault current in typical VSC-based DC microgrids. These make the design of protection system for DC microgrids complex compared to the typical radial AC power systems.

4.3.1 AC side protection

4.3.1.1 Fault detection

In case the DC microgrid has a connection point to an AC grid, AC CBs disconnect the DC microgrid from the AC systems when some abnormalities occur in the AC side. Possible AC side faults include phase-to-ground fault, phase-to-phase fault, phase unbalance, converter transformer differential, transformer short-circuit, AC overvoltage, AC overcurrent, AC filter overload/unbalance, and AC > DC current differential fault. Caution needs to be taken for the DC microgrid protection system not to detect AC side fault as a DC system fault as they would generate a disturbance on DC side voltage and current, which typically appears as oscillations in DC bus voltage and current. A method to use the rate of change of DC bus current (di_{DC}/dt) to differentiate AC side faults from DC faults, because di_{DC}/dt is lower for an AC side fault than that of a DC side fault due to the filtering effect of DC bus capacitors [19]. Once the AC CB connecting a source VSC to AC grid trips, the DC microgrid can experience a substantial power loss that requires coordination between bus voltage controller, energy storage controller, and load side controllers to mitigate the impact and continue the operation [20,21]. The internal faults of the source VSC, e.g., failures on IGBTs or control power supply, that typically trip the associated AC CB will also generate a similar effect.

4.3.1.2 Protection by AC circuit breakers

DC system protection can be achieved by using AC system's CBs as well. The differential protection scheme can be applied in the AC side of two-terminal DC systems as shown in Figure 4.8. This approach can typically be seen in the applications like PtP-type HVDC systems where the DC bus works as a conduit between AC systems, and it needs cooperation between the AC CBs, power converters, and relays monitoring DC side voltage and current. For the DC side fault conditions such as DC under- or overvoltage, DC pole voltage differential, DC overcurrent, DC > AC current differential fault, the converters are blocked and AC CBs are tripped. For a LLC-based HVDC system, in conjunction with the trip of AC CBs, the thyristor rectifier valves can be controlled in the forced retard mode to deenergize the bus, while the inverter operates in bypass mode as shown in Figure 4.9 [22]. The rectifier polarity will be reversed so that the bus current can decrease (retard mode) and the inverter bypasses the DC current through bypass pairs (bypass mode). The protection system

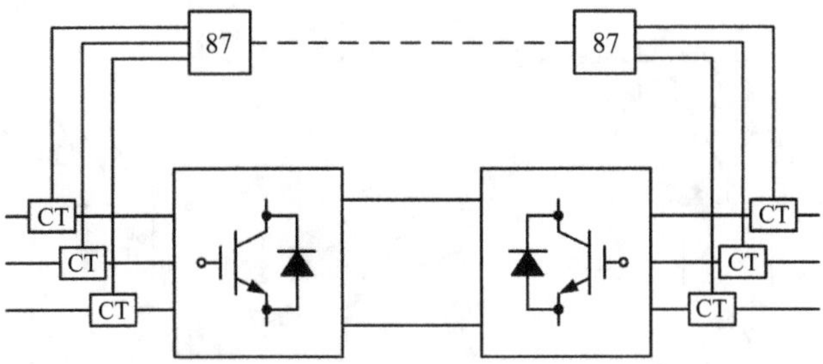

Figure 4.8 Differential protection scheme used in the AC sides of a two-terminal DC system. The ANSI standard device number 87 is for differential protection relay

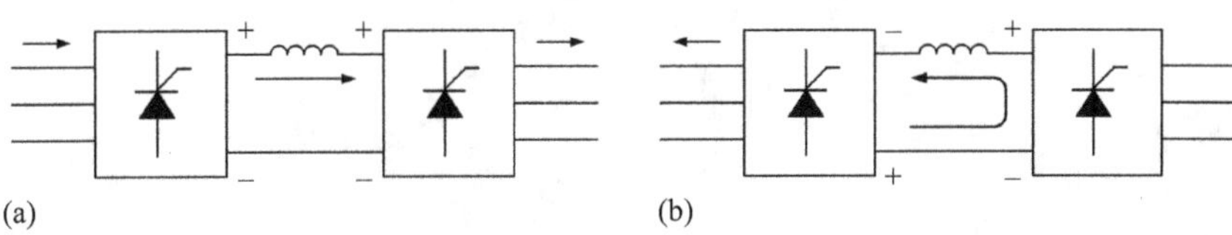

Figure 4.9 LCC operation modes: (a) normal operation, (b) forced retard operation. The rectifier polarity will be reversed so that the bus current can decrease, while the inverter bypasses the DC current through bypass pairs

has to be able to coordinate to properly discharge surge arresters and capacitors in the bus. The system can enter the reclosing cycle, and the voltage level can be coordinated in the restart mode (e.g., full or reduced voltage).

For MTDC systems, a "hand-shaking" method using fast-acting DC switches as well as AC CBs has been proposed [23] (Figure 4.10). Although switches can only open when there is no current, they can provide a lower cost solution compared to those with DC CBs. Downside is that the DC bus has to be deenergized because all of the AC CBs trip when a fault occurs. Using AC CBs would be one of the economical ways to protect DC systems, but the slow interruption speed may be an issue as the fastest AC CBs can interrupt the current in two cycles [24]. Furthermore, it would completely shut down the whole system, which could be inefficient especially for multiterminal systems. However, AC CBs can be readily used as a back-up protection measure.

4.3.2 DC side protection

4.3.2.1 Fault detection

When a short-circuit fault happens in a DC bus, e.g., between line to ground or line to line, the source devices connected to the bus, such as generators, bus capacitors, and energy storage, will experience excessive current, and loads, as well as the sources, will detect bus voltage drop. Typically DC bus fault can be detected by

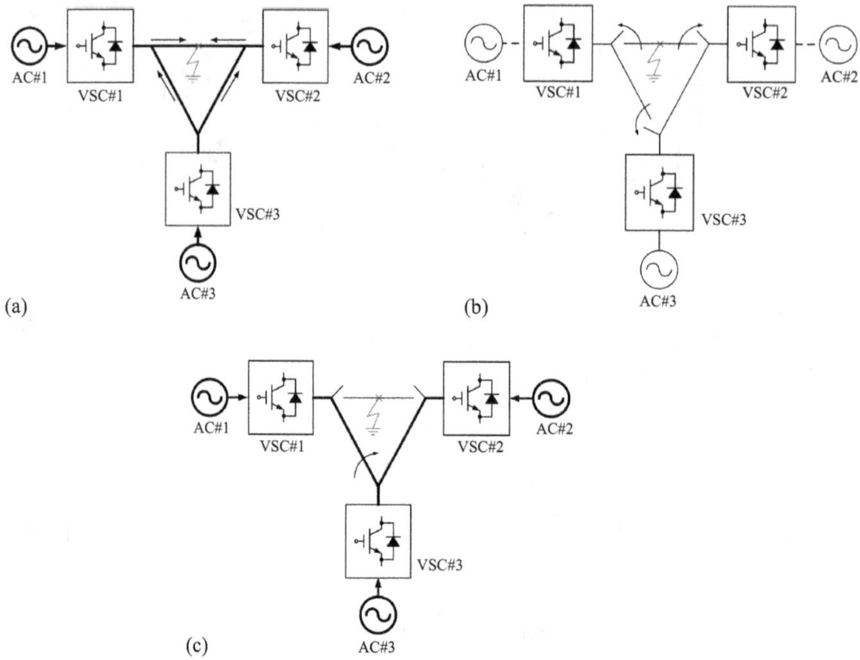

Figure 4.10 Handshaking method [23]: (a) fault current flows to the fault point. (b) All AC CBs open followed by appropriate DC switches after the fault current is extinguished. Each VSC opens one of the associated DC switches that had outgoing current at the time of fault. If all of them are outgoing, the one with highest magnitude is opened. (c) Reclose the switch if it sees the bus voltage after the AC CBs are reclosed

monitoring bus undervoltage and overcurrent, and various detection schemes have been proposed, including large current change, rise time, and oscillation pattern methods [25], dv/dt [5,26], di/dt and d^2i/dt^2 measurement [27], traveling wave [28], wavelet-based [29,30], and model-based [31] detection. Using the first- and second-order differentiation of voltage or current measurement can reduce detection time so that a smaller capacity CB can be used, but it is susceptible to signal noise and filtering could decrease the detection speed.

Short-circuit detection of line-to-ground fault, which is most common, is relatively straightforward unless the bus is ungrounded or high-resistance grounded as can be seen in traction and shipboard power systems, or the fault resistance is large. In the large fault resistance cases, the fault current is difficult to distinguish from a normal load current or detect it with ground fault relay. For an ungrounded bus that is adopted to avoid problems like corrosion by ground leakage current, there is no fault current when just one line is short-circuited to ground. This is an advantage because the system can keep operating with one line short-circuited to ground, but the impact is catastrophic if the first fault is not cleared before the second line-to-ground fault happens in another pole. Although a ground fault can be detected by measuring the voltage between the pole and ground, it is challenging to extract more information for protection because of the low fault current level.

Unit protection techniques that compare the magnitudes or directions of quantities at each end of a defined zone or unit, e.g., current differential protection, can provide faster detection and better selectivity and have been widely used typically for the primary protection method for AC systems. The current differential methods are readily applicable to DC systems and actually suitable to meet the much faster protection requirements. Although it will require communication between relays which increases the cost, the development of modern power systems with increased number of sensors and communication infrastructure can help the expansion of the usage of unit protection schemes. Differential detection schemes have been suggested for various types and applications of DC power systems, and many of them require high-speed data communication between protective devices. Although it provides higher selectivity and flexibility, the communication-based approach needs some independent back-up plan because communication error or glitch needs to be taken into consideration. Distance protection techniques are also widely used for AC systems to detect the fault and coordinate CBs, but small impedance of DC cables would cause significant problems using distance relaying methods for coordination (Figure 4.11) [5].

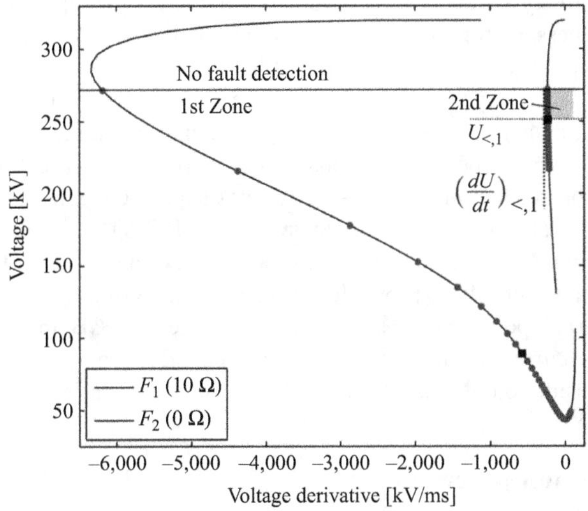

Figure 4.11 A dv/dt-based detection scheme [5]: protection zones and thresholds in the voltage/voltage derivative plane is shown. Locus of the voltage U and voltage derivative dU/dt is plotted for two line-to-ground faults F_1 (near) and F_2 (far) from the measuring point. The system is assumed to be asymmetric monopolar with pole voltage 320 kV. It can be seen that the fault can be detected using the undervoltage threshold, e.g., $U_{<,1}$ and derivative threshold $(dU/dt)_{<,1}$

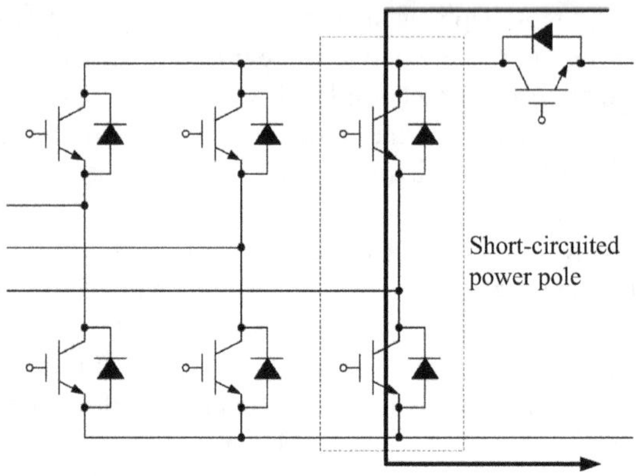

Figure 4.12 Fault current flows through short-circuited power pole via antiparallel diode of a DC CB

Most of the devices participating in a DC microgrid are interfaced with the bus through active power converters, such as LCCs and VSCs, which will disconnect from the bus to protect themselves by blocking all the switches to avoid overload or malfunctions from DC link undervoltage when a bus fault occurs. Diode rectifiers are passive, but they typically have contactors and fuses for separation. Therefore, it would be necessary for the protection system to interrupt the fault current and isolate the faulted section as soon as possible to minimize the impact of the fault and downtime of the system. However, if the DC bus is interfaced to AC grid through a VSC, it should be noted that the VSC cannot block the fault current from AC side. This happens because the bus voltage rapidly decreases due to the discharge of DC bus capacitors and sources cannot supply enough current to sustain the bus voltage. Hence, the protection system in DC side has to manage fault conditions during this quite short critical time period or should coordinate with the AC CBs to achieve a back-up protection from AC side. Similarly, if a fault happens in VSC's internal power poles, a single solid-state DC CB in the line cannot block the fault current coming from the DC bus side through antiparallel diode (Figure 4.12). Bidirectional solid-state CBs can be used, but high conduction loss can be an issue.

4.3.2.2 Postfault processes

The fault detection is typically followed by fault current interruption, fault location, and isolation processes. Fault current interruption in DC side can be implemented using DC CBs of mechanical, solid-state, and hybrid types and fast-acting DC switches. Typically, DC power systems are configured as shown in Figure 4.7, and the CBs can be placed at the terminals of each VSC or at the end of each line. Physical separation of the line can be implemented with additional DC switches. Although having DC CBs in all of the lines would be most efficient for protection, it may not be economical considering the cost of the DC CBs especially for the medium-voltage level. DC/DC converters and choppers have been investigated for DC fault current interruption as well. Conventional protective device coordination techniques in AC systems, such as short-circuit calculations, main/back-up protection, and current-time coordinated settings, can be applied to the solid-state DC CBs, fast-acting fuses, and no-fuse mechanical CBs in the DC system as well. However, the conventional time-current coordination between CBs could be challenging in DC systems due to the fast dynamics, bidirectional nature of power flow, and very short time interval that is allowed between CBs.

To effectively interrupt the bus from the AC and DC feeder, power converters such as thyristor rectifiers and DC/DC converters have been suggested to limit and block the fault current [32,33]. Fault current interruption can be coordinated with current limiter to reduce the interruption time [34]. Modern protection systems have adopted communication-based schemes to increase selectivity and flexibility. It is relatively easy to isolate the faulted segment when the differential or directional detection is used (Figure 4.13) [35].

Fault location is an important task for protection systems to reduce the repair time and system downtime. In AC systems, the fault location has been typically performed by microcontroller-based relays that calculate the location of the fault using short-circuit study data, traveling wave measurements, and impedance-based methods. For the impedance-based algorithms, the related voltage and current phase information is required, and this can be provided by phasor measurement unit. The impedance-based distance calculation concept can readily be applied to

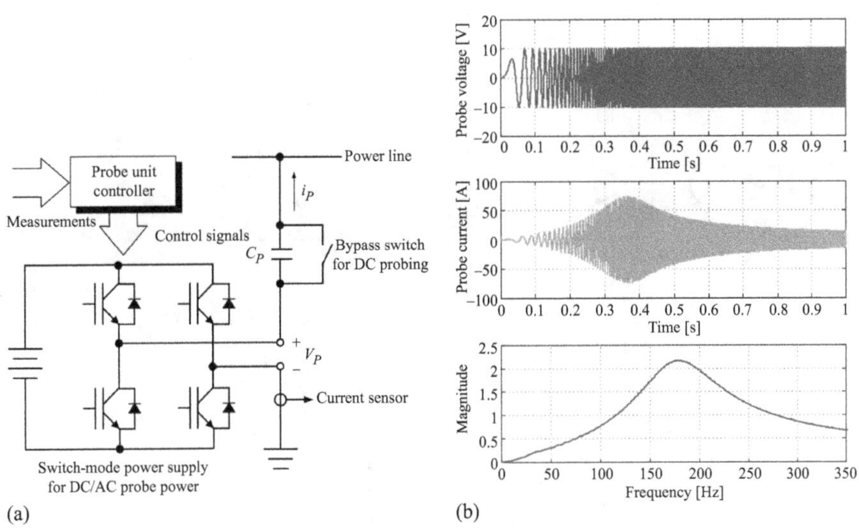

Figure 4.13 Sinusoidal DC fault probe unit [35]: (a) probe unit. (b) Probe voltage, current, and FFT result (from top). The probe unit applies voltage with sweeping frequency and the current response shows the characteristic of fault circuit

the fault location in DC microgrids. Techniques that extract information from the instantaneous measurement have been suggested [36–38]; however, due to the quite lower reactance and shorter length of the cables compared to those of AC systems, it is not trivial to measure from the transient current especially when the fault resistance is low. Furthermore, the transient happens at the top of the fault current; however, the protection system must interrupt the fault current as quickly as possible before the transient can be measured because the converters typically cannot sustain the large fault current. Some off-line measurement approaches using signal analysis have been proposed, e.g., traveling-wave-based methods [39,40], wavelet [30,41], and probe signal injection [3,42]. Although these can concurrently achieve fast fault current interruption and location, they require additional hardware equipment and can be time-consuming or inaccurate for complex configurations.

Isolation of the fault will minimize the affected area and maximize operation continuity once the fault current is interrupted. The isolation of the faulted device, e.g., a converter, from the bus is relatively straightforward because it typically trips itself out, but it can generate some disturbance on the bus. For example, when a source converter is lost from a DC microgrid system, the bus voltage will be maintained by other source converters if the load is sustainable with remaining sources; however, the voltage will overshoot if that converter is back on line. Chopper circuits [43,44], snubber circuits [45,46], and controller-based approaches [47,48] have been suggested to suppress the overvoltage transient (Figure 4.14). Compared to the typical control scheme that a single controller supplies the control actions for normal and fault conditions, these multiple controller approaches have separate controllers

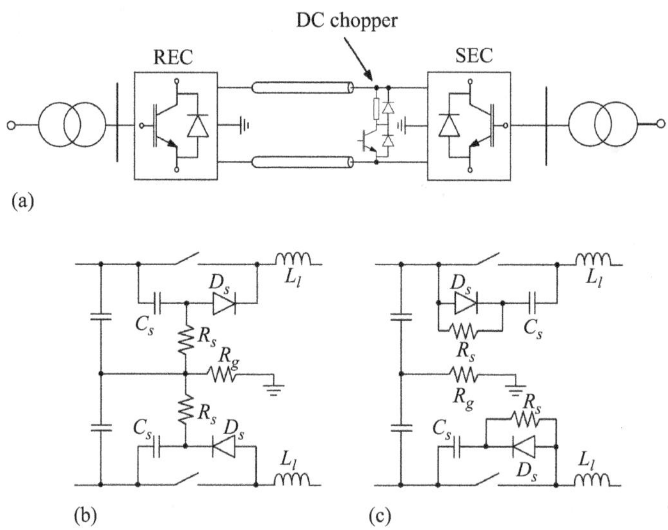

Figure 4.14 DC bus overvoltage suppression: (a) DC chopper, (b) discharge-suppressing RCD snubber, (c) charge–discharge RCD snubber

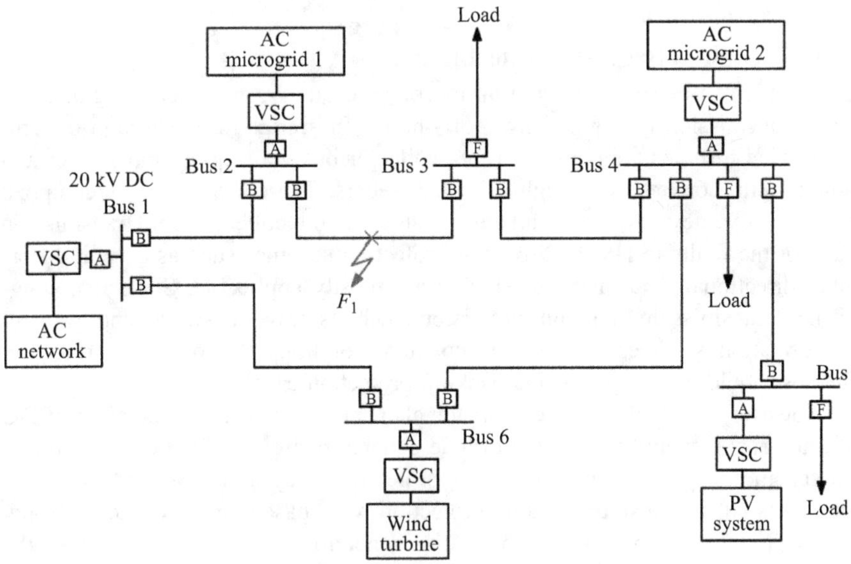

Figure 4.15 Exemplary medium voltage DC (MVDC) network, where boxed A, B, and F denote circuit breaker, static switch, and fuse, respectively [49]. A local protection unit is installed at each connection point of breaker to detect the faults comparing the line current and predetermined threshold value. Once fault F_1 is detected, the circuit breakers will interrupt the fault current, and the line end static switches will isolate the faulty section afterwards

that regulate steady-state variables such as power and bus voltage, and transient disturbances, e.g., loss of a converter, respectively. In a fault condition, the fault controller takes control to maintain the best possible performance and continuous operation by redistributing setpoints or shutting down load converters. The idea of multiple hierarchical controllers is similar to the coordination of protective devices in AC systems.

In case the fault happens in the bus, the faulted segment has to be isolated so that the other intact parts of the microgrid continue to operate. This task is especially important for the mission-critical power systems. For example, the zones of DC shipboard power systems are implemented with DC CBs, static switches and redundant buses, so that the load can be continuously supported after the fault is isolated. The system can be reconfigured to operate without the faulted section and coordination between protection relays and source converters may be required in this reconfiguration or self-healing process. In the system shown in Figure 4.15 [49], a fault can be isolated by stopping the source converters, opening the CBs, locate the fault, and open associated static switches. Then, the system can resume operation by restarting the source converters; however, load shedding or load redistribution would be necessary because of the reduced capacity.

4.3.3 Applications

4.3.3.1 High-voltage DC transmission

An example can be found in wind farms collector bus system where long distance power transmission is required from off-shore wind farms to on-shore grid. Generally, wind turbines are connected to a collector bus via step-up transformer and high-capacity converters through collector feeders. The collector bus is equipped with protective devices to respond during faults in both collection and transmission side. For the faults in HVDC link, unit protection schemes such as current differential, directional, and distance relaying are mostly applicable. Since it is long-distance transmission, nonunit protection such as traveling-wave and wavelet transformation schemes can also be potential solutions. All of these protection schemes should have primary and back-up protection capabilities.

However, any fault in collection side also causes problem in operation of DC collector bus. Generally, the wind turbine generators are protected by unit protection schemes with individual generator protection relay and CBs, and the transformer has either a fuse or separate protection relay based on the capacity. If any fault happens near the wind turbine, the generator relay coordinates with the transformer or converter protection devices to isolate the faulty section from the collector feeder which is connected to the collector bus. If this primary protection fails, then the collector bus relay is activated as a back-up protection measure. However, to enhance the reliability of the protection, effective coordination and communication of the protective devices need to be ensured.

4.3.3.2 Shipboard power system

Conventional AC shipboard power system is moving toward the next generation medium-voltage DC (MVDC) architecture as DC is advantageous over AC power system, e.g., higher efficiency and reliability, no issues with frequency and reactive power regulation, and simpler interface and control. For the MVDC and LVDC shipboard power systems, fault tolerant DC zonal electrical distribution system (DCZEDS) concept has been widely investigated. The DCZEDS allows a number of electrical zones connected to the main bus to improve the survivability of the system during abnormal events. For the protection of DCZEDS, unit-based and breaker-based approaches have been extensively investigated [50].

In the unit-based approach, the DCZEDS allows two separate longitudinal buses along the length of the ship that provides redundant paths to deliver power to the critical loads from both the buses through power converters and no-load electromechanical switches [51]. If fault happens in a certain zone, this approach enables the fault-feeding converters of inherent current limiting and voltage fold back capabilities to decouple the fault effect from upstream generators and coordinate with the downstream no-load contactors to deenergize the faulty zone. If fault happens in one longitudinal bus, the DCZEDS utilizes the other bus to supply the critical loads during system reconfiguration (Figure 4.16).

The breaker-based protection approach relies on the parallel combination of protection devices such as DC CBs and no-load contactors. The DCZEDS form a

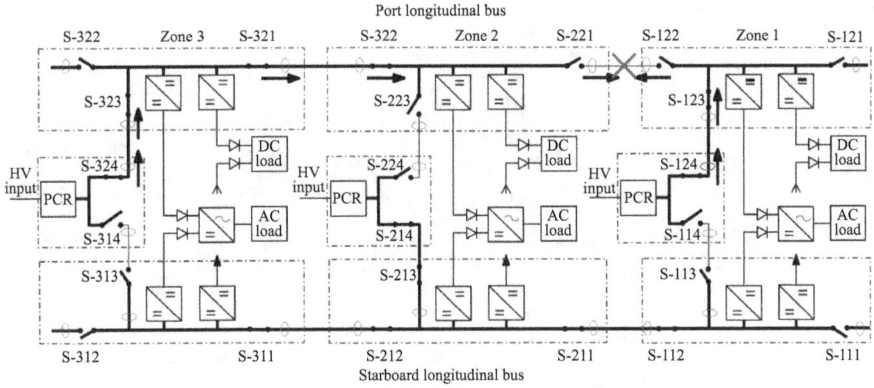

Figure 4.16 An unit-based protection scheme for a shipboard DC zonal electrical distribution system [51]. The adjacent zones consist of switches and phase controlled rectifier (PCR) with communication links that detect the fault from predetermined current threshold value, locate the fault comparing the vector direction of current in each zone, and isolate the faulted section afterwards

ring bus configuration by connecting the longitudinal buses with solid-state CBs which allows to isolate the faulty section without deenergizing the entire bus. In a fault condition, DC CBs should react very fast to extinguish the fault current to zero before the upstream converters shut down by themselves. However, to coordinate the fault isolation and system reconfiguration, effective communication is required between the converters, CBs, and switches, and at the same time, fault location schemes are also required. Traveling-wave-based techniques have been suggested for locating fault as it offers sufficient accuracy [52]. Artificial neural network and wavelets have been suggested as well [53].

4.3.3.3 Traction power systems

Mass transit systems such as light-rail and subway systems are growing rapidly throughout the world due to the urbanization and massive traffic growth. Single-phase AC as well as DC power system has been employed for traction system. A typical substation in DC traction power systems consists of a transformer and rectifiers from high-voltage AC supply which feeds bus bars through high-speed DC CBs and bypasses switches. The bus voltage used in DC traction power systems is relatively low (600/750/1,500 V), and the ground impedance is high [54]; as a consequence, it is challenging to differentiate between the line-to-ground fault current and start-up current drawn from substation which are close in magnitude. It becomes worse if a short-circuit fault occurs at a distant place from substations (Figure 4.17).

di/dt Measurement-based protection methods have been proposed [55]; if current outstrips the setting limit at any instant of time with higher di/dt than the threshold and lasts longer than the setting time interval, then a short-circuit fault is detected, otherwise it is a start-up current. Auto reclosers are also installed at the

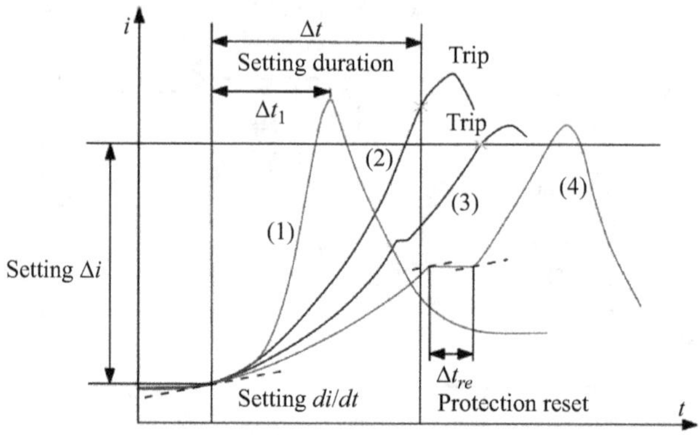

Figure 4.17 A di/dt protection scheme for traction power system [55]. If di/dt rises and exceeds the set point, tripping decision will be taken only when both Δi and Δt are higher than the setting limits [curves (2) and (3)]. However, if there is any momentary reduction in rising di/dt but Δt is higher than the protection reset interval (Δt_{re}), then there will be no tripping [curve (4)]

track feeders to see if the fault is temporary. However, typically, the traction power system is ungrounded or grounded with high resistance to mitigate ground leakage current and corrosion, which makes it difficult to detect line-to-ground fault. The system experiences a line-to-line fault if ground fault happens at the same time on another pole. Hence, it is required to detect and locate the line-to-ground fault quickly to avoid unwanted discontinuity in the service due to a line-to-line fault. Techniques such as off-line impedance probe unit [35], traveling wave-based methods [56], superconducting fault current limiter [57], and high-speed digital relay-based protection [58] have been suggested for traction power system fault protection.

4.3.3.4 Photovoltaic park

A number of parallel strings consist of series connected PV panels form a PV generation system, which is tied to the main bus through power electronics converters and protective devices. Possible faults in a PV power system include arc, open-circuit, line-to-ground, and line-to-line fault. The line-to-ground fault happens most often, and line-to-line fault occurs typically due to water ingress, animal chewing, mechanical damage, or DC junction box corrosion [59].

Considering the short length of bus, an unit protection with DC switches can be utilized for PV parks. Overcurrent protection devices (OCPDs, e.g., fuse) and ground-fault protection devices (GFPDs) are used as well in typical PV systems (Figure 4.18) to detect line-to-line and line-to-ground faults, respectively. A line-to-line fault in PV strings is difficult to detect and protection becomes challenging if solar irradiance is low or a fault occurs between two points of close electrical potential [60]. The OCPDs respond to the fault with typically 2.1 times of short-circuit current at standard test condition (1,000 W m^{-2}, 25 °C, 1.5 air mass) [60], but

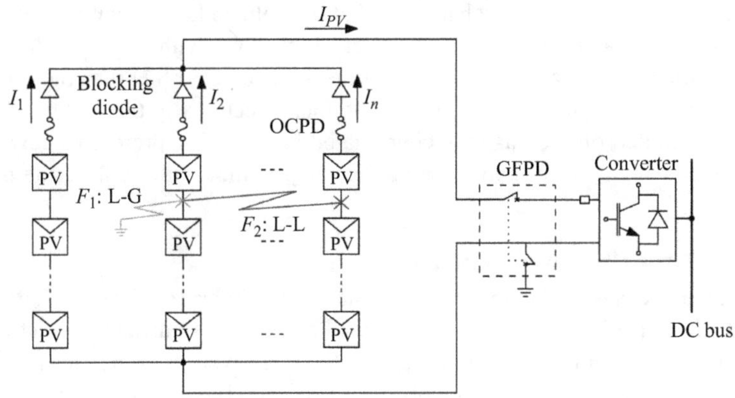

Figure 4.18 A typical PV array protection system. During line-to-ground and line-to-line fault, fault current is interrupted by GFPD and OCPD, respectively

at low solar irradiation specially at cloudy day or evening, fault current level is considerably low. If the fault remains undetected, it could potentially damage the system, and furthermore, it consumes power from the healthy section as a load. Although a blocking diode can be placed on each string to cut off the back-feeding current, it challenges the OCPDs in fault interruption [59]. Furthermore, GFPDs in a traditional PV system may trip the whole system for a line-to-ground fault. Recent studies have been mainly focused on fault location and detection strategies. A voltage transducers-based scheme has been proposed in [61] to protect against line-to-ground as well as line-to-line fault, in which the top and bottom end string voltage is compared with predefined threshold to classify and locate the fault. Differential protection [62] and machine learning-based protection [63] strategies have been suggested as well for PV system fault location and detection.

4.3.3.5 Data centers

In the past decade, there has been a significant advancement in data center from a small server room to large data centers. Besides AC grids, conventional diesel generators and uninterruptible power supply (UPS) are also employed to supply power during grid outage. Data center loads are mostly DC, comprised of IT equipment such as servers, routers, switches, and energy storage devices. From AC grid to the loads, several AC–DC converter stations are installed with associated protective devices typically in radial configuration. However, data center requires highly reliable and accurate fault protection schemes to maintain the service availability.

The unit protection technique with microcontroller-based relay can be applicable for the primary detection and location. If the relays and DC CBs fail to respond to the relay's trip signal, the converters will shut off with self-protection capability as back-up protection. However, it should be noted that the power distribution system in DC data centers is highly capacitive in nature because of UPSs and multiple power supply units in parallel, which causes problem during short-circuit failure [64].

During a fault, the capacitor discharges so that the voltage falls down rapidly and the amplitude of the fault current rises to a very high level within a very short time. Moreover, this current can be detected by protective devices other than those located on the fault path and unnecessarily trips the intact sections in the system. A relay should be installed on the bus with coordinated fast reacting protective devices, so that the faulty segment of the bus can be isolated without interrupting the healthy sections.

4.3.3.6 Aircraft power systems

An aircraft power system is mostly small-scale AC/DC hybrid system with 230 V AC/ 360–800 Hz and ±270 V DC, consists of small generators, transmission–distribution system, electrical components, and onboard loads, as well as some energy storage devices as an emergency power supply for vital loads [66,67]. The common faults in aircraft are arc fault, overloads, and electrical short-circuit fault, and differential

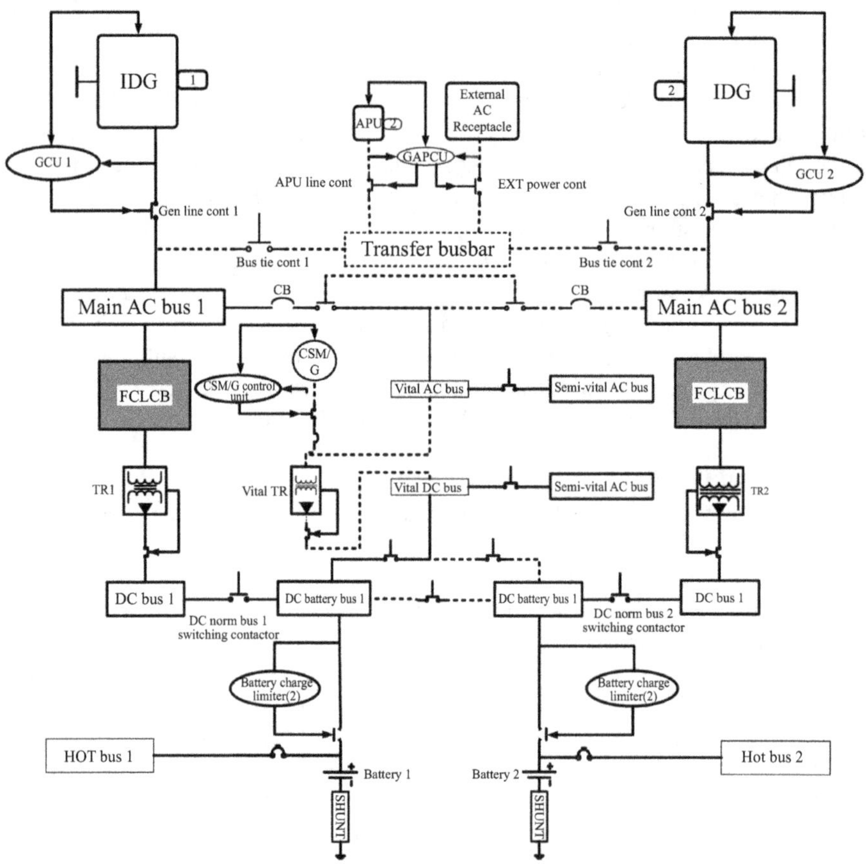

Figure 4.19 Single line diagram of the j150+ aircraft restructured distribution network for a more electric aircraft (MEA) [65]

current protection and OCPDs are mainly used for single DC bus-based system [65]. In case of the ring-bus aircraft, directional protection scheme can be applicable, because once a fault happens at a point, then current will flow to that point from both sides. Fault current can be interrupted by fuses, but replacement would be difficult for aircraft onboard power systems. Resettable CBs can be used, but it may not be an effective solution to arc fault detection and interruption. Semiconductor-based protective devices have been suggested for arc fault, which also provide protection against short-circuit fault and overloads using I^2t operating characteristics [68]. Recently, fault current limiting CBs (FCLCBs) based on series resonance with superior current limiting and breaking advantages have been proposed [65]. A single line diagram of a more electric aircraft power system is shown in Figure 4.19.

4.4 Summary

DC power systems have been investigated for transmission and distribution systems due to the advantages such as improved efficiency, reliability, and simpler configuration, as well as increasing DC-based distributed generation, energy storage, and loads. Along with comparatively short operating history and insufficient guidelines and standards, one of the current challenges is system protection techniques to deal with the highly dynamic and nonzero crossing nature of DC power. In this chapter, various coordinated protection schemes and applications for DC power systems have been reviewed in conjunction with the analysis of DC fault currents dynamics and topological bus configurations.

Acknowledgment

This work was supported by the National Science Foundation (NSF) under Award ECCS-1554626.

References

[1] T. Dragičević and F. Blaabjerg, *Microgrid: advanced control methods and renewable energy system integration*. Waltham, MA: Butterworth-Heinemann, 2016, pp. 263–279.

[2] IEEE Power and Energy Society, "IEEE guide for determining fault location on AC transmission and distribution lines," *IEEE Std C37.114-2004*, pp. 1–36, 2005.

[3] J. D. Park, J. Candelaria, L. Ma, and K. Dunn, "DC ring-bus microgrid fault protection and identification of fault location," *IEEE Transactions on Power Delivery*, vol. 28, no. 4, pp. 2574–2584, October 2013.

[4] A. Berizzi, A. Silvestri, D. Zaninelli, and S. Massucco, "Short-circuit current calculations for DC systems," *IEEE Transactions on Industry Applications*, vol. 32, no. 5, pp. 990–997, 1996.

[5] W. Leterme, J. Beerten, and D. V. Hertem, "Nonunit protection of HVDC grids with inductive DC cable termination," *IEEE Transactions on Power Delivery*, vol. 31, no. 2, pp. 820–828, April 2016.

[6] M. K. Bucher and C. M. Franck, "Contribution of fault current sources in multiterminal HVDC cable networks," *IEEE Transactions on Power Delivery*, vol. 28, no. 3, pp. 1796–1803, 2013.

[7] D. Jovcic and K. Ahmed, *High voltage direct current transmission: converters, systems and DC grids*. Chichester: John Wiley & Sons, 2015.

[8] J. Suonan, J. Zhang, Z. Jiao, L. Yang, and G. Song, "Distance protection for HVDC transmission lines considering frequency-dependent parameters," *IEEE Transactions on Power Delivery*, vol. 28, no. 2, pp. 723–732, 2013.

[9] S. Luo, X. Dong, S. Shi, and B. Wang, "A directional protection scheme for HVDC transmission lines based on reactive energy," *IEEE Transactions on Power Delivery*, vol. 31, no. 2, pp. 559–567, 2016.

[10] E. Kontos, G. Tsolaridis, R. Teodorescu, and P. Bauer, "On dc fault dynamics of MMC-based HVDC connections," *IEEE Transactions on Power Delivery*, vol. 33, no. 1, pp. 497–507, 2018.

[11] D. Pieniazek, "HV substation design: applications and considerations," in *IEEE CED–Houston Chapter*, 2012.

[12] V. Mehta and R. Mehta, *Principles of power system*, New Delhi: S. Chand, 2004.

[13] A. Kwasinski, "Advanced power electronics enabled distribution architectures: design, operation, and control," in *2011 IEEE 8th International Conference on Power Electronics and ECCE Asia (ICPE & ECCE)*, 2011, pp. 1484–1491.

[14] M. K. Bucher, R. Wiget, G. Andersson, and C. M. Franck, "Multiterminal HVDC networks—what is the preferred topology?" *IEEE Transactions on Power Delivery*, vol. 29, no. 1, pp. 406–413, 2014.

[15] J. Yang, J. E. Fletcher, and J. O'Reilly, "Multiterminal DC wind farm collection grid internal fault analysis and protection design," *IEEE Transactions on Power Delivery*, vol. 25, no. 4, pp. 2308–2318, 2010.

[16] P. Bresesti, W. L. Kling, R. L. Hendriks, and R. Vailati, "HVDC connection of offshore wind farms to the transmission system," *IEEE Transactions on Energy Conversion*, vol. 22, no. 1, pp. 37–43, 2007.

[17] O. Gomis-Bellmunt, J. Liang, J. Ekanayake, R. King, and N. Jenkins, "Topologies of multiterminal HVDC-VSC transmission for large offshore wind farms," *Electric Power Systems Research*, vol. 81, no. 2, pp. 271–281, 2011.

[18] Y. Liu, J. Wang, N. Li, Y. Fu, and Y. Ji, "Enhanced load power sharing accuracy in droop-controlled DC microgrids with both mesh and radial configurations," *Energies*, vol. 8, no. 5, pp. 3591–3605, 2015.

[19] L. Tang and B.-T. Ooi, "Protection of VSC-multi-terminal HVDC against DC faults," in *2002 IEEE 33rd Annual IEEE Power Electronics Specialists Conference. Proceedings (Cat. No.02CH37289)*, vol. 2, 2002, pp. 719–724.

[20] D. Chen, L. Xu, and L. Yao, "DC voltage variation based autonomous control of DC microgrids," *IEEE Transactions on Power Delivery*, vol. 28, no. 2, pp. 637–648, April 2013.

[21] N. Eghtedarpour and E. Farjah, "Distributed charge/discharge control of energy storages in a renewable-energy-based DC micro-grid," *IET Renewable Power Generation*, vol. 8, no. 1, pp. 45–57, January 2014.
[22] C. Kim, V. Sood, G. Jang, S. Lim, and S. Lee, *HVDC transmission: power conversion applications in power systems*. Singapore: John Wiley & Sons (Asia), 2009.
[23] L. Tang and B. Ooi, "Locating and isolating DC faults in multi-terminal DC systems," *IEEE Transactions on Power Delivery*, vol. 22, no. 3, pp. 1877–1884, July 2007.
[24] IEEE Power and Energy Society, "IEEE application guide for AC high-voltage circuit breakers rated on a symmetrical current basis," in *ANSI/IEEE C37.010-1979*, pp. 1–54, 1980.
[25] L. Tang and B. Ooi, "Protection of VSC-multi-terminal HVDC against DC faults," in *33rd Annual IEEE Power Electronics Specialist Conference*, vol. 2, pp. 719–724, November 2002.
[26] S. Foster, L. Xu, and B. Fox, "Control of an LCC HVDC system for connecting large offshore wind farms with special consideration of grid fault," in *2008 IEEE Power and Energy Society General Meeting – Conversion and Delivery of Electrical Energy in the 21st Century*, July 2008, pp. 1–8.
[27] A. Meghwani, S. C. Srivastava, and S. Chakrabarti, "A non-unit protection scheme for DC microgrid based on local measurements," *IEEE Transactions on Power Delivery*, vol. 32, no. 1, pp. 172–181, February 2017.
[28] S. Azizi, M. Sanaye-Pasand, M. Abedini, and A. Hasani, "A traveling-wave-based methodology for wide-area fault location in multiterminal DC systems," *IEEE Transactions on Power Delivery*, vol. 29, no. 6, pp. 2552–2560, December 2014.
[29] K. D. Kerf, K. Srivastava, M. Reza, D. Bekaert, S. Cole, D. V. Hertem, and R. Belmans, "Wavelet-based protection strategy for DC faults in multi-terminal VSC HVDC systems," *IET Generation, Transmission Distribution*, vol. 5, no. 4, pp. 496–503, April 2011.
[30] W. Li, A. Monti, and F. Ponci, "Fault detection and classification in medium voltage DC shipboard power systems with wavelets and artificial neural networks," *IEEE Transactions on Instrumentation and Measurement*, vol. 63, no. 11, pp. 2651–2665, November 2014.
[31] Y. Ji and J. Bals, "Multi-model based fault detection for the power system of more electric aircraft," in *2009 7th Asian Control Conference*, August 2009, pp. 93–98.
[32] D. Dong, Y. Pan, R. Lai, X. Wu, and K. Weeber, "Active fault-current foldback control in thyristor rectifier for DC shipboard electrical system," *IEEE Journal of Emerging and Selected Topics in Power Electronics*, vol. 5, no. 1, pp. 203–212, March 2017.
[33] M. Baran and N. Mahajan, "Overcurrent protection on voltage sourced converter based multiterminal DC distribution systems," *IEEE Transactions on Power Delivery*, vol. 22, no. 1, pp. 406–412, January 2007.

[34] C. Zhan, C. Smith, A. Crane, A. Bullock, and D. Grieve, "DC transmission and distribution system for a large offshore wind farm," in *9th IET International Conference on AC and DC Power Transmission (ACDC 2010)*, October 2010, pp. 1–5.

[35] J. D. Park, "Ground fault detection and location for ungrounded DC traction power systems," *IEEE Transactions on Vehicular Technology*, vol. 64, no. 12, pp. 5667–5676, December 2015.

[36] J. Yang, J. E. Fletcher, and J. O'Reilly, "Short-circuit and ground fault analyses and location in VSC-based DC network cables," *IEEE Transactions on Industrial Electronics*, vol. 59, no. 10, pp. 3827–3837, October 2012.

[37] Y. Pan, M. Steurer, and T. L. Baldwin, "Ground fault location testing of a noise-pattern-based approach on an ungrounded DC system," *IEEE Transactions on Industry Applications*, vol. 47, no. 2, pp. 996–1002, March 2011.

[38] J. Cheng, M. Guan, L. Tang, H. Huang, X. Chen, and J. Xie, "Paralleled multi-terminal DC transmission line fault locating method based on travelling wave," *IET Generation, Transmission Distribution*, vol. 8, no. 12, pp. 2092–2101, 2014.

[39] X. Liu, A. H. Osman, and O. P. Malik, "Hybrid traveling wave/boundary protection for monopolar HVDC line," *IEEE Transactions on Power Delivery*, vol. 24, no. 2, pp. 569–578, April 2009.

[40] Y. J. Kwon, S. H. Kang, D. G. Lee, and H. K. Kim, "Fault location algorithm based on cross correlation method for HVDC cable lines," in *2008 IET 9th International Conference on Developments in Power System Protection (DPSP 2008)*, March 2008, pp. 360–364.

[41] K. Jia, T. Bi, B. Liu, E. Christopher, D. W. P. Thomas, and M. Sumner, "Marine power distribution system fault location using a portable injection unit," *IEEE Transactions on Power Delivery*, vol. 30, no. 2, pp. 818–826, April 2015.

[42] E. Christopher, M. Sumner, D. W. P. Thomas, X. Wang, and F. de Wildt, "Fault location in a zonal DC marine power system using active impedance estimation," *IEEE Transactions on Industry Applications*, vol. 49, no. 2, pp. 860–865, March 2013.

[43] C. Feltes, H. Wrede, F. W. Koch, and I. Erlich, "Enhanced fault ride-through method for wind farms connected to the grid through VSC-based HVDC transmission," *IEEE Transactions on Power Systems*, vol. 24, no. 3, pp. 1537–1546, August 2009.

[44] L. Bing, J. Xu, R. Torres-Olguin, and T. Undeland, "Fault mitigation control design for grid integration of offshore wind farms and oil and gas installations using VSC-HVDC," in *International Symposium on Power Electronics*, pp. 792–797, June 2010.

[45] J. D. Park and J. Candelaria, "Fault detection and isolation in low-voltage DC-bus microgrid system," *IEEE Transactions on Power Delivery*, vol. 28, no. 2, pp. 779–787, April 2013.

[46] F. Liu, W. Liu, X. Zha, H. Yang, and K. Feng, "Solid-state circuit breaker snubber design for transient overvoltage suppression at bus fault interruption in low-voltage DC microgrid," *IEEE Transactions on Power Electronics*, vol. 32, no. 4, pp. 3007–3021, April 2017.

[47] W. Lu and B. Ooi, "DC overvoltage control during loss of converter in multiterminal voltage sourced converter based HVDC (M-VSC-HVDC)," *IEEE Transactions on Power Delivery*, vol. 18, no. 3, pp. 915–920, July 2003.

[48] L. Lamont, D. Jovcic, and K. Abbot, "VSC transmission control under faults," *Universities Power Engineering Conference*, vol. 3, pp. 1209–1213, 2004.

[49] M. Monadi, C. Koch-Ciobotaru, A. Luna, J. I. Candela, and P. Rodriguez, "Multi-terminal medium voltage DC grids fault location and isolation," *IET Generation, Transmission Distribution*, vol. 10, no. 14, pp. 3517–3528, 2016.

[50] R. M. Cuzner and V. Singh, "Future shipboard MVdc system protection requirements and solid-state protective device topological tradeoffs," *IEEE Journal of Emerging and Selected Topics in Power Electronics*, vol. 5, no. 1, pp. 244–259, 2017.

[51] M. W. Rose and R. M. Cuzner, "Fault isolation and reconfiguration in a three-zone system," in *2015 IEEE Electric Ship Technologies Symposium (ESTS)*, 2015, pp. 409–414.

[52] E. Schweitzer, A. Guzman, M. Mynam, V. Skendzic, B. Kasztenny, and S. Marx, "Protective relays with traveling wave technology revolutionize fault locating," *IEEE Power and Energy Magazine*, vol. 14, no. 2, pp. 114–120, 2016.

[53] Y. M. Yeap and A. Ukil, "Wavelet based fault analysis in HVDC system," in *2014 40th Annual Conference of the IEEE Industrial Electronics Society (IECON)*, 2014, pp. 2472–2478.

[54] R. Hill, "Electric railway traction. part 3. traction power supplies," *Power Engineering Journal*, vol. 8, no. 6, pp. 275–286, 1994.

[55] M. X. Li, J. H. He, Z.-q. Q. Bo, H. Yip, L. Yu, and A. Klimek, "Simulation and algorithm development of protection scheme in DC traction system," in *PowerTech, 2009 IEEE Bucharest*. IEEE, 2009, pp. 1–6.

[56] C. Chang, T. Feng, A. Khambadkone, and S. Kumar, "Remote short-circuit current determination in DC railway systems using wavelet transform," *IEE Proceedings-Electric Power Applications*, vol. 147, no. 6, pp. 520–526, 2000.

[57] H.-S. Shin, S.-M. Cho, and J.-C. Kim, "Protection scheme using SFCL for electric railways with automatic power changeover switch system," *IEEE Transactions on Applied Superconductivity*, vol. 22, no. 3, pp. 1–4, 2012.

[58] E. Cinieri, A. Fumi, V. Salvatori, and C. Spalvieri, "A new high-speed digital relay protection of the 3-kVdc electric railway lines," *IEEE Transactions on Power Delivery*, vol. 22, no. 4, pp. 2262–2270, 2007.

[59] Y. Zhao, J.-F. de Palma, J. Mosesian, R. Lyons, and B. Lehman, "Line–line fault analysis and protection challenges in solar photovoltaic arrays," *IEEE Transactions on Industrial Electronics*, vol. 60, no. 9, pp. 3784–3795, 2013.

[60] National Fire Protection Association (NFPA), "Article 690—solar photovoltaic systems," *National Electrical Code*, 2011.

[61] K. A. Saleh, A. Hooshyar, E. F. El-Saadany, and H. H. Zeineldin, "Voltage-based protection scheme for faults within utility-scale photovoltaic arrays," *IEEE Transactions on Smart Grid*, 2017. Early Access.

[62] S. Dhar, R. Patnaik, and P. Dash, "Fault detection and location of photovoltaic based DC microgrid using differential protection strategy," *IEEE Transactions on Smart Grid*, 2017. Early Access.

[63] Z. Yi and A. Etemadi, "Line-to-line fault detection for photovoltaic arrays based on multi-resolution signal decomposition and two-stage support vector machine," *IEEE Transactions on Industrial Electronics*, vol. 64, no. 11, pp. 8546–8556, 2017.

[64] G. Wawrzola, "Challenges of DC data center power distribution protection," in *13th International Conference on Development in Power System Protection 2016 (DPSP)*, March 2016, pp. 1–6.

[65] H. Radmanesh and A. Kavousi, "Aircraft electrical power distribution system protection using smart circuit breaker," *IEEE Aerospace and Electronic Systems Magazine*, vol. 32, no. 1, pp. 30–40, 2017.

[66] J. Chen, X. Zhang, and C. Wen, "Harmonics attenuation and power factor correction of a more electric aircraft power grid using active power filter," *IEEE Transactions on Industrial Electronics*, vol. 63, no. 12, pp. 7310–7319, 2016.

[67] M. Terorde and D. Schulz, "New real-time heuristics for electrical load rebalancing in aircraft," *IEEE Transactions on Aerospace and Electronic Systems*, vol. 52, no. 3, pp. 1120–1131, 2016.

[68] D. Izquierdo, A. Barrado, C. Fernández, M. Sanz, and A. Lázaro, "SSPC active control strategy by optimal trajectory of the current for onboard system applications," *IEEE Transactions on Industrial Electronics*, vol. 60, no. 11, pp. 5195–5205, 2013.

Chapter 5

Energy management systems for dc microgrids

Amjad Anvari-Moghaddam[1], Tomislav Dragičević[1], and Marko Delimar[2]

5.1 Introduction

Over the past few years, large-scale utilization of renewable energy sources together with the high penetration of low-to-medium voltage dc sources into the generation mix of energy networks have increased attention toward deployment of dc microgrids for improving system operation, reducing environmental footprints, mitigating congestion and power losses through the network, increasing energy efficiency and potentially boosting system economics. Moreover, with the increasing share of dc appliances and loads, especially in residential and office buildings, as well as the technological advancements in ac-to-dc rectifiers, dc microgrids have become the point of interest for many research scientists and engineers worldwide and have become a hot subject area for a large number of research, development and demonstration activities. On the other hand, average energy consumption distribution in different sectors at worldwide level illustrates that more than one-third of the total consumption in each society is used by the residential sector [1,2]. Although the increasing rate of energy consumption in this sector has decelerated recently thanks to energy-aware solutions, energy prices are still rising [3]. In this regard, many ongoing research activities are directed toward energy efficiency in residential and office buildings and its impact on the environment and climate [4–10]. Accordingly, a large number of initiatives have been taken to promote sustainable consumption and green living patterns as well as to promote delivery of affordable, clean and reliable power to customers that, in turn, necessitates the use of smart automation and energy management solutions. Although "energy management" is a term that could mean differently, in this chapter, we are mainly concerned about the process of energy saving and optimal scheduling of energy sources in a given system here called microgrids.

[1]Department of Energy Technology, Aalborg University, Denmark
[2]Department of Energy and Power Systems, Faculty of Electrical Engineering and Computing, University of Zagreb, Croatia

5.2 DC microgrid operation and control fundamentals

5.2.1 Power/energy management schemes

For optimal operation of a microgrid system in different working modes (i.e. in grid-connected mode or islanded mode), it is crucial to properly design and implement energy management systems (EMSs). These system optimizers normally determine the best possible operating scheme at the supply and demand sides in terms of optimized set-points for controllable units and then send them as control signals into the dedicated control system of interfacing converters. Generally, there are two types of energy/power management strategies which are used in microgrid applications; interactive schemes based on information sharing mechanisms and passive schemes based on self-autonomy [11].

5.2.1.1 Interactive power/energy management strategies

In any given interactive power (IP)/EMS, local and global system information (such as line currents, nodal voltages, frequency and powers) is communicated in the microgrid and exchanged between corresponding nodes in order to determine the operation point of each controllable distributed generation (DG) or consumption unit. These strategies also benefit from a sort of intelligence in the integration of the computing and communications technologies which help them to define and develop the communication structure based on the computation burden of each node and other related system's objectives and constraints [12]. In this regard, three different communication schemes can be realized for an IP/EMS: centralized, decentralized and hybrid. In each of these schemes, different communication technologies such as microwave (μW), power line communication (PLC), fiber-optics, infrared and/or wireless radio networks (such as global system for mobile communications and code division multiple access) can be effectively used and integrated into existing infrastructures [13,14].

- **Centralized P/EMS**
  In a centralized P/EMS, also known as a supervisory scheme, there is a centralized entity or a control center that monitors the system's behavior, collects information from different parts of the network, makes decisions based on the observations and accordingly updates set-points for the controllable units in the supply/demand sides [15–19]. In other words, a centralized P/EMS acts as a master unit while other local controllers (LCs) within the system are treated as slaves to follow the reference signals coming from the master unit, as shown in Figure 5.1. To improve the effectiveness of a P/EMS, it is also very important to clearly define a system's objectives and constraints. These objectives (such as operating cost minimization, emission mitigation, power loss reduction, etc.) together with the constraints might be conflicting in some cases, which, in turn, makes the optimal decision-making process a difficult or even impossible task. Different examples of centralized P/EMS for microgrids can be found in literature [10,20–24]. As an illustration, an expert centralized energy management approach is presented in [22] for optimal operation of a

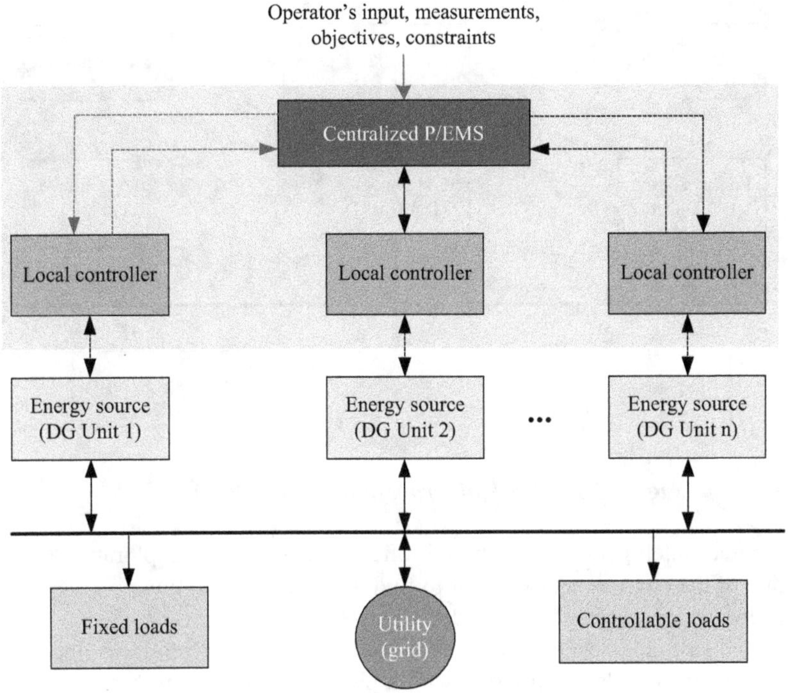

Figure 5.1 Block diagram of a centralized P/EMS

given microgrid with different micro-sources (such as microturbines, fuel cells and batteries) and renewables. In this work, the optimal management problem is formulated as a nonlinear constraint multi-objective optimization problem to minimize the total operating cost and the simultaneous net emissions inside the microgrid.

The advantages of a centralized scheme mainly lie within the simplicity of implementation and globality of optimal solution, however it brings two disadvantages: single point of failure which implies that a centralized P/EMS must be securely designed with appropriate redundancy inbuilt, and massive communication expenditure. The latter is not a challenging problem in small-scale networks but could be problematic for larger systems as the complexity of the centralized optimization grows exponentially with the number of units (control variables) in the system.

- **Distributed P/EMS**
Distributed P/EMS is the second interactive scheme for management of microgrids in which there is no central supervisory unit, but all the LCs are connected and communicate with one another through a communication bus [25,26]. In this way, each controller not only captures local measurements, but also receives information from neighboring nodes which helps in decision-making process according to different optimization objectives [27,28]. In this

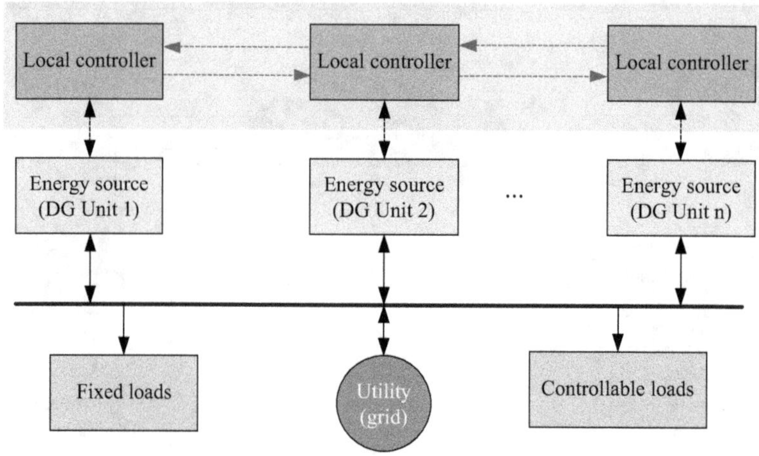

Figure 5.2 Block diagram of a decentralized P/EMS

scheme, intelligent algorithms are often used for better exploration/exploitation of the environment in order to find the optimal operation point. Figure 5.2 shows the block diagram of a decentralized P/EMS.

A distributed scheme has some advantages over a centralized one. Firstly, it supports a scalable structure with a plug-and-play feature for newly added/removed energy sources or load blocks. Secondly, the computation burden of each LC is mitigated which, in turn, reduces the required communication bandwidth. Finally, a distributed P/EMS could improve the redundancy and modularity of the system where this is needed. However, there is still a problem if a communication link fails in the system. This failure would not end in a total system collapse, but the performance of the system would no longer be optimal. Also, a distributed P/EMS suffers from degradation of performance for small/medium networks, increased use of database space, and complex use and administration. A multi-agent system (MAS) is one of the best illustrations of a distributed scheme [28]. Considering a MAS-based structure, there are several interacting intelligent agents that make decisions based on predefined objectives within the environment. The agent's intelligence may include some methodical, functional, procedural approaches, algorithmic searching or reinforcement learning [29–32].

- **Hybrid P/EMS**
  A hybrid scheme for power/energy management in microgrids can be realized as another interactive structure that is based mainly on a combination of centralized and distributed schemes. In a hybrid structure, LCs, which are used for operation management of different energy sources, are divided into groups [33]. Within each group, a centralized scheme is used to control and optimize the performance of LCs. On a higher level, a distributed scheme is utilized to coordinate the operation of centralized controllers in different clusters for global optimization [34]. Such a hybrid strategy can be seen in Figure 5.3.

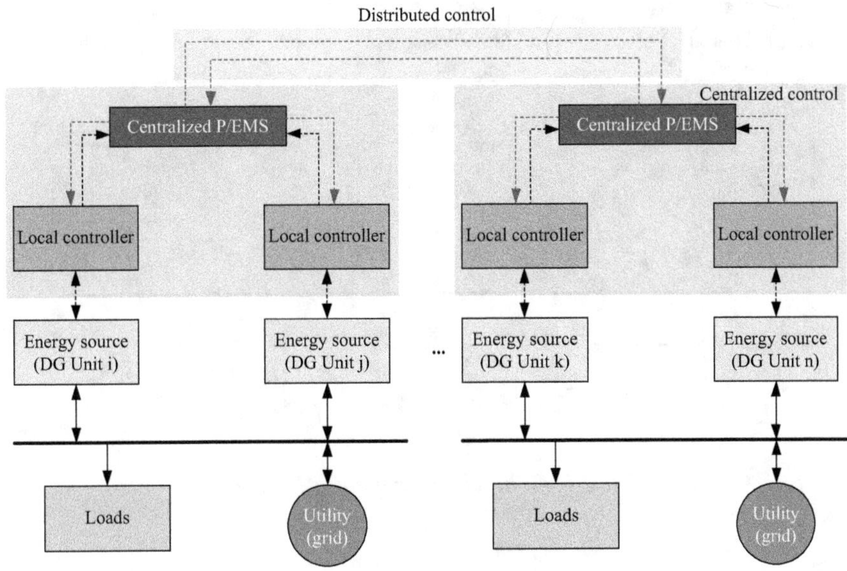

Figure 5.3 Block diagram of a hybrid P/EMS

It is notable that a hybrid P/EMS scheme is normally implemented for large-scale networks such as interconnected energy systems or microgrids, where the optimal operation of the entire system depends on cooperation and coordination of different control layers over a time period. Moreover, by doing this hybridization, it is possible to improve system reliability and resiliency for long-run operations, thanks to the unique features inherently existent in decentralized/decentralized schemes [35,36].

5.2.1.2 Passive power/energy management strategies

Self-autonomy of operation for a LC without having information from neighboring nodes is the main idea of a passive power/energy management scheme (PP/EMS). In this structure, it is assumed that making an information sharing mechanism is too costly or not viable, thus independent operation of energy sources is required. Moreover, it is necessary to clearly define the control objective of each energy source to assure reliable operation of the system. A block diagram for such a power/energy management scheme is shown in Figure 5.4.

Among the existing methods for PP/EMS, droop-based control strategy is regarded as a dominant method [37–43]. This control methodology adopts the behavior of synchronous machines in responding to the changes in voltage and frequency and applies similar rules in the operation management of converters in ac/dc sides. The droop-based control strategy works based on the assumption that the output impedance of a controllable unit (such as a micro-source) is mainly inductive, and it utilizes the droop characteristics of voltage amplitude and

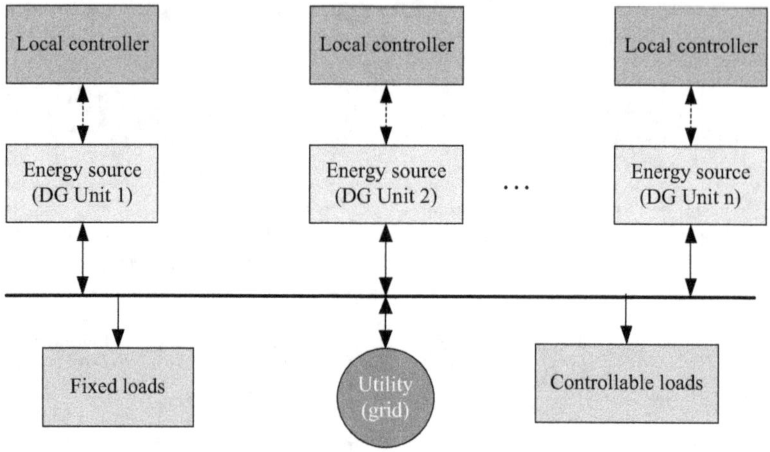

Figure 5.4 Block diagram of a PP/EMS

frequency of each controllable unit to control its output [41]. In the case of a dc microgrid, bus voltages, and in the case of an ac microgrid, the system voltage and frequency are the information sensed by each local droop controller and used subsequently to adjust output active (and/or reactive) power as a generation unit. Since there is no communication requirement for fulfilling the control objectives, this control strategy is highly reliable. Moreover, this control structure could be easily extended to different energy sources while enabling true plug-and-play features [39]. Apart from these benefits, there are several issues in such a power/energy management strategy. First, nonlinear current sharing among parallel units cannot be addressed directly by this method. Second, low X/R line impedance ratio may result in active and reactive power coupling and instability issues in low-voltage microgrid systems and cause power sharing errors for generation units [44,45]. Recently, work has been carried out to improve the performance of a conventional droop-based control method by implementing the droop in virtual frames [44, 46], adding virtual impedance in control loops [47,48], or adjusting the output voltage bandwidth [44]. However, without a coordinating unit, such as a central controller (CC) or a system optimizer, it would be a challenging task to optimally manage the operation of a microgrid system with PP/EMS.

As another type of PP/EMS, maximum power point tracking (MPPT) control methodology is also applied in microgrids to maximize power extraction from renewable energy sources (mainly wind turbines (WTs) and photovoltaic (PV) solar systems) under all conditions [49,50]. In such a power management technique, the unit's voltage and current are sampled frequently, and the duty ratio of the interfaced converter is adjusted accordingly. However, it should be noted that in islanded renewable-based microgrids, which are controlled based on MPPT principles, energy storage systems (ESSs) must also be dispatched to provide voltage and frequency regulation services [51].

Considering the drawbacks of IP/EMS and PP/EMS, it seems that a combined P/EMS structure (e.g. a consensus-based droop framework [52] or a droop-based distributed cooperative control [53]) could not only address reliability issues, but also enhance control performance of the system both in grid-connected and stand-alone modes.

5.2.2 Control schemes

In order to guarantee stable and efficient operation of a dc microgrid, effective control strategies should be developed. The structure of a dc microgrid control system can be depicted as shown in Figure 5.5.

In general, a microgrid consists of several parallel converters that ideally work in harmony. Local control functions of these converters typically cover the following: (1) current, voltage and droop control for each unit; (2) source-dependent functions, e.g. MPPT for PV modules and WTs or a state of charge (SoC) estimation for ESSs; (3) decentralized coordination functions, such as local adaptive calculation of virtual resistances, distributed dc bus signaling (DBS) or power line signaling (PLS).

At a global microgrid level, a dedicated communication-based coordinated control can be implemented to achieve advanced energy management functions. It can be realized either in a centralized or a distributed fashion, via a CC or sparse communication network, respectively [28,38]. In case of a distributed control, variables of interest are only exchanged between LCs. A consensus algorithm can then be used to calculate either the average of all the variable values in distributed LCs (DCLs) or the exact value of any variable present in a specific LC. Some of the functionalities that can be accomplished by using DCLs include secondary/tertiary control, real-time optimization, unit commitment and internal operating mode changing (see Figure 5.6 for more details) [54].

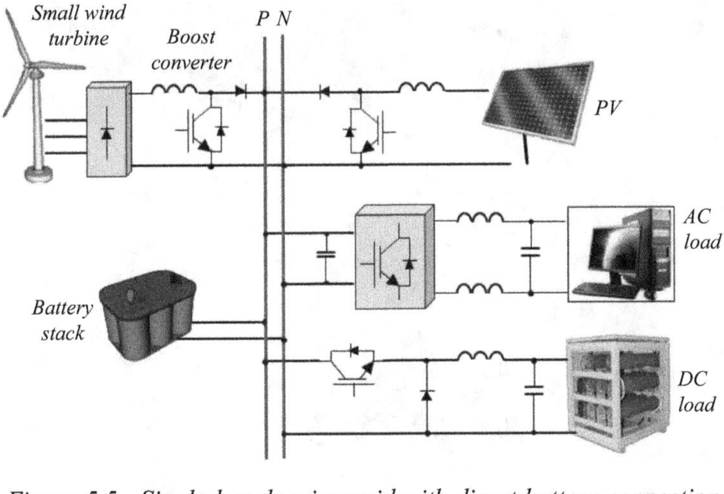

Figure 5.5 Single-bus dc microgrid with direct battery connection

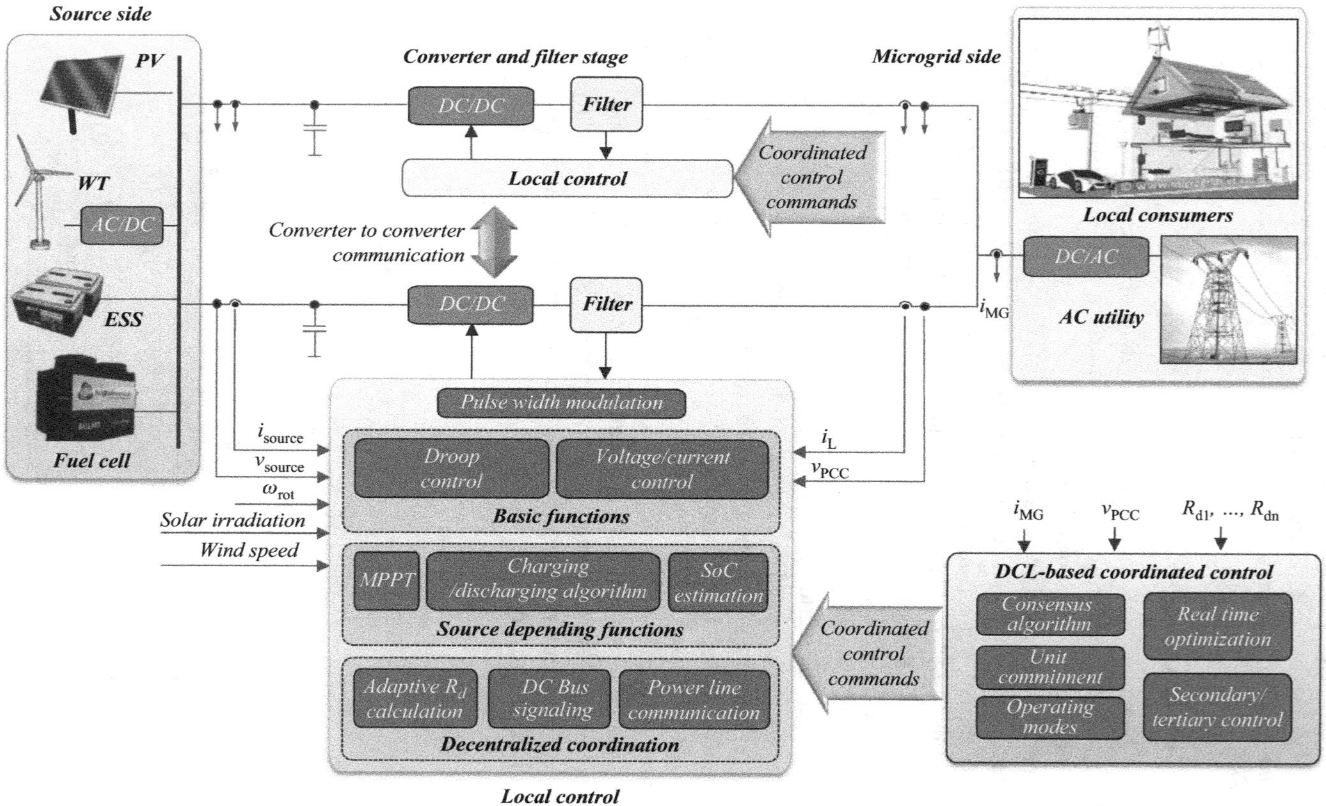

Figure 5.6 General control structures in dc microgrids

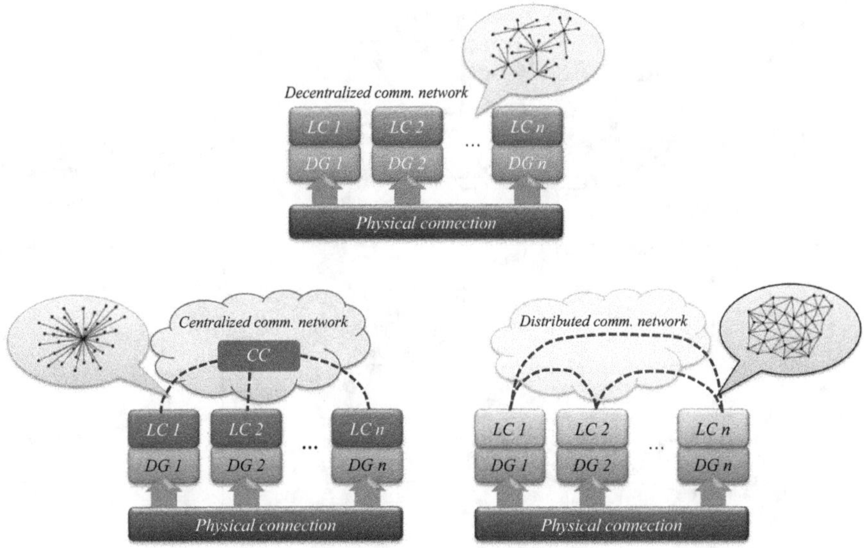

Figure 5.7 Differences between decentralized, centralized and distributed control architectures

From the communication perspective, overall control of dc microgrids can be divided into the following three categories:

- Decentralized control: DCLs do not exist and power lines are used as the only channel of communication.
- Centralized control: data from distributed units are collected in a centralized aggregator, processed and feedback commands are sent back to them via DCLs.
- Distributed control: DCLs exist but are implemented between units and coordinated control strategies are processed locally.

The basic configuration of each of these control structures is depicted in Figure 5.7. A more detailed overview of the significant features of local and coordinated control strategies is provided in the following sections.

5.2.2.1 Local control functionalities

As previously mentioned, the control framework of a dc microgrid consists of local and coordinated control levels. In this section, a local control level is discussed in detail. Basic functions which include current, voltage and droop control are reviewed. A review of MPPT and charging algorithms has been omitted here. More details on charging algorithms for batteries can be found in [55], while an in-depth analysis of MPPT algorithms has been presented in a number of references, e.g. in [56]. As a backbone of a dc microgrid, the interface converters play an important role in efficient and reliable operation of the overall system. In order to ensure not only proper local operation, but also to enable coordinated interconnection between

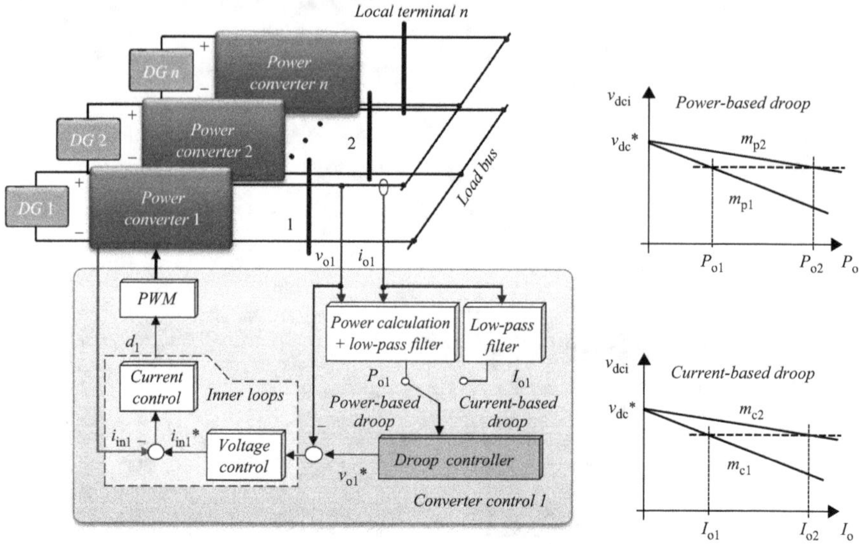

Figure 5.8 Inner control loops for dc microgrids

different modules in a dc microgrid, flexible local current and voltage control should be employed and accurate power sharing among parallel connected converters should be achieved.

The basic local control diagram is shown in Figure 5.8, including local current and voltage controllers, and a droop control loop. For local dc current and voltage control systems in dc microgrids, proportional–integral (PI) controllers are commonly used since they introduce zero steady-state error, can be easily tuned and are highly robust. However, the use of other types of controllers, such as proportional derivative (PD), fuzzy and boundary controllers, has also been reported [57–59]. PD controllers can be used to improve the phase margin (PM) of the system, but they do not eliminate steady-state error and, also, need to have high-frequency poles to attenuate the high-frequency noise. Hence, rather than appearing in a pure PD form, the derivative term in a PD controller is usually replaced by a high-pass digital filter. By combining the beneficial effects of PI and PD controllers, proportional-integral-derivative (PID) controllers can be employed. Fuzzy control is designed to emulate a human being's conclusion deduction process based on the stimulus he/she gets from the environment and his/her own embedded knowledge. In the engineering world, this can be defined as a knowledge-based control method that can simultaneously take advantage of both static and dynamic properties of the system. It has therefore good adaptability in environments with varying conditions.

For the purpose of local voltage and current regulation, fuzzy controllers can either be used as principal regulators that process the error signal or in a series with feedback loops. To ensure fast convergence and extreme robustness, nonlinear control strategies based on state-dependent switching can be employed. They present simple implementation, but their detailed performance analysis can be quite complex. It should be noted that alternative control methods for dc microgrids have recently drawn a lot of attention in academic circles. However, their practical application should be elaborately justified by performing modeling, analysis, simulation, implementation, as well as a full cost-benefit analysis. For instance, increased production cost and lead time often proves to be too large an obstacle for their deployment.

Droop control is commonly installed on top of inner loops, primarily for current sharing purposes. Figure 5.8 demonstrates that either output power or output current can be selected as the feedback signal in droop control [60,61]. For dc microgrids with a power-type load, output power can be used as droop feedback, as shown in (1). On the other hand, when current signal is used, as shown in (2), droop coefficient m_c can be regarded as a virtual internal resistance. In this case, the implementation and design of the parallel converter system in a dc microgrid can be simplified to some extent as the control law is linear [60]. The principle of current-based droop control was also extensively used in distributed power systems for paralleling multiphase converters that supply computer CPUs. Here, droop control is commonly known as adaptive voltage positioning [62–64]. The calculations of references for voltage controller, in the two aforementioned cases, are as follows:

$$v^*_{\mathrm{dci}} = v^*_{\mathrm{dc}} - m_{\mathrm{p}} \cdot P_{\mathrm{oi}} \tag{5.1}$$

$$v^*_{\mathrm{dci}} = v^*_{\mathrm{dc}} - m_{\mathrm{c}} \cdot i_{\mathrm{oi}} \tag{5.2}$$

where v^*_{dci} is the output of the droop controller, i.e. the reference value of the dc output voltage of converter i, v^*_{dc} is the rated value of the dc voltage; m_{p} and m_{c} are the droop coefficients in power- and current-based droop controllers, while P_{oi} and i_{oi} are the output power and current, respectively, of converter i.

5.2.2.2 Coordinated control

Although the local interface converter control is an essential part of a dc microgrid which provides regulation of physical quantities, coordinated control is needed to achieve extended objectives. As already mentioned, depending on the means of communication between the interface converters, coordinated control should be implemented to achieve an intelligent control system with extended objectives.

- **Centralized coordination**
  Centralized coordination control can be implemented in dc microgrids by employing a CC and a dedicated communication network to connect it with sources and loads, as shown in the DCL-based coordination control window of Figures 5.6 and 5.7. For small-scale dc microgrids, each unit can be directly controlled by the CC that employs a high bandwidth communication using a master/slave approach [65,66]. However, for larger scale dc microgrids, hierarchical control is often a preferred choice since it introduces a certain degree of independence between different control levels. This control is more reliable as it can be designed to continue the operation even in a case of failure of

centralized control. Hierarchical control is achieved by simultaneously using local converter control and DCL-based coordinated control, which are separated by at least an order of magnitude in the control bandwidth [60]. Coordinated functions can include secondary/tertiary regulation of the dc voltage, power flow control and different grid-interactive control objectives such as unit commitment, changing operating modes, global optimization aimed at maximizing efficiency, minimizing operating cost, etc.

It should be noted that centralized control provides the best foundation for the employment of advanced control functionalities since all relevant data can be collected and processed in a single controller. However, the most obvious disadvantage of this strategy is that it has a single point of failure. In particular, if the CC or any key communication link fails, the commands from/to the controller will not be transmitted and corresponding control objectives will likely not be achieved. For mission critical applications, redundant communication systems can be installed in order to reduce the possibility of failure, but this needs to be justified by a cost–benefit analysis. Another option to increase the reliability of the system is to combine decentralized and centralized control methods into a hierarchical control structure [67–72]. In this case, basic functions of the dc microgrids can be retained even if the centralized controller fails.

- **Decentralized coordination**

Decentralized coordination strategies are achieved exclusively by LCs, as shown in Figures 5.6 and 5.7. The most common types of decentralized control are DBS, adaptive adjustment of droop coefficients and PLS. While their advantage is in the simplicity of control and independence from dedicated communication technology, they inherently have performance limitations due to lack of information from other units. Moreover, as these methods are invariably based on the interpretation of the voltage in the common dc bus, the accuracy of voltage sensors impacts their effectiveness and reliability.

Originally proposed in [73] and the follow-up in [74], DBS is the most prominent decentralized coordination method for dc microgrids. By using the DBS approach, coordinated operation of different units in dc microgrids is realized by imposing and identifying variations in the common dc bus voltage. The transitions between different modes are triggered by different preset dc bus voltage values.

The method of adaptive droop coefficient calculation has been mainly used for power balancing of distributed ESSs, as shown above. The main limitation of the method is the potential instability induced by improperly designed droop curves. To that end, there always exists a trade-off between the permissible voltage deviation and stability properties of the system, i.e. higher voltage deviation is associated with the higher PM. It should also be noted that some of the SoC balancing methods mentioned above [75,76] were originally proposed for ac microgrids. However, since the same principle can apply to dc microgrids (in ac microgrids, the frequency is normally used as a system level coordinating signal), they are also included here.

PLS is another decentralized method that can be deployed for coordinated control. In particular, sinusoidal signals of a specific frequency are injected through amplifiers into the dc bus, allowing each device to send and receive information on its status, performance, history or internal operational mode. Although PLS relies on dedicated communication, here it is categorized as decentralized since the power network is the only communication medium. It should be noted that in power systems' literature, this particular way of communication is sometimes also referred to as PLC [77]. REbus, an open standard for dc electricity distribution in homes, commercial buildings, campuses and other settings, uses PLS as a primary communication carrier [78]. Alternative methods that exchange information between devices without using dedicated amplifiers were presented in [79, 80]. Signals are generated by the PWM of dc–dc converters in those works. They are injected in open loop in [80], whereas dedicated proportional-resonant controllers are used during injection periods in [79] to avoid the steady-state error in the dc bus [81].

PLS is generally more complex to implement compared to other decentralized methods. Moreover, it is commonly only used for changing operating modes or shutting down corrupted components of the system, and it is not suitable for power sharing. However, as opposed to permanent voltage deviation in the common dc bus which is inherent for DBS and adaptive droop methods, sinusoidal signals are only periodically injected into the system. Therefore, the quality of the voltage waveform can be considered, to some extent, improved compared to other methods.

- **Distributed coordination**
  Distributed control indicates the control principle where a central control unit does not exist and LCs communicate only among themselves through dedicated DCLs, as shown in Figure 5.7. The main advantage of this approach is that the system can maintain full functionality, even if the failure of some communication links occurs, provided that the communication network remains connected.[1] Therefore, distributed control is immune to a single point of failure. The functionalities that can be achieved by this approach resemble those of centralized control and are also represented in the DCL-based coordination control window of Figure 5.6. However, in order to enable these functionalities, the information exchanged through DCLs first needs to be appropriately processed. In particular, information directly exchanged between LCs can contain only locally available variables. In other words, if the two units are not connected by a DCL, they do not have direct access to each other's data and their observation of the system is quite limited. In order to circumvent this problem and to make the level of awareness of an LC similar to that of a CC, a consensus algorithm can be used. In its basic form, a consensus algorithm is a simple protocol installed within every LC which continuously adds up all algebraic differences of some variable(s) of interest

[1] For the exact definition and a more in-depth discussion of the connectivity of communication networks, please refer to [83].

present in a given LC and those present in LCs adjacent to it. If we look at LC i, this definition can also be expressed by the following equation:

$$\dot{x}_i(t) = \sum_{j \in N_i} \left[x_i(t) - x_j(t)\right] + b_i(t) \tag{5.3}$$

where $x_i(t)$ and $x_j(t)$ are the values of variables of interest in LC i and LC j, respectively. Here, j is iterated through all of the neighbors of LC i, represented by N_i. Finally, $b_i(t)$ is an optional input bias of LC, which can be used to declare it as a virtual leader. It can be seen from (5.3) that $x_i(t)$ is interactively adapted with respect to the values of its neighboring units. Likewise, variables in any other controller adapt with respect to the values of their own neighbors. Consequently, it can be analytically proved that, if the communication network is connected, all variable values will converge to a common average after a certain amount of time [82,83]. Another option is to use a nonzero input bias in one of the LCs. In that case, variables of all other LCs will converge to their respective bias [84]. In either case, the ability of a consensus to share information in such a manner has wider applicability than simple data averaging. For instance, if every LC has information on the number of other active LCs, an exact value of any specific variable can be calculated directly from the average.

The collective dynamics of a communication system realized via consensus protocol can be represented by the following equation:

$$\dot{x}(t) = -\Theta \cdot x(t) \tag{5.4}$$

where $\Theta = [\theta_{ij}]$ is the graph Laplacian of the network whose elements are defined as follows:

$$\theta_{ij} = \begin{cases} -1 & j \in N_i \\ |N_i| & j = i \end{cases} \tag{5.5}$$

where $|N_i|$ denotes the number of neighbors of node i. The topology of the communication network is explicitly reflected by graph Laplacian, and it is also possible to design weights of the respective matrix to control the convergence speed [85]. Figure 5.9 shows the configurations of an exemplary power electronics-based microgrid and its sparse communication network

In summary, it can be concluded that distributed control can achieve information awareness comparable to that of centralized control. Therefore, objectives such as output current sharing, voltage restoration, global efficiency enhancement, SoC balancing and others can be easily realized. In this sense, distributed control offers much wider functionalities than decentralized control, but remains protected from the single point of failure. Its main limitation is the complexity of analytical performance analysis, i.e. assessment of convergence speed and stability margins,

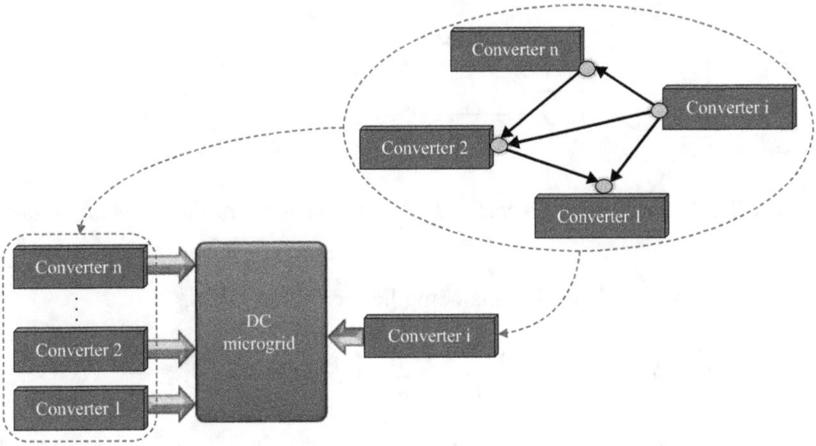

Figure 5.9 Configurations of exemplary physical and communication networks

especially in nonideal environments characterized by communication time delays and measurement errors.

5.3 Interfacing converter control strategies for power/energy management purposes

It was previously discussed that the primary goal of a power/EMS is to determine the optimal operating point of controllable units (such as active/reactive power references) with regard to the system's objective(s) and constraints (e.g. device technical specifications, permissible voltage and frequency deviations, demand–supply balance, etc.). To effectively track optimal set-points (reference signals) coming from a system-level optimizer while regulating current and the voltage, advanced control strategies must be applied in interfacing converters. These control strategies are generally categorized as voltage control mode (VCM) (also known as grid-forming mode) and current control mode (CCM) (also known as grid-following mode) [86]. In the next section, a brief description is presented for each of the interfacing converter control modes.

5.3.1 Voltage control/grid-forming mode

By mimicking a synchronous generator/condenser behavior for adjusting the grid's voltage, VCM can be applied for a converter-based DG unit to regulate the output power of the unit and reach the same objective. Droop control concepts can also be used for efficient implementation of such control strategy. Let us consider the problem of complex power flow through a line as shown in Figure 5.10. Without loss of generality, the line could be considered as an *R–L* circuit ($R+jX$) with the

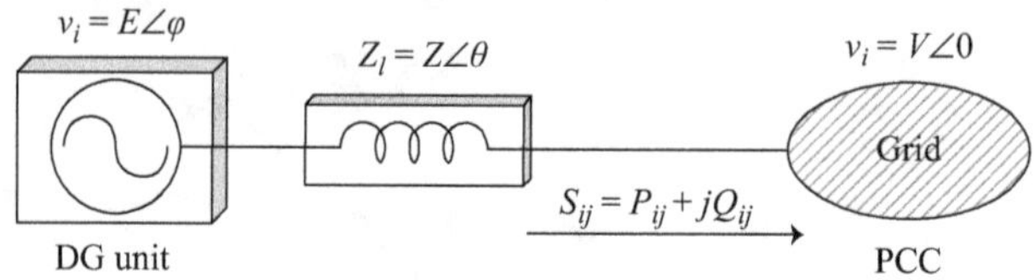

Figure 5.10 Equivalent circuit of DG unit connected to the common ac bus

voltages at the terminals of the line being held constant. Thus,

$$S_{ij} = P_{ij} + jQ_{ij} = v_i\left(\frac{v_i - v_j}{Z_l}\right)^* = \frac{E}{Ze^{-j\theta}}\left(E - Ve^{j\varphi}\right) \tag{5.6}$$

Considering that a converter-based DG unit is connected through a highly inductive line to a node and the phase angle φ between the converter and the point of coupling is small, it can be concluded that

$$P_{ij} \cong \frac{E \cdot V}{X}\sin(\varphi) \rightarrow \varphi \cong \frac{X \cdot P_{ij}}{E \cdot V} \tag{5.7}$$

$$Q_{ij} \cong \frac{E^2}{X} - \frac{E \cdot V}{X}\cos(\varphi) \rightarrow E - V \cong \frac{X \cdot Q_{ij}}{E} \tag{5.8}$$

meaning that the active and reactive power outputs of a given DG unit are proportional to the phase angle difference φ and the voltage magnitude difference $(E-V)$, respectively. In other words, by controlling the real power, it would be possible to adjust the power angle and, through reactive power control, voltage regulation can be achieved. In a droop-based control, the same procedure is adopted with the exception that each unit uses the frequency (instead of the power angle or phase angle), for controlling active power flow.

Figure 5.11 shows the block diagram of the VCM control strategy which can be applied in both grid-tied and islanded operating modes of microgrids and makes the operation mode transition easy and smooth. As observed from the same figure, active and reactive power references are provided by the P/EMS to meet certain goals such as voltage profile smoothing. The lower level controllers embedded in the VCM structure (named as active and reactive power controllers in Figure 5.11) assure that the output voltage phase angle and magnitude are of the right orders and that they can be designed as proportional controllers for realizing active power–frequency droop (P–ω) and reactive power–voltage magnitude droop (Q–E) as follows:

$$\omega = \omega_{ref} - m\left(P_{meas} - P_{ref}\right) \tag{5.9}$$

$$V = V_{ref} - n\left(Q_{meas} - Q_{ref}\right) \tag{5.10}$$

In this strategy, inner current loops can be assigned to the voltage closed-loop control system to improve the transient and stability performance [87].

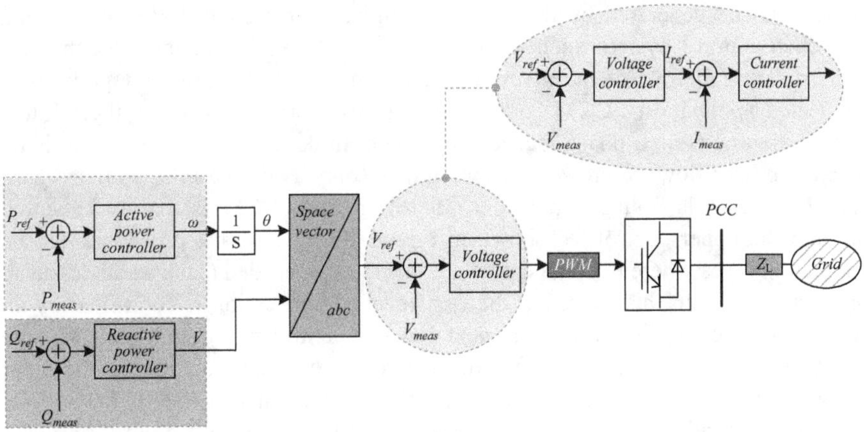

Figure 5.11 Voltage-based real and reactive power control strategy

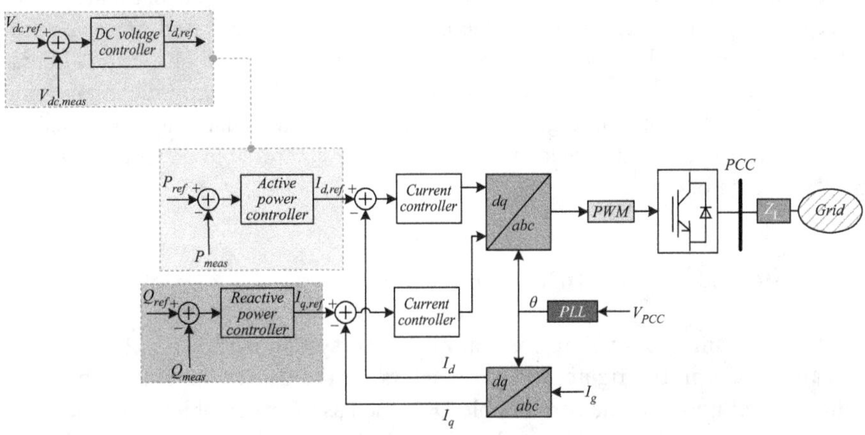

Figure 5.12 CCM/grid-following control strategy

More complex controllers can also be integrated into the VCM to closely mimic the behavior of a synchronous machine equipped with governor and excitation systems [88].

5.3.2 Current control/grid-following mode

Figure 5.12 shows the schematic of a current control/grid-following control strategy for power flow management. As can be observed from the figure, there are several current-based controllers in a CCM that track and regulate the active and reactive powers in the closed-loop manner. The outer active and reactive power control loops are in charge of producing *d*-axis and *q*-axis reference currents (based on the received set-points), respectively. It should be noted that the reference power signals can be generated by the system optimizer (such as an EMS) or through any other control system such as a MPPT unit. It is also noteworthy that for synchronizing the interfacing inverter output current with the grid voltage in grid-connected applications, CCM control strategy uses the grid voltage angle information coming from a phase-locked-loop block. In dc networks where there is no reactive power flow, reference current signals (only available in *d*-axis) are generated using a dc voltage controller. In this regard, CCM is mainly executed through the upper control leg shown in Figure 5.12. The same procedure can be adopted in cases where two-stage power conversion is needed (such as dc/dc–dc/ac or ac/dc–dc/ac). In this situation, the first level of conversion is controlled by an active power controller, while the next level is controlled through a dc voltage controller in which the power difference between the inverter output and the input stage can be used to charge or discharge the dc link capacitor [89]. Taken as a whole, CCM/grid-following mode is mainly used for power flow control purposes in grid-connected conditions where the utility (i.e. main grid) plays the key role in supporting the system frequency and regulating the common bus voltage while other controllable units can be operated/dispatched based on an optimal control strategy enforced by the CCM. On the other hand, the CCM-based power control mode cannot be used as the only control strategy in an off-grid or islanded operation of microgrids as it is not capable of regulating the microgrid voltage and frequency in different working conditions. Thus, complementary actions based on VCM control strategy are needed for at least one controllable generation unit or energy storage option.

5.4 Illustrative example

A block diagram of a given residential microgrid system is shown in Figure 5.13. As can be seen in the figure, the main actors on the generation side include the utility (power grid) and the controllable DGs such as micro-combined cogeneration systems micro-combined heat and power (μCHP), while on the demand side, schedulable appliances serve as actors.

In this structure, the leading actor is the residential smart EMS (RSEMS) that manages the operation of in-home devices and appliances optimally in coordination with the microgrid CC taking into account users' objectives and existing technical/operational constraints. RSEMS also provides required information such as set-points and reference signals for LCs when is needed. While the RSEMS is designed as a soft real-time system to improve and optimize system operation during each decision period, lower level controllers at the supply/demand sides are modeled as hard real-time systems to maintain stability. The two aforementioned systems have different time granularities according to types of controlled objects and performance requirements. In the examined microgrid, sustainable energy sources such as PV systems and WTs are exploited as nondispatchable prime movers of the mentioned system. The power grid is also used as a sustained power supply to support required point of common coupling (PCC) voltage regulation and system

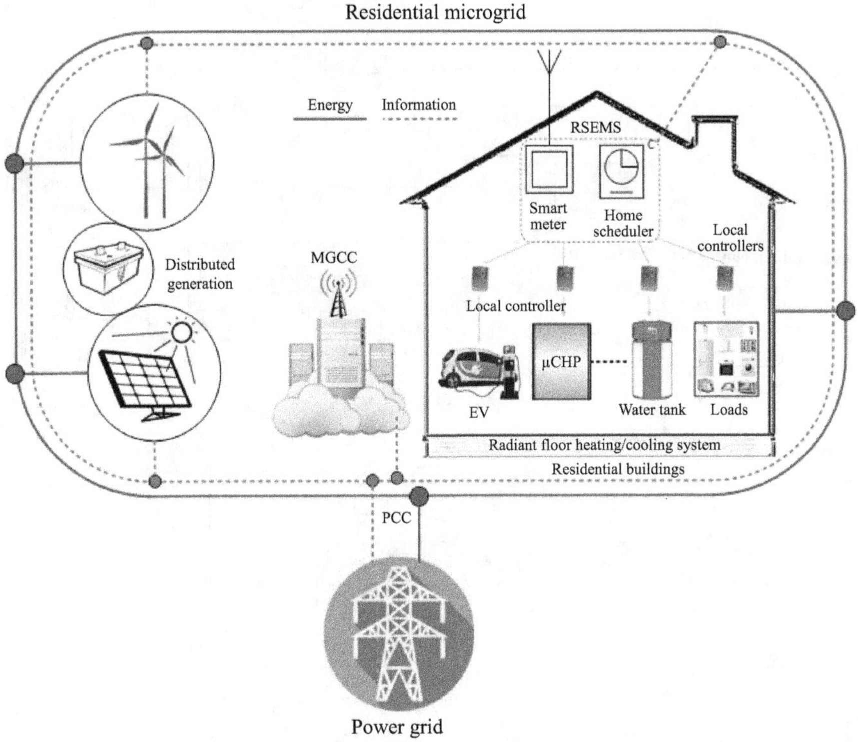

Figure 5.13 Integrated building and microgrid system

frequency. Other dispatchable dispersed generators within the residential section adjust their behaviors according to the power demand and reference signals received from RSEMS [90].

The lower level control structure (here named as the "master-slave") of the existing DG sources can be implemented in a way depicted in Figure 5.14. Based on this control structure, each controllable DG is treated as a "grid-following" unit and operated in CCM/power control mode to follow a power reference (P^*), while the utility (which is the master unit) is operated in grid-forming mode (VCM) to maintain the voltage and the frequency within the range. In this regard, droop control is implemented based on (5.11) and (5.12) in the primary loop for the master unit, and the reverse droop controller is utilized for the slave units as expressed in (5.13):

$$\omega = \omega^* - m_d P \tag{5.11}$$

$$E = E^* - n_d Q \tag{5.12}$$

$$Q^* = \left(\frac{1}{n_r}\right) \cdot (E^* - E_g) \tag{5.13}$$

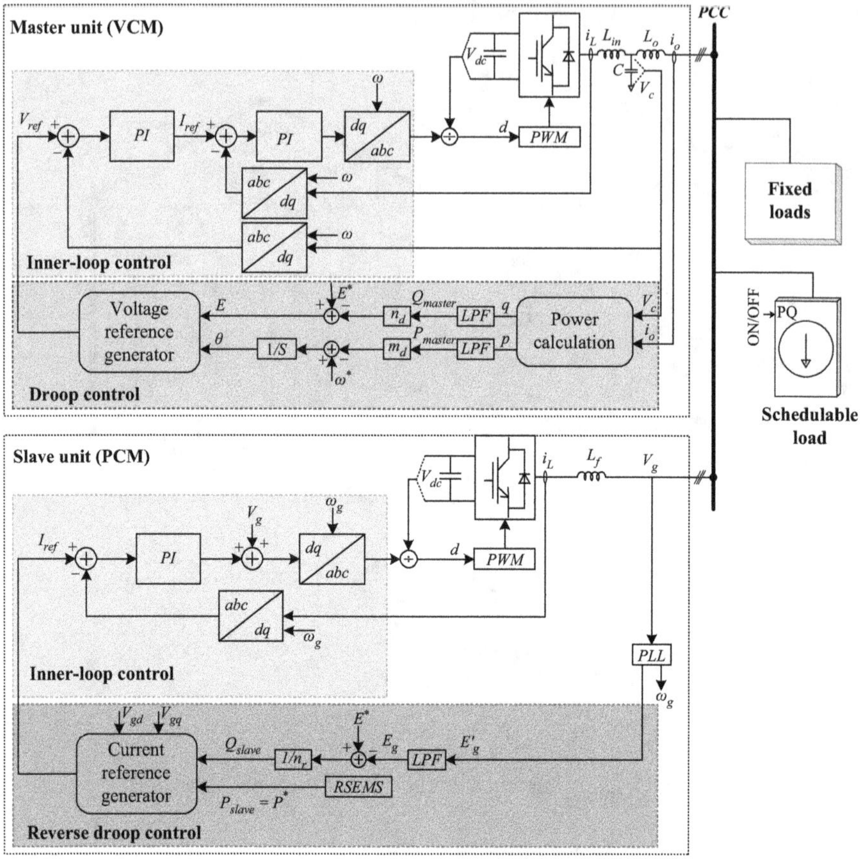

Figure 5.14 Master–slave control of distributed energy sources

It should be mentioned that within the mentioned structure, the inner loop control of the VCM aims to achieve good output voltage regulation with respect to a determined capacitor voltage reference while the inner loop control of the PCM regulates the output power. For both inner loops, typical PI controllers are appropriately designed and used. In the primary control loops, low-pass filters are also applied to limit the loop bandwidth, so that the primary control can be separately designed and the inner loop of the VCM and CCM can be considered as ideal voltage and current sources, respectively.

5.5 Conclusions

With the application of dc microgrids at different scales and topologies, it is possible to integrate different energy sources (such PV systems, fuel cells and batteries) into the energy mix of a larger system with lower conversion requirements.

Moreover, it is much easier and cheaper to convert ac power into dc power which, in turn, helps economic operation of future energy systems. By using dc microgrids, it is also possible to achieve energy savings up to a great extent (~15%) and to improve system reliability by reducing the number of devices required and the total points of failure. The dc architecture also enables cost-effective and green solutions for operation and control of zero-net energy residential/office buildings as well as data centers. However, such optimal performance mainly depends on the proper design and application of EMSs which effectively manage the process of energy production and consumption based on predefined objectives and constraints. There are, definitely, a number of challenges in this regard which have to be suitably addressed. Currently, there is a lack of approved standards and technical codes for dc equipment and distribution networks at low voltage. There is also a lack of approved and recognized dc architectures at low-to-medium voltage levels which in turn necessitates different safety and protection practices in comparison with conventional ac systems. Last but not least, there is a strong need for upgrading the existing infrastructure to accommodate dc systems and interfaces.

References

[1] American Physical Society, "Buildings," in *ENERGY FUTURE: Think Efficiency*. American Physical Society, 2008, pp. 52–85.

[2] International Energy Agency, *Transition to Sustainable Buildings – Strategies and Opportunities to 2050*. Paris: OECD/IEA, 2013.

[3] N. Dempsey, C. Barton, and D. Hough, "Energy Prices," in *The House of Commons Library*, 2016, no. 4153, pp. 1–18.

[4] A. Mousavi, C. W. Yang, C. Pang, and V. Vyatkin, "Energy efficient automation model for office buildings based on ontology, agents and IEC 61499 function blocks," in *19th IEEE Int. Conf. Emerg. Technol. Fact. Autom. ETFA 2014*, 2014, pp. 1–7.

[5] C. Delmastro, G. Mutani, M. Pastorelli, and G. Vicentini, "Urban morphology and energy consumption in Italian residential buildings Urban morphology and energy consumption in Italian residential buildings," in *4th International Conference on Renewable Energy Research and Applications*, 2015, vol. 5, no. Jan. 2016, pp. 1603–1608.

[6] S. I. Symposium, "Simulation-based optimization in energy efficiency retrofit for office building," in *Proc. 2014 IEEE/SICE Int.*, 2014, pp. 222–227.

[7] K. Vatanparvar, Q. Chau, A. Faruque, *et al.*, "Home energy management as a service over networking platforms," in *IEEE PES Conference on Innovative Smart Grid Technologies (ISGT)*, Washington, DC, 2015, pp. 1–5.

[8] A. Gligor, H. Grif, and S. Oltean, "Considerations on an intelligent buildings management system for an optimized energy consumption," in *2006 IEEE Int. Conf. Autom. Qual. Testing, Robot. AQTR*, 2006, no. 1, pp. 280–284.

[9] A. Anvari-Moghaddam, H. Monsef, and A. Rahimi-Kian, "Optimal Smart Home Energy Management Considering Energy Saving and a

Comfortable Lifestyle," *IEEE Trans. Smart Grid*, vol. 6, no. 1, pp. 324–332, 2015.

[10] A. Anvari-Moghaddam, H. Monsef, and A. Rahimi-Kian, "Cost-Effective and Comfort-Aware Residential Energy Management under Different Pricing Schemes and Weather Conditions," *Energy Build.*, vol. 86, pp. 782–793, 2015.

[11] Y. Li and F. Nejabatkhah, "Overview of Control, Integration and Energy Management of Microgrids," *J. Mod. Power Syst. Clean Energy*, vol. 2, no. 3, pp. 212–222, Aug. 2014.

[12] A. Ghasemkhani, A. Anvari-Moghaddam, J. M. Guerrero, and B. Bak-Jensen, "An efficient multi-objective approach for designing of communication interfaces in smart grids," in *IEEE PES Innovative Smart Grid Technologies (ISGT 2016)*, 2016, pp. 1–6.

[13] S. Z. Islam, N. Mariun, H. Hizam, *et al.*, "Communication for distributed renewable generations (DRGs): A review on the penetration to smart grids (SGs)," in *PECon 2012 – 2012 IEEE Int. Conf. Power Energy*, 2012, no. December, pp. 870–875.

[14] K. M. Liyanage, M. A. M. Manaz, A. Yokoyama, Y. Ota, H. Taniguchi, and T. Nakajima, "Impact of communication over a TCP/IP network on the performance of a coordinated control scheme to reduce power fluctuation due to distributed renewable energy generation," in *2011 6th Int. Conf. Ind. Inf. Syst. ICIIS 2011 – Conf. Proc.*, 2011, pp. 198–203.

[15] A. Anvari-Moghaddam, G. Mokhtari, and J. M. Guerrero, "Coordinated Demand Response and Distributed Generation Management in Residential Smart Microgrids," in *Energy Management of Distributed Generation Systems*, 1st ed., Dr. Eng. L. Mihet, Ed. InTech, 2016, pp. 27–57.

[16] M. Sechilariu, B. Wang, and F. Locment, "Power management and optimization for isolated DC microgrid,"in *2014 International Symposium on Power Electronics, Electrical Drives, Automation and Motion*, Ischia, 2014, pp. 1284–1289.

[17] F. Nejabatkhah and Y. W. Li, "Overview of Power Management Strategies of Hybrid AC/DC Microgrid," *IEEE Trans. Power Electron.*, vol. 30, no. 12, pp. 7072–7089, 2015.

[18] D. E. Olivares, S. Member, and C. A. Cañizares, "A Centralized Energy Management System for Isolated Microgrids," *IEEE Trans. Smart Grid*, vol. 5, no. 4, pp. 1864–1875, 2014.

[19] L. E. Zubieta, "Power Management and Optimization Concept for DC Microgrids," in *2015 IEEE First International Conference on DC Microgrids (ICDCM)*, 2015, pp. 81–85.

[20] M. Parvizimosaed, F. Farmani, and A. Anvari-Moghaddam, "Optimal Energy Management of a Micro-Grid with Renewable Energy Resources and Demand Response," *J. Renewable Sustainable Energy*, vol. 5, no. 5, pp. 31–48, 2013.

[21] A. Anvari-Moghaddam, A. Seifi, and T. Niknam, "Multi-Operation Management of a Typical Micro-Grids Using Particle Swarm Optimization: A Comparative Study," *Renewable Sustainable Energy Rev.*, vol. 16, no. 2, pp. 1268–1281, 2012.

[22] A. Anvari-Moghaddam, A. Seifi, T. Niknam, and M. Pahlavani, "Multi-Objective Operation Management of a Renewable MG (Micro-Grid) with Back-up Micro-Turbine/Fuel Cell/Battery Hybrid Power Source," *Energy*, vol. 36, no. 11, pp. 6490–6507, 2011.

[23] Q. Jiang, M. Xue, G. Geng, "Energy Management of Microgrid in Grid-Connected and Stand-Alone Modes," *IEEE Trans. Power Syst.*, vol. 28, no. 3, pp. 3380–3389, 2013.

[24] D. J. Becker and B. J. Sonnenberg, "DC microgrids in buildings and data centers," in *Telecommunications Energy Conference (INTELEC), 2011 IEEE 33rd International*, 2011, pp. 1–7.

[25] A. Tani, M. B. Camara, and B. Dakyo, "Energy Management in the Decentralized Generation Systems based on Renewable Energy—Ultracapacitors and Battery to Compensate the Wind/Load Power Fluctuations," *IEEE Trans. Ind. Appl.*, vol. 51, no. 2, pp. 1817–1827, 2015.

[26] Z. Wang, B. Chen, J. Wang, and J. Kim, "Decentralized Energy Management System for Networked Microgrids in Grid-Connected and Islanded Modes," *IEEE Trans. Smart Grid*, vol. 7, no. 2, pp. 1097–1105, 2016.

[27] G. Mokhtari, A. Anvari-Moghaddam, and G. Nourbakhsh, "Distributed Control and Management of Renewable Electric Energy Resources for Future Grid Requirements," in *Energy Management of Distributed Generation Systems*, 1st ed., no. Jul., Dr. Eng. L. Mihet, Ed. InT, 2016, pp. 1–24.

[28] A. Anvari-Moghaddam, J. J. M. Guerrero, A. Rahimi-Kian, and M. S. Mirian, "Optimal real-time dispatch for integrated energy systems: An ontology-based multi-agent approach," in *7th International Symposium on Power Electronics for Distributed Generation Systems (PEDG'16)*, 2016, pp. 1–7.

[29] C. Dou, D. Yue, S. Member, X. Li, and Y. Xue, "MAS-Based Management and Control Strategies for Integrated Hybrid Energy System," *IEEE Trans. Industrial Informatics*, vol. 12, no. 4, pp. 1332–1349, 2016.

[30] J. Yu, S. Member, C. Dou, and X. Li, "MAS-Based Energy Management Strategies for a Hybrid Energy Generation System," *IEEE Trans. Industrial Electronics*, vol. 63, no. 6, pp. 3756–3764, 2016.

[31] Y. S. F. Eddy, S. Member, H. B. Gooi, S. Member, and S. X. Chen, "Multi-Agent System for Distributed Management of Microgrids," *IEEE Trans. Power Systems*, vol. 30, no. 1, pp. 24–34, 2015.

[32] M. Mao, P. Jin, and N. D. Hatziargyriou, "Multiagent-Based Hybrid Energy Management System for Microgrids," *IEEE Trans. Sustainable Energy*, vol. 5, no. 3, pp. 938–946, 2014.

[33] C. X. Dou, Z. S. Duan, and B. Liu, "Two-Level Hierarchical Hybrid Control for Smart Power System," *IEEE Trans. Autom. Sci. Eng.*, vol. 10, no. 4, pp. 1037–1049, 2013.

[34] C. Dou, X. Jia, Z. Bo, D. Liu, and F. Zhao, "Hybrid Control for Micro-Grid based on Hybrid System Theory," in *2011 IEEE Power and Energy Society General Meeting*, San Diego, CA, pp. 1–11, 2011.

[35] T. Ma, M. H. Cintuglu, S. Member, and O. A. Mohammed, "Control of Hybrid AC/DC Microgrid Involving Energy Storage and Pulsed Loads," *IEEE Trans. Ind. Appl.*, vol. 53, no. 1, pp. 567–575, 2017.

[36] T. Lu, Z. Wang, Q. Ai, and W.-J. Lee, "Interactive Model for Energy Management of Clustered Microgrids," *IEEE Trans. Ind. Appl.*, vol. 53, no. 3, pp. 1739–1750, 2017.

[37] E. Rodriguez, J. C. Vasquez, M. Josep, *et al.*, "Multi-level energy management and optimal control of a residential DC microgrid," in *2017 IEEE International Conference on Consumer Electronics (ICCE)*, Las Vegas, NV, 2017, pp. 312–313.

[38] A. Anvari-moghadam, Q. Shafiee, J. C. Vasquez, and J. M. Guerrero, "Optimal adaptive droop control for effective load sharing in AC microgrids," in *42nd Annual Conference of the IEEE Industrial Electronics Society (IECON'16)*, 2016, pp. 2–7.

[39] A. Tah and D. Das, "An Enhanced Droop Control Method for Accurate Load Sharing and Voltage Improvement of Isolated and Interconnected DC Microgrids," *IEEE Trans. Sustainable Energy*, vol. 7, no. 3, pp. 1194–1204, 2016.

[40] P. H. Huang, P. C. Liu, W. Xiao, and M. S. El Moursi, "A Novel Droop-Based Average Voltage Sharing Control Strategy for DC Microgrids," *IEEE Trans. Smart Grid*, vol. 6, no. 3, pp. 1096–1106, 2015.

[41] A. Maknouninejad, Z. Qu, F. L. Lewis, and A. Davoudi, "Optimal, Nonlinear, and Distributed Designs of Droop Controls for DC Microgrids," *IEEE Trans. Smart Grid*, vol. 5, no. 5, pp. 2508–2516, 2014.

[42] A. P. N. Tahim, D. J. Pagano, E. Lenz, and V. Stramosk, "Modeling and Stability Analysis of Islanded DC Microgrids under Droop Control," *IEEE Trans. Power Electron.*, vol. 30, no. 8, pp. 4597–4607, 2015.

[43] K. J. Bunker and W. W. Weaver, "Multidimensional Droop Control for Wind Resources in dc Microgrids," *IET Gener. Transm. Distrib.*, vol. 11, no. 3, pp. 657–664, 2017.

[44] Y. Li and Y. W. Li, "Power Management of Inverter Interfaced Autonomous Microgrid based on Virtual Frequency-Voltage Frame," *IEEE Trans. Smart Grid*, vol. 2, no. 1, pp. 18–28, 2011.

[45] Y. W. Li and C. N. Kao, "An Accurate Power Control Strategy for Power-Electronics-Interfaced Distributed Generation Units Operating in a Low-Voltage Multibus Microgrid," *IEEE Trans. Power Electron.*, vol. 24, no. 12, pp. 2977–2988, 2009..

[46] Y. Guan, J. M. Guerrero, X. Zhao, J. C. Vasquez, and X. Guo, "A New Way of Controlling Parallel-Connected Inverters by Using Synchronous-Reference-Frame Virtual Impedance Loop – Part I: Control Principle," *IEEE Trans. Power Electron.*, vol. 31, no. 6, pp. 4576–4593, 2016.

[47] J. Matas, M. Castilla, L. G. de Vicuña, J. Miret, and J. C. Vasquez, "Virtual Impedance Loop for Droop-Controlled Single-Phase Parallel Inverters Using a Second-Order General-Integrator Scheme," *IEEE Trans. Power Electron.*, vol. 25, no. 12, pp. 2993–3002, 2010.

[48] Y. Zhang and Y. Wei Li, "Energy Management Strategy for Supercapacitor Virtual Impedance," *IEEE Trans. Power Electron.*, vol. 32, no. 4, pp. 2704–2716, 2017.

[49] I. Tank, "Renewable based DC microgrid with energy management system," pp. 1–5, 2015.

[50] A. Anvari-Moghaddam, A. Seifi, T. Niknam, and M. R. Alizadeh Pahlavani, "Multi-Objective Operation Management of a Renewable MG (Micro-Grid) with Back-up Micro-Turbine/Fuel Cell/Battery Hybrid Power Source," *Energy*, vol. 36, no. 11, pp. 6490–6507, 2011.

[51] A. Anvari-Moghaddam, T. Dragicevic, J. C. Vasquez, and J. M. Guerrero, "Optimal utilization of microgrids supplemented with battery energy storage systems in grid support applications," in *IEEE First International Conference on DC Microgrids (ICDCM) 2015*, 2015, pp. 57–61.

[52] L. Lu and C. Chu, "Consensus-Based Distributed Droop Control of Synchronverters for Isolated Micro-Grids," *IEEE Trans. Power Syst.*, vol. 30, no. 5, pp. 914–917, 2015.

[53] J. Lai, H. Zhou, X. Lu, X. Yu, and W. Hu, "Droop-Based Distributed Cooperative Control for Microgrids With Time-Varying Delays," *IEEE Trans. Smart Grid*, vol. 7, no. 4, pp. 1775–1789, 2016.

[54] T. Dragicevic, J. C. Vasquez, J. M. Guerrero, and D. Skrlec, "Advanced LVDC Electrical Power Architectures and Microgrids: A Step toward a New Generation of Power Distribution Networks," *IEEE Electrif. Mag.*, vol. 2, no. 1, pp. 54–65, 2014.

[55] D. Linden and T. B. Reddy, *Handbook of Batteries*, 3rd ed. New York: McGraw-Hill, 2002.

[56] E. Koutroulis, K. Kalaitzakis, and N. C. Voulgaris, "Development of a Microcontroller-based, Photovoltaic Maximum Power Point Tracking Control System," *IEEE Trans. Power Electron.*, vol. 16, no. 1, pp. 46–54, 2001.

[57] A. Kwasinski and C. N. Onwuchekwa, "Dynamic Behavior and Stabilization of DC Microgrids with Instantaneous Constant-Power Loads," *IEEE Trans. Power Electron.*, vol. 26, no. 3, pp. 822–834, Mar. 2011.

[58] N. L. Diaz, T. Dragicevic, J. C. Vasquez, and J. M. Guerrero, "Intelligent Distributed Generation and Storage Units for DC Microgrids – A New Concept on Cooperative Control Without Communications Beyond Droop Control," *IEEE Trans. Smart Grid*, vol. 5, no. 5, pp. 2476–2485, 2014.

[59] C. N. Onwuchekwa and A. Kwasinski, "Analysis of Boundary Control for Buck Converters With Instantaneous Constant-Power Loads," *IEEE Trans. Power Electron.*, vol. 25, no. 8, pp. 2018–2032, 2010.

[60] J. M. Guerrero, J. C. Vasquez, J. Matas, L. G. de Vicuna, and M. Castilla, "Hierarchical Control of Droop-Controlled AC and DC Microgrids—A General Approach Toward Standardization," *IEEE Trans. Ind. Electron.*, vol. 58, no. 1, pp. 158–172, Jan. 2011.

[61] X. Lu, K. Sun, J. M. M. Guerrero, J. C. C. Vasquez, and L. Huang, "Double-Quadrant State-of-Charge-Based Droop Control Method for Distributed Energy Storage Systems in Autonomous DC Microgrids," *IEEE Trans. Smart Grid*, vol. 6, no. 1, pp. 147–157, 2015.

[62] M. Lee, D. Chen, K. Huang, C.-W. Liu, and B. Tai, "Modeling and Design for a Novel Adaptive Voltage Positioning (AVP) Scheme for

Multiphase VRMs," *IEEE Trans. Power Electron.*, vol. 23, no. 4, pp. 1733–1742, 2008.

[63] C.-J. Chen, D. Chen, C.-S. Huang, M. Lee, and E. K.-L. Tseng, "Modeling and Design Considerations of a Novel High-Gain Peak Current Control Scheme to Achieve Adaptive Voltage Positioning (AVP) for DC Power Converters," *IEEE Trans. Power Electron.*, vol. 24, no. 12, pp. 2942–2950, 2009.

[64] H.-H. Huang, C.-Y. Hsieh, J.-Y. Liao, and K.-H. Chen, "Adaptive Droop Resistance Technique for Adaptive Voltage Positioning in Boost DC–DC Converters," *IEEE Trans. Power Electron.*, vol. 26, no. 7, pp. 1920–1932, 2011.

[65] F. Valenciaga and P. F. F. Puleston, "Supervisor Control for a Stand-Alone Hybrid Generation System Using Wind and Photovoltaic Energy," *IEEE Trans. Energy Convers.*, vol. 20, no. 2, pp. 398–405, Jun. 2005.

[66] F. Valenciaga and P. F. Puleston, "High-Order Sliding Control for a Wind Energy Convers. System Based on a Permanent Magnet Synchronous Generator," *IEEE Trans. Energy Convers.*, vol. 23, no. 3, pp. 860–867, 2008.

[67] T. Dragicevic, J. M. Guerrero, J. C. Vasquez, and D. Skrlec, "Supervisory Control of an Adaptive-Droop Regulated DC Microgrid With Battery Management Capability," *IEEE Trans. Power Electron.*, vol. 29, no. 2, pp. 695–706, 2014.

[68] L. Che and M. Shahidehpour, "DC Microgrids: Economic Operation and Enhancement of Resilience by Hierarchical Control," *IEEE Trans. Smart Grid*, vol. 5, no. 5, pp. 2517–2526, 2014.

[69] X. Lu, J. M. M. Guerrero, K. Sun, J. C. C. Vasquez, R. Teodorescu, and L. Huang, "Hierarchical Control of Parallel AC–DC Converter Interfaces for Hybrid Microgrids," *IEEE Trans. Smart Grid*, vol. 5, no. 2, pp. 683–692, 2014.

[70] B. Wang, M. Sechilariu, and F. Locment, "Intelligent DC Microgrid With Smart Grid Communications: Control Strategy Consideration and Design," *IEEE Trans. Smart Grid*, vol. 3, no. 4, pp. 2148–2156, 2012.

[71] C. Jin, P. Wang, J. Xiao, Y. Tang, and F. H. H. Choo, "Implementation of Hierarchical Control in DC Microgrids," *IEEE Trans. Ind. Electron.*, vol. 61, no. 8, pp. 4032–4042, 2014.

[72] Q. Shafiee, T. Dragicevic, J. C. Vasquez, and J. M. Guerrero, "Hierarchical Control for Multiple DC-Microgrids Clusters," *IEEE Trans. Energy Conversion*, vol. 29, no. 4, pp. 922–933, 2014.

[73] J. Schonberger, R. Duke, and S. D. Round, "DC-Bus Signaling: A Distributed Control Strategy for a Hybrid Renewable Nanogrid," *IEEE Trans. Ind. Electron.*, vol. 53, no. 5, pp. 1453–1460, 2006.

[74] Y. Gu, X. Xiang, W. Li, and X. He, "Mode-Adaptive Decentralized Control for Renewable DC Microgrid With Enhanced Reliability and Flexibility," *IEEE Trans. Power Electron.*, vol. 29, no. 9, pp. 5072–5080, 2014.

[75] A. Gkountaras, S. Dieckerhoff, and T. Sezi, "Performance analysis of hybrid microgrids applying SoC-adaptive droop control," in *Power Electronics and Applications (EPE'14-ECCE Europe), 2014 16th European Conference on*, 2014, pp. 1–10.

[76] E. Liegmann and R. Majumder, "An Efficient Method of Multiple Storage Control in Microgrids," *IEEE Trans. Power Syst.*, vol. 30, no. 6, pp. 3437–3444, 2015.

[77] H. C. Ferreira, L. Lampe, J. Newbury, and T. G. Swart, *Power Line Communications: Theory and Appl. for Narrowband and Broadband Communications over Power Lines*. Hoboken, NJ: John Wiley & Sons, 2010.

[78] "REbus DC Microgrid," 2013. [Online]. Available: http://rebuspower.com/. [Accessed: 08 Jun. 2014].

[79] T. Dragicevic, J. M. Guerrero, and J. C. Vasquez, "A Distributed Control Strategy for Coordination of an Autonomous LVDC Microgrid Based on Power-Line Signaling," *IEEE Trans. Ind. Electron.*, vol. 61, no. 7, pp. 3313–3326, 2014.

[80] W. Stefanutti, S. Saggini, P. Mattavelli, and M. Ghioni, "Power Line Communication in Digitally Controlled DC–DC Converters Using Switching Frequency Modulation," *IEEE Trans. Ind. Electron.*, vol. 55, no. 4, pp. 1509–1518, Apr. 2008.

[81] T. Dragicevic, A. Anvari-Moghaddam, J. C. Vasquez, and J. M. Guerrero, "DC Distribution Systems and Microgrids," in *Large Scale Grid Integration of Renewable Energy Sources*, A. Moreno-Munoz, Ed. IET, 2017, pp. 211–236.

[82] R. Olfati-Saber, J. A. Fax, and R. M. Murray, "Consensus and Cooperation in Networked Multi-Agent Systems," *Proc. IEEE*, vol. 95, no. 1, pp. 215–233, 2007.

[83] R. Olfati-Saber and R. M. Murray, "Consensus Problems in Networks of Agents with Switching Topology and Time-Delays," *IEEE Trans. Autom. Control*, vol. 49, no. 9, pp. 1520–1533, 2004.

[84] Z. Meng, W. Ren, Y. Cao, and Z. You, "Leaderless and Leader-Following Consensus with Communication and Input Delays under a Directed Network Topology," *IEEE Trans. Syst. Man, Cybern. Part B Cybern.*, vol. 41, no. 1, pp. 75–88, Feb. 2011.

[85] L. X. L. Xiao and S. Boyd, "Fast linear iterations for distributed averaging," in *42nd IEEE Int. Conf. Decis. Control (IEEE Cat. No. 03CH37475)*, 2003, vol. 5.

[86] Y. Li and F. Nejabatkhah, "Overview of Control, Integration and Energy Management of Microgrids," *J. Mod. Power Syst. Clean Energy*, vol. 2, no. 3, pp. 212–222, 2014.

[87] L. Yun Wei, "Control and Resonance Damping of Voltage-Source and Current-Source Converters with LC Filters," *IEEE Trans. Ind. Electron.*, vol. 56, no. 5, pp. 1511–1521, 2009.

[88] Q. C. Zhong and G. Weiss, "Synchronverters: Inverters that Mimic Synchronous Generators," *IEEE Trans. Ind. Electron.*, vol. 58, no. 4, pp. 1259–1267, 2011.

[89] F. Blaabjerg, R. Teodorescu, M. Liserre, and A. V. Timbus, "Overview of Control and Grid Synchronization for Distributed Power Generation Systems," *IEEE Trans. Ind. Electron.*, vol. 53, no. 5, pp. 1398–1409, 2006.

[90] A. Anvari-Moghaddam, J. M. Guerrero, J. C. Vasquez, H. Monsef, and A. Rahimi-Kian, "Efficient Energy Management for a Grid-Tied Residential Microgrid," *IET Gener. Transm. Distrib.*, vol. 11, no. 11, pp. 2752–2761, 2017.

Chapter 6

Control of solid-state transformer-enabled DC microgrids

Xu She[1], Alex Huang[2], Xunwei Yu[3], and Yizhe Xu[4]

6.1 Introduction

With the world-wide energy shortage and deterioration of existing power grid, microgrid becomes one of the hottest research directions in the power engineering area. Considering DC nature of many key components in the smart grid, such as photovoltaic (PV), battery, fuel cell, super capacitor, etc., as well as many DC type loads, such as light-emitting diode, DC microgrid has received more attention recently since it brings the opportunity for boosting the efficiency by eliminating the unnecessary power conversion stages. However, the existing DC microgrid can only interface with the distribution system by using a heavy and bulky passive line frequency transformer plus a rectifier, which has large space and heavy weight. Developing a more compact and active grid interface to enable an intelligent DC microgrid system is still a research focus. In this chapter, the solid-state transformer (SST)-enabled DC microgrid is presented. In addition, two system control strategies, namely, the centralized power management and hierarchical power management strategies, are proposed. In addition, an improved control strategy is proposed for increasing the penetration of distributed renewable energy resources (DRER) integration, which controls SST-enabled DC microgrid as a solid-state synchronous machine (SSSM). With the proposed control concept, frequency and voltage stability are improved in case of high-power intermittence at either DC load or DRER side. Design examples are given to illustrate the main characteristics of the presented system and control schemes.

6.2 Solid-state transformer-based microgrid: architecture and benefits

The SST is a power electronic device that replaces the traditional 50/60 Hz power transformer by means of high-frequency transformer isolated AC–AC conversion

[1]Electric Power, GE Global Research, United States
[2]Electric and Computer Engineering, University of Texas Austin, United States
[3]Intersil Corporation, United States
[4]Grid Bridge Inc., United States

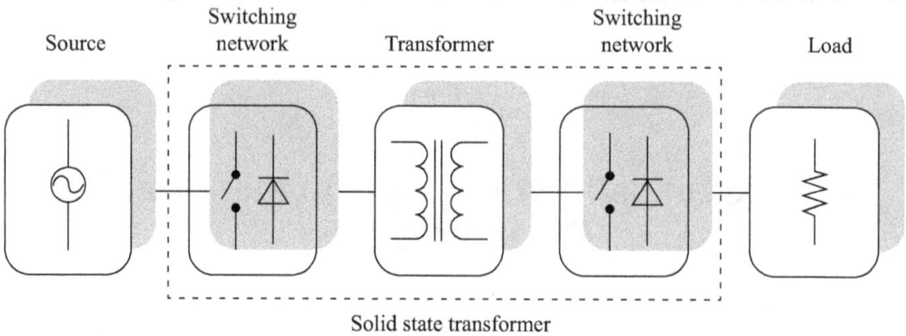

Figure 6.1 Configuration of solid-state transformer (SST) from source to load

technique, which is represented in Figure 6.1 [1]. The basic operation of the SST is firstly to convert the 50/60 Hz AC voltage to a high-frequency one (normally in the range of several to tens of kilohertz), then this high-frequency voltage is stepped up/down by a high-frequency transformer with significantly decreased volume and weight, and finally shaped back into the desired 50/60 Hz voltage to feed the load. Therefore, the first advantage that the SST can offer is its reduced volume and weight compared with traditional transformers. It is further seen from the configuration of the SST that some other potential functionalities that are not owned by the traditional transformers may be obtained. First, the use of solid-state semiconductor devices and circuits makes the voltage and current regulation a possibility. This brings promising features such as power flow control, voltage sag compensation, fault current limitation, and others, which are not possible for traditional transformers. Second, voltage source converters connected from the secondary terminal of the SST could readily support a regulated DC bus, which could be connected to DC microgrid, enabling this new microgrid architecture.

Figure 6.2 demonstrates the proposed SST-based DC microgrid system [2]. The AC loads are connected to the AC mains of SST, and the DC sources and loads are connected to the DC mains for minimizing the conversion stages. The high-voltage side of the SST is connected to the distribution system; thus, the conventional transformer is eliminated. The presence of SST can also isolate the distribution system from the residential side; therefore, the fault on one side will not affect another. Furthermore, the high-voltage AC/DC converter in the distribution system side can realize power factor regulation, enabling the Var compensation capability. Therefore, the SST-based microgrid is more compact and shows superior characteristics over conventional AC and DC microgrids.

Similar like the traditional microgrids, different power management strategies could be applied. In the following sections, both the centralized and hierarchical power management strategies are presented. In addition, SSSM control concepts are also proposed to enhance the system stability at high-power intermittence condition. Design examples are given to demonstrate the feasibility of the proposed SST-enabled DC microgrid.

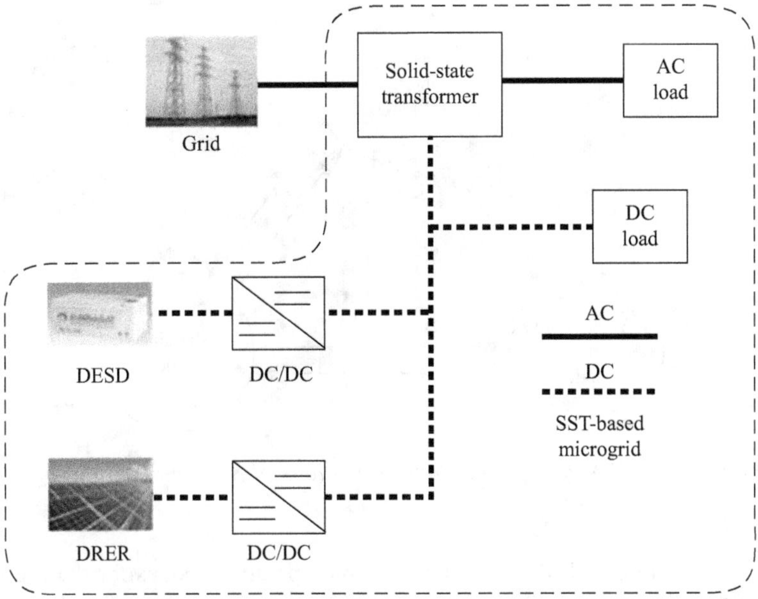

Figure 6.2 SST-based microgrid architecture: DESD and DRER. DESD, distributed energy storage device; DRER, distributed renewable energy resource

6.3 Centralized power management of solid-state transformer-based DC microgrid

6.3.1 Power management strategy

In the centralized power management strategy, an intelligent energy management (IEM) algorithm should be developed. The operating modes of DC microgrid can be defined as shown in Figure 6.3, where three modes are identified, namely, active grid interaction mode, passive grid interaction mode, and islanding mode.

In the passive grid interaction mode, SST controls the DC voltage. Although the DC microgrid is interfaced with the distribution system, it balances the power within the network without any power being exchanged with SST. In the active grid interaction mode, due to the limited power balancing capability of the DC microgrid under certain operating conditions, additional power is transferred between SST and DC microgrid. In the islanding mode, SST stops working and the battery regulates the DC bus. The DC microgrid supplies the additional power for the load at AC side whenever extra power is available. The assumption is made in this chapter that the SST, as the backup of DC microgrid, can always supply enough power under the grid interaction mode by an optimized design. The whole power management algorithm aims at maximizing the utilization of PV and battery and minimizing the burden of existing AC grid, as well as ensuring a high reliability of the power system.

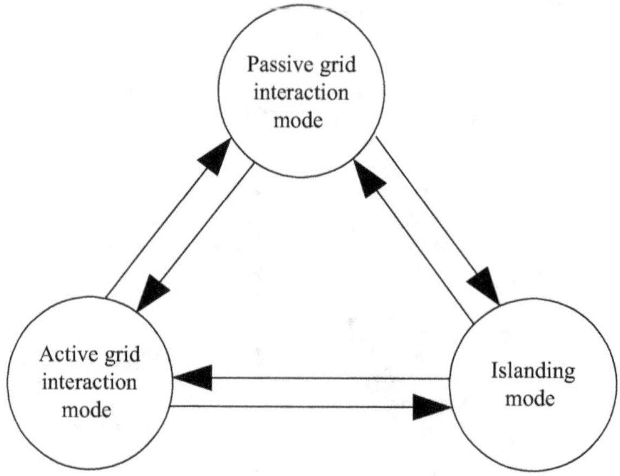

Figure 6.3 Operating modes and transitions of SST-enabled microgrid

If divided in detail, there are ten possible operating modes for the presented system, and they are shown in Figure 6.4. The nomenclature in the flowchart is defined in Table 6.1.

The division of the modes mainly depends on the operating status of the distribution system, battery state of charge (SOC), and battery power. These modes are summarized and explained as below.

Mode 1: PV operates in maximum power point tracking (MPPT) mode, battery stops working, and SST supplies the additional power for DC load.

Mode 2: PV operates in MPPT mode, battery operates in discharging power limitation mode, and SST supplies the additional power for DC load.

Mode 3: PV operates in MPPT mode, battery balances the power within DC microgrid, and SST only supplies power for AC load.

Mode 4: PV operates in MPPT mode, battery operates in charging power limitation mode, and SST absorbs the additional power from DC microgrid.

Mode 5: PV operates in MPPT mode, battery stops working, and SST absorbs the additional power from DC microgrid.

Mode 6: Part of loads is shed and the system operates in either Mode 8 or Mode 9, depending on the operating condition.

Mode 7: Part of loads is shed and the system operates in Mode 8, Mode 9, or Mode 10, depending on the operating condition.

Mode 8: PV operates in MPPT mode and battery operates in voltage tracking mode.

Mode 9: PV operates in power tracking mode for supplying the total load power. In this mode, the power reference of PV converter is set to

$$P_{pv} = P_{load\_DC} + P_{load\_DC} + P_{b\,\max} \tag{6.1}$$

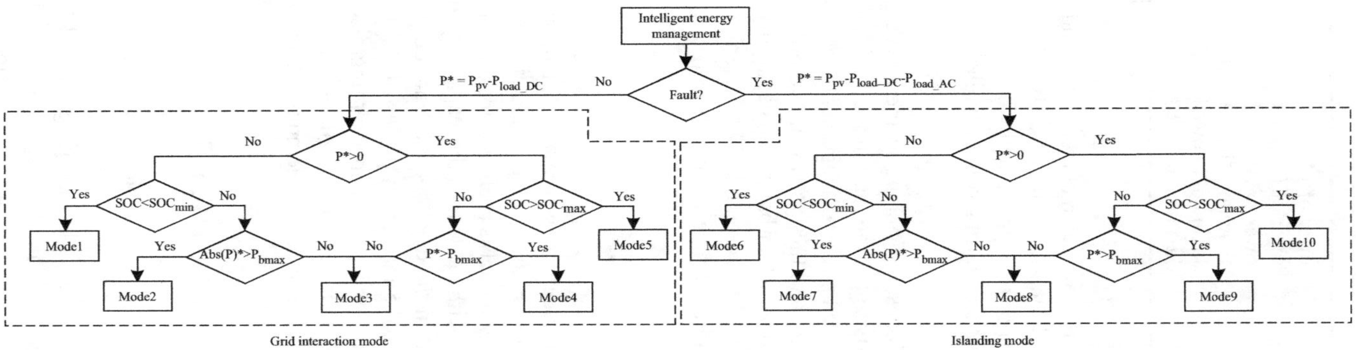

Figure 6.4 Intelligent energy management system diagram of SST using PV and battery

Table 6.1 Nomenclature for IEM system

P_{pv}	Power of PV
P_{load_DC}	Power of DC load
P_{load_AC}	Power of AC load
SOC_{min}	Minimum allowed SOC
SOC_{max}	Maximum allowed SOC
P_{bmax}	Maximum allowed battery power

The battery operates in voltage tracking mode, and its absorbed power is fixed to charging power limitation automatically.

Mode 10: PV operates in power tracking mode for supplying the total load power and the battery controls the DC bus voltage while no power is delivered to the load.

As it can be seen above, Modes 1–5 belong to the active grid interaction mode. Mode 3 belongs to the passive grid interaction mode. Modes 6–10 belong to the islanding mode. In the presented power management strategy, the matched battery capacity needs to be designed, and the system should operate in Mode 3 for most of the time. Only when the local power balancing capability is limited, a transition to other modes occurs. Thus, the presented IEM system can maximize the utilization of PV and battery and minimize the effect to the existing AC power grid architecture.

6.3.2 Case study

To verify the presented intelligent power management algorithm, a simulation platform is established by using the average modeling technique. The average model instead of the switching model is preferred for this large power system since it speeds up the simulation without losing the capability of capturing the main characteristics of the system. In the developed simulation platform, the power rating of SST is set to 20 kVA, the maximum power of PV under the irradiation of 1 kW/m^2 is set to 8 kW, and the maximum charging/discharging power of the battery pack is set to 6 kW. Different load profiles are adopted to test different cases. Only parts of the cases are tested for demonstrating the main characteristics of the system due to page limitations.

6.3.2.1 Passive grid interaction (Mode 3)

To emulate the operating Mode 3, which is the passive grid interaction mode, the AC load for SST is set to 10 kW, and DC load is set to 4 kW. The key operating waveforms are shown in Figure 6.5.

The irradiation of PV is gradually changing from 300 W/m^2 to 1 kW/m^2, and then back to 300 W/m^2 to emulate the irradiation of a day, as shown in Figure 6.5(a). Figure 6.5(b) shows the power distribution of the presented system.

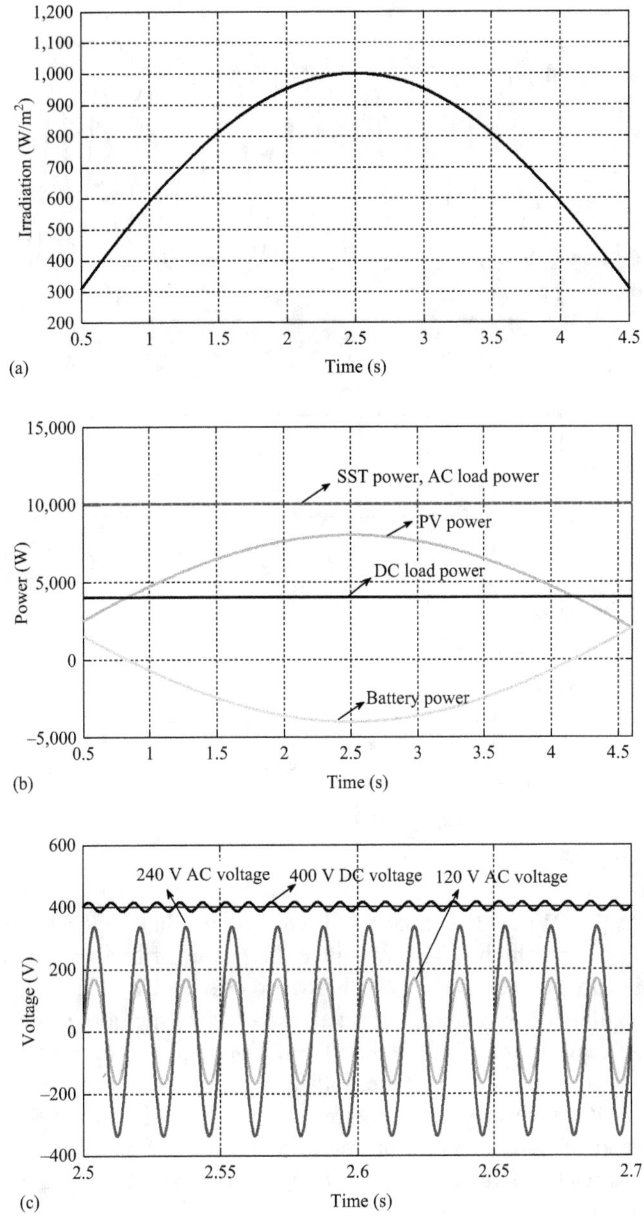

Figure 6.5 Key operating waveforms in Mode 3 of the system in Figure 6.2: (a) irradiation of PV panel, (b) power distribution of the system, (c) low-voltage terminal waveforms, (d) high-voltage terminal waveforms, (e) SOC of the battery

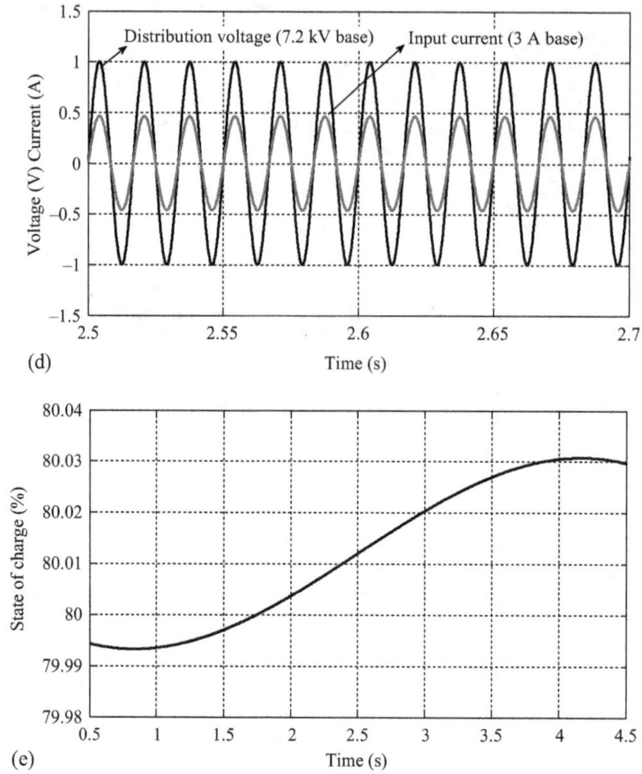

Figure 6.5 (Continued)

The power of the PV has the same trend with the irradiation. SST only supplies the AC load power, which is 10 kW. Since the power of the battery is within its limitation, it can balance the power between DC load and PV. Figure 6.5(c) shows the regulated low-voltage terminal waveforms, including 400 V DC and 120/240 V AC. In Figure 6.5(d), the SST input voltage and current are depicted. The current is in phase with voltage, indicating a unity power factor operation. The battery SOC is described by Figure 6.5(e), where the initial SOC is set to 80%. The trend of SOC depends on the direction of current. A positive battery current discharges the battery and a negative battery current charges the battery in the presented system in Figure 6.2.

6.3.2.2 Transition from passive grid interaction mode to active grid interaction mode (Mode 2 to Mode 3)

The transition from passive grid interaction mode to active grid interaction mode is demonstrated. Mode 2 and Mode 3 are chosen to verify the operation, in which the AC load is set to 10 kW and the DC load is set to 1 kW. Key operating waveforms

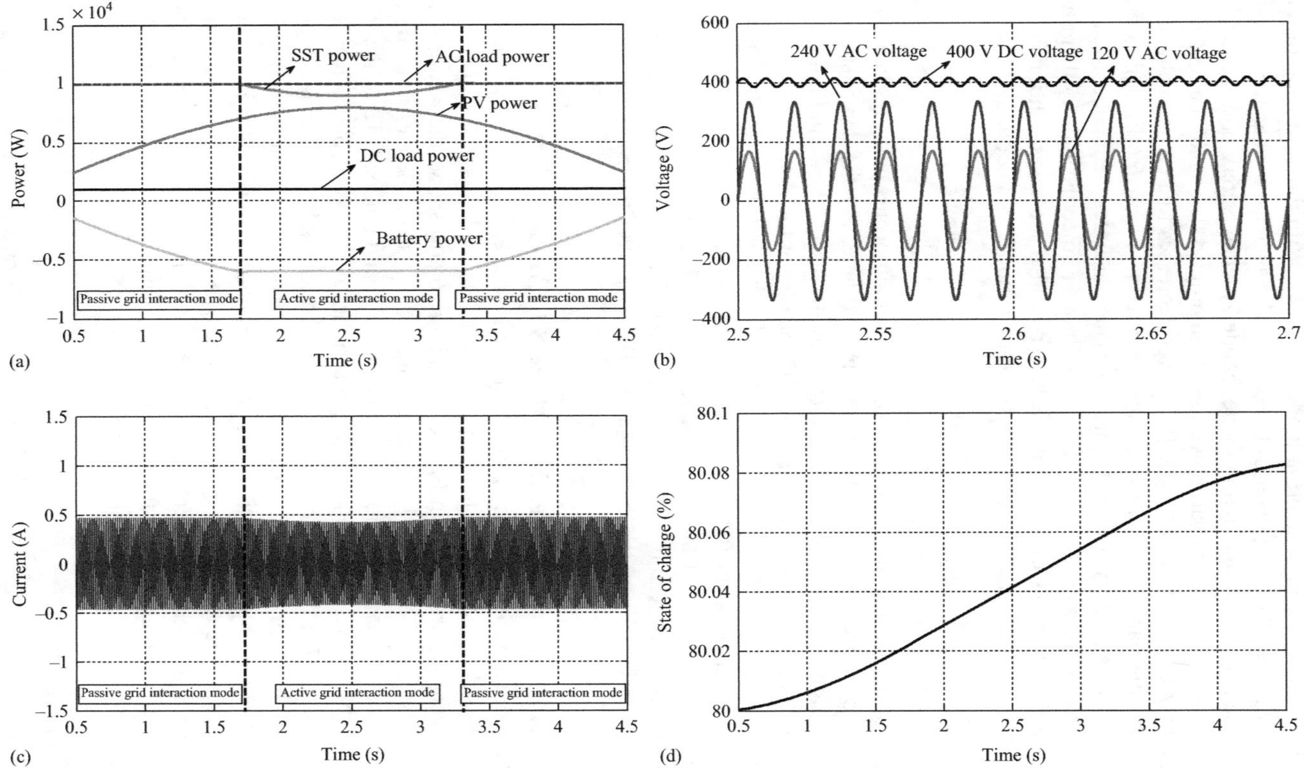

Figure 6.6 Key operating waveforms during mode transition (Mode 2 to Mode 3): (a) power distribution of the system, (b) low-voltage terminal waveforms, (c) high-voltage terminal current (3 A base), (d) SOC of the battery pack

are shown in Figure 6.6. Figure 6.6(a) shows the power distribution of the system. At the beginning, the power of battery pack is within the limitation and the system operates in the passive grid interaction mode (Mode 3). With the irradiation of PV panel increasing, the charging power of the battery also increases. Then the system transits into the active grid interaction mode (Mode 2) when battery power reaches the 6 kW limitation. The additional power of DC microgrid flows into the SST and supplies to the AC load. After that, the PV power decreases to a certain value and the system operates back to passive grid interaction mode. Figure 6.6(b) depicts the well-regulated AC and DC voltages. Figure 6.6(c) demonstrates the input current of SST, which shows the same trend with the SST power since the input voltage is fixed. Figure 6.6(d) shows the SOC of the battery and it increases continuously. As it can be seen, the transition between these two modes is smooth and occurs automatically.

6.3.2.3 Islanding mode (Mode 8)

The microgrid transits into the islanding mode when fault occurs in the distribution system. How to detect the grid fault is not addressed here. For the demonstration purpose, operation of Mode 8 is chosen and the waveforms are shown in Figure 6.7, where the AC load is set to 4.5 kW and the DC load is set to 1 kW.

Power distribution of the system is shown in Figure 6.7(a), where the battery balances the power between PV and load (both AC and DC load). Figure 6.7(b) shows the key waveforms at the low-voltage terminal. It can be seen that the DC bus voltage is regulated well by the battery converter. Figure 6.7(c) shows the SOC of the battery pack. When the battery current is positive, the SOC decreases. Otherwise, the SOC increases.

6.3.2.4 Islanding mode transition (Mode 8 to Mode 9)

Finally, the mode transition in the islanding mode is studied. In the study, the AC load is set to 1 kW, and the DC load changes from 2 to 0 kW at 2.7 s. The key operating waveforms are shown in Figure 6.8. Figure 6.8(a) shows the power distribution in this case. Before the DC load changes, the system operates in mode 8 and the battery can balance the power between the PV and the load. Thus, the PV operates in MPPT mode for extracting the maximum power available. When the DC load is cut at 2.7 s, the power command of the battery is larger than 6 kW and battery then operates in charging power limitation mode. The PV no longer operates in MPPT mode and transits to power tracking mode. Figure 6.8(b) illustrates the 400-V DC voltage during the mode transition. Due to the cutting down of the DC load, a voltage overshoot occurs and recovers within 0.05 s. The SOC of the battery is given in Figure 6.8(c), and the battery is charged continuously. Clearly, the transition between different islanding modes is smooth.

6.3.3 Summary

A centralized power management strategy is presented for SST-enabled microgrid system with PV and battery. It renders the coordinate management of power

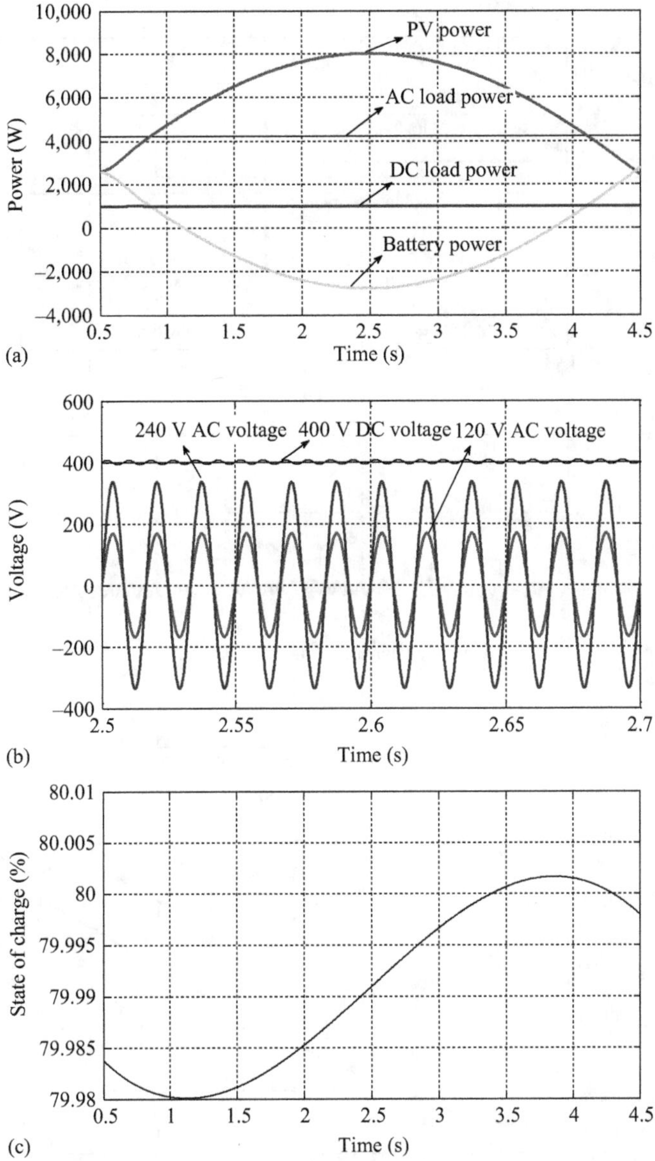

Figure 6.7 Mode 8 key operating waveforms: (a) power distribution of the system, (b) low-voltage terminal waveforms, (c) SOC of battery pack

between DC microgrid and distribution system. In addition, the impact of the established DC microgrid to the existing power grid can be minimized, which promises a better reliability and stability. The presented strategy is verified through some simulation studies.

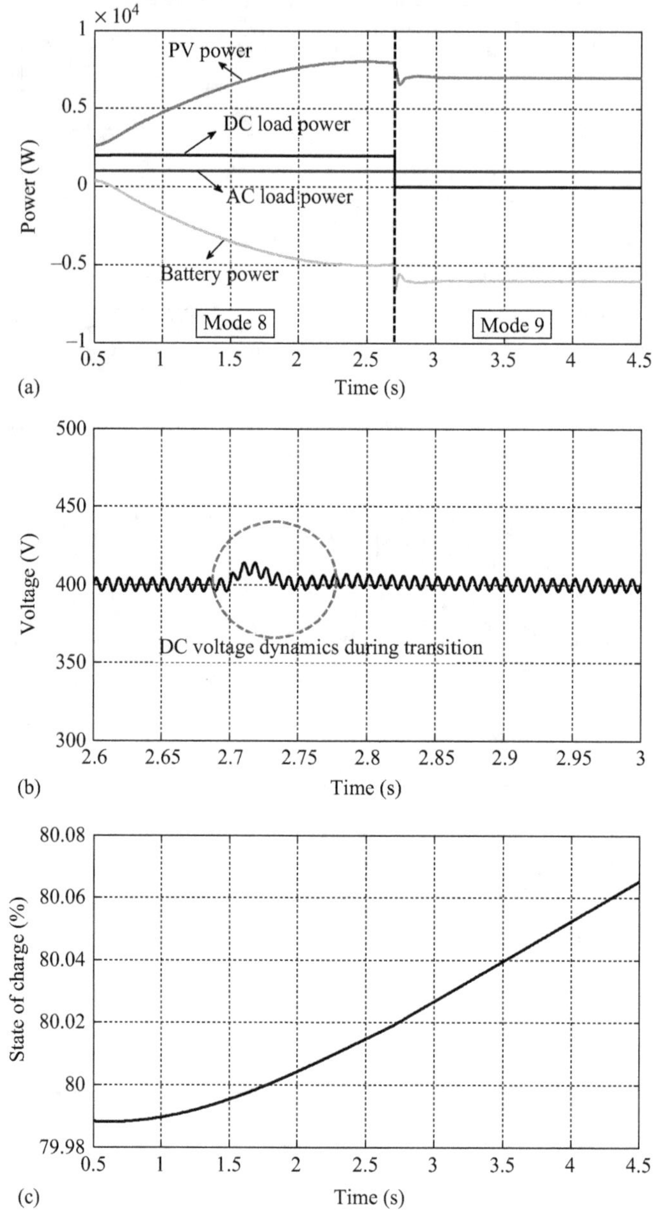

Figure 6.8 Key operating waveforms during mode transition (Mode 8 and Mode 9): (a) power distribution of the system, (b) 400-V DC bus voltage dynamics, (c) SOC of battery pack

6.4 Hierarchical power management of solid-state transformer-enabled DC microgrid

6.4.1 Power management strategy

The hierarchical control is a hybrid method, which combines the advantages of both centralized and distributed control and can be served as standard method for the microgrid [3]. As shown in Figure 6.9, the system control strategy frame is categorized into three layers: primary control, secondary control, and tertiary control. The primary control is the distributed control, which ensures that the microgrid system can operate without communication. Therefore, the primary control usually takes effect at the micro-second level, which is basically the same level as the converter control. All the local information, including the voltage, current, SOC, etc., are sent to the upper controller, which implements the tertiary control and secondary control through a bidirectional communication link. Here, the DC microgrid is enabled by the SST, and therefore, the SST controller is used as the upper controller [4]. The objective of the secondary control is to recover the microgrid bus voltage to achieve seamless transfer as the system switches from islanding mode to SST-enabled mode. The time scale of the secondary control is on the order of milliseconds to seconds.

6.4.1.1 Primary control algorithm

When the system operates in islanding mode, the battery needs to regulate the DC bus voltage alone. Therefore, (6.2) can be obtained.

$$C_{eq}V_{DC}\frac{dV_{DC}}{dt} = (P_{DC\_Load} - P_{Battery} - P_{PV}) \tag{6.2}$$

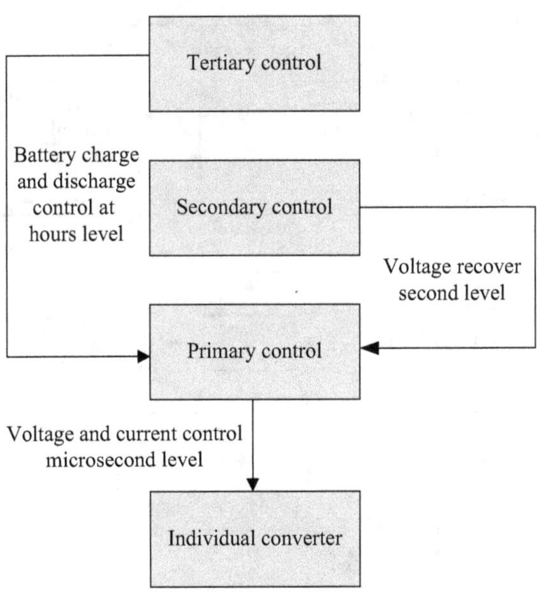

Figure 6.9 Hierarchical control frame

Since P_{PV} is equal or larger than zero, the power direction of battery is determined by the difference between the P_{PV} and P_{DC_Load}. For example, the power of the battery will be larger than zero when P_{DC_Load} is larger than P_{PV}, while the power of the battery will be less than zero when P_{DC_Load} is smaller than P_{PV}.

Basically, the primary control algorithm is based on DC bus signal control, and therefore, the distributed control can be achieved. In this control algorithm, the DC bus voltage is no longer a fixed value, but varies in a certain range in the islanding mode.

The DC bus voltage range is defined as from 360 to 400 V in the presented system design. In islanding mode, only the battery is used to regulate the DC bus voltage. Droop control is adopted for battery control with the expression shown in (6.3):

$$V_b = V_0 - R_b I_b \tag{6.3}$$

where V_b is the reference for the DC bus voltage, V_o is bus voltage value without load and it is set to 380 V, R_b is the virtual output impedance, and I_b is battery output current. For the PV module, it always operates in MPPT mode to deliver power to the system. Noted here, the load management is involved in the primary control. The load is divided into the critical load and the noncritical load. When the PV has no output power to the system, the bus voltage will drop. The battery has to source more power to the load and the battery SOC will decrease rapidly. To ensure that the DC microgrid can operate as long as possible in islanding mode, the battery SOC has to be considered. Therefore, to avoid battery SOC dropping fast, the noncritical load will be shed from the system when the bus voltage decreases below some preset threshold (370 V). When the bus voltage recovers to a certain value, 380 V for instance, the noncritical load can be reconnected to the system. The primary control diagram is shown in Figure 6.10.

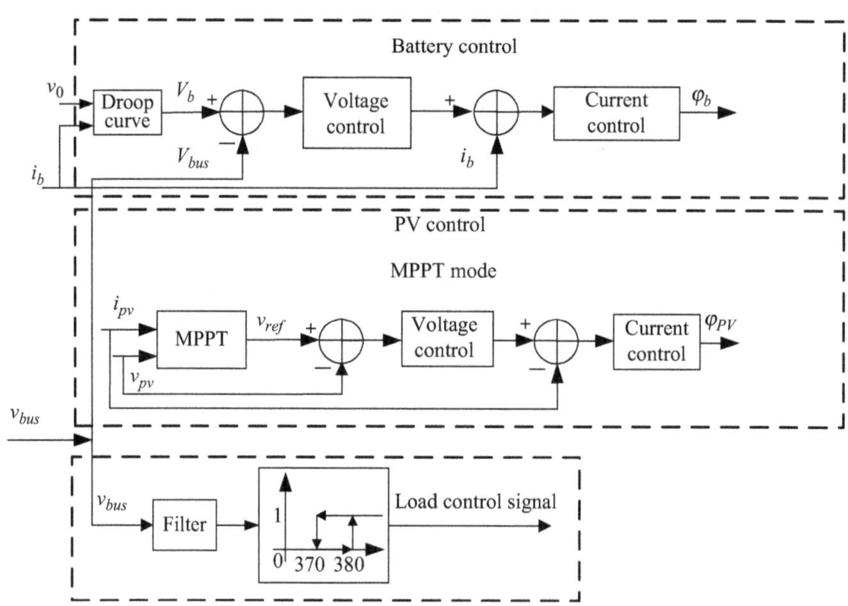

Figure 6.10 Primary control diagram of the DC microgrid system

6.4.1.2 Secondary control algorithm

Since the DC bus voltage is regulated by droop control in islanding mode, the bus voltage might deviate from the SST DC terminal output voltage. To achieve the seamless transfer from islanding mode to SST-enabled mode, the secondary control is adopted. Specifically, when the DC microgrid needs to connect to the SST, the secondary control-enabled signal is triggered. The SST DC terminal voltage is sensed and compared with the DC microgrid bus voltage. Then the voltage difference is processed by a compensator and the output is sent to the battery unit via the low-bandwidth communication to restore the DC microgrid bus voltage (see Figure 6.11). Since the PV and load control do not contribute to regulating the DC bus voltage, their control remains unchanged as shown in Figure 6.10. Due to the space limitations, only the battery module control is depicted in Figure 6.11.

The controller can be represented as

$$V_s = k_{sp}(V_{SST\_dcout} - V_{bus}) + (k_{si}(V_{SST\_dcout} - V_{bus})/s) \tag{6.4}$$

where k_{sp} and k_{si} are the control parameters for the secondary control. Therefore, the droop control of the battery becomes

$$V_b = V_0 - R_b I_b + V_s \tag{6.5}$$

Based on (6.5), the key function of the secondary control is to shift the droop curve (see Figure 6.12). When the SST's DC output voltage is higher than DC microgrid

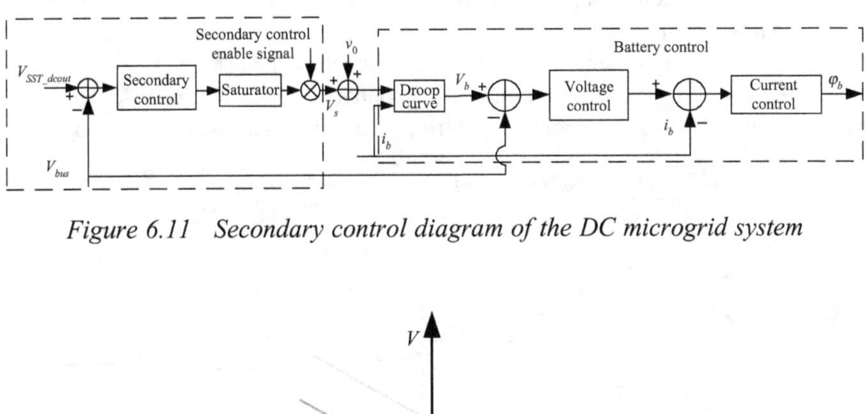

Figure 6.11 Secondary control diagram of the DC microgrid system

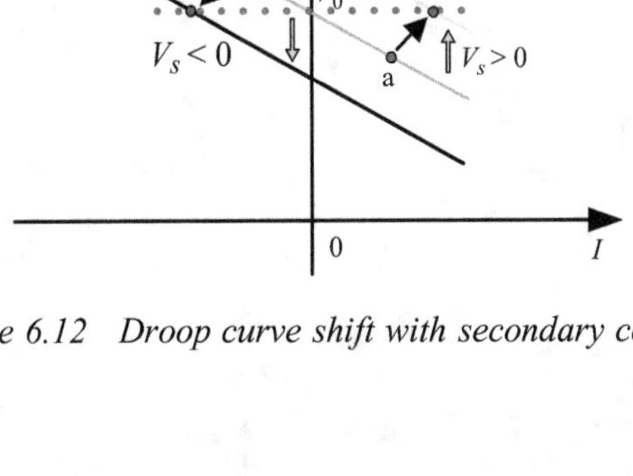

Figure 6.12 Droop curve shift with secondary control

bus voltage, V_s is larger than zero and the droop curve is shifted upward (battery operation point from a to b), and vice versa (battery operation point from c to d).

It is noticed that since the low bandwidth communication is adopted, the secondary controller's time step should agree with that of the communication to prevent secondary controller saturation.

6.4.1.3 Tertiary control algorithm

An appropriate SOC value is necessary for the battery to supply the DC bus voltage while operating in islanding mode. Therefore, SOC is included in the tertiary control. When the system operates in SST-enabled mode, the DC bus voltage will be regulated by the SST, and therefore, SST DC output is a stiff bus as seen from the DC microgrid. Then the battery charge or discharge can be determined by the SST DC voltage and battery droop curve. There are two options for managing the battery SOC in SST-enabled mode. The first is to change the SST DC terminal output voltage, and the second is to shift the battery droop curve, both of which are detailed below.

Change SST DC terminal output voltage

Since the PV and load control do not contribute to regulate the DC bus voltage, their control is the same as shown in Figure 6.10. Only the battery module control is depicted in Figure 6.13. As shown in Figure 6.13, the SOC is evaluated and processed by the current reference selection block. If the SOC reaches the H_{SOC} (upper limit for SOC) or L_{SOC} (lower limit for SOC), I_{ref} equals I_d (discharging current for battery) or I_c (charging current for battery). As soon as the SOC reaches the limit value, the tertiary control-enabled signal is triggered and tertiary control starts to operate. However, based on Figure 6.12, the secondary control achieves voltage recovery by sourcing or sinking more current, which will cause the battery SOC to reach the SOC limit faster. Therefore, the secondary control will be disabled when tertiary control begins to operate. The tertiary controller can be represented as

$$V_c = \frac{k_{cp}(I_{ref} - I_b) + k_{ci}(I_{ref} - I_b)}{s} \tag{6.6}$$

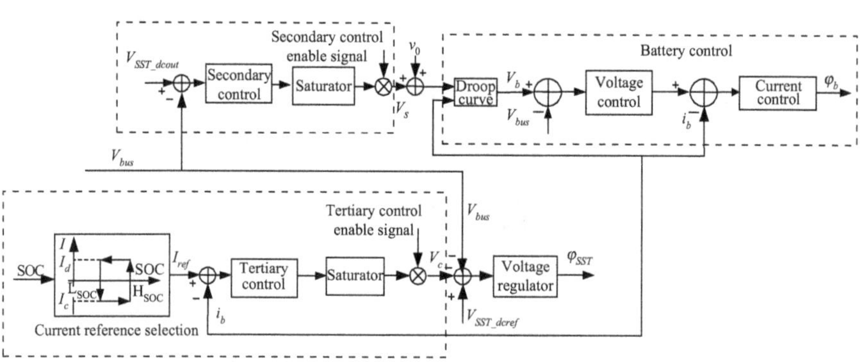

Figure 6.13 First tertiary control diagram of the DC microgrid system

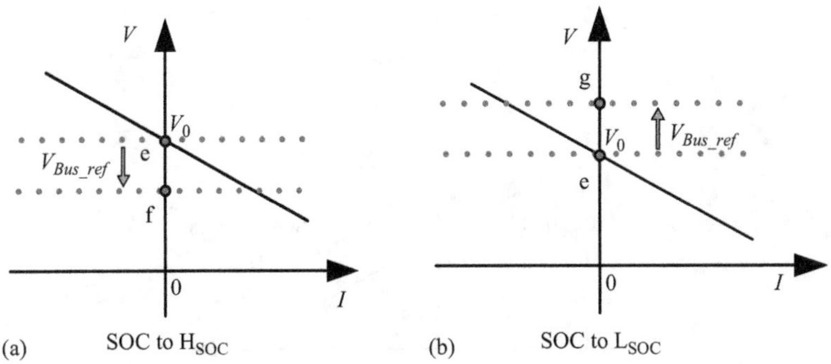

Figure 6.14 SST DC voltage shift in the tertiary control shown in Figure 6.13: (a) SOC to higher boundary and (b) SOC to lower boundary

where k_{cp} and k_{ci} are the tertiary control parameters. The bus voltage reference can be as shown in (6.7):

$$V_{Bus\_ref} = V_{SST\_dcref} - V_c \tag{6.7}$$

As shown in Figure 6.14(a), for example, if I_{ref} equals I_d, the bus voltage operation point will shift from point e to point f. Here, the bus voltage is less than the V_0 for battery droop curve. Thus, the battery will switch to discharging mode automatically. When SOC reaches L_{SOC}, the SST DC output voltage will be higher than V_0 (point e to point g) and the battery will be charged [see Figure 6.14(b)].

Shifting battery droop curve

As in the previous tertiary control scheme, the current reference is determined by the current reference selection block based on the SOC. If the SOC reaches the H_{SOC}, I_{ref} equals I_d. Or if the SOC reaches the L_{SOC}, I_{ref} equals I_c. In contrast to the aforementioned tertiary control scheme, the battery droop curve is changed instead of the SST DC bus voltage. Then the tertiary controller can be represented as following:

$$V_t = \frac{k_{tp}(I_{ref} - I_b) + k_{ti}(I_{ref} - I_b)}{s} \tag{6.8}$$

where K_{tp} and k_{ti} are the control parameters for the tertiary controller. Therefore, for battery droop control, (6.5) becomes

$$V_b = V_0 - R_b I_b + V_t \tag{6.9}$$

The secondary control will be disabled at the time which or before the tertiary control starts. The key point for this tertiary control scheme is to shift the droop curve as shown in Figure 6.15. When the battery SOC is low, the droop curve will shift down, and vice versa. Since the PV and load control do not contribute to regulate the DC bus voltage, their control is the same as shown in Figure 6.10. Due to the space limitations, only the battery module control is depicted in Figure 6.16.

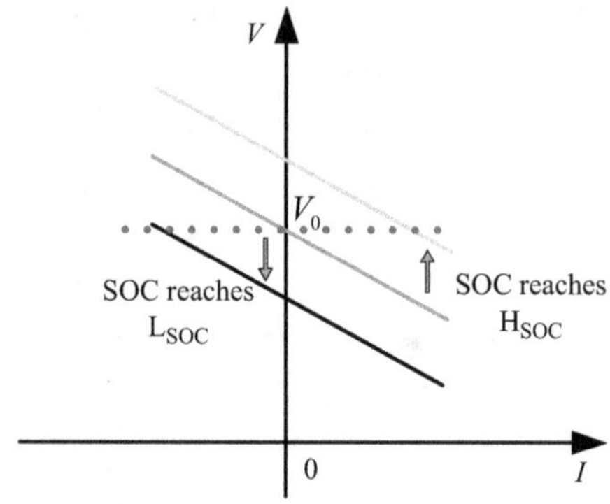

Figure 6.15 Droop curve shift with second tertiary control shown in Figure 6.13

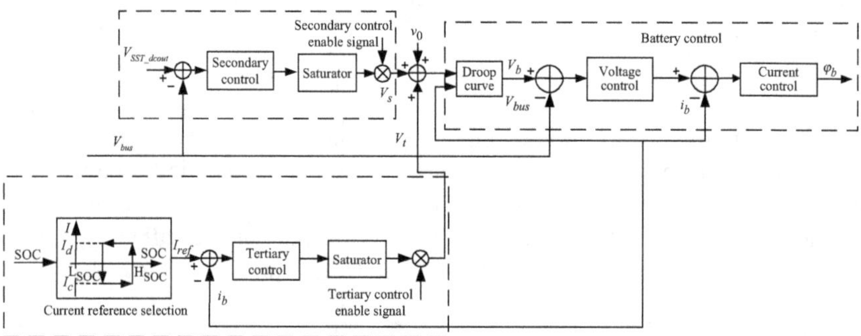

Figure 6.16 Second tertiary control diagram shown in Figure 6.13

6.4.2 *Case study of a small-scale DC microgrid*

To verify the primary control algorithm shown in Figure 6.10, an experiment is carried out on a lab test-bed. A 500-W battery module and a 200-W PV module are constructed to constitute the DC microgrid. The input voltage of the SST is 1.2 kV, the low voltage DC link is 380 V, and the low voltage AC link is 240-V AC. When the PV generation unit is operating with MPPT control, the maximum power is 100 W and the corresponding MPPT voltage is about 48 V. The nominal battery voltage is 48 V and the capacity is 20 A h.

6.4.2.1 Case I: Primary control of the SST-enabled DC microgrid system

To verify the primary control algorithm, an experiment is carried out in the test-bed. The DC microgrid operates in islanding mode and the result is depicted in Figure 6.17. Initially, only the battery supplies power to the load (2 kΩ). Then the PV connects to the system and supply the power to the load. Since the PV's output

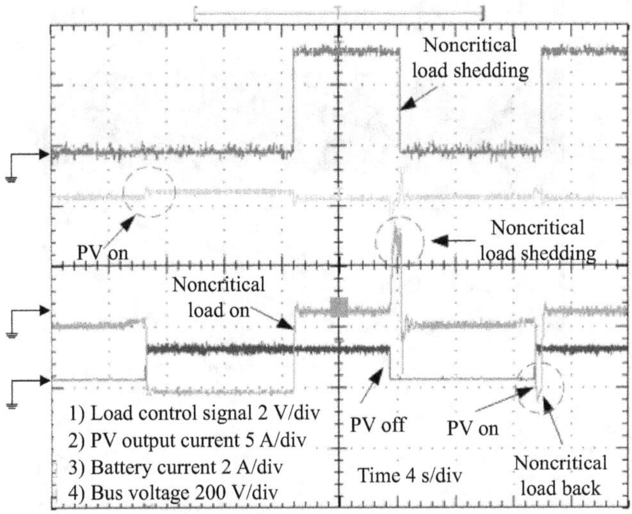

Figure 6.17 Waveforms from the primary control

power is larger than the load, the battery switches to the charging mode automatically and the bus voltage is higher than 380 V. When the noncritical load (1 kΩ) connects to the system, the bus voltage drops and the battery switches back to the discharging mode. Since the bus voltage exceeds the 370 V (value for load shedding), the noncritical load remains connected to the bus. After the PV disconnects from the system, the battery has to supply the power to the load alone and the bus voltage decreases. After 0.5 s when the bus voltage drops below 370 V, the noncritical load is shed from the system, and the bus voltage can recover. When the PV is connected back to the system the bus voltage increases. After 0.5 s when the bus voltage again exceeds 380 V (the value for noncritical load back), the noncritical load is reconnected to the system automatically.

6.4.2.2 Case II: Secondary control of the SST-enabled DC microgrid system

To verify the secondary control algorithm, experimental results are shown in Figure 6.18. The communication ports are not included in this chapter, so a software trigger is used to start the secondary control; the time step for secondary control is 1 ms.

Figure 6.18(a)–(c) shows the waveforms with the secondary controller. Figure 6.18(a) shows the SST waveforms, including the input voltage and current, high voltage DC of one AC/DC power stage, and low-voltage DC. In Figure 6.18(b), before the secondary control loop starts, the DC microgrid bus voltage is less than 380 V because the battery and the PV have been supplying the power to the load. As soon as the secondary control loop begins to operate, the microgrid bus voltage increases with the secondary control time step until it reaches 380 V. Then the DC

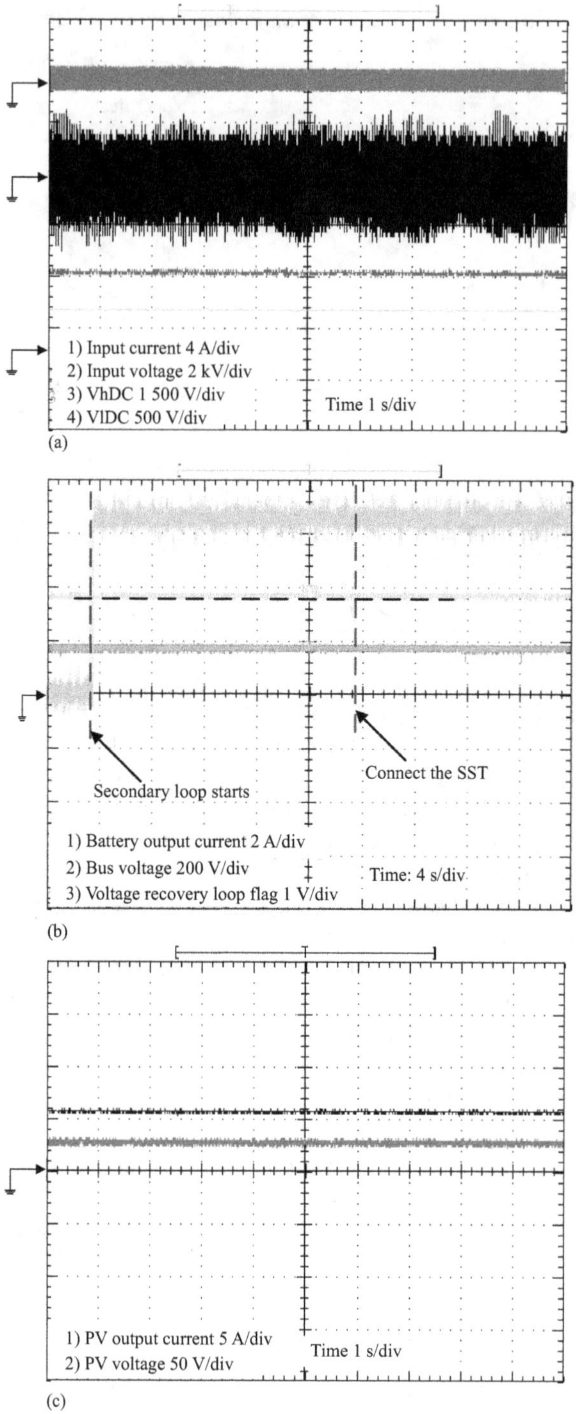

Figure 6.18 Waveforms with secondary control for the system in Figure 6.2: (a) SST waveforms, (b) battery current, DC microgrid bus voltage, and SST AC output, (c) PV waveform

microgrid connects to the SST, and a seamless transfer is achieved. Furthermore, the battery droop curve is lifted based on the secondary control loop as previously described. Therefore, when the system is in SST-enabled mode, the battery still outputs current to the load. For the PV module, there is no change in its output voltage and current because it always operates in the MPPT mode as shown in Figure 6.18(c).

6.4.2.3 Case III: First tertiary control

To verify the first tertiary control algorithm, experimental results are shown in Figure 6.19. SOC estimation is beyond the focus of the chapter, and therefore, it is not covered here. A software trigger is used to start the tertiary control, emulating the time when the battery SOC reaches its upper limit.

Figure 6.19(a) shows the SST waveforms, including the input voltage and current, high-voltage DC of one AC/DC power stage, and low-voltage DC. When the tertiary control starts, the SST output DC bus voltage increases because the V_c in (6.7) is less than zero. Figure 6.19(b) shows the SST AC-side voltage, DC microgrid bus voltage, and the battery current. It is noted that the secondary control is disabled before the tertiary control starts. Therefore, the battery module's output current is almost zero before the tertiary control starts. When the tertiary control starts, the DC bus voltage increases until the battery output current reaches its current reference (−2 A). Since the PV is still in the MPPT mode, its curve is same as the previous case, which is shown in Figure 6.19(c).

6.4.2.4 Case IV: Second tertiary control

To verify the second tertiary control algorithm, experimental results are shown in Figure 6.20. Similarly, a software trigger is used to start the tertiary control to emulate when the SOC reaches the L_{SOH}, and the time step for tertiary control is 0.1 s.

Figure 6.20(a) shows the SST waveforms, including the input voltage and current, high-voltage DC, and low-voltage DC. When the tertiary control starts, in contrast to the previous tertiary control, the SST output DC bus voltage remains constant at 380 V. Figure 6.20(b) shows the SST AC-side voltage, DC microgrid bus voltage, and the battery current. When the secondary control loop stops, the battery output current drops to zero because the V_0 equals 380 V (the SST DC output voltage) without the voltage recovery item (i.e., V_s is zero). After the battery's output current reaches zero, the tertiary control loop is triggered and begins to operate (the green line goes to 3.3 V). Then the battery starts to charge until the current reaches the I_c value (−2 A). As with the first tertiary control method, the PV is still in MPPT mode, and there is no change in its waveform whether the secondary control stops or tertiary control starts, as shown in Figure 6.20(c).

6.4.2.5 Comparison of the two tertiary control methods

The first tertiary control has the following disadvantages: (1) when the DC bus voltage changes, the load power changes; (2) the DC bus voltage range is limited

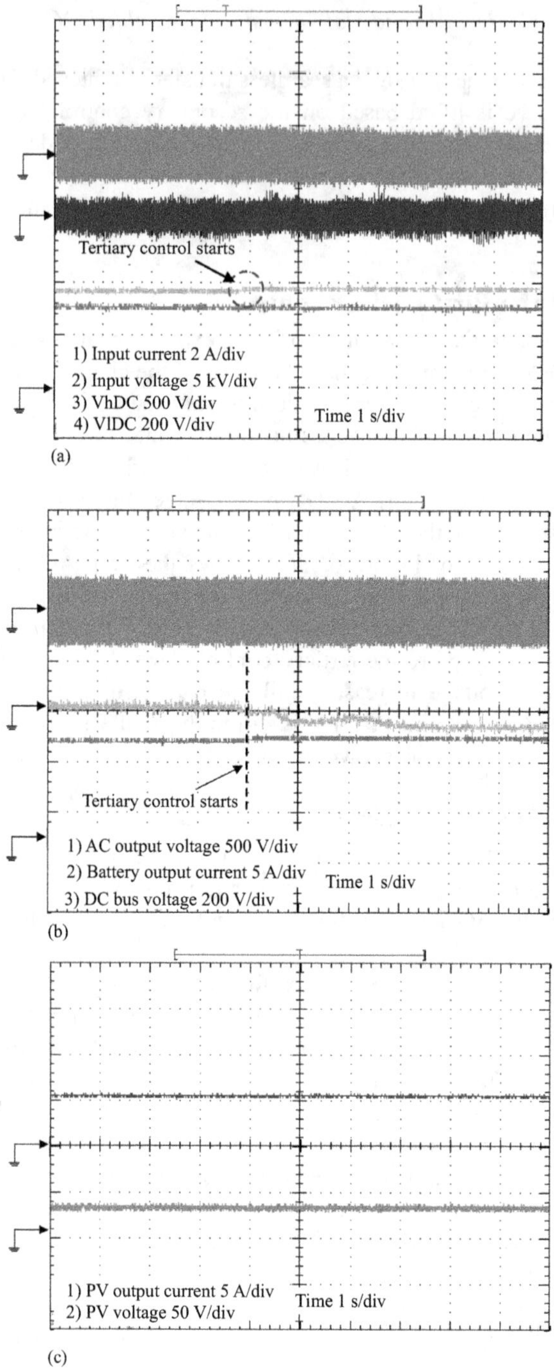

Figure 6.19 Waveforms with first tertiary control principal: (a) SST waveform, (b) battery current, DC microgrid bus voltage, and SST AC output, (c) PV current and voltage

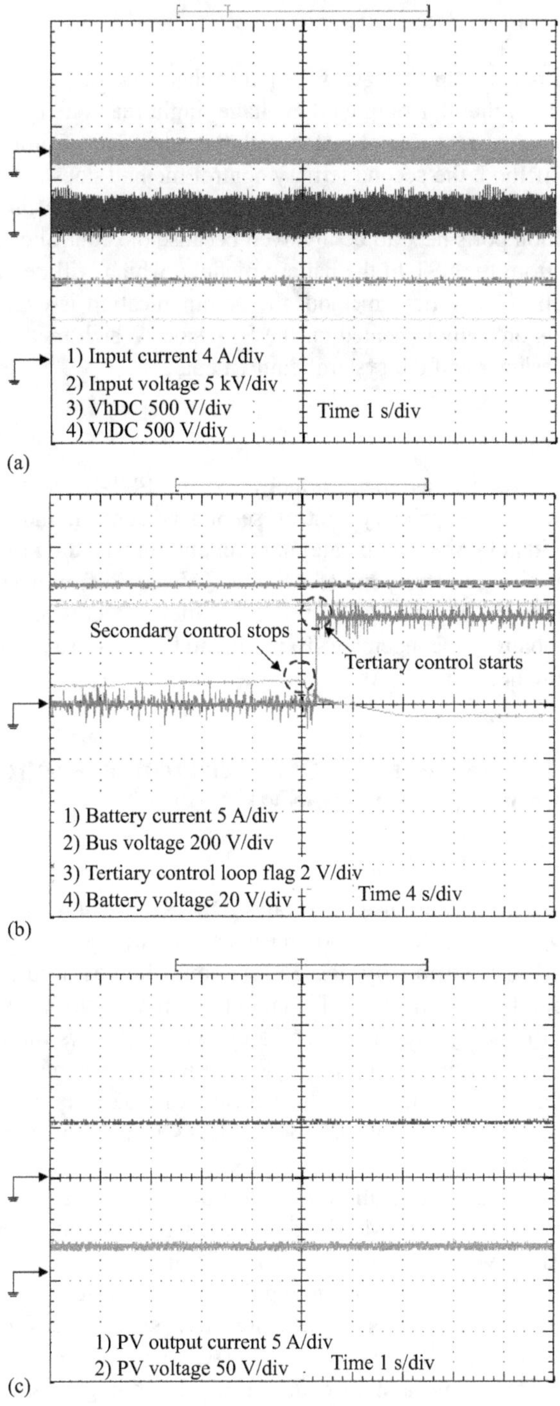

Figure 6.20 Waveforms with second tertiary control: (a) SST waveform, (b) battery current, DC microgrid bus voltage, and control flag, (c) PV waveform

by the SST output voltage range. When the charging and discharging current references are large, the SST output DC voltage might reach its limitation and then the battery's output current cannot match with the references. Compared to the first tertiary control method, the second tertiary control method does not have these two problems since it just changes the battery's droop curve. However, its disadvantage is that communication ports need to be involved because the control output commands need to be sent from the SST to the battery module, which will degrade the system reliability. For the first tertiary method, the communication is exclusive since the control method is only implemented on the SST side. Which method to select will depend on the designer and the system requirements.

6.4.3 Summary

A hierarchical power management strategy for the SST-based DC microgrid is proposed, which includes primary control, secondary control, and tertiary control. The DC microgrid not only can operate more reliably in islanding mode by primary control, but can also seamlessly transfer the DC microgrid from islanding mode to SST-enabled mode via secondary control. The battery SOC is involved in the tertiary control for battery management. Experimental results are presented to verify the proposed schemes.

6.5 Control of SST-enabled DC microgrid as a solid-state synchronous machine (SSSM)

6.5.1 Concept of the SSSM

The SST-enabled DC microgrid integrates DRER, distributed energy storage devices (DESD) and DC loads, and interfaces to the legacy grid by a phase locked loop (PLL). Traditionally, the SST is controlled as a current source and from the grid point of view, the SST-enabled DC microgrid lacks the rotational inertia of a traditional synchronous generator, which follows the grid frequency and the high-power intermittence of DRER, and load is instantaneously reflected at the point of common coupling. The concept of SSSM is proposed to overcome the frequency and voltage instability issues in a high DRER penetration system [5–7].

Figure 6.21 shows the overall architecture of an SST-enabled DC micro-grid when it is controlled as an SSSM. The detailed control diagram applied in single-phase inverter is shown in Figure 6.22. The SST interfaces the DC microgrid to legacy grid with frequency and voltage regulation capabilities. The microgrid acts as a voltage source which mimics the behavior of the synchronous machine to form the grid. It improves the aforementioned SST-enabled DC microgrid with (1) a larger inertia; (2) a simple and extendible model for large-scale power system analysis; (3) no use of PLL during normal operation; (4) seamless transition from grid-connection mode to islanding mode; (5) up/down reserve for frequency regulation with the support of energy storage (battery).

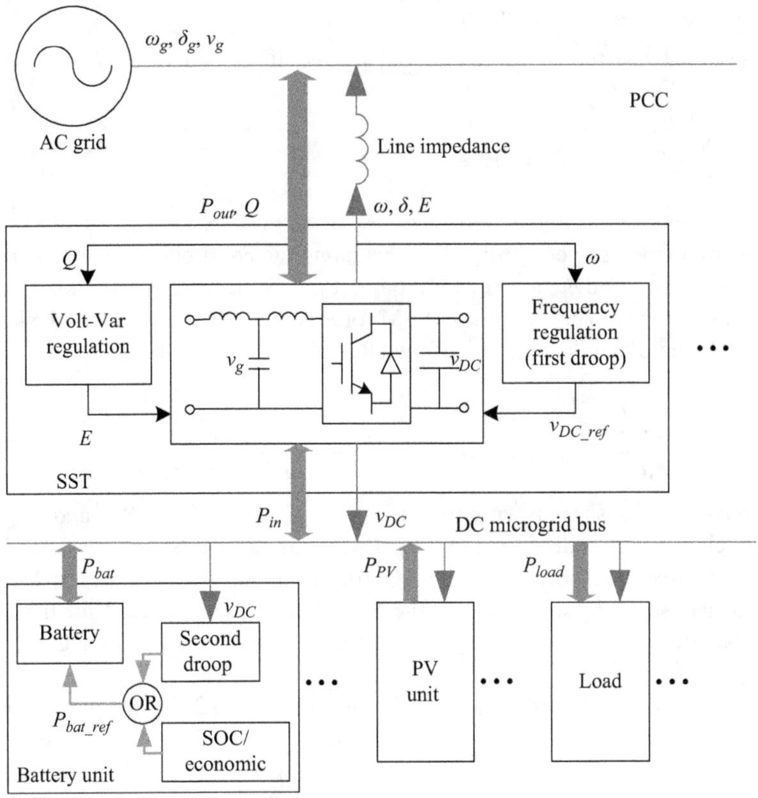

Figure 6.21 Overall architecture of the SST-enabled DC microgrid as an SSSM with the associated control strategy

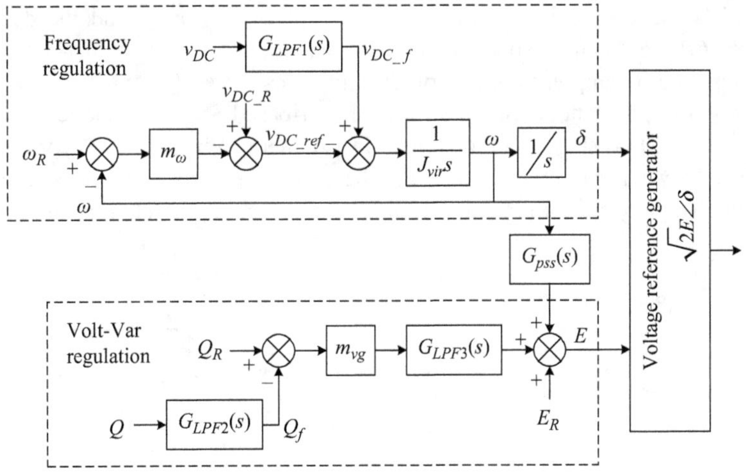

Figure 6.22 Detailed control strategy diagram of the SST controlled as SSSM

6.5.2 Frequency regulation

The traditional SM rotor's mechanical characteristic is called the swing equation and expressed as

$$J\frac{d\omega_r}{dt} + D(\omega_r - \omega_g) = T_m - T_e \tag{6.10}$$

where ω_r is the rotor angular frequency, ω_g is the grid frequency, J is the angular momentum inertia of the rotor, D is the damping coefficient generated by the damper windings, and T_m and T_e are the input mechanical torque and load electric torque, respectively. The dynamics of SM rotor is normally slow due to the large J. In the SST, the dynamic equation of the DC voltage with a DC capacitor C_{DC} can be derived as

$$C_{DC}v_{DC}\frac{dv_{DC}}{dt} = i_{in}v_{DC} - i_{out}v_{DC} = P_{in} - P_{out} \tag{6.11}$$

Equation (6.11) is similar with (6.10), where $C_{DC}v_{DC}$, V_{DC}, P_{in}, and P_{out} are, respectively, corresponding to J, ω_r, T_m, T_e. Assuming that SST is lossless, P_{out} is also the output active power into the AC grid. Equation (6.11) shows that the DC capacitor has similar dynamics with the SM rotor and can be used for the inertia emulation. But in order to get the same inertia as the SM rotor, the DC capacitor will be very large.

The equation of the frequency regulation in Figure 6.23 is derived as

$$J_{vir}\frac{d\omega}{dt} + m_\omega(\omega - \omega_R) = G_{LPF1}(s)v_{DC} - v_{DC\_R} \tag{6.12}$$

Equation (6.12) is also similar with (6.10), where the virtual inertia parameter J_{vir} and damping coefficient m_ω are, respectively, corresponding to J and D. So this regulation can also emulate the SM rotor's characteristic without physical component. The difference is that the torque is replaced by V_{DC}, and the dynamic equation (6.11) is then inserted in the swing equation. G_{LPF1} is a first-order low-pass filter used to suppress the low-order harmonics for V_{DC}.

The virtual rotor model of the SSSM concept-based SST is developed based on (6.11) and (6.12), as shown in Figure 6.23 in s-domain. When P_{in} or P_{out} varies, the difference between them will change V_{DC} based on C_{DC}, and then ω based on J_{vir}. This will change δ and then P_{out} according to the power–angle relationship. Finally,

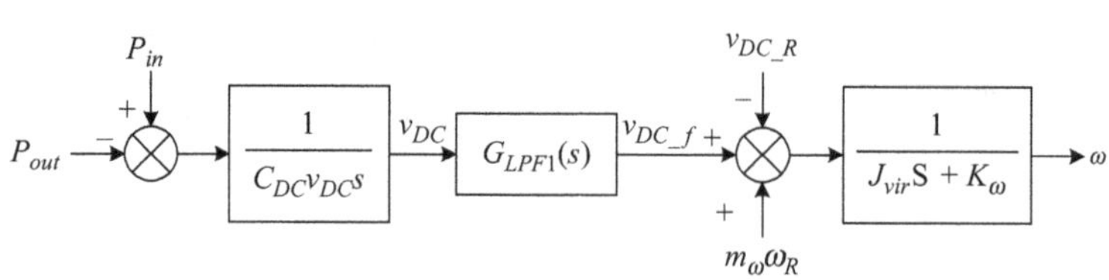

Figure 6.23 Block diagram of the SST's virtual rotor model

a new balanced working point for P_{in}, P_{out}, and δ is reached, while ω synchronizes with ω_g again. The microgrid's total inertia is determined by C_{DC}, V_{DC}, ω_{c1}, and J_{vir}. When C_{DC} is designed normally, J_{vir} will be an extra control parameter to adjust the inertia to meet the stability requirement in the power system.

According to the analysis above, any power from the DC microgrid can be transmitted to the AC grid through the SST at a given AC frequency. The dynamics of AC output active power is slowed down and completely decoupled from the intermittent, variable DRERs and loads at the DC microgrid. So this microgrid's model is simple and promising for studying the scalability and stability of the paralleled systems. Due to its swing characteristic, the microgrid is able to respond to the short-term requirements of the grid frequency regulation.

6.5.3 Power up/down reserve support

Different from the traditional SST control that synchronizes with power grid with no frequency response, the SSSM-based SST will supply the power system with up/down power reserve with its emulated angular speed ω and support of energy storage units such as supercapacitor or battery. Dual droop control, as shown in Figure 6.24 is proposed to coordinate the SST with DESD and implement the up/down reserve for primary frequency regulation, similar to the governor control in synchronous machine.

1. *First droop in SST:* Since the energy storage units in the DC microgrid only react to V_{DC}, ω in the SST must be linked to V_{DC} through the active power. The traditional droop controls for active power in AC and DC system, respectively, are

$$\omega - \omega_R = -k_\omega (P - P_{rate}) \tag{6.13}$$

$$v_{DC} - v_{DC\_R} = -k_{vDC} (P - P_{rate}) \tag{6.14}$$

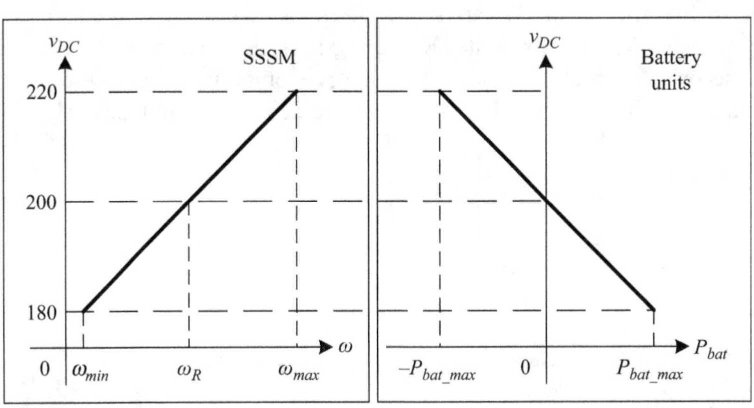

Figure 6.24 Detailed diagram of dual droop control for SSSM and battery units' coordination

where P_{rate} is the converters' rated power, and k_{ω} and k_{vDC} are the droop slopes for AC and DC systems. Combining (6.13) and (6.14), the first droop in the SST can be rewritten as

$$v_{DC} - v_{DC\_R} = \frac{k_{vDC}}{k_{\omega}}(\omega - \omega_R) = m_{\omega}(\omega - \omega_R) \tag{6.15}$$

In (6.15), m_{ω} is determined by the allowable ranges of ω and V_{DC}. Typically, $1/k_{\omega}$ is proportional to the SST's rated power. The DC voltage ranges are designed to be the same for different SSTs, then $1/k_{vDC}$ is also proportional to their rated power, and the resulting m_{ω} will be the same for different SSTs. For the frequency regulation control in Figure 6.23, this droop control is naturally integrated in it, and m_{ω} is also the coefficient for the part of the damping.

2. *Second droop in energy storage units:* The second droop in the energy storage units is a traditional *V–P* droop, and m_{vDC} is its droop slope. According to the first droop in the SST, P_{bat} should be controlled based on V_{DC}. This control is shown in the right column in Figure 6.24. The hierarchical power management discussed in Section 6.4 for battery is still applicable as an offset power reference.
3. *Coordination of dual droop:* Figure 6.25 shows the coordination diagram of the dual droop control. V_{DC} is controlled according to ω through the first droop. This voltage is a command, which indicates the energy storage units how much power SST requires corresponding to the second droop curve. Then energy storage units will output the corresponding power through the second droop and inject it into the AC grid via the SST. This is similar to the traditional active power control in the SM, and finally, the SST with the DC microgrid will have the ω–P droop characteristic.

The dual droop control is distributed and autonomous without communication. In steady state, the SSTs connected to the same AC grid will synchronize with it, share and balance the power via the AC frequency. So the SST can provide up/down reserve to the grid when ω_g varies in a desirable region, with the support from energy storages. The power in the DC microgrid is also shared and balanced based on the second droop among paralleled energy storage units. So the power management of the AC grid and DC microgrid are achieved simultaneously either in

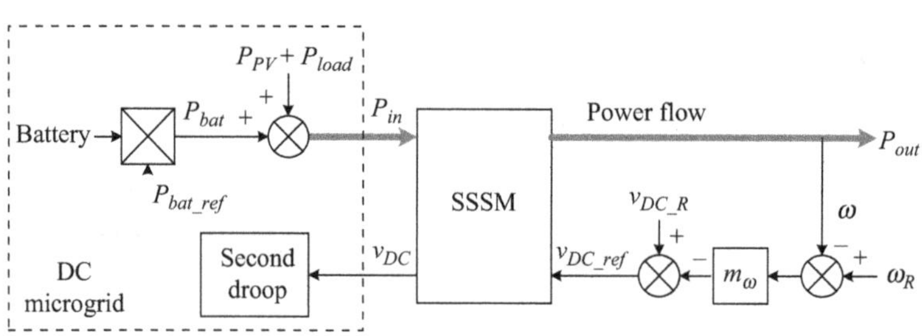

Figure 6.25 Coordination diagram of a dual droop control

grid-connection or islanding modes. During the dynamics, the dual droop control will also support the demanded inertia at a given ω_g. It is achieved by commanding energy storages to provide temporary energy to the SST through V_{DC}, when ω and V_{DC} vary due to the disturbances from the AC grid or DC microgrid.

6.5.4 Voltage regulation

In Figure 6.22, a simple *Volt-Var* droop control (shown in Figure 6.26) is adopted to mimic the excitation system of SMs, where m_{vg} is its droop slope. In SMs, the exciter is usually a proportional controller, and the field circuit flux is produced based on it after a long-time delay. To emulate this, a first-order LPF can be adopted, shown as $G_{LPF3}(s)$ in Figure 6.22.

6.5.5 Case study

To verify the primary control algorithm, a low-power experiment is carried out in the lab test-bed. A 500-W battery module, a 500-W PV module, and up to 500-W DC load are constructed to constitute the DC microgrid. The grid voltage of the 1,200-W SST is 120 V, and the DC link voltage is 200 V. When the PV generation unit is operating with MPPT control, the maximum power is 500 W, and the corresponding MPPT voltage is about 28 V. The nominal battery voltage is 12 V and the capacity is 20 A h. $m_\omega = 31.83$ is chosen so that V_{DC} works from 180 to 220 V, when ω works from 59.9 to 60.1 Hz; $m_{vg} = 0.005$ is chosen so that E will work from 114 to 126 V when the virtual synchronous machine based interface (VSMBI) provides 100% to -100% reactive power to the grid. The constant $m_{vDC} = 25$ is chosen, so that battery will operate from 180 to 220 V with 500-W discharge to 500-W charge power.

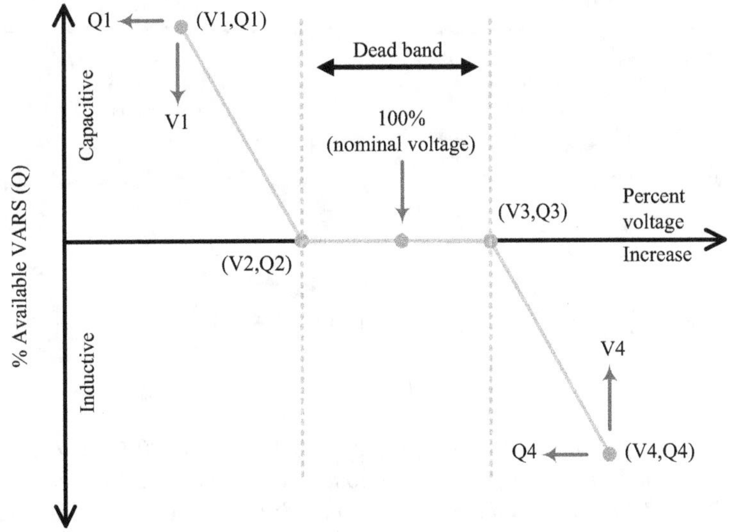

Figure 6.26 Volt-Var droop control of SSSM applied in Figure 6.22

6.5.5.1 Case I: Load change

In this case, an 80-Ω resistance load is disconnected first, and then reconnected. Figure 6.27 shows the results when J_{vir} is separately set at 1.06 and 4.24. The system will take about 1.7 and 2 s to reach new operating points in Figure 6.27(a) and (b), compared with 2 and 2.1 s in Figure 6.27(c) and (d). These results verify the effectiveness and functions of the SST with the DC microgrid as SSSM in generator mode.

6.5.5.2 Case II: Source change

In this case, PV output current changes from 2.5 to 0 A first, and then changes back. Figure 6.28 shows the results when J_{vir} is set at 1.06 and 4.24. The system takes about 1.6 and 1.2 s to reach new stable working points in Figure 6.28(a) and (b), compared with 2.6 and 2 s in Figure 6.28(c) and (d). The system's functions in motor mode are verified.

6.5.5.3 Case III: Power up/down reserve

Figure 6.29 shows the results for 200 s when everything is in default settings. ω varies with the time, because ω_g is varying and ω is synchronizing with it. The frequency variation is normal in utility grid due to the customers' load change, and

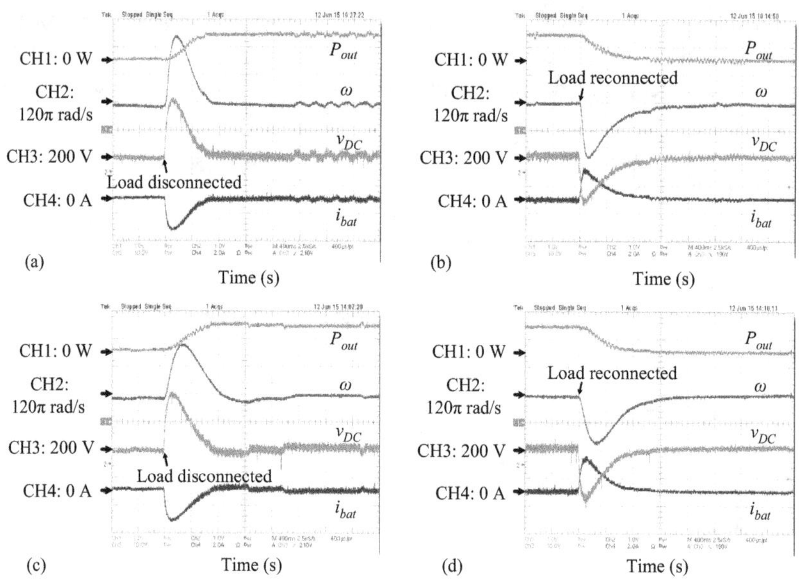

Figure 6.27 Experimental results of Case I: (a) $J_{vir} = 1.06$, 80-Ω load changes to no load; (b) $J_{vir} = 1.06$, no load changes to 80-Ω load; (c) $J_{vir} = 4.24$, 80-Ω load changes to no load; (d) $J_{vir} = 4.24$, no load changes to 80-Ω load. CH1: P_{out} (500 W/div); CH2: ω (0.08 π rad/s/div); CH3: v_{DC} (10 V/div); CH4: i_{bat} (2 A/div); X-axis: time t (400 ms/div)

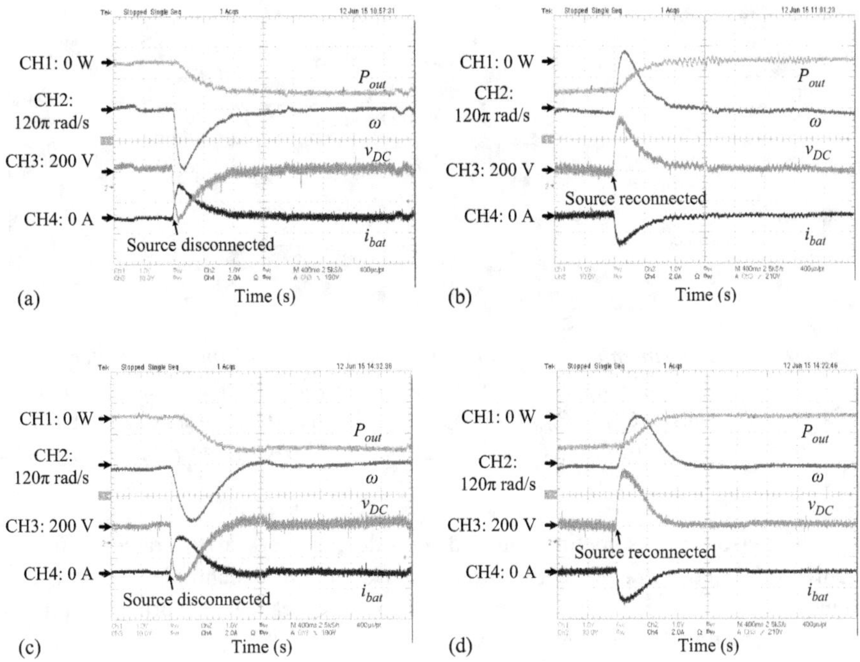

Figure 6.28 Experimental results of Case II: (a) $J_{vir} = 1.06$, 2.5-A source changes to 0 A; (b) $J_{vir} = 1.06$, 0-A source changes to 2.5 A; (c) $J_{vir} = 4.24$, 2.5-A source changes to 0-A; (d) $J_{vir} = 4.24$, 0-A source changes to 2.5 A. CH1: P_{out} (500 W/div); CH2: ω (0.08 π rad/s /div); CH3: v_{DC} (10 V/div); CH4: i_{bat} (2 A/div); X-axis: time t (400 ms/div)

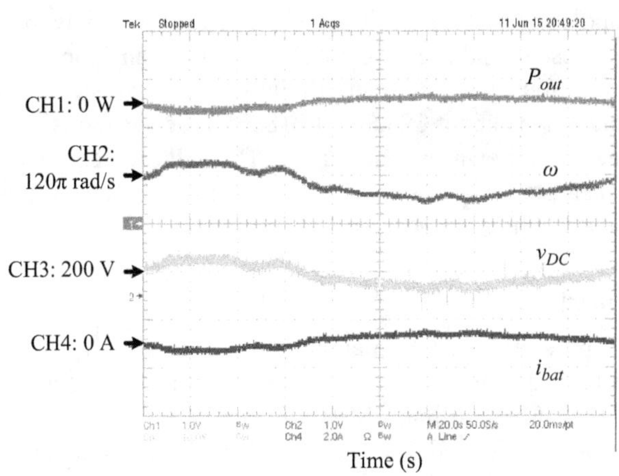

Figure 6.29 Experimental results of Case III: CH1: P_{out} (500 W/div); CH2: ω (0.08 π rad/s /div); CH3: v_{DC} (10 V/div); CH4: i_{bat} (2 A/div); X-axis: time t (20 s/div)

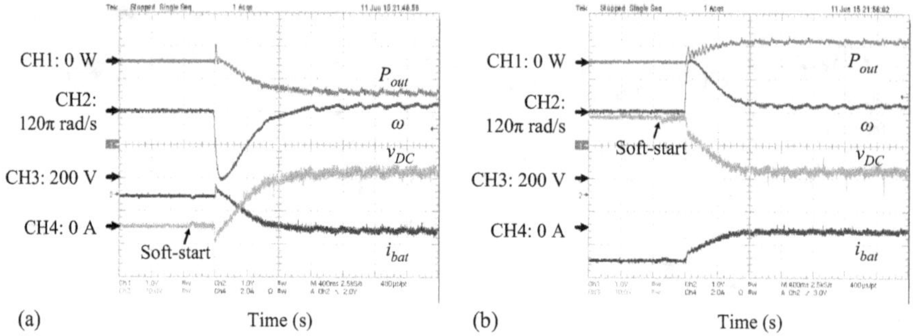

Figure 6.30 Experimental results of Case IV: (a) reconnecting with battery unit and load and (b) reconnecting with battery unit and current source. CH1: P_{out} (500 W/div); CH2: ω (0.08 π rad/s /div); CH3: V_{DC} (10 V/div); CH4: i_{bat} (2 A/div); X-axis: time t (400 ms/div)

it is the reason why sometimes ω and V_{DC} do not work at the rated values in Figures 6.27 and 6.28. Since the proposed system can synchronize its own frequency with ω_g with a high degree of accuracy, it can also work as a phasor measurement unit in the power systems. In Figure 6.29, V_{DC} and i_{bat} vary and follow ω according to the dual droop control. This verifies the effectiveness and functions of the dual droop control in the proposed concept, supplying the power system with up/down reserve when the frequency deviates from the rated 60 Hz.

6.5.5.4 Case IV: Islanding and reconnection

In this case, at first the AC grid is disconnected, SST stops working and the DC microgrid keeps working. Figure 6.30(a) shows the transition for the system reconnecting and synchronizing with the AC grid when only battery unit and 400-W load are present, and Figure 6.30(b) shows the transition when only battery unit and 2-A DC source are present. Initially, i_{bat} is about 1.95 and −2.1 A, respectively, and V_{DC} is about 185.5 and 218.5 V; the system takes about 1.4 and 1.3 s to reconnect, as shown in Figure 6.30. The achievement of power management in DC microgrid when SST is not working and the proposed concept's synchronization capability are verified.

6.5.6 *Summary*

An SSSM concept is proposed to integrate the DC microgrid into legacy AC grid through an SST. The benefits are (1) integrates the DERs, loads, and energy storages in the DC microgrid into SSSM; (2) achieves power management for the AC grid and DC microgrid, and enable the DERs and loads to output any power at a given AC frequency; (3) introduces inertia, provides up-reserve and down-reserve for the grid with support from practical energy storages, and decouples the AC output from the fast responding DERs and loads. Experimental results verify the effectiveness and the benefits of the presented method.

6.6 Conclusion

The SST-enabled DC microgrid concept is introduced in this chapter. Compared to the traditional microgrids, the presented system enables a compact and integrated interface for the DC and AC grids. Centralized power management strategy, hierarchical power management strategy, and SSSM concept are proposed to control the operation of the presented system. Simulation and experimental results are given to demonstrate the feasibility of the proposed system.

References

[1] X. She, A. Q. Huang, and R. Burgos, "Review of Solid-State Transformer Technologies and Their Application in Power Distribution Systems," *IEEE J. Emerg. Sel. Topics Power Electron.*, vol. 1, no. 3, pp. 186–198, September 2013.

[2] X. She, A. Q. Huang, S. Lukic, and M. Baran, "On Integration of Solid-State Transformer with Zonal DC Microgrid," *IEEE Trans. Smart Grid*, vol. 26, no. 12, pp. 3778–3789, June 2012.

[3] J. M. Guerrero, J. C. Vasquez, J. Matas, L. Vicuna, and M. Castilla, "Hierarchical Control of Droop-Controlled ac and dc Microgrids – A General Approach Toward Standardization," *IEEE Trans. Ind. Electron.*, vol. 58, no. 1, pp. 158–172, May 2011.

[4] X. Yu, X. She, X. Ni, and A. Q. Huang, "System Integration and Hierarchical Power Management Strategy for a Solid-State Transformer Interfaced Microgrid System," *IEEE Trans. Power Electron.*, vol. 29, no. 8, pp. 4414–4425, August 2014.

[5] M. Ashabani, and Y. A.-R. I. Mohamed, "Novel Comprehensive Control Framework for Incorporating VSCs to Smart Power Grids Using Bidirectional Synchronous-VSC," *IEEE Trans. Power System*, vol. 29, no. 2, pp. 805–814, March 2014.

[6] Q. Zhong, and G. Weiss, "Synchronverters: Inverters that Mimic Synchronous Generators," *IEEE Trans. Ind. Electron.*, vol. 58, no. 4, pp. 1259–1267, April 2011.

[7] D. Chen, Y. Xu, and A. Q. Huang, "Integration of DC Microgrids as Virtual Synchronous Machines into the AC Grid," *IEEE Trans. Ind. Electron.*, vol. 64, no. 9, 7455–7466, September 2017.

Chapter 7

The load as a controllable energy asset in dc microgrids

Robert S. Balog[1,2]*, Morcos Metry*[1,2]*, and Mohammad Shadmand*[1,2]

7.1 Introduction

The electric power grid is the backbone of modern societies and their economies by interconnecting electrical energy generation to loads via a vast transmission and distribution network. Heralded as the greatest engineering achievement of last century [1], the power grid in the United States and other countries is deteriorating due to age and usage stresses [2]. In fact, it experiences a major outage every decade that costs more than $2 billion, and on any given day, 500,000 customers are without power for 2 h or more [3]. The traditional grid was not designed with the renewable energy sources in mind and the reverse power flows that occur and thus does not easily accommodate the transformation from *consumer* to *prosumer*. Simply patching the old system beyond its original design and safety margins will not alleviate the problems of power delivery congestion, feeder system voltage profile regulation, reactive power, harmonic power, and the other challenges of the way in which the grid is now being used.

One solution that does not require a $5 trillion complete overhaul of the existing $1.5–$2 trillion [4] US electricity grid is to separate the functions of bulk power delivery from point-of-load (POL) control. In this paradigm, we interconnect loads and distributed energy sources and storage into a local area power and energy system (LAPES) [5], sometimes called a microgrid [6], which is then connected to the utility power grid or to other microgrids [7]. The Department of Energy defines a microgrid as "a group of interconnected loads and distributed energy resources within clearly defined electrical boundaries that acts as a single controllable entity with respect to the grid" [8]. A classical perspective of ac power systems is a top-down, generator-to-load flow of power in which utilities control the generation and infrastructure and are required by a regulatory body to deliver power to the load.

[1]Department of Electrical and Computer Engineering Texas, A&M University, USA
[2]Department of Electrical and Computer Engineering, Kansas State University, USA

The utility has no control of the load, which imposes operational constraints on the systems. In a LAPES, the load as well as distributed sources and storage can all be controlled, which changes the paradigm of generation following load into a paradigm in which control of the load itself can be thought of as a controllable energy asset [9]. The paradigm further changes as the control moved from centralized to distributed.

In the past, some utilities have had load-shed programs in which customers relinquish some limited control of their load. Such a system was most commonly used for peak load shaving in which customer-owned air-conditioning systems were fitted with cut-off switches that responded to a command from the grid or system operator, as explained by Commonwealth Edison Company of Illinois [10]. Thus, shedding individual load to match available generation capabilities or congestion constraints to avert a blackout is one way the load can be considered as an energy asset [11]. Load shed has also been considered for ac systems where both voltage and frequency stability were a concern and implemented with frequency-activated load shedding [12]. Less aggressive approaches that have been considered include conservation voltage reduction in which the voltage on a feeder system is lowered to reduce power consumption, which can work to some degree when the loads are mainly resistive in nature [13]. Market-based incentive programs take a slightly different approach and attempt to encourage customers to modify their power consumption based on adjusting the price of electricity. In some cases, a smartphone app is used to signal cost savings that may be possible by voluntarily reducing consumption, or offer an incentive, such as a coupon, for volunteering to shed load [14,15]. However, such initiatives depend on the active participation of the customer. While it raises public awareness of energy consumption, it may not be reliable for peak-time load shaving to avert a crisis if the customer does not behave as expected.

Fundamentally, the challenge in any electrical power system is that of just-in-time delivery: supply must balance consumption. If the system is congested, or generation is already at peak, one way to relax the constraint is to introduce energy storage. Pumped hydro can be cost effective but depends on geography and thus is not a universal solution, just like compressed air storage which depends on vast underground storage. Battery-based storage can be more arbitrarily located in the utility system as needed. Examples of grid-scale storage include the Younicos cube or the Tesla power pack [16,17]. Another approach is to tap into a growing market of residential-scale storage element such as the Tesla power wall located in individual houses [18]. However, while adding storage can decouple supply from demand, it also adds cost.

7.1.1 Local area power and energy system

Consider the LAPES microgrid shown in Figure 7.1. From the perspective of the legacy power system, a LAPES would exist at the grid-edge and would incorporate distributed renewable energy sources such as solar photovoltaics and wind, energy storage, and load. The LAPES would also provide a point of common coupling (PCC) for all of the individual components to the grid-connection. The LAPES

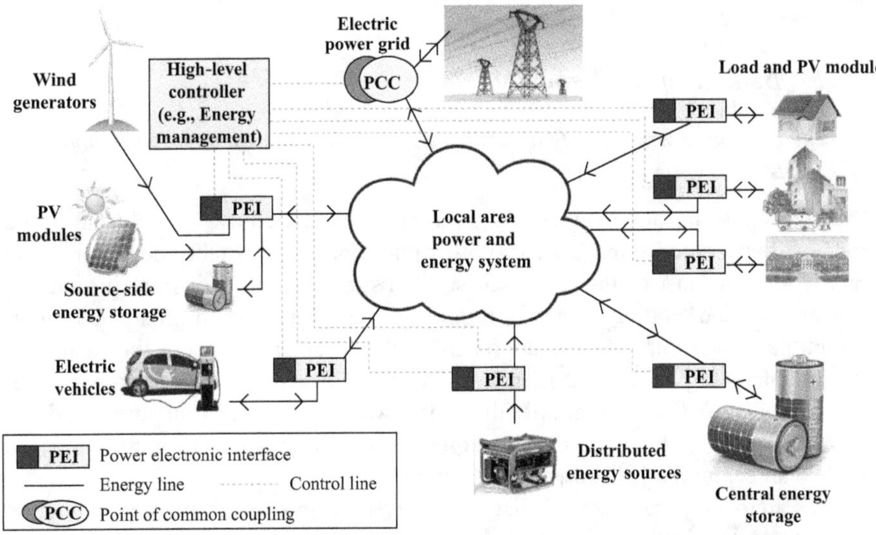

Figure 7.1 One-line system diagram for a local area power electronic system (LAPES) that integrates community-scale energy storage with distributed solar generation colocated with the electrical load. The system is interconnected with the electric utility at a single point of common coupling (PCC). Sizing and control of the energy storage unit, the "Central Energy Storage" can enable arbitrary power flow profiles at the PCC to the grid

would not merely supply power to the load but would also interact with the loads, sources, and storage. In such a paradigm, the operation of the load, distributed sources, and distributed storage can respond to the LAPES system to ensure stable operation and even participate in energy market opportunities upstream in the utility power grid by controlling power flow at the PCC. Without loss of generality, examples of a LAPES include a neighborhood or subdivision of houses, an industrial park, a shopping center, a school campus. In these examples, the central energy storage shown in Figure 7.1 would be a common resource used by the entire system, much like a community lake is a common resource used for firefighting or a retention basin is used for flood control.

In this chapter, we consider a LAPES as a dc microgrid system. Virtually, all modern loads are dc, and dc systems offer efficiency and control advantages over ac for distribution networks and integration of renewable sources and distributed storage [5]. The focus of this chapter is on control of the load so that the system can view the load as an energy asset rather than a liability. In other words, view the load as an integral participant in the overall power and energy balance of the microgrid. This is done by exploring how POL controllers combined with modulation of the bus voltage can control the power and energy consumption of the load in the system without needing expensive communications.

7.2 Why control the load?

7.2.1 Benefit of load control

> *The challenge with quality assurance is that everyone wants it but nobody wants to pay for it.* [19]

The practicality for any engineering solution is determined by the ability to find a trade-off between cost and benefit. Power interruptions inevitably do occur in the microgrid, and most of the discussed solutions to these problems are profoundly centered around adding generation and storage capacity to guarantee uninterruptible power supply. The approach in this chapter considers the load as a controllable energy asset that is included within the cost and benefit trade-off for system design and operation. Are all loads always necessary? Can some loads be reduced instead of being turned off completely? The answers to these questions provide an economic justification and benefit.

A Pareto Frontier is a tool that is usually used in engineering and economic applications to visualize the most efficient set of solutions to a multiobjective optimization problem [20]. Consider a design in which there is a trade-off between two parameters, such as converter efficiency and cost. When we consider all of the possible designs in the design space of these two parameters, we find the Pareto Frontier as the set of nondominated designs. In other words, for a give value of parameter 1 (efficiency, for example), the nondominated design is the one that achieves this efficiency at the lowest cost. Thus, all points on a Pareto Frontier are considered to be Pareto Efficient for the particular objective function. Pareto Efficient points are the only solutions that guarantee satisfying all the assigned objectives without making any particular objective worse off [21].

The benefits of load control can be observed by considering the set of equally good solutions, the Pareto Frontier, for the optimal multicriteria design of the hybrid microgrid energy system in Figure 7.1. The Pareto Frontier can be generated by using optimization methods for hybrid energy systems such as nondominated sorting genetic algorithm (NSGA-II) [22]. Figure 7.2 illustrates the set of optimal solutions given by the Pareto Frontier to maximize the energy availability and minimize the cost associated with energy storage requirement of the hybrid microgrid. The energy availability, the fraction of the time when energy is available, is a key figure of merit of a hybrid microgrid system. It is important to make a clear distinction between reliability and energy availability. The reliability is the ability of the system to operate without failure; energy availability is the ability of the system to supply power to the load. As an example, a highly reliable renewable energy system, where the components are not prone to failure, can have low energy availability if there is insufficient energy storage to support the power requirements of the load during the night or during an overcast day. Thus, a specified level of energy availability can be achieved with many configurations of a microgrid system while minimizing the system size to have a financially feasible system. This is the equilibrium point of the techno-economic optimization.

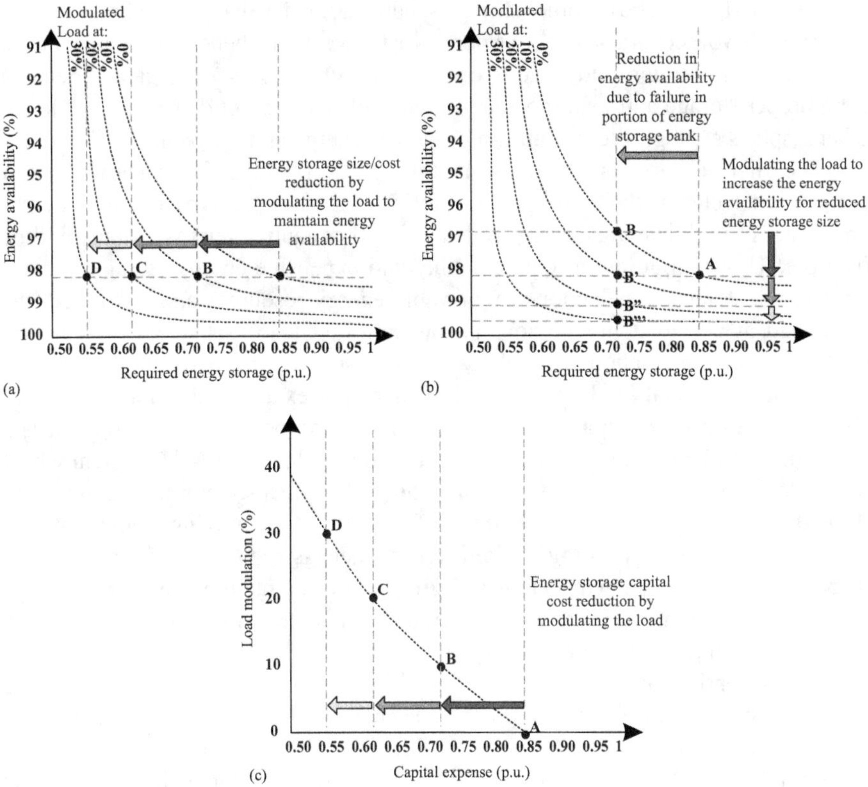

Figure 7.2 Pareto-frontier reveals the set of optimal solutions for the hybrid system [22]. (a) Load control benefits to design: energy storage size/cost reduction by moving from point A to D (modulating the load from 0% to 30% while maintaining the energy availability of the microgrid), (b) load control benefit to control: portion of the energy storage bank is failed, thus energy availability will be decreased by moving from point A to B due to this failure; the energy availability for the reduced available energy storage size can be increased by moving from point B to B''' (load modulation from 0% to 30%), (c) load control benefit to capital cost of the energy storage system

Consider a LAPES that contains PV generation and storage. The amount of PV generation and energy storage determines the energy availability for particular load [23]. Figure 7.2 illustrates the trade-off between the cost and energy availability. Consider that individual loads can be controlled to reduce power consumption. Without loss of generality, one may think of dimming a light or slowing a heating, ventilation, and air conditioning (HVAC) blower motor. The base-case is the microgrid system without load control, illustrated as 0% load modulation. This is compared to systems in which the load can be modulated to reduce power consumption from

10% to 30%. This can be accomplished by sending a signal to each dc/dc or dc/ac POL converter. Two scenarios are considered that reveal the benefit of load control (modulation) in a microgrid system: one scenario encompasses the gained benefits from proper storage and load modulation ratios of the microgrid systems, while the other displays strategies for failures in portion of energy storage bank.

The first scenario is illustrated in Figure 7.2(a). This figure shows that if point A is selected as the final optimal solution of the base-case based on particular preferences to achieve 98% energy availability, the required energy storage will be 0.85 p.u. This configuration requires bulky and expensive energy storage to compensate the stochastic behavior of renewable energy resources and/or to meet the peak load at 0% load modulation (an uncontrolled load). The load modulation could refer to the fraction of total load that is shed or the percentage of power reduced in individual load for a period of time. For example, dimming a light or changing the set point on a thermostat represents load power modulation that is a different control paradigm than simply turning off the load or HVAC system which may balance providing benefit to stabilize the grid while also minimizing impact to the end-user. By modulating the load from 0% to 30% (i.e., moving from point A to D), the required energy storage system requirement is reduced significantly from 0.85 to 0.55 p.u. as shown in Figure 7.2(a), while maintaining the same energy availability levels. This scenario demonstrates the benefits of the load control to the design and sizing for the microgrid system.

The second scenario is illustrated in Figure 7.2(b) where the system was initially designed for operation at point A. Then, portion of the energy storage bank is failed, causing reduction of energy availability from 98% to 97% (i.e., moving from point A to B). The advantage of the load modulation comes into play in this scenario; by modulating the load at 10%, the same energy availability of 98% could be maintained with a failed storage element. Further energy availability improvement could be achieved by modulating the load further. For example, at 30% modulation, the energy availability reaches 99.99% (shown by moving from point B to B''') at an energy storage size of 0.72 p.u. Thus, the function of load modulation, in this example, increased the energy availability by 1.9% and decreased the energy storage size by 0.13 p.u. in comparison with the originally designed system that did not have controllable load considered initially. Thus, the load control can improve the system controllability for maximizing the energy availability as well as compensating the consequences of energy storage device failures in a typical microgrid system.

It is worth mentioning that moving from point A to D in Figure 7.2(a) is a capital expense saving at the design stage because less energy storage is required. The effect of load modulation on the capital cost of the energy storage system is illustrated in Figure 7.2(c). As is illustrated, for example, if lights are dimmed to modulate the load to 30%, the capital cost of the system will be reduced to 0.55 p.u. at point D. However, if the loads are not modulated, the capital cost of the system is 0.85 p.u. at point A. The capital cost of 0.85 p.u. demonstrates the cost of an optimal system design at 0% load modulation, and the capital cost of 0.55 p.u. demonstrates the cost of an optimal system design at 30% load modulation. This demonstrates the impact of load modulation on the end user in terms of productivity, capability, and economic model because the end user would expect financial compensation for the reduced performance or reduced capability.

7.2.2 Is load modulation practical?

Figure 7.3 shows the generation inertia of the ERCOT (Electric Reliability Council of Texas) utility grid for the state of Texas during the month of April 2014 along with the corresponding load demand [24]. For the most part, the inertia seems to meet and surpass the actual load for that month. This is because nuclear and coal generation inertia is scheduled based on historical demand and forecasted weathers and fast peaking supply, along with spinning reserves ensure rapid response to unpredictable changes in load. However, microgrids that rely on renewable energy sources, particularly solar energy, along with other sources may not have such flexibility, particularly during times of peak load. During peak loads, microgrids have multiple options: generating more energy, importing energy from the main grid or from other microgrids, shedding the load, or modulating the load. There are many scenarios; the first option of generating more energy is not possible physically because the system is already at maximum generation capacity. The second option, importing energy from the grid or other microgrids, may be unfavorable economically because of the exorbitant cost from purchasing power on the volatile spot market. The third option, load shed, may be unfavorable since customers would be inconvenienced from the power interruption. Load modulation, on the other hand, requires a closer look at the characteristics of loads in the microgrids. Let us consider lighting, HVAC and electric/plug-in hybrid vehicles as examples of controllable loads.

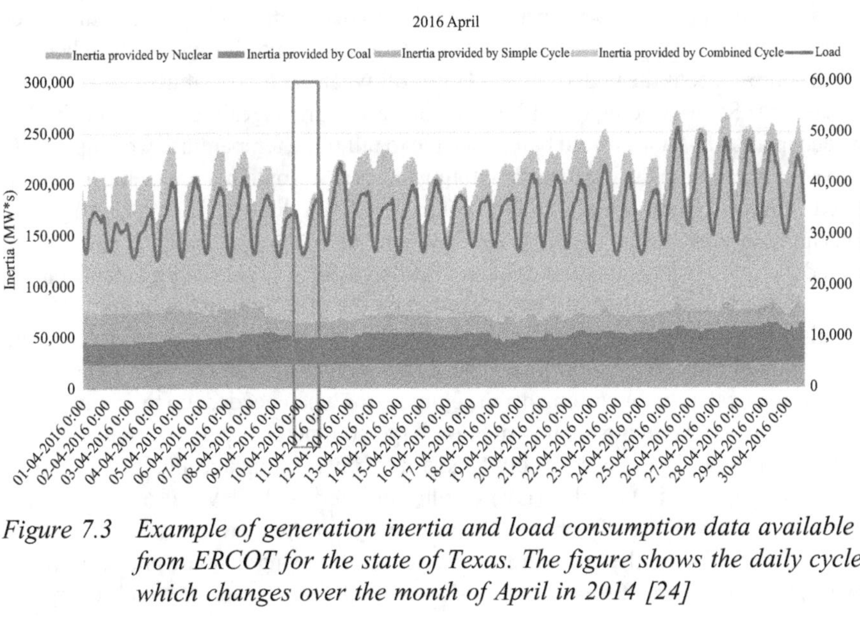

Figure 7.3 Example of generation inertia and load consumption data available from ERCOT for the state of Texas. The figure shows the daily cycle which changes over the month of April in 2014 [24]

Consider the peak-time load consumption which occurs during daylight hours. Since lighting loads represent 30% of the electrical consumption in commercial buildings, modern buildings are designed with controllable lighting systems for energy conservation and adjust artificial lighting based on the amount of sunlight entering the building [25], occupancy, and usage requirements. More windows unshaded means a lower illuminosity foot-candles (ftcd) is required from the artificial lighting system. Standards recommend office spaces to have an illuminosity of 46 ftcd [26]. Hence, sunlight exposure from the office windows could reduce light consumption. While this might seem like a good idea in principle, tenants of the office complain about sudden switch off of lighting during direct sunlight, delayed turn on of the lighting when the room is darker, and a continuous switching on and off of lights on cloudy days [27]. Such concerns mean that in many workspaces, these "intelligent" systems are bypassed; hence, lights remain at their full illuminosity the whole day, impeding the energy saving opportunity and missing the opportunity to control the load as an energy source. As many lightbulbs come now with dimming features, lighting modulation could be achieved by adjusting the brightness rather than turning on or off certain light fixtures. This is achieved by giving some load modulation control to the grid. After all, a user can barely distinguish the difference between a 46-ftcd and a 35-ftcd room [28] yet when taken at the scale of an entire building over a period of time can provide enough energy to have a meaningful impact on the stability and energy availability of the microgrid. Hence, lighting in commercial buildings is well suited as a controllable energy asset.

Consider a building heating, ventilation, and air conditioning system (HVAC). A building has a large thermal mass so adjusting blower speeds or temperature set points on a short enough time bases may have minimal impact on the occupancies perception but frees up energy. Typical HVAC systems have a control approach similar in principle to the bang-bang controller, in that there is a 2°C upper and lower limits around the set point and the HVAC compressor and fan will work when the temperature exceeds the upper limit and turns off when the temperature goes under the lower limit. Such systems could be thought of as a manageable load asset, in that the occupants of a room would barely be able to tell if the temperature went up to 3°C above the upper limit during times of contingencies and would not notice the lower fan speed intensity. Such initiative takes place in the middle of the day, during peak consumption time, assuming the HVAC system has been on all day and has already cooled the space. Many companies have considered demand side energy management in which the HVAC unit could be controlled by the grid. Edison Commonwealth of Illinois [10] has the initiative of Central AC Cycling for homeowners, which offers the installation of direct load control switches or utilizing a smart thermostat to cycle the load during peak time by turning off the compressor when needed, up to 15 min every 30-min interval. Applying such system on large commercial and government buildings such as schools and offices could add a flexibility of manageable load assets.

Consider electric vehicles (EV) or plug in hybrids (PHEV): if all the owners care about is having a fully charged battery ready for the next trip, he or she could set the expected time the car will next be needed by using a smart phone app [29] when parking the car. From the driver's perspective, the rate at which energy is delivered to the car battery is meaningless as long the battery is fully charged and ready for service when the user returns [30]. Thus, EVs become a controllable energy asset by scheduling unnecessary EV charging off peak times. In 20 years or so, when EV become the mainstream technology, utilizing such load could be significant. For example, a parking garage next to a hospital with a 1,000 cars capacity could have enough energy storage, using vehicle-to-grid protocols [31], for one of the hospital's building segments (e.g., operation rooms or intensive care units) during emergency power deficiencies.

Numerous application examples can be envisioned, ranging from civilian to military, terrestrial to outer space, to personal power networks, in which load can participate in the overall control and operation of power and energy management of the dc microgrid system. It may be easy to visualize the imperative for load control if the microgrid is stand-alone and energy constrained. The need also exists, however, for grid-connected dc microgrids, like the one in Figure 7.1, in which the operator desires to shape the load profile at the PCC for reasons of economics or utility-side stability. For example, the microgrid can take advantage of market clearing prices to control local energy assets [32]. This may be obvious for the cycling of energy stored in battery, but we can also modulate a load to achieve similar effects. Therefore, there are many considerations to be taken into account during the design of the microgrid. During the design stage for the energy storage, sizing of the storage is determined by assuming fully modulated load to size the storage based on worst case availably requirement. During operation, energy storage must be first used as primary energy asset and then the load modulation is secondary. Energy tapping opportunities from the load should be categorized in a way that ensures the minimum cost. For example, tapping into EV batteries could potentially result in higher battery maintenance cost than reducing the fan speed of an HVAC system which could only cost a little convenience.

7.3 Time-scale of energy requirements

Not all power interruptions have the same impact and may require different strategies to mitigate or minimize disturbance to the system. Figure 7.4 illustrates a time continuum of events that may be experienced on an electrical grid The timescale could be as fast as thousandths of a second for power electronics stability, subsecond for electrical network stability, tens of seconds for automatic generator control, minutes to hours for economic dispatch/optimal power flow, diurnal (daily) for solar and wind power, and seasonal for heating and air-conditioning systems. Such continuum of interruption timescales makes it impossible to tackle such variability with only one strategy. Hence, it is crucial to divide the different interruptions based on the range of their time scale (long or short) and plan for the appropriate contingency plan.

7.3.1 Short-term transients

One method to maintain stability through short-term disturbances is to reduce the power to the load—appropriate for lighting or inertial loads where the momentary

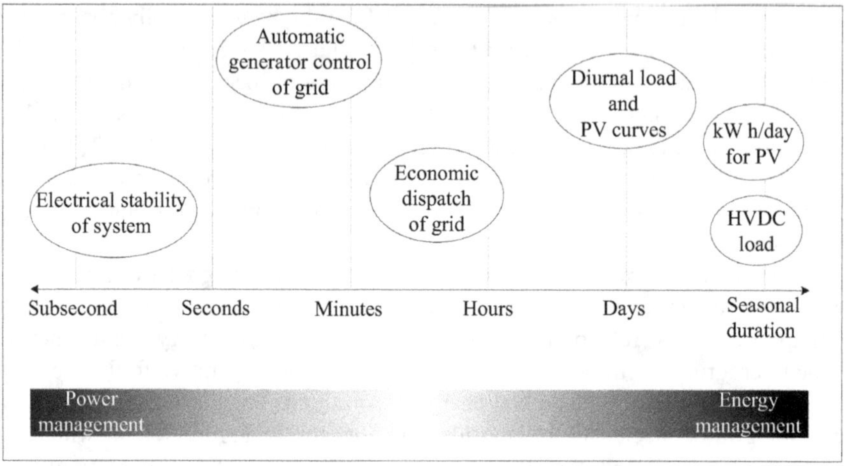

Figure 7.4 An illustration of the timescales of potential load interruptions

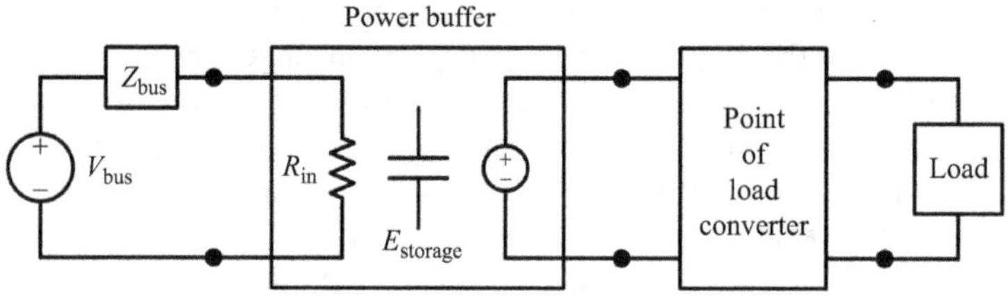

Figure 7.5 Power buffer for a dc system

slowing of fans and pumps or dimming of lights may be an acceptable alternative to system-wide instability. However, many modern electronic loads and other sensitive loads do not lend themselves to this technique. A more general method is to implement an active dynamic buffer as an interface between the power system and the POL converter as shown in Figure 7.5.

A power buffer decouples the load from the bus dynamics [33]. The rating of the local energy storage device provides the designer with a degree of freedom to choose the extent of the transients through which the load can be sustained. Once the local energy has been depleted, however, continued operation of the load is no longer possible. This gives rise to the notion of time scales based on the energy storage in the power buffer.

The power buffer has several modes of operation. During normal system operation, the power buffer supplies the load by drawing power directly from the bus. When a system transient occurs, the buffer senses the system voltage sag and presents constant impedance to the bus while continuing to supply the load with constant power. Since less power is drawn from the bus during the constant impedance mode, internal storage is required to maintain the power requirements of the load. After the transient passes, the buffer returns to a power regulation mode and draws additional incremental power to recharge the buffer capacitor. In effect, a power buffer stretches the time scale of the transient, diminishing the impact of tight converter regulation.

A power buffer is limited by the amount of stored energy available in the bus capacitor. The time that a power buffer can maintain constant input impedance while continuing to supply full power to the load is called its sustaining time and is defined as

$$T_{sustain} = \frac{C_{buffer}\left(V_{load0}^2 - V_{load1}^2\right)V_{bus0}^2}{2P_{laod}\left(V_{bus0}^2 - V_{bus1}^2\right)} \tag{7.1}$$

where V_{load0} is the nominal load voltage, V_{bus0} is the nominal input voltage, V_{bus1} is the sag voltage, and P_{load1} is the load power. The buffer design parameters in the sustaining time are V_{load1} (minimum allowable load voltage) and C_{buffer} (energy storage capacity).

A plot of a typical sustaining time versus voltage sag for a buffer supplying a 100 W, 400 V load from a 100 V distribution system is shown in Figure 7.6. As long as the voltage sag magnitude and duration fall above the curve, the power buffer can successfully ride through the transient, presenting constant impedance to the power system bus while maintaining constant power to the load. If the transient event begins to approach the sustaining time limit, then the local load control needs to switch strategies to maintain stability.

7.3.2 Long-term transients

System transients that exceed the sustaining time of the power buffer (or that are caused by topological failures such as loss of generation or a bus fault) require an alternative technique to mitigate system instability and voltage collapse. In this

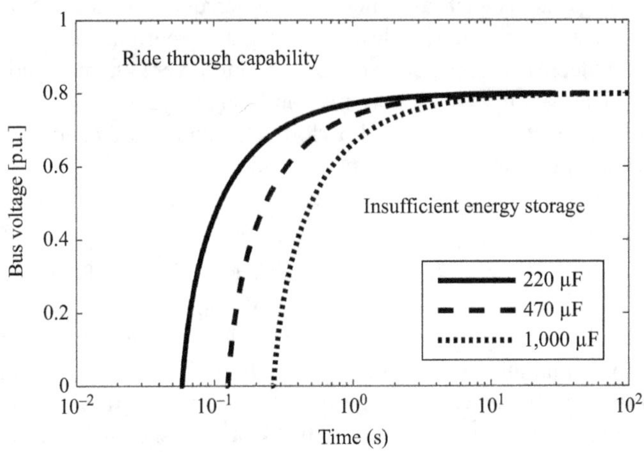

Figure 7.6 Sustaining time capability

case, the only long-term strategy to stabilize an energy-constrained dc system is to switch to another supply or to load shed.

In practice, dc–dc regulators have minimum allowable input voltages that prevent them from operating as true constant power loads (CPLs). As the system bus voltage decreases due to increased loading or loss of generation, these converters will turn off to self-protect as the bus voltage decreases below this specified voltage limit. In a radial system, the voltage drop along the bus due to bus impedance automatically gives rise to a notion of priority to the loads located closest to the source and hence with the highest bus voltage. Thus, the first converters shed due to under-voltage protection (UVP) are the ones physically farthest from the source. This behavior directly couples the system topology to the load priority. In general, it is not desirable that the topology dictate priority. Instead, a supervisor control at each POL converter monitors the bus voltage and turns off the converter based on the priority of the load. Mapping the priority setting to a particular bus voltage forces loads with the lowest priority to be shed first and allows higher priority loads to remain energized, regardless of location in the system. Thus, system operation is decoupled from system topology. A similar technique has been proposed for ring-bus architectures.

7.4 Autonomous load control

Direct current power systems have long been the standard architecture for many applications where reliability is critical, like telecommunication and naval ships. dc Power systems allow for more straightforward integration of renewable energy sources into the grid. Modern high frequency dc–dc converters have enabled more efficient voltage conversion at higher power densities than the AC 60-Hz converters could [34]. dc Systems are better fitted for energy storage integration to improve reliability and energy availability of distributed systems. Also, many concerns that accompanied ac systems such as synchronization requirements for multiple small generation sources, reactive power flow, and circulating current due to differences in voltage magnitude, phase angle, or dc offset in a multilayer system, do not exist in dc systems [35]. The paradigm shift from the common ac system to a dc system facilitates easier control of individual load performance and coordination, especially as energy allocation priorities change to match load priority requirements.

7.4.1 Control

The generally adopted control hierarchy of the dc microgrid is of close resemblance to that of the conventional grid, in that it is stratified into tertiary, secondary, and primary control levels [36]. The highest level controller, the tertiary level, responsible for economic dispatch and coordination, assigns the microgrid voltage that is appropriate for power exchange between the microgrid and the main grid [37]. Such function is communicated to the PCC, as shown in Figure 7.1. The secondary controller updates the voltage set points for the primary controllers to match the voltage demands of the secondary controller. Like the tertiary controller, the secondary controller is carried out at the high-level controller. Then, the signals from the secondary controller are sent to all the power electronics interface (PEI) units as illustrated in Figure 7.1.

Implemented locally at each PEI, the primary controller regulates the voltages for the individual converters to facilitate load sharing among sources. The control structure of the primary controller based on distributed control [35] is inherently different from the secondary and tertiary controllers that are based on centralized control. Accordingly, secondary and tertiary controller units are prone to higher risks of failure and instability in the event of losing any links in the topology [34]. On the other hand, distributed control provides higher reliability as each PEI unit controller is independent in operation from failures in other units. It also means that systems based on distributed control are easily scalable [38]. A detailed illustration on the differences between centralized and distributed control is illustrated when studying the difference between active current sharing (centralized control) and droop control (distributed control) in Table 7.2.

7.4.2 Architecture

As has been shown in Figure 7.1, a LAPES microgrid system incorporates renewable energy generation systems such as solar photovoltaics and wind along with storage elements and EV. A LAPES microgrid is not merely a supplier of load to its load but goes beyond by using residential scale generation on the load when available. Without loss of generality, the loads could be residential houses, office buildings, hospitals, university campuses, etc.; within each building, there could be several distribution and load systems as illustrated in Table 7.1.

A high-level energy management system provides signals to the buildings on power and energy usage/setting, and then the PEI for the building turns that setting into a control response. A case scenario could be to balance generation to meet the load; however, there is a point when generator control is no longer possible to supply loads from generation alone. Hence, the need to tap into the energy resources of the load itself. Consider a case study of a residential house with an ac power system and that has different types of load as shown in Figure 7.7. One could observe the energy storage potential in some of the loads (i.e., the UPS of a computer system) and utilize that for the overall system stability. Moreover, a load like residential lighting may be modulated/dimmed to reduce load demands while keeping the lights to a comfortable illumination level. Such control actions may

Table 7.1 Summary of the different loads used in different buildings

Residential	Office space	Retail outlets	Hospitals	University campus/schools	Public spaces
Lighting	Lighting	Lighting	Lighting	Lighting	Street lighting
HVAC	HVAC	HVAC	HVAC	HVAC	Security
Consumer electronics	Consumer electronics	Refrigeration	Medical devices	Security and safety	
Kitchen devices	Telecomm/ datacom	Telecomm/ datacom	Telecomm/ datacom	Telecomm/ datacom	

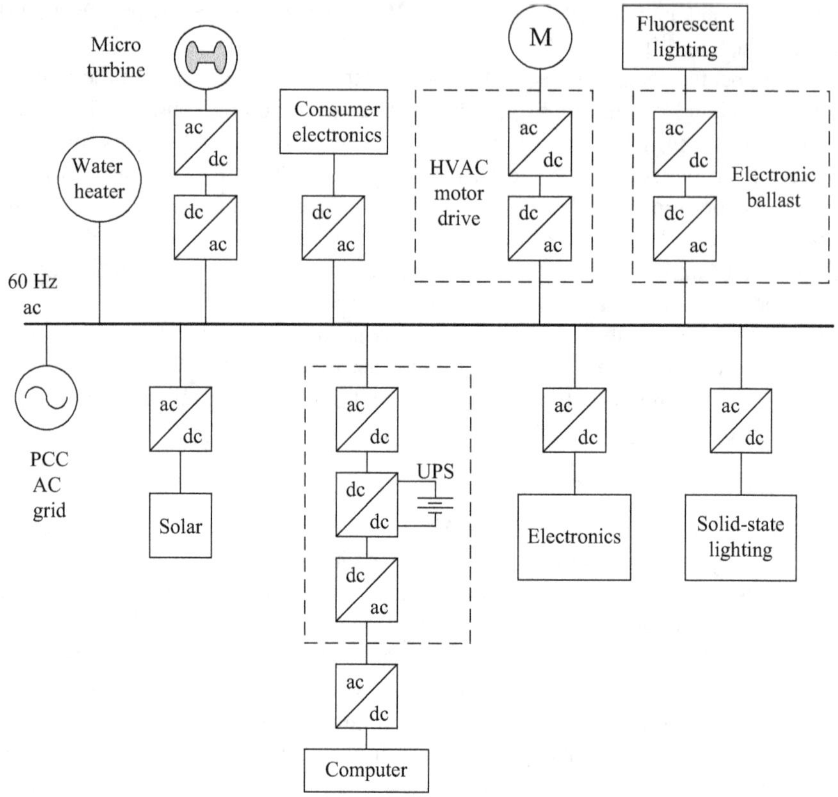

Figure 7.7 An example of a residential ac building with different loads. Utilizing different energy storage potential in the load could be used for load modulation

be achieved using a configuration like that of the single bus radial system of Figure 7.8 connected to the LAPES microgrid of Figure 7.1. The details of the control methods are explained in the following subsections.

7.4.3 Strategy for controlling the load to be an energy asset

Control strategy for each POL bus interface when the system is stressed depends on the load priority [9]. High-priority loads are required to continue operation while lower priority loads can be turned off to preserve the systems voltage stability. Also, the timeframe of the need for energy determines the type of control used: short term or long term. For short term, a power buffer decouples the dynamics of the load from the dynamics of the dc bus. For long term, modulating the load consumption balances energy in the system. A power buffer on the higher priority loads eases the burden on the bus by presenting constant input impedance instead of a CPL during a buffering event. In this mode, current is still drawn from the bus and internal energy storage satisfies the load requirements. Figure 7.9 illustrates the decision logic for a complete POL power unit.

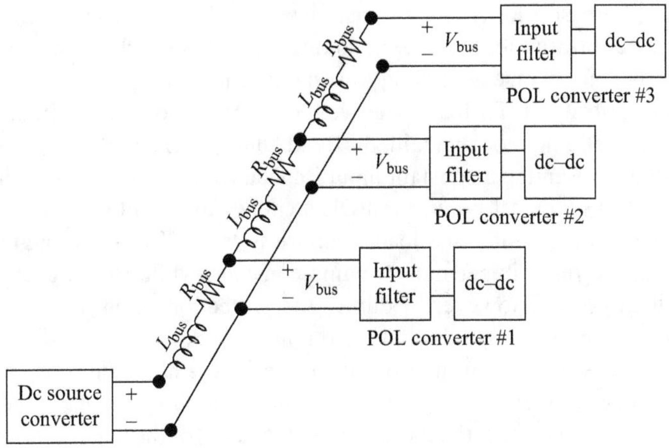

Figure 7.8 Single bus radial system with three loads [39]

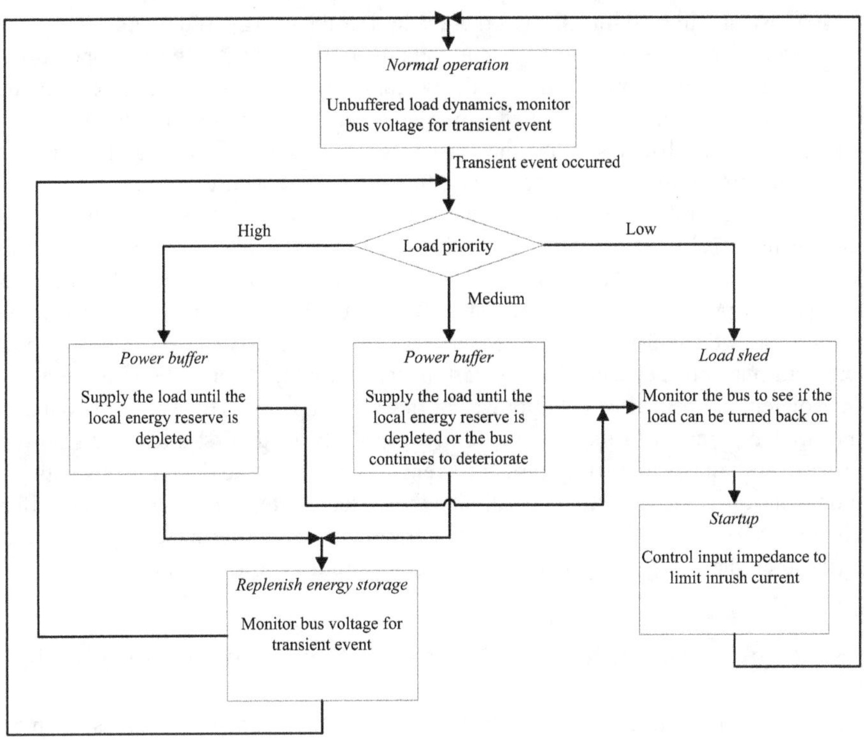

Figure 7.9 Local control strategy for each POL supervisor

During normal operation, the buffer transfers the load power demand from the bus, while maintaining its internal energy storage. This mode completely couples the load and bus dynamics. When the bus experiences a transient that triggers a protection event, load priority determines the course of action to maintain system stability. The highest priority loads remain connected to the bus until the energy is depleted. Lesser priority loads continue to monitor the bus and disconnect if the bus is sensed to become worse according to some metric.

For high-priority buffer loads, the wellbeing of the load is favored; therefore, the buffer will attempt to maintain the load until its internal energy storage is depleted, while presenting a constant input impedance to the bus. When the internal energy reaches a given set-point, the POL switches to a load-shed strategy. For a medium-priority buffer load, the loads welfare is favored less. Throughout a protection event, the input impedance remains constant, while stored energy supplements the load power. However, if during the protection event, the bus condition becomes worse, the strategy is switched to a load shed.

When the bus recovers from a transient, the high and medium priority buffers change to replenish energy storage mode in anticipation of the next event. While drawing full load power from the bus to supply the needs of the load, additional power is drawn to recharge the buffer energy storage capacitor. If during the replenish cycle the bus experiences another transient, then the buffer reenters a constant input impedance mode. Since the replenish cycle was interrupted, the sustaining time for the latest transient will be diminished as there is less stored energy. If the replenish cycle finishes uninterrupted by a bus transient, then the buffer returns to normal operation.

When the load is low priority, or the buffer energy storage has been depleted in a high or medium buffer, load shed is implemented. This entails shutting down the load in a manner that will cause the least inconvenience for startup and minimize the impact on the load. The strategy is a function of the nature of the load.

When the protection event has cleared, a load-appropriate startup strategy is implemented. To minimize the chance of triggering further system transients, the bus power to the load should be minimized during startup, so-called soft-start. When the load is high priority, power is immediately delivered to the load. When a low or medium-priority load is started, the buffer energy storage capacitor is precharged before turning on the load. During startup, load priority determines if the load is immediately connected or if the power buffer starts up first. Highest priority loads immediately connect to the bus, while lower priority loads first allow their power buffer to soft-start, eliminating inrush current and gracefully loading the bus with an initial constant impedance to help stabilize the system. After a time, the power buffer will revert to normal operation. If a bus transient occurs during charging of the buffer, then the charging cycle ends until the bus returns to its nominal state.

7.5 Droop control for stability and information communication

It is common in dc microgrid systems to have multiple generation sources, such as photovoltaics, and multiple storage elements, such as batteries, supplying one dc bus. Thus it is critical to have techniques to manage current sharing. Ideally, supply converters will identically share the current amongst them; however, mismatches in components and feedback networks as well as different impedances at different locations on the dc bus can cause imbalance in current sharing. Such imbalance, when significant, could result in overload and thermal stresses that jeopardize the system reliability [40]. The common methods of current sharing used in the literature fall into two general groups: active sharing and droop control [41].

Active current-sharing techniques involve a control structure and a method of programming individual converters with a reference current. One implementation is to use a master/slave configuration such that one dc source is designated as the master and is used to control the bus voltage. The remaining dc sources, designated as slaves, operate as current sources. This strategy produces a stiff bus voltage and controlled load dispatch at each source. There are two main limitations of this technique: high-speed communication is required and a single point failure can disable the entire system. In practice, active current sharing techniques are best suited for physically small systems, such as paralleled voltage regulator module applications. If the topology were fixed and known a priori, more sophisticated controls such as interleaving can be used to reduce ripple. In droop control [42], the output voltage of the source drops as current increases. This is a form of local control since converters autonomously share load current by sensing the local bus voltage. Droop control can be as simple as a series resistance or a more efficient closed-loop controller such as a phase-angle controller in a rectifier source converter [32]. This scheme has been proposed for use in large-scale distributed systems with dynamically changing topologies since it supports plug-and-play reconfiguration and system scaling and is robust to component failures. A summary of the differences between active current sharing and droop control techniques is as shown in Table 7.2.

7.5.1 Constant power load and its deleterious effect on dc systems

Similar to the ac grids, dc microgrids encounter stability issues that arise from the voltage regulation of PEIs which have become a necessity in the integration of

Table 7.2 Comparison between load sharing techniques: active current sharing and droop control

Active current sharing	***Droop control***
Involves a current structure and a method of programming individual converters with a reference current	The output voltage of the source drops as current increases, making it a local autonomous control
Requires communication channels	Does not require communication channels
Principle: Master/slave	Principle: Virtual resistance
One dc source acts as a master to control bus voltage, while the rest act as slaves operating as current sources	Implemented as a series resistance or a more efficient closed loop controller
Plug and play reconfiguration not possible	Plug and play reconfiguration possible
Single point of failure	Robust distributes system
Suitable for small-scale embedded systems	Suitable for large-scale distributed systems

sources, loads, and storage devices. In fact, switching dc power converters is inherently nonlinear, that interconnecting a large number of them poses stability issues to the whole microgrid. POL converters act as instantaneous CPLs, as they regulate the load voltage too tightly [43]. The tight POL voltage regulation results in negative dynamic input resistance which leads to destabilizing effects during a voltage sag. CPLs have become very common in various applications related to microgrid integration such as in data centers, telecommunication, wireless communication base stations. Hence, the need for an effective solution to the destabilizing effect during voltage sags, which are quite common due to the events of loss of generation or increase in load. Since finite dc systems are inherently weak, such CPL systems are subject to extreme voltage sags or even voltage collapse.

Many solutions are proposed in the literature [43,44], which include the addition of filters, the addition of bulk energy storage devices that are directly connected to the system main bus, load shedding, the addition of linear controllers and the addition of boundary controllers. This section reviews the benefits of droop control and load modulation to remedy the destabilizing effects that could accompany CPLs.

7.5.2 Steady state stabilization

In droop control, the output voltage drops as the current increases. Converters share load current by sensing the bus voltage and increasing current as the bus voltage drops. The model for droop-control based on sensed bus voltage, taken from, is shown in Figure 7.10 and incorporates the effects of finite bus impedance Z_{bus} and source output-capacitance C_s. Variables I_s and V_s are the current and voltage at the output terminals of the converter, respectively. The voltage at the output, the bus voltage, is low-pass filtered and used to close a feedback loop. The droop gain K converts the voltage error into a current command for the source converter. Assuming the converter current perfectly tracks the reference current, the steady-state droop relationship is

$$I_{s,ref} = K\left(V_{ref} - v_{s,sense}\right) \tag{7.2}$$

This scheme has been proposed for use in large-scale distributed systems since it does not require any communication between the dc sources. The distributed system is inherently robust because droop control automatically shares current among

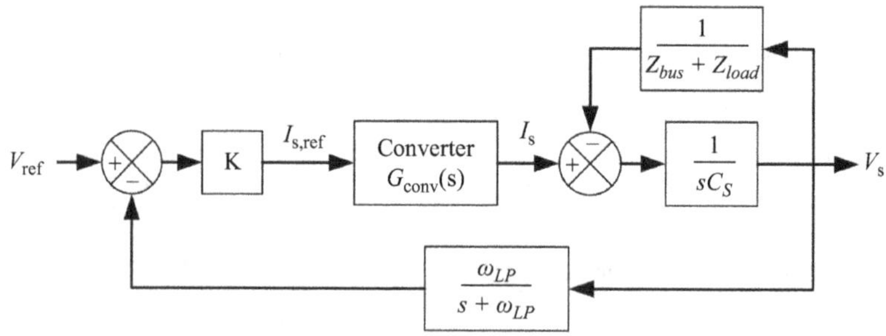

Figure 7.10 Droop-controlled voltage source

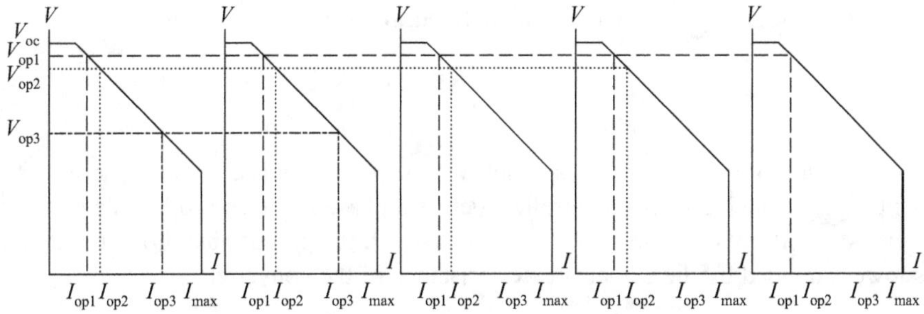

Figure 7.11 Current sharing using droop-control

Table 7.3 Table of operating points

	Number of sources		
	5	**4**	**2**
V_{op} (V)	47.75	46.54	38.97
I_{op} (A)	5.672	7.273	17.38
I_{bus} (A)	28.36	29.09	34.75
P_{load} (W)	1,354	1,354	1,354

the available converters without the need for a central controller to redispatch the source converters. If a converter turns off or fails, the remaining converters sense a decrease in bus voltage and increase their respective output current to compensate for the lost source.

Consider a system with five sources on a common dc bus supplying 1,354 W of total load. Each dc source converter has a load-line that describes the *v–i* terminal characteristics, as shown in Figure 7.11. Assuming negligible bus impedance between the five converters, the solution to the base case (where all converters are operational) results in the bus voltage V_{op1} with each converter supplying I_{op1} current. The analytical solution for the operating point is found by solving the load-flow equations for n source converters and m constant-power loads (Table 7.3):

$$V_{oc,n} - I_n \frac{1}{K_n} = V_{bus}, \forall n \qquad \sum_n I_n = \sum_m \frac{P_m}{V_{bus}} \tag{7.3}$$

It is observed that the droop gain is the slope of the *v–i* curve and modifies the actual source impedance. Thus, a simple model for a source converter under droop control is a constant voltage behind impedance:

$$V_s = V_{oc} - \frac{1}{K} i_s \tag{7.4}$$

where K is the droop gain and can be defined in terms of a resistance:

$$K = \frac{1}{R_{droop}} \tag{7.5}$$

Although droop control can be as simple as a series resistance, a more energy efficient choice is a closed-loop controller such as a phase-angle controller in a rectifier source converter. For an arbitrary source converter, the permissible droop resistance is lower bounded by the actual source resistance of the converter:

$$R_s \leq R_{droop} \tag{7.6}$$

In the previous example, the source converters are assumed to be identical with identical droop characteristics. Thus, the total load current is shared equally. In general, however, each converter can have an arbitrary droop characteristic representing its operating parameters, power limits, or preferred dispatch:

$$R_{droop} = \frac{1}{1-\lambda} R_s, \quad \text{where } 0 \leq \lambda < 1 \tag{7.7}$$

Thus, droop control programs the effective output impedance of the source-converter and has been shown to result in current sharing. It also has direct implications to system dynamic behavior.

7.5.3 *Dynamic stabilization*

The closed-loop reference-to-output transfer function for the converter model in Figure 7.10, assuming an ideal and lossless bus, is

$$G_{vg}(s) = \frac{K(s+\omega_{lp})Z_{load}}{s^2 + \left(\left(1 + C_s\omega_{lp}Z_{load}\right)/C_sZ_{load}\right)s + \left(\left(\omega_{lp}(KZ_{load}+1)\right)/C_sZ_{load}\right)} \tag{7.8}$$

After writing the characteristic polynomial in the usual way and substituting (7.5), the expression for damping is

$$\zeta = \frac{C_s\omega_{lp}Z_{load} + 1}{2\sqrt{C_s\omega_{lp}Z_{load}\left(\left(Z_{load}/R_{droop}\right)+1\right)}} \tag{7.9}$$

This analysis suggests that droop control can be used to dynamically stabilize a system [45]. Consider the three-bus system from Figure 7.8. The system is simulated in DYMOLA using linearized averaged-model buck converters with damped input filters. The source is modeled as an ideal voltage. The system is stable, albeit with very light damping, $\zeta = 107 \cdot 10^{-6}$, as shown by a transient response in Figure 7.12. The system eventually reaches steady state with each load-converter supplying 144 W of load (about 3 A of current from the bus).

The system has reached steady state prior to 5 s when the load on bus 1 experiences a step-change in output power to 576 W. Figures 7.13 and 7.14 show that the increase in current at bus 1 depresses the voltages at each bus in the system

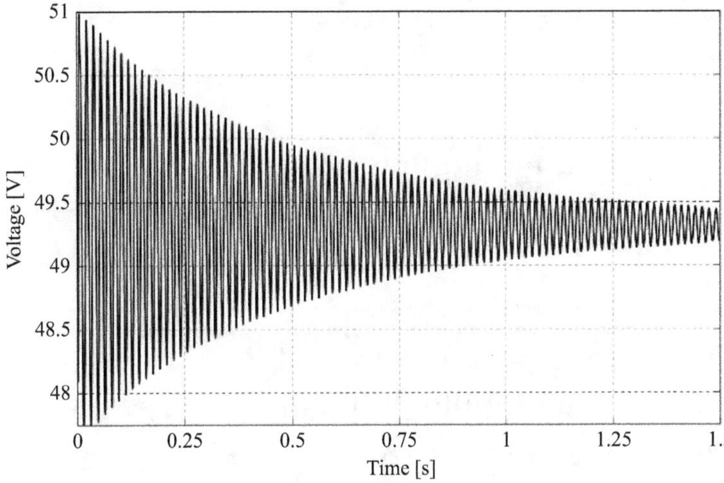

Figure 7.12 Initial system transient response for a lightly damped system

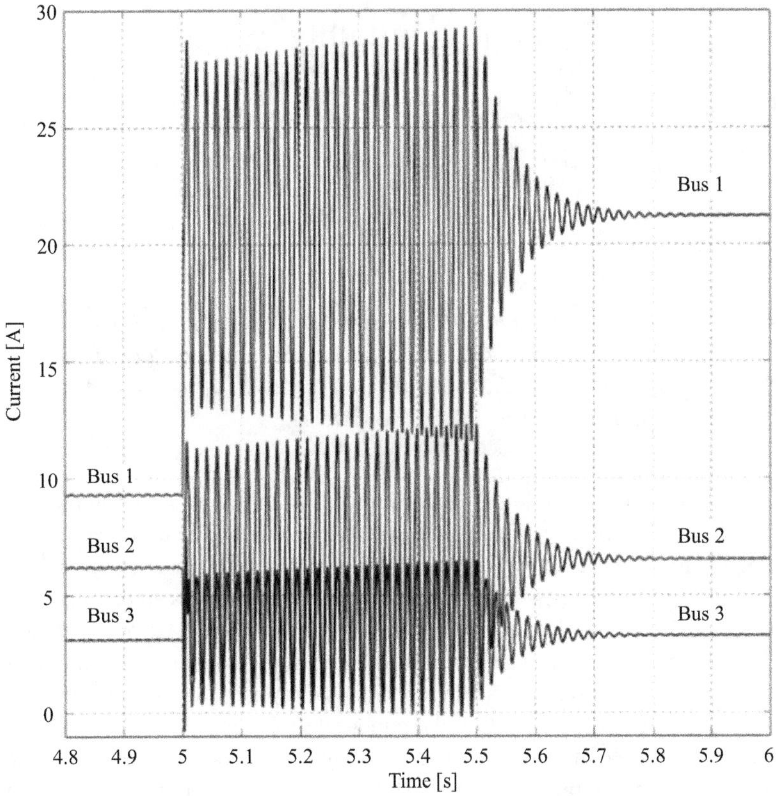

Figure 7.13 Bus current of each POL converter. At 5 s, the load increases at bus 1. The previously stable system becomes unstable with growing oscillations. At 5.5 s, the droop resistance at the source converter is adjusted to restabilize the system

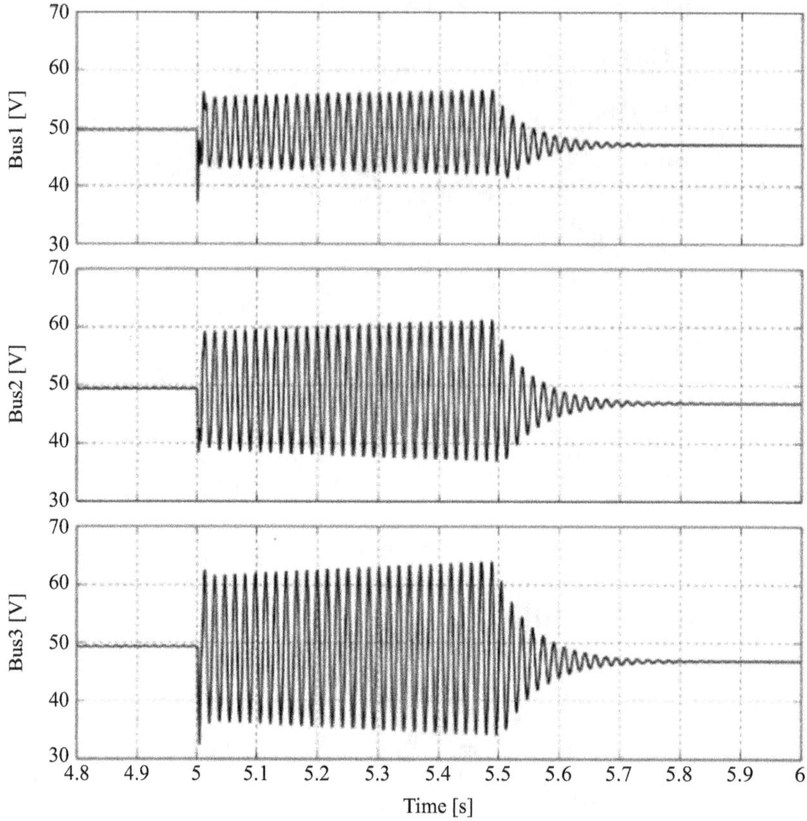

Figure 7.14 Bus voltage at each POL converter. At 5 s, the load increases at bus 1. The previously stable system becomes unstable with growing oscillations. At 5.5 s, the droop resistance at the source converter is adjusted to restabilize the system

and that the previously stable system becomes unstable with growing oscillations. At 5.5 s, droop control is enabled on the source converter. The droop control results in an equivalent resistance of 0.10 Ω. The droop control increases the effective system damping to $\zeta = 5.64 \cdot 10^{-3}$ and the unstable system is stabilized.

A potential drawback to droop control is that there is a stationary error introduced in the bus voltage as seen in Figure 7.14. In the literature, a second control loop with low-pass filtering and PI control is proposed to increase the internal open-circuit source voltage VOC and compensate for this stationary error. Since in this example, droop control is only used to stabilize the system, not share load, the stationary error can be minimized by adaptively programming the minimum droop necessary to achieve the desired damping.

The drop in bus voltage, however, carries useful information about the health of the system. The drop can be due to a partial loss of generation and indicate that the system may be energy constrained, as shown in Figure 7.11, or the drop can be a result of needing increased damping which would indicate that the system is nearing a region of instability as shown in Figure 7.14. Load-side controls that sense this bus voltage drop will not know why the drop occurred but can take mitigating action such as switching supply buses, activating a power buffer, or invoking load interruption to alleviate system stress.

7.6 Voltage-based load interruption

Autonomous local control performs dynamic load interruption based on the sensed bus voltage at the POL converter and the priority of that load. Each of the loads in the radial system shown in Figure 7.8 is assigned a priority based on the explanation of Section 7.4.3. Table 7.4 gives an example of the priority assignment that will be used in this section. By selecting the highest priority load to be the one furthest from the source, the example will demonstrate that priority need not be constrained by the topology.

The control strategy in (7.10) is to turn off a particular load when the sensed bus voltage drops below a lower threshold and turn the converter back on when the sensed voltage exceeds an upper threshold. The control is complicated when the on and off control action of the converter causes its input voltage to drop below lower threshold during turn-on and to rise above the upper threshold during turn-off. Hysteresis is useful to provide a dead band to prevent this type of chattering. In a larger system, however, the possibility arises of not having a large enough hysteresis band. In this situation, additional information and controls are needed to prevent the chattering (Figure 7.15).

$$\begin{aligned} &\text{Turn on if}: V_{input} > V_{\text{upper limit}} \\ &\text{Turn off if}: V_{input} < V_{\text{upper limit}} \end{aligned} \tag{7.10}$$

One method to obtain these set points is to perform an exhaustive contingency load flow analysis. Each contingency represents a particular combination of operating points for the system POL converters. For a small system, this offline process is straightforward and computationally fast. The resulting set of power flow solutions is processed by a search algorithm to find the upper and lower voltage limits for each load converter.

Table 7.4 Load priority assignment in the three-bus test-bed

Converter	Priority
POL 1	Semiessential (SE)
POL 2	Nonessential (NE)
POL 3	Essential (E)

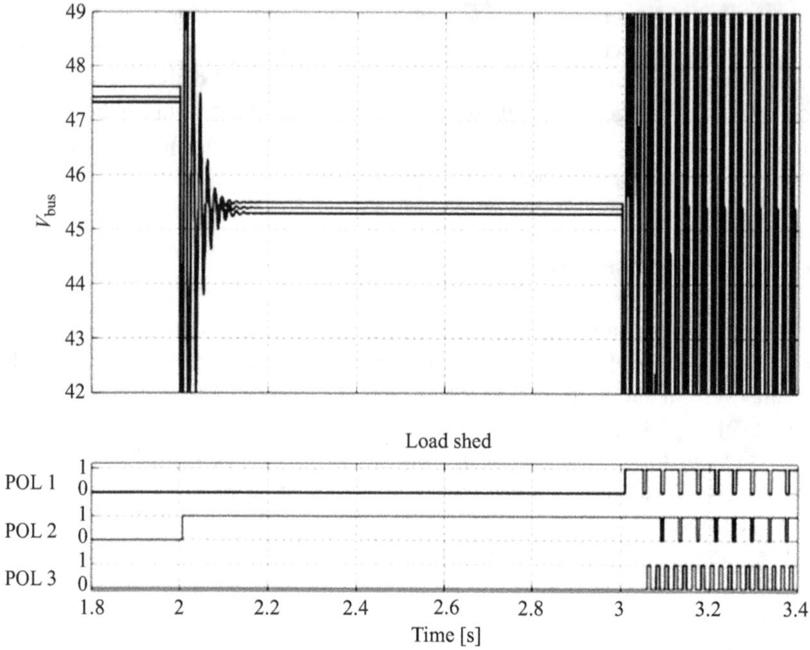

Figure 7.15 Control introduces chattering into the system

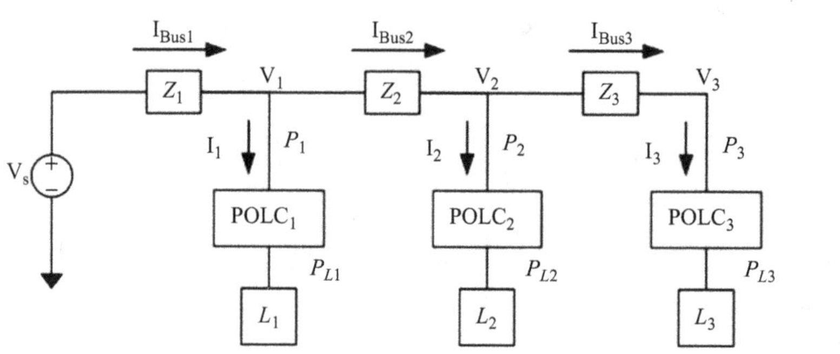

Figure 7.16 Model of a radial dc power system with a single source and three point-of-load converters

7.6.1 Power flow analysis

Power flow analysis is used to determine a priori the values of the voltage set points for under-voltage load interruption and restoration. The MATLAB® Power System Toolbox contains a suitable power flow algorithm for the test system, redrawn in Figure 7.16. The radial design results in rapid convergence of the power flow algorithm. The POL converter is a closed-loop buck converter designed to have

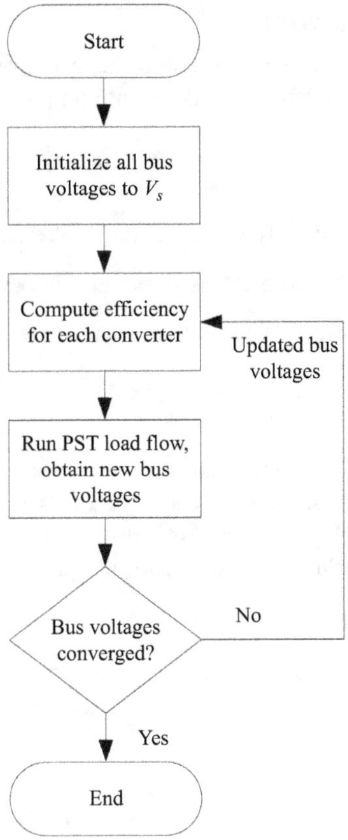

Figure 7.17 Outer loop of power flow algorithm incorporating converter efficiency

good line and load regulation to ensure constant output power given a fixed resistive load. Input power to the converter depends on the efficiency of the converter, which is a function of the input voltage and the output power. Thus, the power flow equations are complicated by the nonlinear efficiency of the dc-to-dc converters.

To improve the power flow results, the efficiency of the converters was included in the power flow algorithm, shown in Figure 7.17. This added an outer iteration loop to the power flow algorithm although it could be incorporated directly into a power flow routine to increase computation efficiency. The power flow algorithm begins by assuming that there is no voltage drop in the system and that the voltage at each node on the bus is identical to the open-circuit source voltage. These node voltages are then used to calculate the initial efficiency for each dc–dc converter and hence the initial power withdrawn from each node on the bus. The PST power flow program is then run. The result of the power flow, the new system node voltages, is used to update the efficiency of each converter, and the process is repeated until node voltages converge to within an acceptable bound.

7.6.2 Contingency analysis

Contingency analysis is performed to obtain the system voltages for each combination of POL converter operating points. Notation that links the power flow result to the configuration of the converter is

$$V(converter)^{\text{state}}_{\text{limit}} \tag{7.11}$$

The parameter *converter* uniquely identifies the converter and takes values

$$converter \in \{1, 2, \ldots, N\}, \text{where } N \text{ is the total number of converters} \tag{7.12}$$

The parameter *state* identifies the current operating state of the converter and takes values

$$state \in \{\text{off, on, base, high}\} \tag{7.13}$$

The states base and high describe the output power of the converter as either nominal rated base power or overload power. The state on is the compliment of off and includes both the base and high states. This is useful when the actual output power is unimportant. Thus, the cardinality of the set of unique output-power levels is

$$|state| = 3 \tag{7.14}$$

The number of power flows that are needed to perform an exhaustive contingency analysis for all possible configurations for N POL converters is

$$N^{|state|} \tag{7.15}$$

The three-converter example in Figure 7.16 will therefore require 27 power flow computations. The parameter limit identifies the type of voltage *limit.* It takes the value

$$\text{limit} \in \{\text{min, max}, UVP\} \tag{7.16}$$

The minimum limit is the lowest allowed input voltage before the converter turns off, maximum is the highest input voltage before a converter turns on, and *UVP* refers to the hardware UVP limit designed to self-protect.

The results of the exhaustive contingency analysis for the system in Figure 7.16 are graphically presented in Figure 7.18. The results of the 27 power flows are shown for each of the three POL converters. Horizontal lines indicate the open circuit supply voltage, VS, and the UVP limit, VUVP, where the converter turns off to self-protect. Since POL3 is furthest from the source, it has the lowest bus voltages for each contingency. Solid bars indicate contingencies of interest where one or more converters in that contingency would trip off-line due to UVP.

7.6.3 Search algorithm

The upper-limit and lower-limit voltages for each POL converter can be found from the maximum and minimum values for sets of specific contingencies. This optimization is done by a search algorithm that parses the results of the contingency

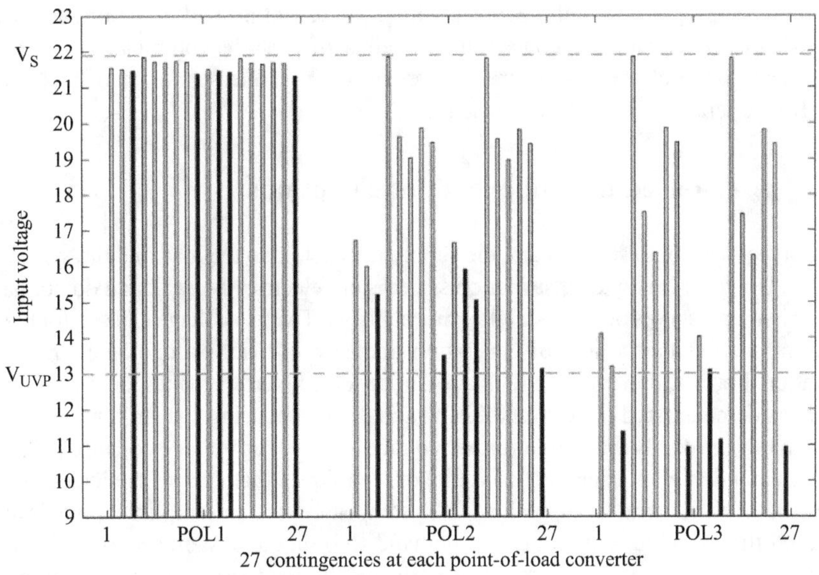

Figure 7.18 Results of exhaustive contingency analysis on the radial test system with a single source and three point-of-load converters. The load on each converter can be either off, base load, or overload. Solid bars indicate contingencies where a bus voltage in the system is below the UVP for that converter

analysis. The process starts with the lowest priority converter, POL2. The lower bound on the lower-limit is

$$V(NE)^{on}_{\min} > \max\left\{\begin{array}{c} V(NE)_{UVP}, \\ V(NE)^{on}; V(E) \leq V(E)_{UVP} \\ V(NE)^{on}; V(SE) \leq V(SE)_{UVP} \end{array}\right\} \tag{7.17}$$

The algorithm finds the voltage POL2 for all contingencies where the higher priority loads are on and their input voltage is below the *UVP* limit. In this example, this occurs for five contingencies at POL3, shown as the dark bars in Figure 7.18. The algorithm also considers the possibility that the voltage at POL2 can fall below the *UVP* limit.

The upper-limit has both an upper and lower bound. Only, the essential and nonessential loads are considered here for simplicity.

$$V(E)^{high}; V(NE)^{off} < V(NE)^{off}_{\max} < \min\left\{\begin{array}{c} V(E)^{off}, \\ V(E)^{base}; V(NE)^{off} \end{array}\right\} \tag{7.18}$$

The lower bound for restoring the nonessential load occurs when the essential load is at the maximum power and the nonessential load is *off*. The upper bound is found by minimizing a set of contingencies where the essential load is off or supplying base load power and the nonessential load is off. These combinations of contingencies were chosen to ensure that restoring the nonessential load never jeopardizes operation of the essential load.

7.7 *dv/dt*-Based dynamic load interruptions

In a power system in steady state, the supply is matched to the load and the system is stable. However, many dc distribution systems are electrically weak and do not have the spinning reserves or other stability mechanisms. The bus voltage can sag for many reasons such as partial loss of generation, increase in load, or topological reconfiguration. Further, tight voltage regulation in dc–dc converters makes them operate as CPL, as demonstrated in Section 7.5.1, which draw increasing current for decreasing bus voltage, possibly leading to further voltage sag or even voltage collapse [43].

Demand-side management is a suite of techniques that control the loads so that they become integral components in system stability. Interruptible load is one method that provides curtailment of demand to promote system security. Autonomous local control is investigated to perform this load-side control and improve system reliability.

The *P–V* curve is a useful tool to visualize the operation of a power system. Figure 7.19 illustrates a family of the familiar *P–V* system curve. Maximum power transmission occurs at the nose were the source impedance and load impedance are equal. In a dc system, the bus voltage drops as the load increases due to voltage-divider action of the source impedance and the load impedance.

A system is initially in steady state with voltage $V(t_1)$ delivering total load power of $P(t_1)$. The system impedance suddenly increases, perhaps due to a partial loss of generation or topological reconfiguration, and the operating point moves to a new *P–V* curve at time t_2. However, the voltage $V(t_2)$ is below the under-voltage

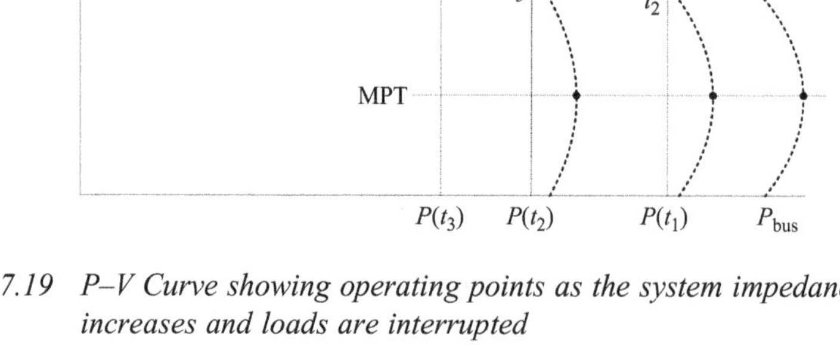

Figure 7.19 P–V Curve showing operating points as the system impedance increases and loads are interrupted

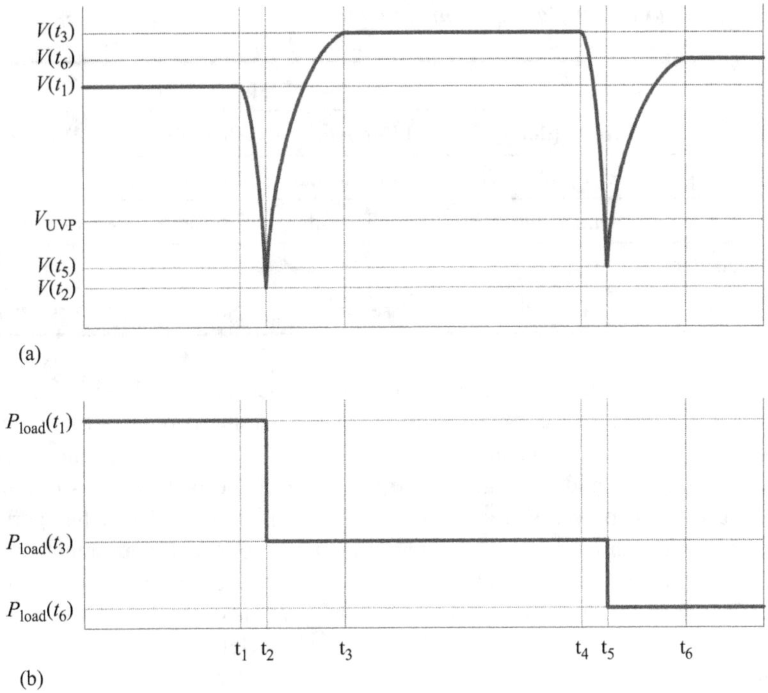

Figure 7.20 Ideal bus voltage and load power as system impedance increases and loads are interrupted to prevent voltage collapse. (a) Bus voltage decreases in response to increased system impedance at t_1 to reach the operating point on the new P–V curve at t_2. The new bus voltage is below the UVP limit, so control action causes load to be shed, moving to a mew operating point on the same P–V curve at t_3 with a higher bus voltage. The cycle repeats at t_4. (b) Load power in the system changes as point-of-load converters are turned-off to reduce total system load when the bus voltage drops below the UVP

limit and load is shed, moving to a new operating point on the same *P–V* curve at t_3. The time-domain waveforms in Figure 7.20 reveal that these changes in operating points do not occur instantaneously. The trajectories on the two figures, however, are idealized to improve clarity of the system response and do not include the dynamics associated with the inductance of the bus, the input filter, and the constant-power dc–dc converters.

7.8 Load prioritization and scheduling

In an energy-constrained system, load prioritization is critical for system stability and control because it provides a structured approach to the decision and control process. It is likely that a particular load's priority may need to change depending

Table 7.5 Load priority-assignment example

Status \ Load type	Hotel (domestic)	Lifeboats	Propulsion	Weaponry
In port	Essential	Nonessential	Nonessential	Nonessential
Patrol	Semiessential	Semiessential	Essential	Semiessential
General quarters	Nonessential	Nonessential	Essential	Essential
Abandon ship	Nonessential	Essential	Nonessential	Essential

on the operation of the system. In a naval ship, each load is classified as either nonessential, semiessential, or essential. A general framework for organizing the load priorities is a two-dimensional matrix as shown in Table 7.5. In one example, propulsion is considered the highest priority, while hotel loads such as lighting in crew quarters and power in the galley are less important and can be sacrificed depending on the threat level. In another example, the launch equipment for lifeboats has a higher priority under "Patrol" status for safety reasons but yields priority to other systems such as weapons and propulsion under "General Quarters" status. If local energy is available in a power buffer, load priority is useful to determine if the load simply turns off or the power buffer operates when trouble is sensed on the dc bus.

Fine-tuning the performance of the system requires that each POL converter has some information about the entire system. Low-bandwidth signaling from the command and control center can broadcast the current state of the system, but each POL converter ultimately decides how to use the information—unlike in a centralized control scheme where each load is directly controlled. The distributed control strategy is inherently fault-tolerant because each controller acts independently. If the low-bandwidth communication is compromised, each controller can continue to operate using the last-known state or revert to fail-safe operation as determined by the load-priority table.

7.9 Summary

In a traditional power grid system, the operator had total control of generation and distribution assets while the load was viewed as a disturbance. Thus, planning and operation necessitated always being prepared for unforeseen changes in the load consumption. The result is that the US power grid is amazingly resilient, robust, and expensive. As we consider a new paradigm of dc microgrid systems, overcapacity may not be feasible for technological and/or economic reason. Yet, high power quality and availability is more important than ever particularly to support the digital economy and information age. This is compounded as renewable sources become increasingly utilized, and the system operator no longer has total and arbitrary control of the generation.

In this chapter, we introduced a framework for load control in a LAPES. In this paradigm, the load is considered to be an energy asset which can be controlled not just by the end user but also by the system. As such, a new degree of freedom is introduced in the control problem of balancing electrical supply and demand. Key to implementing any of these concepts is to strike the balance between the opportunity and cost, which would provide the techno-economic trade-off needed to implement these concepts practically. Important application-specific design considerations are the desired energy availability and the price willing to be paid for that availability as well as the tolerable amount of control relinquished. Optimizing this is not straightforward and may lead to solutions that change with time and conditions. To anticipate this, the chapter discussed the concept of load prioritization and a method to change the allowable control of the load.

To understand the benefits of load control, we introduced the use of a multi-objective optimization problem, which is concerned with energy availability and reliability. The Pareto Frontier technique was used as a way to visualize the engineering trade-off and assist with the optimization. Within this chapter, we were able to see the application and feasibility of load modulation for different time-scale transients. Modulation was explored as an alternative to load shed for situations that could tolerate a reduction of power, such as building lighting and HVAC systems, and PHEV charging. Within a LAPES, POL converters can be controlled, hence modulate the load, to achieve the objective of energy availability. These techniques range from the use of dc/dc level power buffers, load scheduling, replenishing energy storage, and a strategy for prioritized load shedding. Using droop control, the system can realize distribute control to eliminate a single-point of failure of a central control architecture. A highly desirable benefit is that droop control can also improve system stability by introducing extra damping which is important as a power system nears a boundary of stability. Other techniques like voltage-based interruptions and load prioritization were discussed in this chapter.

Acknowledgments

The authors wish to acknowledge the various funding sources which supported the research behind this content of this chapter. Dr. Balog acknowledges support received from the Qatar Foundation. This publication was made possible by NPRP grant no. 9-204-2-103 from the Qatar National Research Fund (a member of Qatar Foundation). The statements made herein are solely the responsibility of the authors. This work was also supported in part by the Grainger Center for Electric Machinery and Electromechanics at the University of Illinois. Dr. Mohamad Shadmand acknowledges support from the Qatar Foundation. This publication was made possible by NPRP-EP grant no. X-033-2-007 from the Qatar National Research Fund (a member of Qatar Foundation). The statements made herein are solely the responsibility of the authors. Mr. Morcos Metry acknowledges the sponsorship of the Qatar Research Leadership Program (QRLP) under the Qatar National Research Fund.

References

[1] N. A. Armstrong, "The Engineered Century: A Century Hence, 2000 May Be Viewed as Quite a Primitive Period in Human History. It's Something to Hope For," *The Bridge*, vol. 30, no. 1, pp. 14–18, Spring 2000.

[2] C. W. Gellings, "New Products and Services for the Electric Power Industry," *The Bridge*, vol. 40, no. 1, pp. 21–28, Spring 2010.

[3] S. M. Amin, "Securing the Electricity Grid," *The Bridge*, vol. 40, no. 1, pp. 13–20, Spring 2010.

[4] J. D. Rhodes and The Conversation. (May 2017). *The Outdated US Electric Grid is Going to Cost $5 Trillion to Replace*. Available: http://www.businessinsider.com/replacing-us-electrical-grid-cost-2017-3.

[5] A. Kwasinski, W. Weaver, and R. S. Balog, *Microgrids and Other Local Area Power and Energy Systems*. Cambridge University Press, Cambridge, UK, 2016.

[6] R. H. Lassetter, "Smart Distribution: Coupled Microgrids," *Proceedings of the IEEE*, vol. 99, no. 6, pp. 1074–1082, Jun 2011.

[7] Y. Zhang, L. Xie, and Q. Ding, "Interactive Control of Coupled Microgrids for Guaranteed System-Wide Small Signal Stability," *IEEE Transactions on Smart Grid*, vol. 7, no. 2, pp. 1088–1096, Nov 2015.

[8] D. T. Ton and M. A. Smith, "The U.S. Department of Energy's Microgrid Initiative," *The Electricity Journal*, vol. 25, no. 8, pp. 84–94, Oct 2012.

[9] R. S. Balog, W. W. Weaver, and P. T. Krein, "The Load as an Energy Asset in a Distributed DC SmartGrid Architecture," *IEEE Transactions on Smart Grid*, vol. 3, no. 1, pp. 253–260, March 2012, Art. no. TSG-00146-2011.

[10] Commonwealth Edison Co. (May 2017). *Central AC Cycling*. Available: https://www.comed.com/WaysToSave/ForYourHome/Pages/CentralACCycling.aspx.

[11] J. O. Swanson and J. P. Jolliffe, "Load Shedding Program in the Pacific Northwest," *Transactions of the American Institute of Electrical Engineers. Part III: Power Apparatus and Systems*, vol. 73, no. 2, pp. 1655–1668, Jan 1954.

[12] R. M. Maliszewski, R. D. Dunlop, and G. L. Wilson, "Frequency Actuated Load Shedding and Restoration Part I—Philosophy," *IEEE Transactions on Power Apparatus and Systems*, vol. PAS-90, no. 4, pp. 1452–1459, Jul 1971.

[13] Z. Wang, B. Chen, J. Wang, and M. M. Begovic, "Stochastic DG Placement for Conservation Voltage Reduction Based on Multiple Replications Procedure," *IEEE Transactions on Power Delivery*, vol. 30, no. 3, pp. 1039–1047, Jun 2015.

[14] H. Chen, Y. Li, R. H. Y. Louie, and B. Vucetic, "Autonomous Demand Side Management Based on Energy Consumption Scheduling and Instantaneous Load Billing: An Aggregative Game Approach," *IEEE Transactions on Smart Grid*, vol. 5, no. 4, pp. 1744–1754, Jun 2014.

[15] H. Zhong, L. Xie, and Q. Xia, "Coupon Incentive-Based Demand Response: Theory and Case Study," *IEEE Transactions on Power Systems*, vol. 28, no. 2, pp. 1266–1276, May 2013.

[16] Younicos. (May 2017). *Y.Cube: Our Plug-and-Play Energy Storage Solution* [Online]. Available: https://www.younicos.com/products/y-cube/.

[17] Tesla. (May 2017). *PowerPack: Utility and Business Energy Storage*. Available: https://www.tesla.com/powerpack.

[18] Tesla. (May 2017). *Powerwall: Reliable Power Day and Night*. Available: https://www.tesla.com/powerwall.

[19] L. Poulin, *Reducing Risk with Software Process Improvement*. Boca Raton, FL: Auerbach Publications, Taylor & Francis Group, 2005.

[20] A. V. Lotov and K. Miettinen, "Visualizing the Pareto Frontier," in *Multiobjective Optimization: Interactive and Evolutionary Approaches*, J. Branke, K. Deb, K. Mirettinen, and R. Slowinski, Eds.: Springer-Verlag Berlin Heidelberg, 2008, pp. 213–243.

[21] N. Barr, *Economics of the Welfare State*, 5 ed. United Kingdom: Oxford University Press, 2012.

[22] M. B. Shadmand and R. S. Balog, "Multi-Objective Optimization and Design of Photovoltaic-Wind Hybrid System for Community Smart dc Microgrid," *IEEE Transactions on Smart Grid*, vol. 5, no. 5, pp. 2635–2643, Sep 2014.

[23] J. W. Kimball, B. T. Kuhn, and R. S. Balog, "A System Design Approach for Unattended Solar Energy Harvesting Supply," *IEEE Transactions on Power Electronics*, vol. 24, no. 4, pp. 952–962, Apr 2009.

[24] ERCOT. (Sep 2016, May 2017). *Renewable Integration at ERCOT*. Available: http://www.cigre.cl/seminarios/wp-content/uploads/2016/09/ERCOT-DAN-WOODFIN.pdf.

[25] Lighting Research Center (LRC). (April 2010, May 2017). *Demonstration and Evaluation of Lighting Technologies and Applications: Daylight-Harvesting Switch*. Available: http://www.lrc.rpi.edu/programs/DELTA/pdf/FTDELTA_DaylightHarvestingSwitch.pdf.

[26] The National Optical Astronomy Observatory. (May 2017). *Recommended Light Levels (Illuminance) for Outdoor and Indoor Venues*. Available: https://www.noao.edu/education/QLTkit/ACTIVITY_Documents/Safety/LightLevels_outdoor+indoor.pdf.

[27] Occupational Safety and Health Branch. (Dec 2008, May 2017). *Lighting Assessment in the Workplace*. Available: http://www.labour.gov.hk/eng/public/oh/Lighting.pdf.

[28] R. J. Sledz, "Control Room Lighting: An Application of Human Factors Engineering," *IEEE Power Engineering Review*, vol. PAS-101, no. 8, pp. 2755–2761, Aug 1982.

[29] J. Voelcker. (Oct 2007, May 2017). *Can Plug-In Hybrid Electric Vehicles Keep the Electric Grid Stable?* Available: http://spectrum.ieee.org/transportation/advanced-cars/can-plugin-hybrid-electric-vehicles-keep-the-electric-grid-stable.

[30] A. Vaughan. (Mar 2017, May 2017). *Charge Electric Cars Smartly to Take Pressure off National Grid—Minister*. Available: https://www.theguardian.

com/environment/2017/mar/20/electric-cars-uk-power-grids-charging-peaks-sse-demand-side-response.

[31] S. Chakraborty, W. Kramer, B. Kroposki, *et al.*, "Interim Test Procedures for Evaluating Electrical Performance and Grid Integration of Vehicle-to-Grid Applications," National Renewable Energy Lab (NREL), Golden, Colorado. Jun 2011, Available: http://www.nrel.gov/docs/fy11osti/51001.pdf.

[32] I. U. Nutkani, P. C. Loh, P. Wang, and F. Blaabjerg, "Cost-Prioritized Droop Schemes for Autonomous AC Microgrids," *IEEE Transactions on Power Electronics*, vol. 30, no. 2, pp. 1109–1119, Feb 2015.

[33] L.-L. Fan, V. Nasirian, H. Modares, F. L. Lewis, Y.-D. Song, and A. Davoudi, "Game-Theoretic Control of Active Loads in DC Microgrids," *IEEE Transactions on Energy Conversion*, vol. 31, no. 3, pp. 882–895, Sep 2016.

[34] J. M. Guerrero, P. C. Loh, T. L. Lee, and M. Chandorkar, "Advanced Control Architectures for Intelligent Microgrids Part II: Power Quality, Energy Storage, and ac/dc Microgrids," *IEEE Transactions on Industrial Electronics*, vol. 60, no. 4, pp. 1263–1270, Apr 2013.

[35] J. M. Guerrero, M. Chandorkar, T.-L. Lee, and P. C. Loh, "Advanced Control Architectures for Intelligent Microgrids—Part I: Decentralized and Hierarchical Control," *IEEE Transactions on Industrial Electronics*, vol. 60, no. 4, pp. 1254–1262, Apr 2013.

[36] V. Nasirian, A. Davoudi, F. L. Lewis, and J. M. Guerrero, "Distributed Adaptive Droop Control for DC Distribution Systems," *IEEE Transactions on Energy Conversion*, vol. 29, no. 4, pp. 944–956, Dec 2014.

[37] M. D. Cook, G. G. Parker, R. D. Robinett, and W. W. Weaver, "Decentralized Mode-Adaptive Guidance and Control for DC Microgrid," *IEEE Transactions on Power Delivery*, vol. 32, no. 1, pp. 263–271, Feb 2017.

[38] J. He, Y. W. Li, J. M. Guerrero, F. Blaabjerg, and J. C. Vasquez, "An Islanding Microgrid Power Sharing Approach Using Enhanced Virtual Impedance Control Scheme," *IEEE Transactions on Power Electronics*, vol. 28, no. 11, pp. 5272–5282, Nov 2013.

[39] R. S. Balog, "Autonomous Local Control in Distributed DC Power Systems," PhD Dissertation, Department of Electrical and Computer Engineering, University of Illinois at Urbana-Champaign, 2006.

[40] F. Gao, S. Bozhko, G. Asher, P. Wheeler, and C. Patel, "An Improved Voltage Compensation Approach in a Droop-Controlled DC Power System for the More Electric Aircraft," *IEEE Transactions on Power Electronics*, vol. 31, no. 10, pp. 7369–7383, Oct 2016.

[41] M. B. Shadmand, R. S. Balog, and H. Abu-Rub, "Model Predictive Control of PV Sources in a Smart dc Distribution System: Maximum Power Point Tracking and Droop Control," *IEEE Transactions on Energy Conversion*, vol. 29, no. 4, pp. 913–921, Dec 2014.

[42] X. Lu, J. M. Guerrero, K. Sun, and J. C. Vasquez, "An Improved Droop Control Method for dc Microgrids Based on Low Bandwidth Communication with dc Bus Voltage Restoration and Enhanced Current Sharing

Accuracy," *IEEE Transactions on Power Electronics*, vol. 29, no. 4, pp. 1800–1812, Apr 2014.

[43] W. W. Weaver and P. T. Krein, "Optimal Geometric Control of Power Buffers," *IEEE Transactions on Power Electronics*, vol. 24, no. 5, pp. 1248–1258, May 2009.

[44] A. Kwasinski and C. N. Onwuchekwa, "Dynamic Behavior and Stabilization of DC Microgrids With Instantaneous Constant-Power Loads," *IEEE Transactions on Power Electronics*, vol. 26, no. 3, pp. 822–834, Mar 2011.

[45] N. Bottrell, M. Prodanovic, and T. C. Green, "Dynamic Stability of a Microgrid With an Active Load," *IEEE Transactions on Power Electronics*, vol. 28, no. 11, pp. 5107–5119, Nov 2013.

Chapter 8
Electric vehicle charging infrastructure and dc microgrids

Srdjan Srdic[1] and Srdjan Lukic[1]

8.1 Overview of EV and EVSE markets and trends

With substantial growth in sales of electric vehicles (EVs) globally, there is a push for expansion of the recharging infrastructure to service these vehicles. Over 2 million of electric cars (battery-electric and plug-in hybrid electric), 200 million electric motorcycles and 345 thousand buses (primarily in China) were deployed worldwide by the end of 2016 [1], and over 1.2 million of electric cars were sold globally in 2017 alone [2]. However, the global electric car stock made only a 0.2% of the total number of passenger cars globally in 2017. Assessments of country targets, original equipment manufacturer announcements and deployment scenarios for electric cars indicate that the number of EVs will range between 9 and 20 million by 2020 and between 40 and 70 million by 2025 [1]. Furthermore, a number of countries have decided to end the sales of fossil-fuel-powered cars in the near future (Norway by 2025, India and Netherlands by 2030, Scotland by 2032, France and rest of the United Kingdom by 2040), further accelerating the shift to electric transportation. The electric vehicle supply equipment (EVSE) is closely following the EV stock growth, with 2.3 million EVSE outlets (including 110,000 publicly available fast-charging outlets) available globally in 2016, and predicted six-fold increase in the available outlets by 2025 [1]. The fastest growing EVSE market is the Chinese market, with over 88,000 publicly available fast-charging outlets in 2016.

With the advancement and maturity of batteries used in EVs, the storage capacity and, therefore, the range of EVs are increasing, while the vehicle cost is dropping. Major vehicle manufacturers are already producing EVs with range exceeding 200 mi (320 km) at prices competitive with conventional vehicles. With these trends in place, the power of EV chargers needs to increase in order to reduce the recharging time and provide the convenience inherent in refueling conventional vehicles at the gas stations. According to the US Environmental Protection Agency, the fuel economy of electric cars produced from 2016 to 2018 ranges from 72 to

[1]Department of Electrical and Computer Engineering, North Carolina State University, USA

136 mi/gal equivalent [3]. Expressed in terms of energy consumption, modern electric cars consume between 25 and 47 kWh of energy (stored in the battery) per 100 mi (15.6–29.4 kWh per 100 km). Therefore, a charger with a power between 300 and 564 kW would be needed to add 200 mi of range to a modern electric car in 10 min. However, number of barriers related to battery technology cost, vehicle electrical architecture and power electronics subsystems, and infrastructure issues related to high power demand, would still need to be overcome before the ultrafast charging (350 kW or higher) can become a commercially viable EV recharging option [4]. To support the need for shorter recharge times, dc fast chargers rated 150 kW have been recently deployed across the world and the new high power units (350 kW or higher) have been announced for 2018 by major EV and EVSE makers.

8.2 dc Fast charging systems and requirements

8.2.1 dc Fast charging systems and standards

To ensure charging compatibility, a number of governing bodies, including the Society of Automotive Engineers (SAE) and International Electrotechnical Commission (IEC), have developed standardized protocols and couplers that interface a plug-in EV with the EVSE. The five standard dc fast charging systems that are in use are listed in Table 8.1. The IEC 62196-3 Standard defines four different vehicle coupler configurations for dc fast charging: Configuration AA [proposed and implemented by CHAdeMO (CHArge de MOve) Association], Configuration BB (Also known as GB/T and available only in China), Configuration EE [Type 1 Combined Charging System (CCS), adopted in North America] and Configuration FF (Type 2 CCS, adopted in Europe and Australia). There is also a proprietary system developed by Tesla Inc. and used exclusively for Tesla vehicles.

With over 18,000 charging outlets deployed worldwide throughout the end of first quarter of 2018, CHAdeMO interface has the largest global coverage of all fast charging systems [5]. The system was proposed by several Japanese companies who formally established CHAdeMO Association in March 2010. The current version (v1.2) of the CHAdeMO charging protocol, published in March 2017, supports dc fast charging with power levels up to 200 kW (500 V/400 A), and the 400-kW (1,000 V/400 A) version of the protocol is announced for 2018 [6]. The new CHAdeMO high-power charging system was demonstrated in March 2017, with 150 kW units charging modified Nissan Leaf at 100 kW peak power.

The GB/T is a fast charging standard used predominantly in China, and it originates from a Chinese national standard (GuoBiao) GB/T 20234.3 [7], issued by the Standardization Administration of the People's Republic of China. Since 2017, this standard is mandatory for all new EVs and fast charging infrastructure in China [8]. At the beginning of 2018, the dc fast chargers for passenger cars with GB/T coupler were available at power levels up to 120 kW in single-port configuration and 160 kW (2 $\times$ 80 kW) in dual-port configuration.

Table 8.1 Standard dc fast charging systems

dc Fast charging system	**CHAdeMO IEEE 2030.1.1, IEC 62196-3 (Configuration AA)**	**GB/T (China) GB/T 20234.3, IEC 62196-3 (Configuration BB)**	**CCS Type 1 (US) SAE J1772, IEC 62196-3 (Configuration EE)**	**CCS Type 2 (EU) IEC 62196-3 (Configuration FF)**	**Tesla (US, EU)**
Charge coupler inlet					 Supercharger  Modified Mennekes Type 2 (EU)
Standard max. rated voltage	600 V	750 V (1,000 V per GB/T 20234.3-2015)	600 V	1,000 V	410 V
Standard max. rated current	200 A (400 A per CHAdeMO v1.2)	250 A	200 A	200 A	330 A continuous shared, or 2 × 200 A continuous in Urban Superchargers
Charging power available in 2017	50 kW (150 kW per CHAdeMO v1.2)	120 kW	150 kW	175 kW	135 kW shared (2 posts), or 2 × 72 kW (2 posts) in Urban Superchargers
Communication protocol	Controller area network (CAN)	Controller area network (CAN)	Power line communication (PLC)	Power line communication (PLC)	Proprietary

CharIN (Charging interface) initiative is established in 2015 with the aim to globally promote the CCS for EVs, to define requirements for standards and certification of the CCS and to prepare global introduction of the CCS [9]. The CharIN initiative had over 100 members at the beginning of 2018, including some of the major automobile manufacturers. Approximately 7,000 CCS charging outlets (both Type 1 and Type 2) were available globally at the end of 2017 [10]. The CharIN initiative promotes two systems that support dc fast charging: CCS Type 1 (also known as Combo 1) system, which is mainly in use in North America and supports dc fast charging with power up to 120 kW (per CCS v1.0), and a CCS Type 2 (also known as Combo 2) system, which supports dc fast charging with power up to 200 kW (per CCS v1.0), and is predominantly used in Europe [11]. Both systems also support ac charging up to 43 kW, and both are derived by adding two additional high-current-carrying pins to the existing Type 1 and Type 2 ac charging couplers defined by IEC 62196-2 and SAE J1772. The announced future version of the standard (CCS v2.0) will support dc fast charging with power up to 350 kW. Even though high-power charging is not supported by the CCS v1.0 specification, a few high-power charging stations advertised as 350 kW-capable have already been opened for public: a 150-kW prototype station in the United States [12] and a 4 × 175-kW station in Germany [13]. The future version of the standard (CCS v2.0), which is announced for 2018, will support dc fast charging with power up to 350 kW at voltages up to 1,000 V [11,14].

Tesla Inc. has developed a proprietary charging system for their Superchargers, which differs from the systems described above. More than 1,000 Supercharger stations with almost 7,500 outlets (stalls) have been available globally at the beginning of 2018 [15]. Tesla started deploying their superchargers in October 2012, with 120 kW output power (up to 210 A continuous, and 50–410 V at the output) shared between two stalls. The new Gen. II units with 135 kW of power shared between two stalls were introduced at the beginning of 2014. In September 2017, Tesla has introduced Urban Superchargers with two compact stalls per charger, delivering up to 72 kW (200 A, 50–410 V) each [16].

8.2.2 State-of-the-art EV dc fast chargers

The majority of EV dc fast chargers available on the market at the beginning of 2018 are rated at 50 kW and supplied from a three-phase low-voltage distribution grid (up to 480 V line-to-line, depending on the region). The technical specifications of these dc fast chargers with different output power levels are listed in Table 8.2.

Most of the modern dc fast chargers support more than one charging standard, and some of them support the ac charging as well. The peak efficiency of most of the dc fast chargers is below 94%. The efficiency of Tesla's 135 kW Supercharger was assumed to be the same as for Tesla's onboard charger [17], since Superchargers are made by combining 12 onboard chargers [18].

The EV charging process is a sequence of events defined by the charging protocol. The charging sequence typically starts with signal handshaking and

Table 8.2 Technical specifications of state-of-the-art dc fast chargers

Manufacturer and model	ABB Terra 53	Tritium Veefil-RT	PHIHONG integrated type 120 kW	Tesla supercharger	EVTEC espresso and charge Delta Stromtankstelle 4.0
Maximum power (kW)	50	50	120	135	150
Supported standards	CCS Type 1 CHAdeMO 1.0	CCS Type 1 and 2 CHAdeMO 1.0	GB/T	Supercharger	SAE Combo-1 CHAdeMO 1.0
Input voltage	480 V ac	380–480 V ac 600–900 V dc	380 V ac ± 15% 480 V ac ± 10%	200–480 V ac	400 V ac ± 10%
Output dc voltage (V)	200–500 50–500	200–500 50–500	200–750	50–410	170–500
Output dc current (A)	120	125	240	330	300
Peak efficiency (charger only) (%)	94	>92	93.5	91	93
Volume (L)	758	495	591	1,047	1,581
Weight (lb)	880 (400 kg)	364 (165 kg)	529 (240 kg)	1,320 (600 kg)	880 (400 kg)

exchange of maximum charging parameters between a car and a charger. The car would also perform the insulation test before the charging can begin. If all the required criteria are met, the car closes its dc contactor and the charging may begin. During the charging process, the charger follows the voltage and current commands from the car's Battery Management System (BMS). Each charging event typically consists of two charging modes (regimes): a "Constant Current" (CC) mode followed by a "Constant Voltage" (CV) mode. During the CC mode, the charger follows a slow varying current reference set by the BMS. After a certain time spent in the CC mode (i.e., when the battery voltage reaches a certain value), the BMS switches to CV mode and sends a voltage reference command, which increases slowly as the charging progresses. The charging voltage increases much faster in the CC mode, than in the CV mode. During the CV mode, the charging current decreases. The time spent in CC and CV modes depends mainly on the applied charging current and the initial battery State of Charge (SOC). When the battery reaches a certain preset SOC, the car signals the charger to terminate the charging by reducing the charging current to zero. The car then disconnects itself from the charger by opening its dc contactor. The example of a charging profile of a Tesla model S85 and the corresponding charging power and the battery SOC are shown in Figure 8.1 [19]. The battery was charged from 12% SOC to 94% SOC, with a peak power of 117.2 kW occurring at 16% SOC (2 min after the charging was started). In this case, the CC mode lasted for approximately 2 min. During the charging process, the car and the charger typically exchange the information on current and voltage reference, the battery SOC and some status information. In case of Tesla Superchargers, the car will also periodically send its Vehicle Identification Number (VIN) to identify itself as a Tesla vehicle.

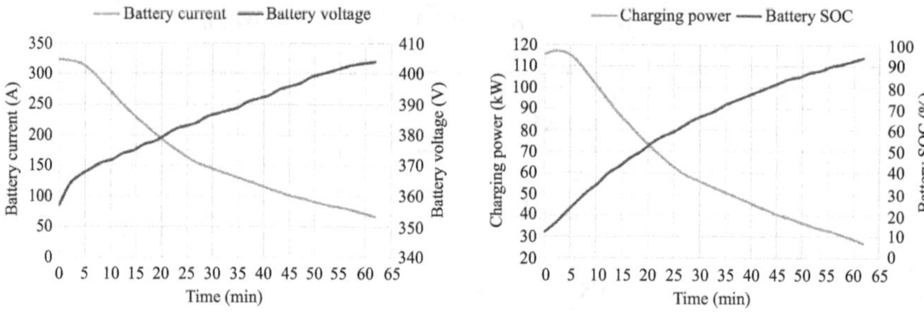

Figure 8.1 The charging profile of a Tesla model S85 and the corresponding charging power and the battery SOC

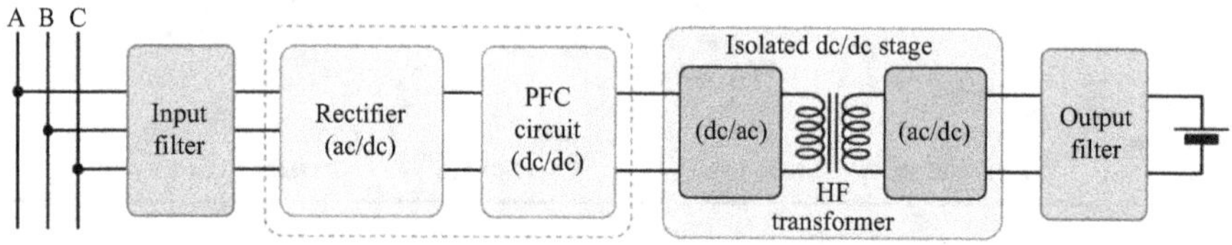

Figure 8.2 The simplified block diagram of a conventional dc fast charger power conversion system

8.2.3 dc Fast charger power converter topologies

The conventional dc fast chargers typically have two power electronics conversion stages: ac/dc rectification stage with power factor correction (PFC), which converts three-phase input ac voltage to an intermediate dc voltage, and an isolated dc/dc stage, which converts the intermediate dc voltage into regulated dc voltage required to charge the EV and provides galvanic isolation from the ac grid. The fast charger power conversion system is typically based on well-known converter topologies used in motor drives. The simplified block diagram of a conventional dc fast charger power conversion system is shown in Figure 8.2 [20].

Even though the input ac/dc stage can be made by implementing separate converters for the rectifier and the PFC, a single-stage converter is preferred at higher power levels due to its simplicity and higher reliability. An LLC filter and a three-phase active pulse width-modulated (PWM) rectifier (shown in Figure 8.3) are often used as an input stage in conventional dc fast chargers. This input stage is usually based on converter modules from commercial motor drives [20,21], which are typically made using the reverse conducting IGBTs, and operate at relatively low switching frequencies (typically up to 10 kHz). The more efficient and more compact solution can be achieved using the silicon carbide (SiC) or gallium nitride (GaN) power MOSFETs (packaged as discrete devices or more often as power modules) which can operate at higher switching frequencies (typically several tens of kHz) and have substantially lower total losses, while the amperage rating for SiC devices is less than half of the required amperage for Si IGBTs [22]. Alternatively, a cheaper and lighter LLCL filter (which has additional inductors connected in

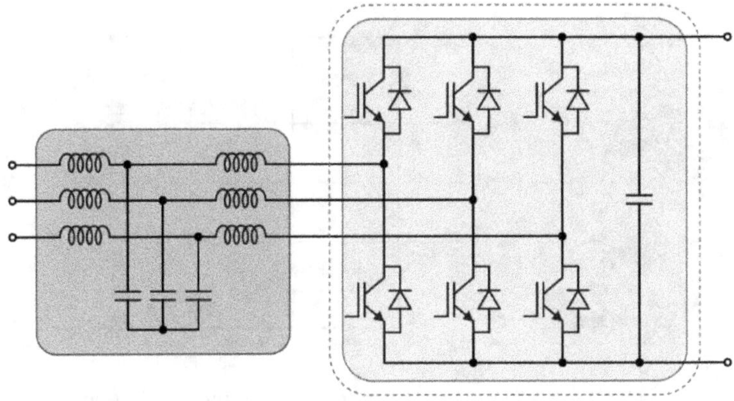

Figure 8.3 The boost-type six-switch PFC converter using IGBTs with input LLC filter

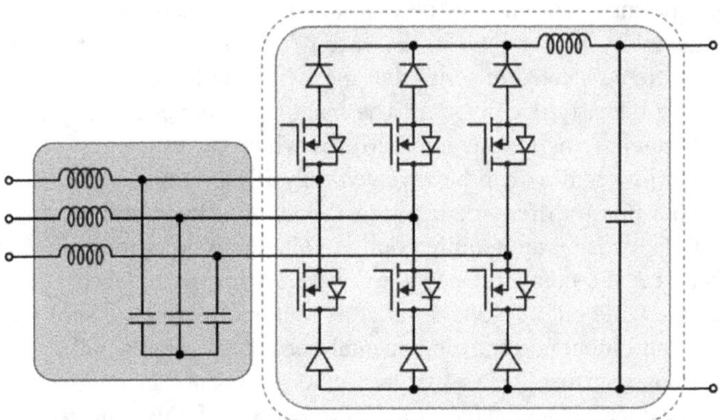

Figure 8.4 The buck-type six-switch PFC converter with input and output filters

series with the capacitors) can be used instead of the LLC filter to make the system more efficient [23,24]. The six-switch PWM converter shown in Figure 8.3 has a boost-type characteristic (the output dc voltage must be higher than the peak input ac line-to-line voltage), enables sinusoidal input currents, bidirectional power flow and can operate with high efficiency and arbitrary phase difference between the input voltage and the input current fundamental [25,26].

The sinusoidal waveform of the input current is achieved by a closed-loop current control, which requires sensing of the input ac currents. The high turn-on switching loss in each IGBT as a result of a reverse recovery of the antiparallel diode of the other IGBT in the same phase leg can be avoided by using SiC MOSFETs with antiparallel Junction Barrier Schottky (JBS) diodes which practically do not exhibit any reverse recovery.

If only unidirectional power flow is required, a buck-type six-switch converter shown in Figure 8.4 can be used. This converter has some advantages over the

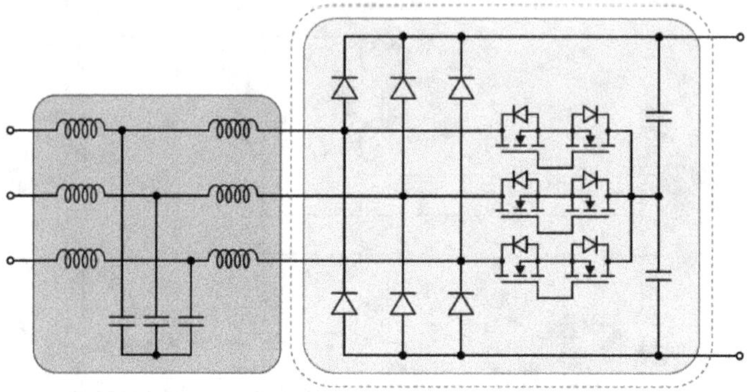

Figure 8.5 The Vienna rectifier with input LLC filter

boost-type rectifier from Figure 8.3, such as inherent short-circuit protection, simple inrush current control and lower output voltage [25]. An additional advantage is that the input current can be controlled in an open loop. The conduction losses are generally higher than in the case of the boost-type converter, because more devices are connected in series, but the switching losses can be lower than in the case of the boost-type converter. The buck-type converter can still operate at very high efficiency, as reported in [26], in which an efficiency of 98.9% was achieved. The power flow can be reversed only if the output voltage is reversed. The achievable phase difference between the input voltage and the input current fundamental depends on the required output voltage. In order to achieve the higher phase difference, the converter needs to operate with the reduced output voltage range (i.e., if a wide output voltage range is required, the phase shift between the input voltage and input current fundamental needs to be kept small).

The Vienna rectifier [27] shown in Figure 8.5 is another boost-type topology that can be used as an input stage of a dc fast charger if only unidirectional power flow is required. Being a three-level converter, the Vienna rectifier offers advantages of 50% lower voltage stress on the used semiconductors switches, lower input current ripple and less-generated EMI. Its limitations are the unidirectional power flow, need for a dc-link capacitor voltage balancing and narrow range of phase difference between the input voltage and the input current which depends on the output voltage (the range is $-30^\circ < \phi < 30^\circ$ when the output voltage is higher than twice the peak input ac line-to-line voltage, and it is reduced to $\phi = 0$ if the output voltage is equal to the peak input ac line-to-line voltage) [26,28,29].

When it comes to the isolated dc/dc stage, several topologies can be used [30]. If only unidirectional power flow is required, a simple topology shown in Figure 8.6 with the full-bridge inverter and the diode bridge rectifier can be used. The full zero-voltage switching (ZVS) operation of the active switches can be achieved when phase-shift PWM control is applied [31]. The main disadvantages of this topology are relatively high switching losses in the output diodes (even when the SiC JBS diodes are used), the high voltage spike due to the current–source character of the rectifying part of the converter (the diode bridge is connected between two current

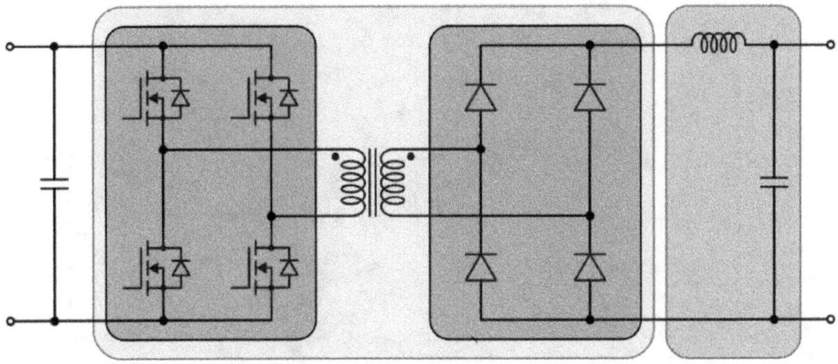

Figure 8.6 Four-switch unidirectional dc/dc converter with output LC filter

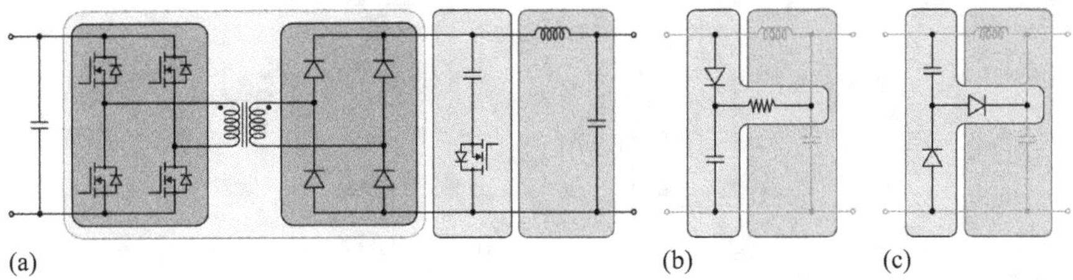

Figure 8.7 (a) Unidirectional dc/dc converter with active snubber and LC output filter, (b) passive RCD snubber and (c) passive CDD snubber

sources: the transformer leakage inductance and the output filter inductance) and severe ringing at the diode bridge output due to the interaction of the transformer leakage inductance and the parasitic capacitance of the reverse biased diodes. To reduce the voltage overshoot and the ringing, the different types of snubber circuits are proposed and applied. The ones that are typically used due to their simplicity and effectiveness are shown in Figure 8.7. The active snubber circuit shows the best performance since it can completely eliminate the voltage ringing and the overshoot [31]. However, that comes with a price of the increased control complexity. The passive RCD snubber is a simple but effective alternative, and the CDD snubber is a cost-effective and low-loss version of the passive snubber with better clamping ability than the RCD snubber [32].

If bidirectional power flow is required, an active rectifier can be used instead of the diode bridge at the transformer's secondary. The snubber would still be required to reduce the ringing across the secondary-side switches.

A dual active bridge (DAB) converter (shown in Figure 8.8) is another bidirectional converter that has been proposed for EV charging applications mainly due to its high power density and high efficiency, buck-boost operation, low device stress, small filter components and low sensitivity to system parasitics [33–40].

When introduced in 1991 [40], the DAB converter initially did not receive a wider adoption due to the high power losses and relatively low switching frequency

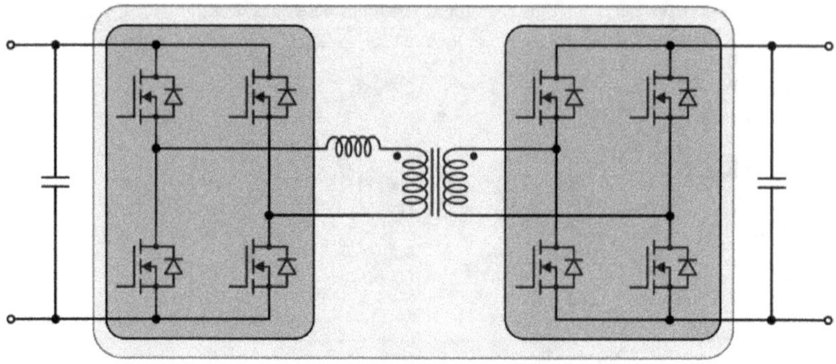

Figure 8.8 Dual active bridge converter

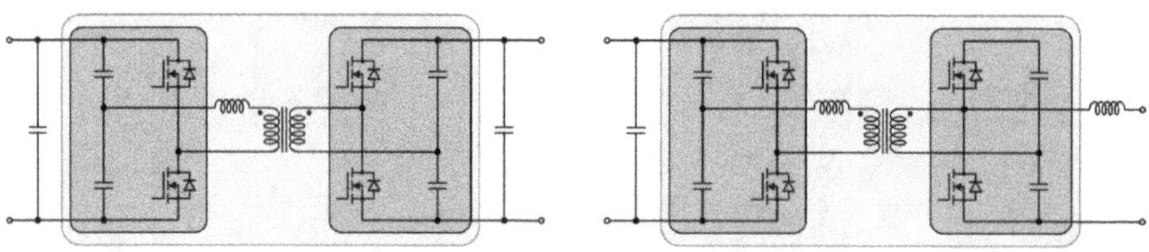

(a) (b)

Figure 8.9 Dual half-bridge converter: (a) with voltage-fed output and (b) with current-fed output

operation of the power semiconductor devices at that time. However, it started regaining the researchers' attention with the introduction of new SiC- and GaN-based power semiconductor devices and the advances in nanocrystalline soft magnetic materials, which enabled the converter efficiency and power density improvements. All the switches operate at 50% duty cycle producing square voltage waveforms at both primary and secondary of the transformer. Using transformer leakage inductance as an energy transfer element, the power flow is controlled by adjusting the time delay (also called "phase shift") between the primary and the secondary voltage waveforms. The active power will be transferred from primary to secondary if the primary waveform is leading the secondary waveform, and vice versa. The ZVS operation of all switches can be achieved only for a narrow range of input-to-output voltage ratio, and it cannot be maintained at light load operation. The DAB can be controlled either by using the voltage mode, current mode or by "mode-hopping" control approach (at light load operation, the system is cycled between the forward and reverse power flow to maintain the required average output voltage) [33].

Several other converter topologies were proposed for the isolated dc/dc stage, including dual half-bridge (DHB) with voltage-fed output [41] and with current-fed output [42], which use only four active switches (two half-bridges, as shown in Figure 8.9), feature ZVS and bidirectional operation, but have the same high device-stress problem as DAB when the input voltage range is wider than 2:1 (i.e., wider than V–$2V$).

8.3 Microgrid topologies for EV charging

Larger scale penetration of EVs is expected to have significant impacts on power grids. Since current EVs act as electric loads, the main expected impacts are increased power demand and energy consumption, increased grid power losses, more prominent load and voltage profile changes along the power grid and eventually requirements to reinforce the grid. With the wider adoption of vehicle-to-grid (V2G) technology, EVs would be able to act as distributed energy storage and provide power to the grid while parked. Based on considerations in [43], the EVs with V2G functionality would not be suitable for baseload power generation (minimum power that needs to be generated at any given time), since baseload power can be generated more cheaply by large nuclear or coal-fired power plants. However, EVs could provide peak power and ancillary services (to serve spinning reserve and regulation markets) which typically account for 5%–10% of the total electricity cost. The EVs with V2G technology could be very effective in isolated power grids (e.g., on islands), where they could enable higher penetration of renewables [44] and, if EV penetration is high enough, significantly help to reduce the variations in daily power demand [45]. When it comes to the integration of EVs into the electric power system, several approaches are proposed. An approach that includes a local entities called Aggregators, which serve as an interface between the EVs and the electricity markets, is proposed in [46]. The Aggregators would group EVs based on their owners' willingness to participate in the electricity markets and perform load shifting to provide financial benefits to their clients. Another approach, proposed in [47], features coordinated charging that can help lower power losses and voltage deviations by flattening the grid output power peaks. The approach proposed in [47] features a more integrated way of managing technical and commercial aspects of EV integration, which requires implementation of smart metering to coordinate charging and communication between the EVs, the distribution system operator and the transmission system operator. If no EV charging coordination is applied, the typical semiurban 15 kV MV grid could accept up to 10% of EVs (out of total number of EVs calculated as 1.5 EV per household), charged at home at the same time, without any significant changes to the grid operation and planning procedures [46].

Since EV chargers present a large electric load, it is conceivable to form a microgrid that would be able to supply multiple EVs simultaneously. Such a system would try to replicate the design of a conventional gasoline refueling station, where the same infrastructure is able to supply multiple vehicles by using the same reservoir, and multiple pumps. By sharing a large common power supply, the system would benefit from load diversification. The work in [48] explored the design of the power supply for a system with large number of Level 2 chargers and showed that for 50 Level 2 chargers, when accounting for diversity, the system would require a power supply of about 60% of the worst-case (rated power) scenario. Similarly, in case of a system with a single ac/dc conversion stage and a dc distribution to dc/dc fast charging units, it was found that the average power load would be substantially lower than the peak power load, and that the grid tie could be sized for the average power demand, rather than the peak demand [49]. In this configuration, an energy storage system is connected to the dc bus to supply power when the demand exceeds the average that can be provided from the grid. It is shown that over 98% of power demand could be satisfied even with a relatively small storage system that is rated based on the average value of the needed storage power. Even though the charging would be delayed for over 60% of the customers, maximum delay time would be approximately 2 min, and the average delay time would not be longer than 10 s.

8.3.1 State-of-the-art dc fast charger installation

A typical dc fast charger station consists of an MV/LV service transformer, switchgear, low-voltage metering and dc fast charging units. State-of-the-art dc fast chargers are typically rated at 50 kW, installed as single units, and can add up to 100 mi (161 km) of range in about 30 min. Higher power Tesla's stations are equipped with multiple 135 kW units (each 135-kW unit serves two charging stalls), which can add up to 170 mi (273 km) in 30 min. As an example, the circuit diagram and the rendering of an actual installation with four Tesla Supercharger units (eight stalls) is illustrated in Figure 8.10 [50].

The dc fast charger and the MV/LV transformer have substantial size and weight which add cost and complexity to the installation, as both the charger and the transformer are commonly installed on a concrete foundation. Additionally, installation in seismically active areas would require seismically restrained concrete foundations for both the transformer and the fast charger, further increasing the installation costs. The single-port state-of-the-art dc fast charger unit cost ranges from $10,000 to $40,000 [51], and the installation costs for a state-of-the-art dc fast charger unit are in the range from $8,500 to $50,820 [based on the data from 111 units deployed until 2013 and including estimates for another 50+ fast charger sites, obtained by the Idaho National Laboratory (INL)] [52]. According to the INL white paper, the main cost drivers include requirements for new electric service, type of surface material under which the electrical wiring/conduits were installed, distance from the power source to the service transformer and from the service transformer to the charger, materials, permits and engineering drawings. The installation of a charging station with multiple charging stalls is even more costly, since it involves the installation of the switchgear (shown in Figure 8.10), more boring/trenching operations and more materials. The result can be a complex and expensive system that requires significant infrastructure.

Considering that the next generation dc fast chargers will have significantly higher power ratings to enable ultrafast charging, designing the dc fast charger stations for future upgradeability may be more cost-effective in a long term, as concluded from a case study performed in [53], where the cost of community and corridor dc fast charging complexes was analyzed. If the system is designed with upgradeability in mind, the initial system cost would be approximately $25,000 higher (for both the community and corridor dc fast charging complexes), but up to $120,000 would be saved in the long run.

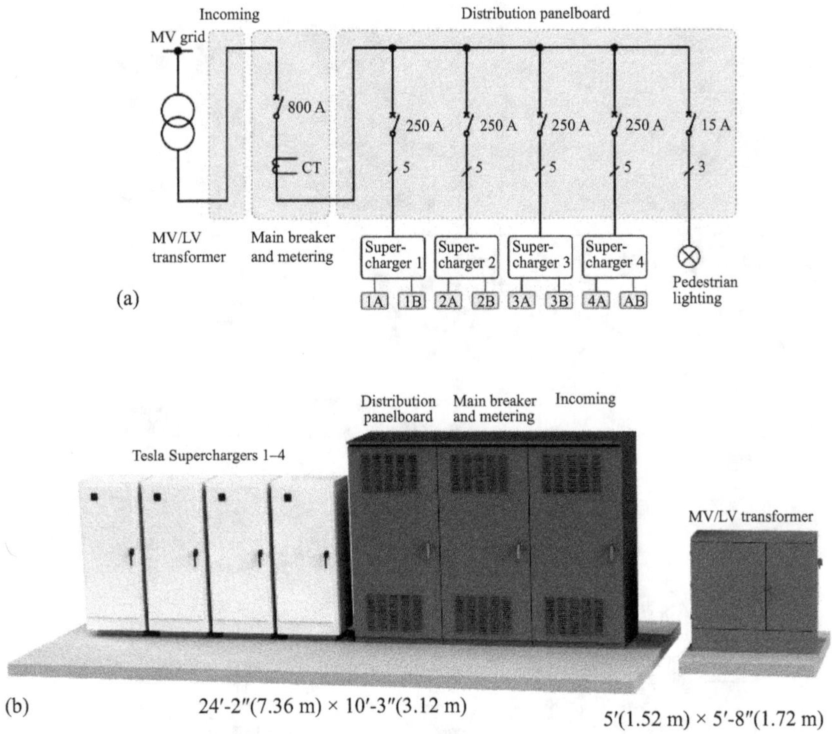

Figure 8.10 State-of-the-art dc fast charger installation with four Tesla Supercharger units: (a) one-line circuit diagram, (b) rendering of the system (charging stalls and seclusion composite fence surrounding the equipment are not shown)

As already mentioned [49], adding the energy storage to the dc fast charging system could minimize the peak demand charges and therefore reduce the operating costs. The added energy storage needs to be sized properly to be capable of buffering the power and energy demands of the charging station during the peak demand period. During the off-peak hours, the stored energy can be used to provide ancillary services for improved grid stability and security. The onsite photovoltaic energy generation system can also be added to the charging station to improve decoupling of power provided to the vehicles from the power drawn from the grid. State-of-the-art systems are typically coupled on the LV ac side through ac-to-dc or dc-to-ac converters, as illustrated in Figure 8.11(a). Even though bidirectional operation is enabled by charging standards, majority of modern dc fast chargers operate in unidirectional mode.

One significant advantage of using the ac bus is the availability and maturity of the rectifier and inverter technology, availability of ac switchgear and protection devices, and well-established standards and practices for the ac power distribution systems. On the other hand, having more conversion stages in the ac-coupled

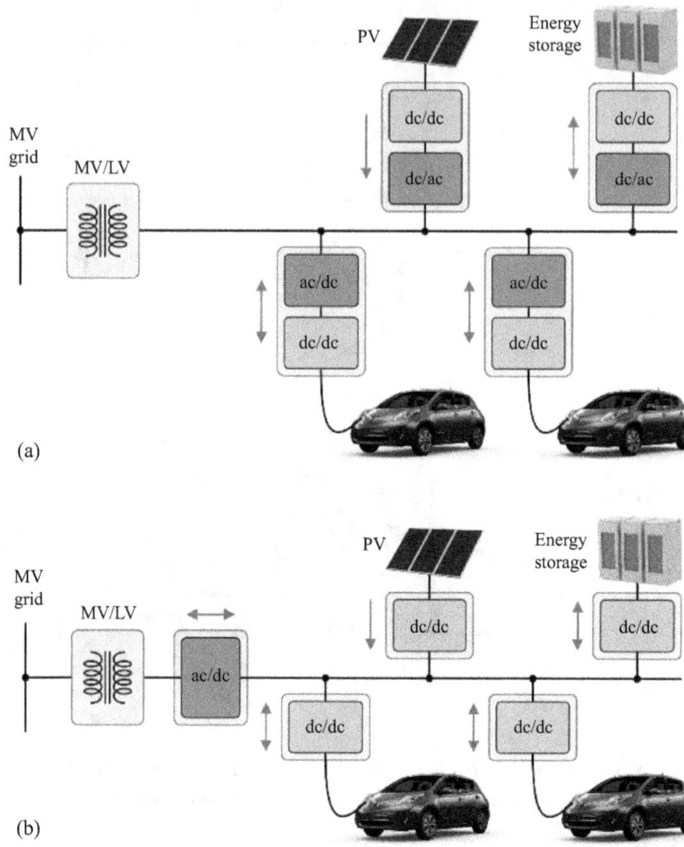

Figure 8.11 Microgrid configurations for EV charging applications: (a) typically used ac microgrid and (b) dc microgrid

system (to interface dc generation, energy storage and dc loads to the ac system) increases the system complexity and cost and decreases the system efficiency. Also, dealing with reactive power, inverter synchronization, and voltage and frequency control during islanded operation of the ac-connected systems makes the ac systems more complicated to control than the dc-connected systems. A dc-connected system, illustrated in Figure 8.11(b), features one centralized ac/dc converter and provides more convenient and more energy efficient way of connecting the energy storage and the renewable energy sources to the central dc bus.

The test results of a simple ac-coupled system with 20 kW (16 kWh) lithium–polymer battery storage and an EV charger commercial prototype with maximum 50 kW dc and 22 kW ac output power are reported in [54]. The battery voltage was first boosted to 600 V and then fed to an inverter which was connected to the 400-V ac grid through the low-frequency isolation transformer. The peak power shaving ability is demonstrated for two different cases: charging Nissan Leaf from 65%

SOC for approximately 10 min, and from 40% to 100% SOC. During the second test, the energy delivered from the energy storage was 4.5 kWh, which was 36% of the total energy absorbed by the charger (12.4 kWh). An implementation of dc-coupled system is reported in [55], where 288 V/40 Ah (11.5 kWh) lead-acid battery is used as a storage to reduce the power drawn from the ac grid during the fast charging of the 50 V/40 Ah (2 kWh) $LiFePO_4$ battery used in two wheelers.

8.3.2 Medium-voltage dc fast chargers: a new approach to high-power EV fast charging

Another approach to EV fast charging is to connect the fast charger directly to MV distribution line and eliminate the low-frequency MV/LV distribution transformer. This approach is based on solid-state transformer (SST) technology and wide bandgap (WBG) semiconductor power devices (typically 1.2 kV SiC MOSFETs) and offers several benefits, such as higher efficiency, higher power density and system-level cost savings, compared to the SOA transformer-and-charger approach. Eliminating low-frequency distribution transformer can be particularly beneficial if the SST-based dc fast charger is installed in densely populated areas. The SST-based MVAC/LVDC (or MVAC/MVDC) converter provides either unidirectional or bidirectional power flow, regulates active and reactive power, performs PFC function and provides isolation and protection features. The typical implementation of a system that uses this approach is illustrated in Figure 8.12 [56,57]. The system uses a common dc bus to interface renewable energy generation systems and energy storage to EVs via dc/dc converters only, ensuring reduced number of conversion stages and high efficiency. Several studies and SST-based EV fast charger prototypes demonstrating these benefits have been reported in the literature.

The Utility Direct Fast Charger (UDFC) has been developed and demonstrated by the Electric Power Research Institute in 2013 [58,59]. Two prototype dc fast chargers were demonstrated: a 50-kW prototype that connects to the 2.4-kV ac

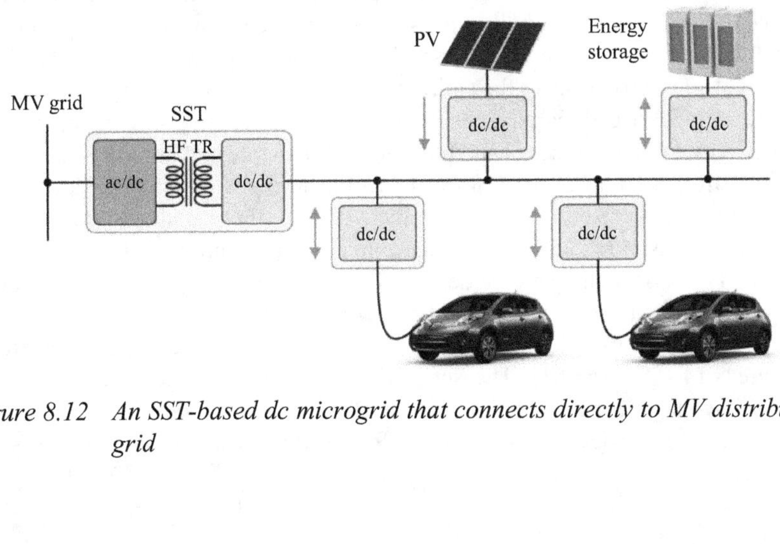

Figure 8.12 An SST-based dc microgrid that connects directly to MV distribution grid

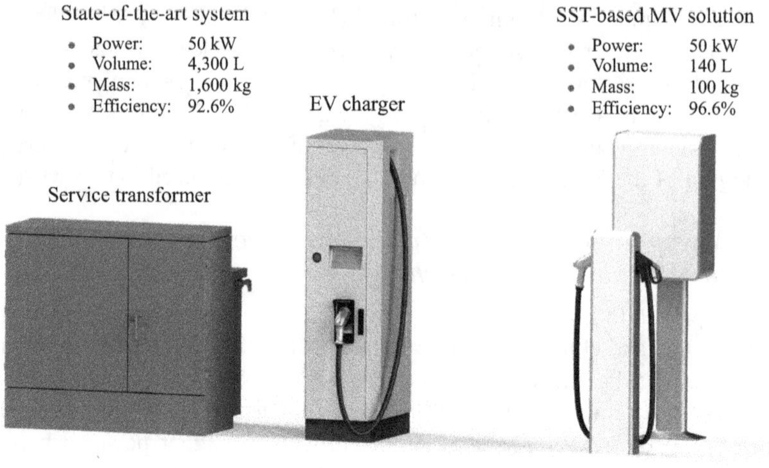

Figure 8.13 Comparison of the SOA EV charger system and the SST-based MV solution from [60]

single-phase line and a 25-kW prototype that connects to the 7.2-kV ac single-phase line. In both the cases, the rated output voltage was 400 V and the grid-to-vehicle efficiency was higher than 96%. Another, more recent effort, has demonstrated a modular 50-kW dc fast charger that uses 1.2 kV SiC devices, connects to 2.4 kV ac single-phase line, outputs 400 V and has the efficiency of over 96% [60]. The benefits of the MV approach with SiC devices are illustrated in Figure 8.13.

As previously mentioned, a dedicated service transformer is typically required to step down the distribution system medium voltage and provide three-phase supply to the commercial dc fast charger. With the service transformer's efficiency of 98.5% (at 50 kVA level), the grid-to-battery efficiency of the 50-kW transformer-and-charger systems is less than 93%. The efficiency of higher power chargers (135 kW Tesla Supercharger or 150 kW Delta Stromtankstelle 4.0) do not exceed 93% while still requiring bulky low-voltage service transformers to provide 208/480 V three-phase input, which itself has an efficiency below 99% (at 150 kVA power level). The efficiency of the entire system with the 150-kV charger is therefore approximately 92%. Compared to the SOA system, the MV dc fast charger prototype reported in [60] reduces the size and the weight by ten-fold and improves the system efficiency from 93% to over 96%—a reduction in losses by approximately 50%. Additionally, by connecting directly to the MV line, the installation costs can be reduced by at least 40%, compared to the conventional solution [61].

The MV EV fast chargers are designed to be modular, using identical modules as building blocks to reach the desired power and voltage levels. The modules are typically connected in series at the input and in parallel at the output, as illustrated in Figure 8.14 [58,60,62,63]. The series connection at the input allows the relatively low-voltage off-the-shelf silicon IGBTs or SiC MOSFETs to serve the medium-voltage application. The use of SiC devices substantially increases the

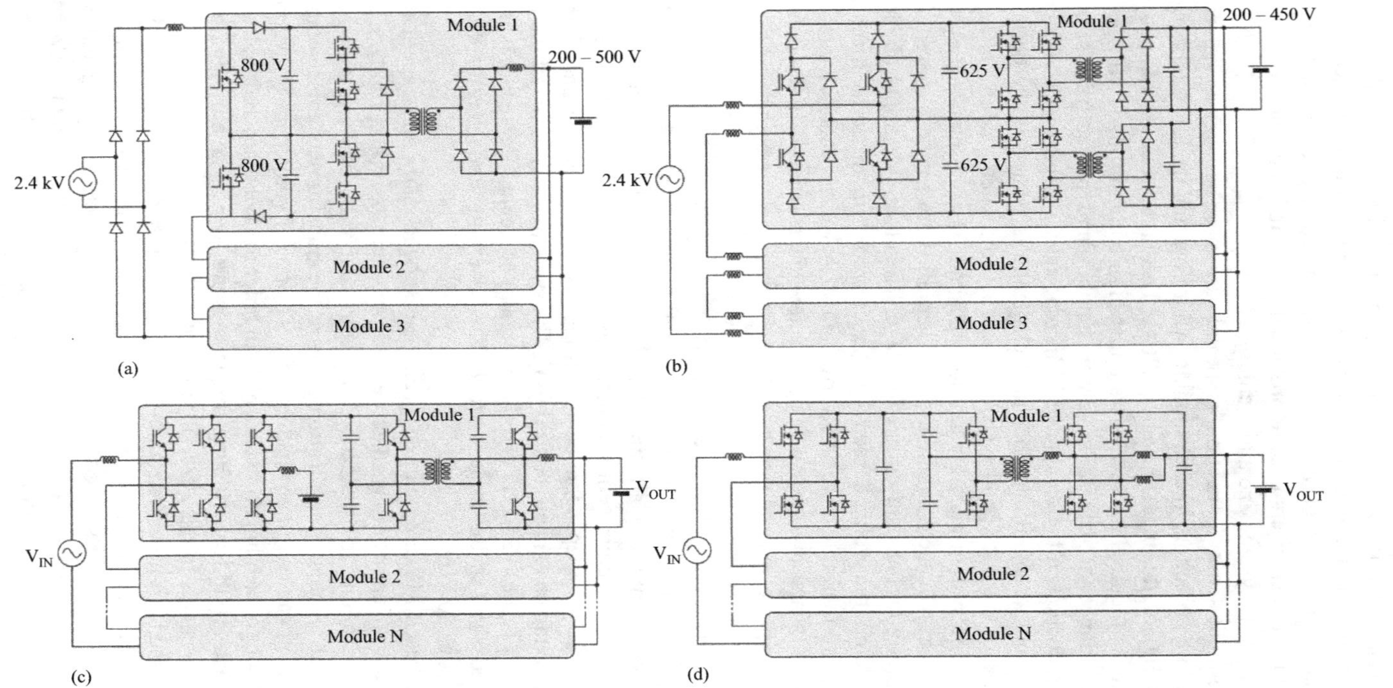

Figure 8.14 *Input-series–output-parallel MV charger topologies: (a) multicell boost (MCB) topology [60], (b) UDFC topology [58], (c) cascaded H-bridge topology with integrated split battery [62], (d) cascaded H-Bridge with current-fed DAB [63]*

efficiency of the system compared to the use of the silicon equivalents, due to the lower switching and conduction losses. Additionally, the high switching frequency operation of SiC MOSFETs enables significant reduction in size of the inductors and the transformers used to provide the required galvanic isolation, thus reducing the overall system size and weight.

With the input-series connection, a special care must be taken to balance the voltage of dc link capacitors in each module, while ensuring the low total harmonic distortion (THD) of the input current [63,64]. The high isolation of gate drive power supplies is required for safe operation at high common-mode voltage. Low coupling capacitance of gate drive power supplies and power transformers is essential for reduction of the conducted EMI, especially if high-speed WBG devices are used. Another significant technical challenge is related to the design and implementation of the fast-acting system protections and safety features at medium voltage level. The system must be able to withstand Basic Insulation Level tests and low-frequency-voltage insulation tests, where the converter is exposed to voltages significantly higher than its rated voltage. Therefore, additional circuitry is required to clamp the voltage seen by the converter in order to meet these requirements. Furthermore, the converter must have an internal protection mechanism that clears any system fault in a few microseconds if the converter devices are to be protected and the system is to continue to operate after the fault is cleared. Such fast clearing of a fault cannot be achieved using conventional MV mechanical switches, and solid-state solutions are required.

The design of a 16.7-kW isolated dc/dc MCB module used in a 50-kW MV fast charger prototype from [60] is illustrated in Figure 8.15 [65]. The module topology uses low number of active semiconductor switches, has high utilization of the switches, and enables independent control of input and output stages and simple dc bus capacitor voltage balancing control [66]. The switching frequency of the three-level boost stage is set to 25 kHz and the minimum switching frequency is limited by the required THD of the input current. The switching frequency of the NPC converter is set to 50 kHz.

The SST-based MV EV charger system can service many other applications, including industrial applications, data centers, and aerospace and military applications. As an example, the reduction of weight and volume is particularly significant in naval shipboard applications, where the transformers for a redundant propulsion system can weigh as much as 35 t. The power converters for shipboard applications need to have high power density in order to save space and fuel, and dc MV systems with 100 kHz switching frequencies are seen as an emerging benchmark for these systems in the 2025–30 time frame [67].

The proposed high-power medium-voltage modular rectifier can also serve as a power supply for data centers. According to DOE's Center of Expertise for Energy Efficiency in Data Center's recent report [68], data centers in the United States consumed an estimated 70 billion kWh in 2014, representing about 1.8% of the total US electricity consumption. Additionally, based on current trends, US data centers are projected to consume approximately 73 billion kWh in 2020. Most of these data centers (except high-end and hyperscale centers) operate with the Power

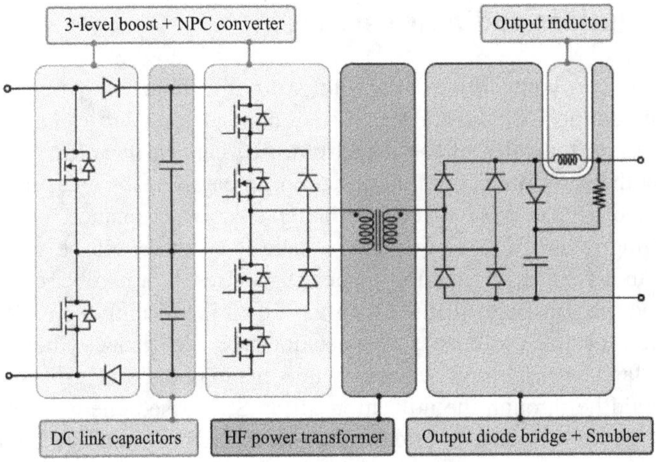

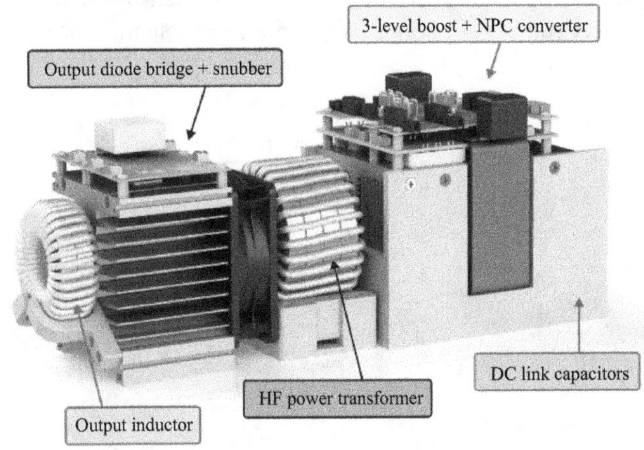

Figure 8.15 16.7 kW isolated dc/dc module as a building block for MV EV fast charger [65]

Usage Effectiveness of approximately 2, which means that half of the used energy is consumed by the data center's supporting infrastructure (cooling equipment, uninterrupted power supplies, lighting, etc.). As a result, any power savings that can be made in the operation of the data center will have significant environmental implications as well as impacts on the system cost of operation. Recently, data centers with 380 V dc distribution systems have been proposed, with their inherently higher efficiency due to the fewer conversion stages. The SST-based topology used for MV EV chargers can be used to efficiently supply 380 V dc to the data centers. The benefits of higher system efficiency, better power conditioning and better quality power can all be achieved with the SST-based design [69].

8.4 Conclusions and future trends

A number of cost-competitive EVs with a driving range of over 200 mi have recently entered the EV marketplace. Also, a number of countries have stated that they plan to end the sales of fossil-fuel-powered cars in the near future, further accelerating the shift to electric transportation. In parallel, advances in lithium ion battery technology promise to deliver better charge acceptance, allowing even faster charging rates. To serve this new vehicle fleet, there will be a need for fast chargers with a very large charging capacity. As a result, a number of automakers have reached the consensus that the long-term goal is to supply up to 350 kW to a single vehicle in order to enable a gas station-like experience. One way of providing the fast EV charging services while maintaining the grid stability and security would be through the utilization of the SST-based energy routers, which would provide both ac and dc grid interconnectivity, grid intelligence and necessary protection features, as illustrated in Figure 8.16. This is part of the Future Renewable Electric Energy Delivery and Management visions to enable the seamless interaction of distributed energy resources, distributed intelligence and advanced power electronics [70]. The SST provides the potential for revolutionizing

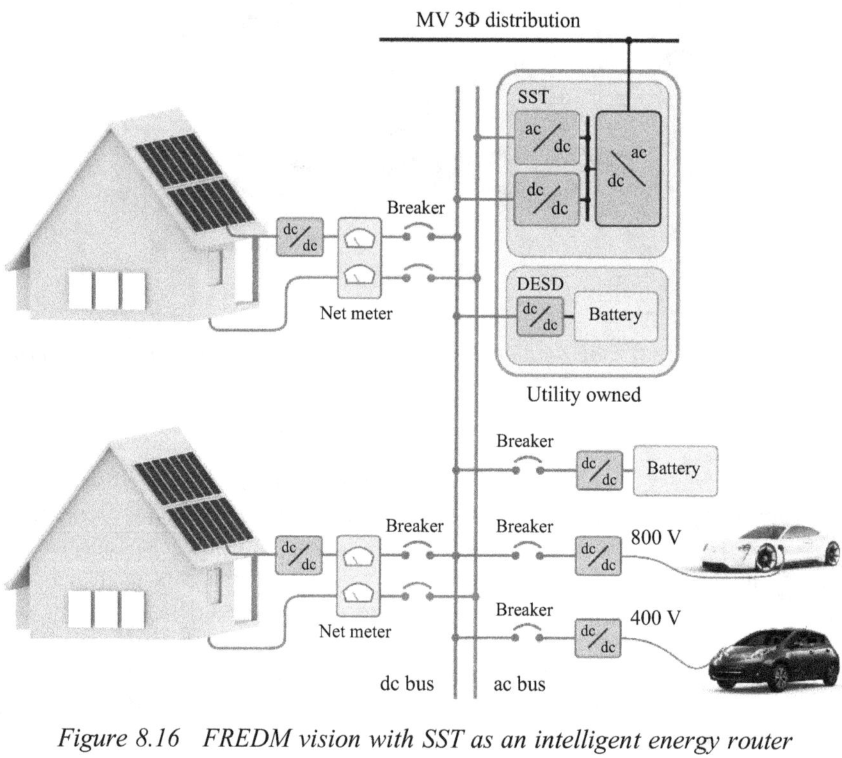

Figure 8.16 FREDM vision with SST as an intelligent energy router

the power distribution system by completely decoupling the medium voltage distribution grid from the low voltage feeder sections. The bidirectional power flow capability of the SST provides possibilities to integrate the distributed energy storage devices and to feed locally generated power back to the grid in a controlled way. Recent technology advances in WBG semiconductor devices, such as high-voltage SiC MOSFETs and diodes, serve as the enabling technology for the SST.

Despite its unparalleled capabilities, significant challenges stand in the way of the complete adoption of the SST by the electric utilities today. The key barriers include (1) inherent reliability concerns of replacing the passive transformer with a power electronic device, (2) relatively low penetration of renewable resources seen in the system today, which can mostly be handled by the existing infrastructure, (3) the large inertia still present in the distribution system, supplied by the legacy generators and (4) limited ability to monetize the perfect power quality supplied by the SST. As a result, the other venues to demonstrate the SST technology can be explored and showcase the benefits of direct connection of power electronics to the MV grid. Medium-voltage SSTs using WBG devices can serve many other emerging and established applications including EV fast chargers, data center power supplies and military power conversion stages, to name a few. The EV fast chargers present the largest opportunity, given the size and the rate of growth of the market. Further, the use of a medium-voltage dc fast charger has clear monetary advantages since this solution will substantially reduce the installation cost, which makes up to half of the price of deploying a commercial dc fast charger, and it is particularly suitable for deployment in densely populated urban areas. Overall, the SST-based approach has potential to help reduce the EV charging time and bring a gas station experience to the EV users, while providing the integration of renewable resources and energy storage in a cost-competitive way.

References

[1] International Energy Agency (IEA) and the Electric Vehicles Initiative of the Clean Energy Ministerial (EVI), "Global EV Outlook 2017," OECD/IEA, Paris, France, Jul. 2017.

[2] "Monthly Plug-In Sales Scorecard". [Online]. Available: https://insideevs.com/monthly-plug-in-sales-scorecard/. Accessed on Jan. 5, 2018.

[3] "Compare Electric Cars Side-by-Side". [Online]. Available: https://www.fueleconomy.gov/feg/evsbs.shtml. Accessed on Jan. 5, 2018.

[4] D. Howell, S. Ahmed, R. B. Carlson, *et al.*, "Enabling Fast Charging: A Technology Gap Assessment," U.S. Department of Energy, Office of Energy efficiency & Renewable Energy, Report INL/EXT-17-41638, Oct. 2017. [Online]. Available: https://energy.gov/sites/prod/files/2017/10/f38/XFC%20Technology%20Gap%20Assessment%20Report_FINAL_10202017.pdf.

[5] "Largest Global Charger Coverage". [Online]. Available: https://www.chademo.com/. Accessed on Jan. 5, 2018.

[6] CHAdeMO, "2016 Activity Report (1 April 2016–31 March 2017)", CHAdeMO Association. [Online]. Available: http://www.chademo.com/wp2016/wp-content/japan-uploads/2017GA/Activityreport2016en.pdf.
[7] Chinese National Standard, "Connection Set for Conductive Charging of Electric Vehicles – Part 3: DC Charging Coupler," National Standard of the People's Republic of China, 2015, GB/T 20234.3-2015.
[8] D. Hall, N. Lutsey, "Emerging Best Practices for Electric Vehicle Charging Infrastructure," International Council on Clean Transportation, Washington, DC, Oct. 2017. [Online]. Available: https://www.theicct.org/sites/default/files/publications/EV-charging-best-practices_ICCT-white-paper_04102017_vF.pdf.
[9] "Articles of Association of Charging Interface Initiative e.V.," Berlin, May 19, 2015. [Online]. Available: http://charinev.org/fileadmin/Downloads/Association_Documents/ArticlesOfAssociation_CharIN.pdf.
[10] "DC-CCS-Charging Stations in the Main Markets". [Online]. Available: http://charinev.org/ccs-at-a-glance/infrastructure/. Accessed on Jan. 5, 2018.
[11] Combined Charging System Standard Specification 1.0, "Combined Charging System 1.0 Specification – CCS 1.0," Coordination Office Charging Interface, c/o innos-Sperlich GmbH, Jan. 2017.
[12] "EVgo and ABB to Deploy Nation's First High-Power Electric Vehicle Fast Charging Station," Press Release, Feb. 27, 2017. [Online]. Available: https://www.evgo.com/about/news/evgo-abb-deploy-nations-first-high-power-electric-vehicle-fast-charging-station/.
[13] "Europe's First Public Ultra-Fast Charging Station Now Operational," Press Release, Dec. 20, 2017. [Online]. Available: https://www.allego.eu/wp-content/uploads/2017/12/ENG-20171221_Ultra-E_PM.pdf.
[14] "Charging Interface Initiative e.V. – Industry Statement on Future Charging Infrastructure," Coordination Office Charging Interface, c/o innos-Sperlich GmbH. Feb. 20, 2017. [Online]. Available: http://www.charinev.org/fileadmin/Downloads/Papers_and_Regulations/CharIN_industry_statement.pdf.
[15] "Superchargers". [Online]. Available: https://www.tesla.com/supercharger. Accessed on Jan. 5, 2018.
[16] The Tesla Team, "Supercharging Cities," Sep. 11, 2017. [Online]. Available: https://www.tesla.com/blog/supercharging-cities.
[17] "2014 Tesla Model S 85 kWh Advanced Vehicle Testing – Baseline Vehicle Testing Results", Idaho National Laboratory, Factsheet INL/MIS-15-34211, Mar. 31, 2016. [Online]. Available: https://avt.inl.gov/sites/default/files/pdf/fsev/fact4500tesla2014.pdf.
[18] E. Loveday, "Rare Look Inside Tesla Supercharger". Jun. 24, 2014. [Online]. Available: https://insideevs.com/rare-look-inside-tesla-supercharger/.
[19] "Supercharge Data from Cross-Country Road Trip". Dec. 8, 2014. [Online]. Available: https://teslamotorsclub.com/tmc/threads/supercharge-data-from-cross-country-road-trip.39184/.
[20] T. Anegawa, "Safety Design of CHAdeMO Quick Charging System," in World Electric Vehicle Journal, vol. 4, no. 4, pp. 855–859, 2010.

[21] D. Aggeler, F. Canales, H. Zelaya, *et al.*, "Ultra-fast DC-charge infrastructures for EV-mobility and future smart grids," 2010 IEEE PES Innovative Smart Grid Technologies Conference Europe (ISGT Europe), Gothenburg, 2010, pp. 1–8.

[22] J. Casady, "Power products commercial roadmap for SiC from 2012–2020," Cree Power – Sept 2014 HMW Direct-Drive Motor Workshop, 2014. [Online]. Available: https://www.nist.gov/document-2374.

[23] W. Wu, Y. He and F. Blaabjerg, "An LLCL Power Filter for Single-Phase Grid-Tied Inverter," in IEEE Transactions on Power Electronics, vol. 27, no. 2, pp. 782–789, Feb. 2012.

[24] W. Wu, M. Huang, F. Blaabjerg, "Efficiency Comparison between the LLCL and LCL-Filters Based Single-Phase Grid-Tied Inverters," in Archives of Electrical Engineering, vol. 63, no. 1, pp. 63–79, 2014.

[25] H. Mao, C. Y. Lee, D. Boroyevich and S. Hiti, "Review of High-Performance Three-Phase Power-Factor Correction Circuits," in IEEE Transactions on Industrial Electronics, vol. 44, no. 4, pp. 437–446, Aug. 1997.

[26] J. W. Kolar, T. Friedli, "The Essence of Three-Phase PFC Rectifier Systems—Part I," in IEEE Transactions on Power Electronics, vol. 28, no. 1, pp. 176–198, Jan. 2013.

[27] J. W. Kolar and F. C. Zach, "A novel three-phase three-switch three-level PWM rectifier," Proc. 28th Power Conversion Conference, PCIM'94, Jun. 28–30, 1994, pp. 125–138.

[28] R. Lai, F. Wang, R. Burgos, D. Boroyevich, D. Jiang and D. Zhang, "Average Modeling and Control Design for VIENNA-Type Rectifiers Considering the DC-Link Voltage Balance," in IEEE Transactions on Power Electronics, vol. 24, no. 11, pp. 2509–2522, Nov. 2009.

[29] J. Liu, W. Ding, Han Qiu, C. Zhang and B. Duan, "Neutral-point voltage balance control and oscillation suppression for VIENNA rectifier," 2017 IEEE 3rd International Future Energy Electronics Conference and ECCE Asia (IFEEC 2017 – ECCE Asia), Kaohsiung, 2017, pp. 1275–1279.

[30] Y. Du, S. Lukic, B. Jacobson and A. Huang, "Review of high power isolated bi-directional DC–DC converters for PHEV/EV DC charging infrastructure," 2011 IEEE Energy Conversion Congress and Exposition, Phoenix, AZ, 2011, pp. 553–560.

[31] J. A. Sabate, V. Vlatkovic, R. B. Ridley and F. C. Lee, "High-voltage, high-power, ZVS, full-bridge PWM converter employing an active snubber," Conference Proceedings of the Sixth Annual Applied Power Electronics Conference and Exposition, APEC'91, Dallas, TX, 1991, pp. 158–163.

[32] J.-G. Cho, J.-W. Baek, C.-Y. Jeong and G.-H. Rim, "Novel Zero-Voltage and Zero-Current-Switching Full-Bridge PWM Converter Using a Simple Auxiliary Circuit," in IEEE Transactions on Industry Applications, vol. 35, no. 1, pp. 15–20, Jan./Feb. 1999.

[33] M. N. Kheraluwala, R. W. Gascoigne, D. M. Divan and E. D. Baumann, "Performance Characterization of a High-Power Dual Active Bridge DC-to-DC Converter," in IEEE Transactions on Industry Applications, vol. 28, no. 6, pp. 1294–1301, Nov./Dec. 1992.

[34] S. Han and D. Divan, "Bi-directional DC/DC converters for plug-in hybrid electric vehicle (PHEV) applications," 2008 Twenty-Third Annual IEEE Applied Power Electronics Conference and Exposition, Austin, TX, 2008, pp. 784–789.

[35] C. Mi, H. Bai, C. Wang and S. Gargies, "Operation, Design and Control of Dual H-Bridge-based Isolated Bidirectional DC–DC Converter," in IET Power Electronics, vol. 1, no. 4, pp. 507–517, Dec. 2008.

[36] R. J. Ferreira, L. M. Miranda, R. E. Araújo and J. P. Lopes, "A new bi-directional charger for vehicle-to-grid integration," 2011 2nd IEEE PES International Conference and Exhibition on Innovative Smart Grid Technologies, Manchester, 2011, pp. 1–5.

[37] H. van Hoek, M. Neubert and R. W. De Doncker, "Enhanced Modulation Strategy for a Three-Phase Dual Active Bridge—Boosting Efficiency of an Electric Vehicle Converter," in IEEE Transactions on Power Electronics, vol. 28, no. 12, pp. 5499–5507, Dec. 2013.

[38] L. Xue, Z. Shen, D. Boroyevich, P. Mattavelli and D. Diaz, "Dual Active Bridge-Based Battery Charger for Plug-in Hybrid Electric Vehicle with Charging Current Containing Low Frequency Ripple," in IEEE Transactions on Power Electronics, vol. 30, no. 12, pp. 7299–7307, Dec. 2015.

[39] L. Xue, M. Mu, D. Boroyevich and P. Mattavelli, "The optimal design of GaN-based Dual Active Bridge for bi-directional Plug-IN Hybrid Electric Vehicle (PHEV) charger," 2015 IEEE Applied Power Electronics Conference and Exposition (APEC), Charlotte, NC, 2015, pp. 602–608.

[40] R. W. A. A. De Doncker, D. M. Divan and M. H. Kheraluwala, "A Three-Phase Soft-Switched High-Power-Density DC/DC Converter for High-Power Applications," in IEEE Transactions on Industry Applications, vol. 27, no. 1, pp. 63–73, Jan./Feb. 1991.

[41] H. Li, D. Liu, F. Z. Peng and G.-J. Su, "Small signal analysis of a dual half bridge isolated ZVS Bi-directional DC–DC converter for electrical vehicle applications," 2005 IEEE 36th Power Electronics Specialists Conference, Recife, 2005, pp. 2777–2782.

[42] M. Vasiladiotis, B. Bahrani, N. Burger and A. Rufer, "Modular converter architecture for medium voltage ultra fast EV charging stations: Dual half-bridge-based isolation stage," 2014 International Power Electronics Conference (IPEC-Hiroshima 2014 – ECCE ASIA), Hiroshima, 2014, pp. 1386–1393.

[43] W. Kempton, J. Tomić, "Vehicle-to-Grid Power Fundamentals: Calculating Capacity and Net Revenue," in Journal of Power Sources, vol. 144, no. 1, pp. 268–279, 2005.

[44] J. A. P. Lopes, P. M. R. Almeida and F. J. Soares, "Using vehicle-to-grid to maximize the integration of intermittent renewable energy resources in islanded electric grids," 2009 International Conference on Clean Electrical Power, Capri, 2009, pp. 290–295.

[45] P. Kadurek, C. Ioakimidis and P. Ferrao, "Electric vehicles and their impact to the electric grid in isolated systems," 2009 International Conference on Power Engineering, Energy and Electrical Drives, Lisbon, 2009, pp. 49–54.

[46] J. A. P. Lopes, F. J. Soares and P. M. R. Almeida, "Integration of Electric Vehicles in the Electric Power System," in Proceedings of the IEEE, vol. 99, no. 1, pp. 168–183, Jan. 2011.
[47] K. Clement-Nyns, E. Haesen and J. Driesen, "The Impact of Charging Plug-In Hybrid Electric Vehicles on a Residential Distribution Grid," in IEEE Transactions on Power Systems, vol. 25, no. 1, pp. 371–380, Feb. 2010.
[48] S. Hutchinson, M. Baran and S. Lukic, "Power supply for an electric vehicle charging system for a large parking deck," 2009 IEEE Industry Applications Society Annual Meeting, Houston, TX, 2009, pp. 1–4.
[49] S. Bai, D. Yu and S. Lukic, "Optimum design of an EV/PHEV charging station with DC bus and storage system," 2010 IEEE Energy Conversion Congress and Exposition, Atlanta, GA, 2010, pp. 1178–1184.
[50] "Tesla Supercharging station Madison, WI009 Madison, 3801 E Washington Ave, Madison, WI 53704", Tesla, Inc. Jun. 2017. [Online]. Available: https://www.cityofmadison.com/dpced/planning/documents/3801ewa_site.pdf.
[51] M. Smith and J. Castellano, "Costs Associated with Non-Residential Electric Vehicle Supply Equipment – Factors to Consider in the Implementation of Electric Vehicle Charging Stations," U.S. Department of Energy Vehicle Technologies Office, no. DOE/EE-1289, Nov. 2015. [Online]. Available: https://www.afdc.energy.gov/uploads/publication/evse_cost_report_2015.pdf.
[52] Idaho National Laboratory (INL) 2015. "What were the Cost Drivers for the Direct Current Fast Charging Installations?" INL/MIS-15-35060, Mar. 2015. [Online]. Available: http://avt.inel.gov/pdf/EVProj/WhatWereTheCostDriversForDCFCinstallations.pdf.
[53] J. Francfort, S. Salisbury, J. Smart, T. Garetson and D. Karner, "Considerations for Corridor and Community DC Fast Charging Complex System Design," INL/EXT-17-40829, Idaho National Laboratory, Mar. 2017. [Online]. Available: https://avt.inl.gov/sites/default/files/pdf/reports/DCFCChargingComplexSystemDesign.pdf.
[54] D. Sbordone, I. Bertini, B. Di Pietra, M. C. Falvo, A. Genovese and L. Martirano, "EV Fast Charging Stations and Energy Storage Technologies: A Real Implementation in the Smart Micro Grid Paradigm," in Electric Power Systems Research, vol. 120, pp. 96–108, 2015.
[55] O. Veneri, C. Capasso and D. Iannuzzi, "Experimental Evaluation of DC Charging Architecture for Fully-Electrified Low-Power Two-Wheeler," in Applied Energy, vol. 162, pp. 1428–1438, 2016.
[56] X. Yu, X. She, X. Ni and A. Q. Huang, "System Integration and Hierarchical Power Management Strategy for a Solid-State Transformer Interfaced Microgrid System," in IEEE Transactions on Power Electronics, vol. 29, no. 8, pp. 4414–4425, Aug. 2014.
[57] R. Crosier, S. Wang and M. Jamshidi, "A 4800-V grid-connected electric vehicle charging station that provides STACOM-APF functions with a bi-directional, multi-level, cascaded converter," 2012 Twenty-Seventh Annual IEEE Applied Power Electronics Conference and Exposition (APEC), Orlando, FL, 2012, pp. 1508–1515.

[58] EPRI, Utility Direct Medium Voltage DC Fast Charger Update: DC Fast Charger Characterization. EPRI, Palo Alto, CA: 2012. 1024106.

[59] S. Rajagopalan, A. Maitra, J. Halliwell, M. Davis and M. Duvall, "Fast charging: An in-depth look at market penetration, charging characteristics, and advanced technologies," 2013 World Electric Vehicle Symposium and Exhibition (EVS27), Barcelona, 2013, pp. 1–11.

[60] S. Srdic, X. Liang, C. Zhang, W. Yu and S. Lukic, "A SiC-based high-performance medium-voltage fast charger for plug-in electric vehicles," 2016 IEEE Energy Conversion Congress and Exposition (ECCE), Milwaukee, WI, 2016, pp. 1–6.

[61] A. Maitra, "EPRI's utility direct DC fast charger–development, testing, demonstration," EPRI IWC Meeting, Atlanta, GA, Mar. 28, 2012.

[62] M. Vasiladiotis and A. Rufer, "A Modular Multiport Power Electronic Transformer with Integrated Split Battery Energy Storage for Versatile Ultrafast EV Charging Stations," in IEEE Transactions on Industrial Electronics, vol. 62, no. 5, pp. 3213–3222, May 2015.

[63] D. Sha, G. Xu and Y. Xu, "Utility Direct Interfaced Charger/Discharger Employing Unified Voltage Balance Control for Cascaded H-Bridge Units and Decentralized Control for CF-DAB Modules," in IEEE Transactions on Industrial Electronics, vol. 64, no. 10, pp. 7831–7841, Oct. 2017.

[64] X. Liang, C. Zhang, S. Srdic and S. Lukic, "Predictive Control of a Series-Interleaved Multi-Cell Three-Level Boost Power Factor Correction Converter," in IEEE Transactions on Power Electronics, vol. 33, no. 10, pp. 8948–8960, Oct. 2018.

[65] S. Srdic, C. Zhang, X. Liang, W. Yu and S. Lukic, "A SiC-based power converter module for medium-voltage fast charger for plug-in electric vehicles," 2016 IEEE Applied Power Electronics Conference and Exposition (APEC), Long Beach, CA, 2016, pp. 2714–2719.

[66] D. Rothmund, G. Ortiz and J. W. Kolar, "SiC-based unidirectional solid-state transformer concepts for directly interfacing 400 V DC to medium-voltage AC distribution systems," 2014 IEEE 36th International Telecommunications Energy Conference (INTELEC), Vancouver, BC, 2014, pp. 1–9.

[67] F. Wang, Z. Zhang, T. Ericsen, R. Raju, R. Burgos and D. Boroyevich, "Advances in Power Conversion and Drives for Shipboard Systems," in Proceedings of the IEEE, vol. 103, no. 12, pp. 2285–2311, Dec. 2015.

[68] A. Shehabi, S. Smith, D. Sartor, *et al.*, "United States Data Center Energy Usage Report," Lawrence Berkeley National Laboratory, Jun. 2016. [Online]. Available: https://eta.lbl.gov/sites/all/files/publications/lbnl-1005775_v2.pdf.

[69] G. Zhabelova, A. Yavarian, V. Vyatkin and A. Q. Huang, "Data center energy efficiency and power quality: An alternative approach with solid state transformer," IECON 2015 – 41st Annual Conference of the IEEE Industrial Electronics Society, Yokohama, 2015, pp. 1294–1300.

[70] A. Huang, "FREEDM system – a vision for the future grid," IEEE PES General Meeting, Minneapolis, MN, 2010, pp. 1–4.

Chapter 9

Overview and design of solid-state transformers

Levy Costa[1], *Marco Liserre*[1], *and Giampaolo Buticchi*[1]

9.1 Solid-state transformer: concept

The concept of electronic transformer was first introduced by [1] in the end of 1960s, where the main concept would be to adjust the output voltage by using the power electronics, while contributing to size and volume reduction. The apparatus in question isolated the primary and secondary side in frequency higher than the grid frequency and for that reason the size, weight and volume could be considerably reduced. At that time, there was no real applicability for such apparatus, mainly because of the performance limitation of the available semiconductors on the market.

However, with the advancement of the power electronics and semiconductors technology, new faster devices with lower switching energy were developed, enabling faster switching frequency operation. As a result, the volume and size of the power converters were considerably reduced, bringing advantages for those applications in which the power density is very important. At the same time, the classic solution used in traction for the electric locomotive was very heavy, bulky and inefficient, because of the low frequency transformer (LFT) employed [2] associated to the power electronics converters required for providing the suitable dc voltage to the variable speed drive on the locomotive. In this sense, the power electronics transformer or solid-state transformer (SST), as usually called, could provide the suitable controlled and isolated dc voltage in a unique solution, offering either volume and weight reduction (around 20%–50%) and efficiency improvement (from 93% to 96%) [2,3]. Thus, the SST became a reality in traction applications.

Figure 9.1 illustrates the conventional solution (state of the art) used on the locomotives, as well as the modern solution based on SST. The benefits offered by the SST in traction application are mainly on volume and weight reduction. Thus, the advantages of the system are only related to its hardware. The SST has been frequently conceptualized as presented below.

[1]Chair of Power Electronics Faculty of Engineering, Christian-Albrechts-University of Kiel, Germany.

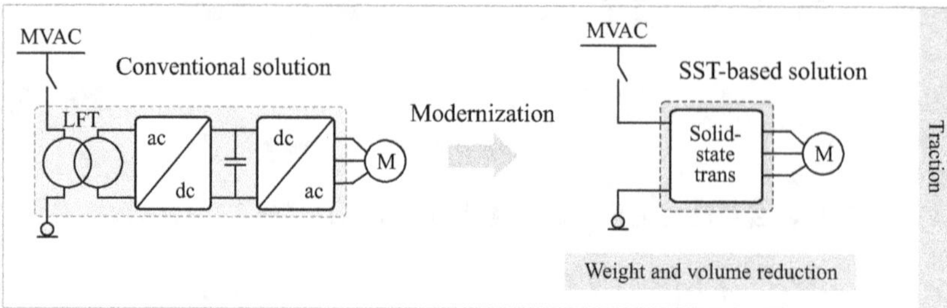

Figure 9.1 Solid-state transformer in traction application. MVAC, medium voltage ac; LFT, low frequency transformer

Solid-state transformer (SST):
Power electronics-based system with galvanic isolation in medium or high frequency between the input and output having waveform conditioning capability. It is normally employed to replace the traditional transformer, offering advantages in volume and weight reduction. Thus, the advantages offered by the SST are related to its hardware stage.

9.2 SST in electric distribution grid application

Apart from the advantages achieved in terms of hardware, the power electronics involved in the SST enables the implementation of an advanced control strategy with multiple control loops (e.g., power, voltage and current). This extra ability enlarges the SST application to other fields, such as electric distribution system, in which power and voltage control is the central requirement. Thus, the SST is very suitable for electric distribution system application, because it can perform as an LFT (replacing it), while providing ancillary services to the grid through its high-advanced control platform, improving the overall power quality of the distribution grid. Therefore, this technology has received a lot of attention in the research community (even academia and industry), and it was cited as one of the most emerging technologies by Massachusetts Institute of Technology Review in 2010 [4–6].

The new functionalities in the form of ancillary services to the electric grid are discussed in [3,7]. Among them, the reverse power flow condition, storage integration, management of hybrid grids (dc and ac) and power quality improvement are highlighted. Then, in this application, the SST is not only supposed to replace the conventional LFT, connecting the medium-voltage (MV) grid to the low-voltage (LV) grid, but also to offer dc connectivity and services to both LV and MV grids, as highlighted in Figure 9.2. In this case, weight and volume advantages have a limited impact, while the efficiency, reliability and functionalities are the primordial requirements. The aforementioned functionalities, enabled by the advanced software communication and control system, make the SST a smart

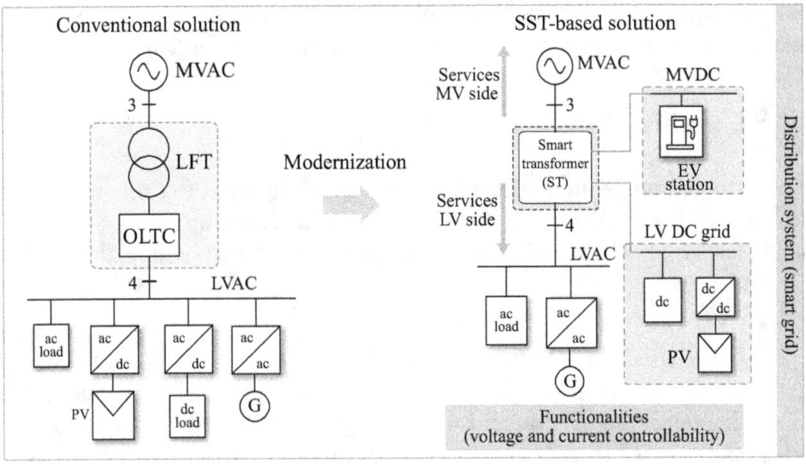

Figure 9.2 Solid-state transformer in electric distribution grid application, including the connection to dc grids

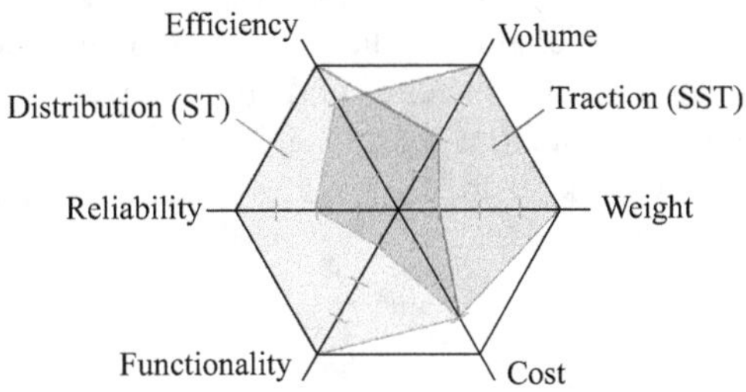

Figure 9.3 Requirements and performance of the SST and ST for traction and distribution grid applications

device, leading to the concept of smart transformer (ST). Then the ST can be conceptualized as described below.

Smart transformer (ST):
Power electronics based system with a galvanic isolation in medium or high frequency between the input and output with extra control functionalities, obtained through advanced communication and control system. In addition to the hardware advantage, its enhanced control makes the ST a powerful technology able to provide ancillary services to the distribution grid, solving most of the problems posed by the grid modernization. This system is a remarkable technology able to fulfill most of the requirements for the implementation of a smart grid, either ac or dc.

Figure 9.3 shows the main requirements and performance of the SST for both described applications. While the volume and weight reduction are the most important requirement for traction application, the functionalities, reliability and efficiency play the most important roles in the electric distribution grid application.

9.3 ST architecture classification

The need to guarantee high efficiency and high reliability, offering new functionalities, has a direct impact on the ST architecture selection. To assist a proper structure selection, Figure 9.4 provides an overview of the possible architectures.

9.3.1 Power conversion stages

The most discussed classification is regarded the number of power conversion stages, that can be single-stage, two-stages or three-stages. The single-stage is normally implemented by a direct isolated matrix converter, and it is characterized to have high power density, because of the absence of the dc-link capacitors. However, the LV and MV are not decoupled, limiting the control capability. The two-stage topology is usually based on the indirect matrix converter and an inverter. This structure might present up to one constant dc-link, improving the controllability in respect to the single-stage structure. However, the two-stage structure offers limited functionalities. The three-stage is implemented by an MV ac–dc converter, isolated dc–dc converter and an LV inverter. This topology has normally two dc-links, in which at least one is available for connectivity. The decoupling between MV and LV sides provides more degree of freedom for the

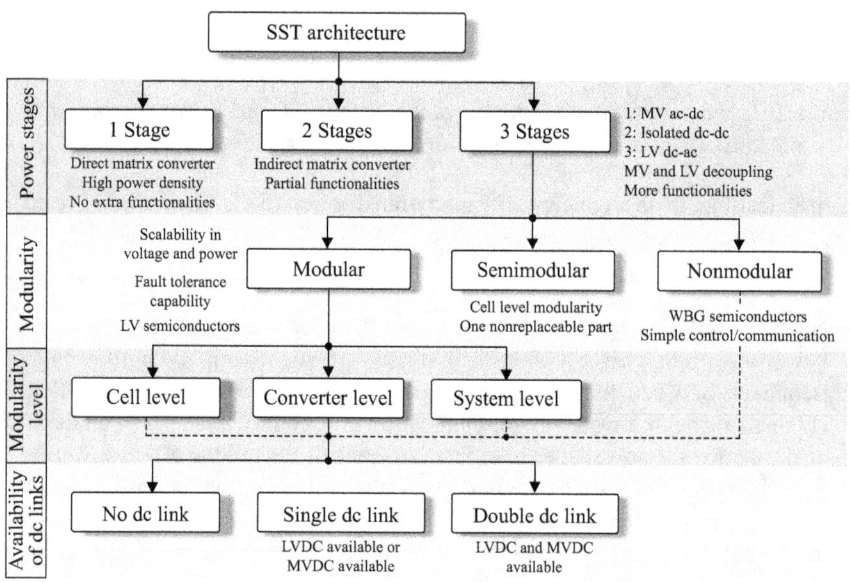

Figure 9.4 Classification of the ST architectures according to the number of power conversions, modularity and modularity level

system control, allowing the implementation of functionalities. Hence, this is the preferred architecture for the ST implementation.

9.3.2 Modularity

The next classification for the ST architecture concerns the degree of modularity. It can be classified as modular, nonmodular or semimodular. The nonmodular system is based on a single power converter, as exemplified in Figure 9.5(a), and usually takes the advantage of high voltage wide-bandgap (WBG) semiconductors [4,5,8]. Special devices with high blocking voltage capabilities, such as 10 kV Silicon Carbide (SiC) MOSFETs or 15 kV SiC IGBTs [9–11] are needed to handle the MV level in the power converter. Because these devices are not commercially available, but only for research purposes, no currently available products are using this technology. A standard approach to handle the MV in a nonmodular solution used in industrial applications is to employ semiconductor connected in series; however, this demands an additional control effort to balance the voltage over the semiconductors.

Conversely, the modular approach consists of several basic converters rated for LV or low current, which are used as building blocks for the entire system, as depicted in Figure 9.5(b). In this solution, the basic building block shares the voltage

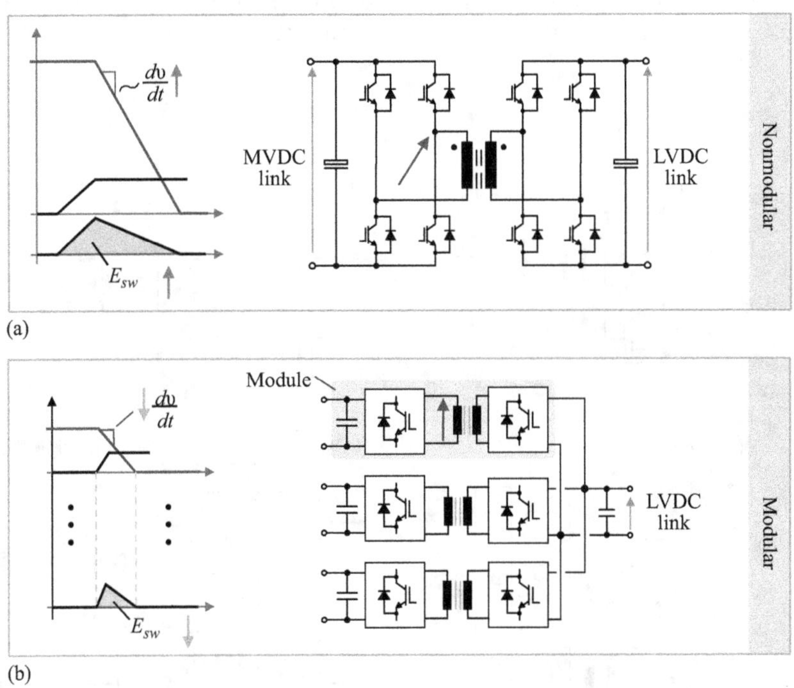

Figure 9.5 Possible implementation approaches of the ST architecture to handle the MV level involved in the power conversion: (a) nonmodular and (b) modular

on the MV side and the current in the LV among them, enabling the use of LV rated devices. In spite of the fact that more components are used compared to the non-modular solution, the modular approach is more economically advantageous, as demonstrated in [12]. Additionally, modular architectures bring several advantages to the power and voltage scalability, maintenance and the implementation of fault-tolerant strategies. In comparison to the nonmodular, the modular architectures have reduced electromagnetic interference (EMI) emissions [due to the low dv/dt and di/dt, as illustrated in Figure 9.5(a) and (b)] and the possibility to use standard LV rated devices that perform well, benefiting both efficiency and cost.

Some features from the modular and nonmodular concepts can be combined to form the semimodular concept, as an alternative for both previous approaches. In this case, the architecture is implemented using basic modules sharing the power, similar to the modular approach, but a nonreplaceable component processing the total system power is part of the architecture. It can also be called as a partially modular approach, in which the modules on the primary side divide the power among them, while the high frequency transformer (HFT) and the secondary side bridge processes all the power.

9.3.3 Modularity level

The modular approach can be implemented in different level, as shown in Figure 9.6, and they are classified as cell level, converter level and system level.

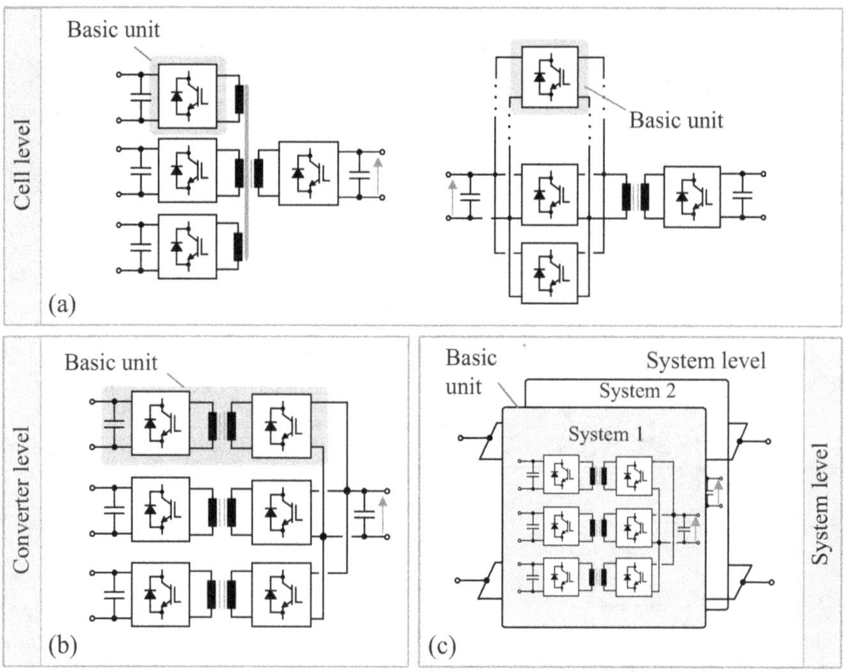

Figure 9.6 Modularization level: (a) cell level, (b) converter level and (c) system level

Cell level:
This is the most basic level, in which the basic unit is the cell used inside the converter, as exemplified in Figure 9.6(a). Another example of a cell-level modular system is the modular multilevel converter that is a single converter using several cells as basic units.

Converter level:
The next level is the converter level. In this case, the power converter is the basic unit, and they are combined in series and/or parallel to implement the entire system, as shown in Figure 9.6(b). This is the most employed modularization level and the input-series output-parallel connection is very popular. Interleaved converter is another example of the converter level modular system, in which several converters are connected in parallel.

System level:
Finally, the system level is presented in Figure 9.6(c). In this case, the full system is used as the basic unit. It is often used in application where fault-tolerance or scalability is highly desired.

All these levels can be freely combined, resulting in a modular system in multiple levels. In modular ST architectures, the converter level is the most common one, although the cell level can also be used.

9.3.4 dc Link voltage availability

The conceptualization of the ST is an important task during its conceivement, because it affects and constraints the further design phases. Besides the number of stages and the modularity level, another very important decision must be taken during the conceiving of the ST architecture: the availability of the dc link. Recently, the dc microgrid technology has been developed significantly, and many solutions for issues related to the grid stability and protection have been proposed in literature. This sort of grid is usually integrated to the ac grid through an LFT associated to a front-end rectifier for forming the required dc voltage. Alternatively, the dc microgrid can be connected directly to the dc link of the ST, so that its integration into the conventional ac grid happens through the ST. Hence, the ST technology enables efficiently the integration of the dc microgrid, and this feature play an important role during the conceptualization of the ST architecture.

The modular three-stage architecture has enough degree of freedom to provide one, two or even no connectivity to the dc links. It depends on the configuration of the power converters, as depicted in Figure 9.7 for a simplified single-phase structure. The availability of one dc link on the LV side (i.e., LVDC link) is the state-of-the-art and the most adopted option, enabling the dc microgrid connection [3–5].

The configurations without the LVDC link connectivity, as seen in Figure 9.7(a) and (c), have distributed dc links on the LV side, instead of having them

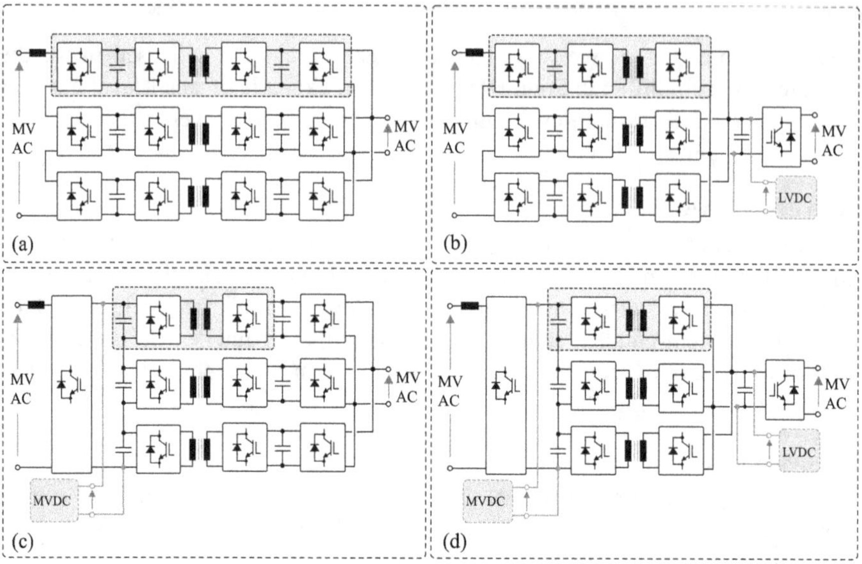

Figure 9.7 Different ST architecture configuration regarding the dc link connectivity for a simplified single-phase structure: (a) without dc link availability, (b) availability of the LVDC link, (c) availability of the MVDC link, (d) availability of both LVDC and MVDC links

connected. This strategy does not bring any advantages, and for this reason it has not being adopted [3]. On the other hand, the distributed dc links on the MV side provides advantages to the system, because the MV level involved in this side can be distributed among the modules. For this reason, this concept has been intensively used [3,4,12–16] and the most adopted structure is the one presented in Figure 9.7(b).

Recently, the availability of the MVDC link has been discussed by the research community, but there is no solution in this regard. The few amount of dc loads applicable for MV level compared to the LV one (which is already well established) makes the MVDC availability still controversial. Nevertheless, new dc loads in LV level are emerging (e.g., electric fast charging station), supporting research on the architectures with MVDC link. Consequently, the architecture with MVDC link as shown in Figure 9.7(c) and (d) is not well adopted yet, but it is a high potential structure and deserves attention.

In summary, the availability of the LVDC link is of paramount importance, and then two configurations with respect to the dc link connectivity are highlighted: the available LVDC link [Figure 9.7(b)] and the available LVDC and MVDC links [Figure 9.7(d)].

9.4 Solid-state transformer and smart transformer architectures overview

In the last two decades, many different SST and ST architectures have been proposed and discussed in literature for traction and distribution grid applications. The most relevant architectures used in both application are discussed.

In traction applications, the architectures are characterized to use commercialized HV Si-IGBT, two power processing stages, as well as modular approach. The efficiencies for each architecture for traction system are presented in Table 9.1. All of them provide higher efficiency and power density, when compared to the conventional solution using the LFT.

The three-stage architecture offers decoupling between the input and output, besides the availability of at least one dc link. Consequently, it is the preferable structure for grid applications, as seen in Table 9.1. To handle the MV level, two approaches are commonly used: modular with commercial LV devices and nonmodular with customized HV devices. Regardless the adopted concept, most of these architecture takes advantages of the high performance WBG SiC devices. They offer lower conduction and switching losses compared to the standard devices, and they are very promising in ST application.

Figure 9.8(a) shows the efficiency values available in the literature according to power of the system, as denoted in Table 9.1. For distribution system, very few ST demonstrators with solid results have been presented in the literature so far. The

Table 9.1 Comparison of the ST and SST architectures from different research groups/companies

			Distribution application		
Name	**Project/group**	**Stages**	**Concept**	**DC link**	**References**
A	FREEDM	3	Modular	LV	[13]
B	FREEDM	3	Nonmodular	MV/LV	[11,17]
C	FREEDM	3	Nonmodular	MV/LV	[9,10]
D	ETH	3	Modular	LV	[18]
E	Uniflex	3	Modular	No	[19–21]
F	GE	2	Modular	No	[22]
			Traction application		
Name	**Company**	**Stages**	**Concept**	**Efficiency (%)**	**References**
G	ABB	1	Modular	94.5	[23,24]
H	ABB	2	Modular	96.2	[14,25]
I	Alstom	2	Partially modular	94	[26,27]
J	Bombardier	2	Modular	96.2	[28]
K	Siemens	2	Partially modular	97	[29,30]

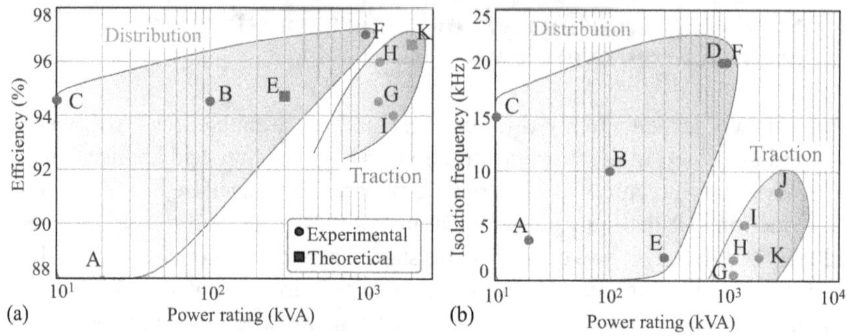

Figure 9.8 Performance comparison of the most relevant ST and SST architectures developed by industries and universities (see Table 9.1): (a) achieved efficiency compared to the power rating and (b) isolation frequency as a function of the power rating

available data shows that the efficiency achieved by this kind of system is around 94.5%–97%. For traction systems, the ST efficiency is in the range of 94.5%–96%, according to the available data provided experimentally (excluding the theoretical one). Although the ST architectures in the distribution use more power stages compared to those used in Traction, their performance is potentially higher, because the utilized WBG semiconductors that plays a very important role in this regards. Figure 9.8(b) shows the isolation frequency for each ST architecture shown in Table 9.1, according to the power level. In traction, the isolation frequency is normally below 10 kHz, whereas in distribution grid is mostly around 10–20 kHz. The limitation of switching frequency is ST for traction is explained by the fact of HV Si-IGBT are normally used to save cost. It results in more switching losses or switching frequency limitations.

9.5 dc–dc Stage: power converter topologies

The dc–dc stage is considered as the core element of the ST architecture, because it is responsible to connect the MV side to the LV side providing galvanic isolation in medium or high frequency among them. In consequence of its strict requirements discussed next, this stage is the most challenging for the implementation [3,16,18, 31,32]. The high voltage and current levels involved on the power conversion makes the dc–dc stage responsible for the majority of the system losses; hence, it deserves more attention during the ST design. Moreover, this stage controls the LVDC link, in which the dc microgrid is connected to. The requirements and a review of the possible solutions for the dc–dc stage is discussed next.

9.5.1 Requirements

The requirements for the dc–dc stage of the ST, independently of the adopted approach (modular or nonmodular), are summarized as follows:

- **High voltage capability in the MV side:** Handle voltage level of around 6–25 kV.
- **High current in the LV side:** Current level in the range of 100–2,000 A is expected [3].
- **High voltage isolation:** The MF/HF transformer employed has an isolation requirement defined by the own MVAC level, i.e., between 10 and 15 kV, regardless of the solution adopted for the dc–dc stage.
- **Decoupling degree between the MV dc side and the LV dc side:** Ability to provide total decoupling between the MVAC grid and the LVAC grid, so that the MV stage and LV stage converters can operate independently from each other. It means that the MV ac–dc stage can operate and provide the required ancillary services to the MV grid (e.g., provide reactive power), without disturbing or being disturbed by the LV dc–ac stage operation. The same remark is also valid for the LV dc–ac stage operation. To achieve such decoupling degree, both dc links (MV and LV) must be maintained constant with minimal oscillation voltage (i.e., 5% of the LVDC [3]), and the dc–dc converter should guarantee the voltage regulation.
- **Control of the power flow:** Ability to control the power flow between the LVDC link and MVDC link and manage the connection of loads of both sides [4,5,8].
- **Bidirectionality:** Many publications agree that the bidirectional power flow capability is required for the dc–dc stage [3–5,8,33]. On the other hand, this requirement is questionable, because the reverse power is an exceptional case in the distribution system. If economically viable, unidirectional solution could also be accepted. Despite that, bidirectionality is considered as a requirement in modern ST.
- **dc Breaker function:** Overload and short-circuit protection (working as a dc breaker) of the possible load/source/microgrid connected to the LVDC link) [3–5,8,33].
- **High efficiency:** The ST is not only intended to replace the LFT of the distribution system but also solve problems coming from the grid modernization [3]. However, high efficiency is expected for the ST, in order to compete with the LFT. The highest efficiencies obtained from an SST in traction applications are around 95%–97%, for two processing stages (i.e., ac–dc and dc–dc) [14]. Taking this fact into account, an expected efficiency of 96% is defined for the whole ST system, as shown in Figure 9.9(b). The MV stage can achieve an efficiency around 98.6%–99.3% [14,34], when multilevel converters are used, while the LV stage can offer an efficiency in the range of 98.8%–99.2% [35–37]. The dc–dc stage should provide an efficiency between 97.5% and 98.6%, so that the entire system achieves the desired efficiency, as shown in Figure 9.9. Thus, an efficiency goal above 97.5% can be defined.

9.5.2 Basic module topologies

From the dc–dc converter requirements, the high efficiency is the most severe one. The modular approach contributes a lot in terms of performance improvement of the ST, because this approach allows the use of lower voltage rated devices, which performs significantly better than those rated for higher voltage [3]. As described

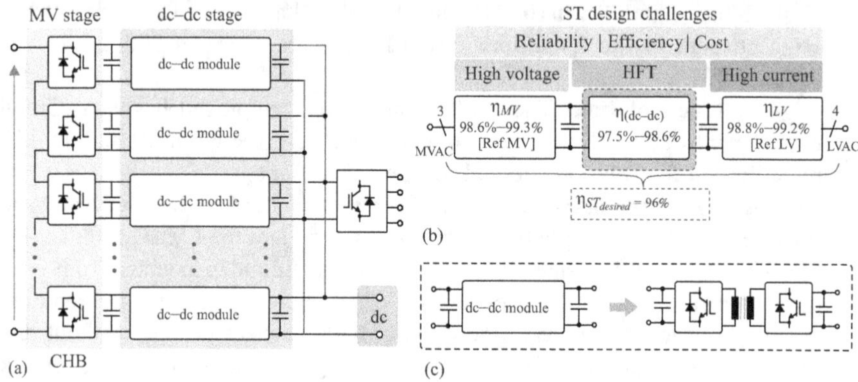

Figure 9.9 (a) Equivalent single phase modular ST architecture adopting the defined CHB, (b) expected efficiency for the whole system, as well as for each power conversion stage ([Ref MV] = [14,34], [Ref LV] = [35–37]), (c) implementation of the basic module of the modular dc–dc stage

Table 9.2 Specification range of the basic module of the dc–dc stage

Power level	5–100 kW
Input voltage	0.6–1.7 kV
Output voltage	600–800 V
Isolation frequency	1–40 kHz
Desired efficiency	>97.5%

before, this approach consists of several basic modules combined to build the entire architecture, as shown in Figure 9.9(a). The basic modules are isolated dc–dc converters, as depicted in Figure 9.9(c), and many topology options are available. The topology choice of the basic module plays an important role in the ST design, once the performance of the entire system depends on the efficiency of the individual module. The specification of the dc–dc basic module is usually in the range shown in Table 9.2. Considering this and the requirements described earlier, an overview of the most suitable topologies for the basic modules is presented as follows.

Among several options of the dc–dc converters, two families of converters are pointed out as the most promising: the SRCs and the active bridges converters.

9.6 Series resonant converter

The series resonant converter (SRC) is the most adopted converter in ST, considering the family of resonant converters, and its topology is presented in

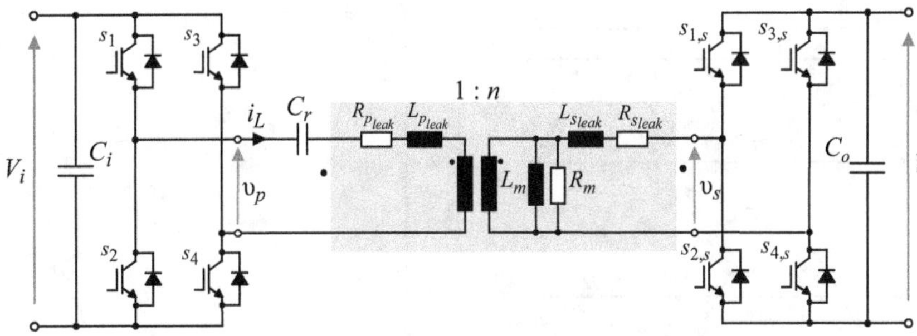

Figure 9.10 Topology of the series-resonant dc–dc converter

Table 9.3 Main parameters used on the analysis of the series resonant converter in Figure 9.10

Parameter	Definition
Angular switching frequency	$\omega_s = 2\pi f_s$
Angular resonant frequency	$\omega_o = 2\pi f_o$
Relative angular frequency	$\omega = \omega_s/\omega_s$
Resonant frequency	$f_o = 1/2\pi\sqrt{L_r C_r}$
Quality factor	$Q = 2\pi f_o L_r/R_a = 1/2\pi f_o C_r R_a$

Figure 9.10. The SRC converter is composed by a full-bridge converter on the primary and secondary sides, connected through a resonant tank circuit and a transformer. The tank circuit is composed of a resonant capacitor (C_r) connected in series to the resonant inductor (L_r). The SRC has different operation modes, according to the load and to the relationship between the switching frequency (f_s) and resonance frequency (f_o). The resonance frequency depends only on the pair L_r and C_r, and it is defined in Table 9.3. With respect to the load, the SRC can operate in continuous conduction mode (CCM) or discontinuous conduction mode (DCM). Regarding the switching frequency, it can operate below the resonant frequency, equal to the resonant frequency and above the resonant frequency [38].

Independent of the operation mode, the SRC converter can generally be analyzed according to [38–40]. In those works, the input–output voltage relation is derived, and it described in (9.1). Note that the variables presented in this equation are described in Table 9.3.

$$\frac{V_o}{V_i}(j\omega) = \frac{1}{1 + jQ((\omega_s/\omega_o) - (\omega_o/\omega_s))} \tag{9.1}$$

Figure 9.11(b) shows the normalized voltage gain (V_o/V_i) of the SRC as a function of the normalized operation frequency (f_o/f_s), according to the quality factor Q. According to this figure, the gain of the SRC varies with the frequency,

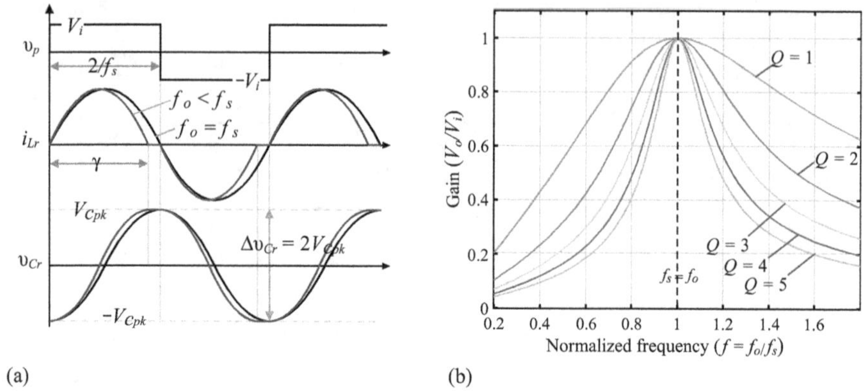

Figure 9.11 Series resonant converter: (a) main waveforms of the SRC and (b) normalized voltage gain (V_o/V_i) of the SRC as a function of the normalized operation frequency (f_o/f_s), according to the quality factor Q

and usually the switching frequency is used to control the output voltage of the SRC. However, this strategy implies a complex control system and poor utilization of the HFT, once it must be designed to operate for the minimum frequency. Thus, it is not suitable for ST applications, in which the power level is relatively high.

The most efficient operating point of the SRC is when it operates in the DCM with the switching frequency (f_s) equal or slightly below the resonance frequency (f_o), i.e., with unity gain. In this operation mode, the primary side switches achieve ZVS and the output diodes achieve zero-current-switching. Moreover, the converter has good transformer utilization and also low EMI emission due to the smooth current shape (low di/dt). In this operation point, the power flow is not directly controlled, and the power conversion between input and output is defined by (9.2), where n is the transformer turns ratio. On the other hand, it is not a disadvantage in ST application, once the system has enough degrees of freedom to control the input voltage by using the front-end rectifier (first stage).

$$V_o = n \cdot V_i \tag{9.2}$$

9.6.1 Current stress

To design the SRC converter, mapping the stresses on the semiconductors and transformers are required. They are normally calculated using waveforms illustrated in Figure 9.11(a) and using the concept of average and rms value. Using this approach, the mean value and rms value of the current in the semiconductors can be calculated using the equations shown in Table 9.4. Likewise, the current stresses on the transformer are calculated by the equations given in Table 9.5.

Table 9.4 Current stresses on the components S_{1b} (MV side) of the SRC [16]

Semiconductor	AVG	RMS
MV side	nP_o/V_o	$nP_o/8V_o\sqrt{2\pi/(f_s\sqrt{L_rC_r})}$
LV side	P_o/V_o	$P_o/8V_o\sqrt{2\pi/(f_s\sqrt{L_rC_r})}$

Table 9.5 Current stresses on the transformer of the SRC [16]

MV side	LV side
$nP_o/4V_o\sqrt{\pi/(f_s\sqrt{L_rC_r})}$	$P_o/4V_o\sqrt{\pi/(f_s\sqrt{L_rC_r})}$

9.6.2 Efficiency expectation

A high efficiency SRC is reported in [41] for high power and LV level applications (5 kW, 600 V). In that paper, SiC MOSFETs are used, presenting a peak efficiency of 97.6%, which is one of the highest efficiencies reported in the literature, considering the power/voltage level. Similarly, the SRC has been adopted in SST for traction applications in [14,25], where an efficiency of 98% was reported for a converter of 160 kW, employing special IGBTs devices.

Similarly, an optimum design of an SRC in ST application has been reported in [16]. In this paper, the authors proposed a design methodology to improve the efficiency and lifetime of the SRC, and the performance of the converter was evaluated considering different semiconductors. The converter provides an efficiency of 98.61%, when SiC-MOSFETs are used. Consequently, a very high efficiency can be expected for this kind of converter using WBG devices.

The main drawback of the SRC for ST application is the lack of controllability. The converter operates in open-loop, and it is not able to control the power flow. Consequently, the control of the LVDC link for dc microgrid connectivity must be shifted to other converter, when the SRC is used in the ST.

9.7 Dual active bridge (DAB) converter

The most popular converter of the active bridge family is the dual active bridge (DAB) and its topology is presented in Figure 9.12. The DAB converter is composed of two active bridges, connected through the inductance L_{DAB} and a transformer. Even though the DAB topology is similar to the SRC, its operation is totally different. Other topologies of the active bridge converter family, such as the three-phase DAB and the multiple-active-bridge (MAB), can also be adopted as a basic module of the ST. In fact, the MAB converter has been intensively

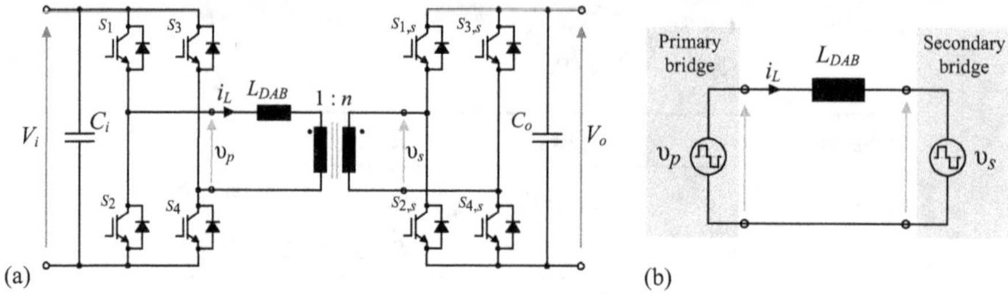

Figure 9.12 Topology of the dual active bridge (DAB) dc–dc converter (a) and its equivalent circuit (b)

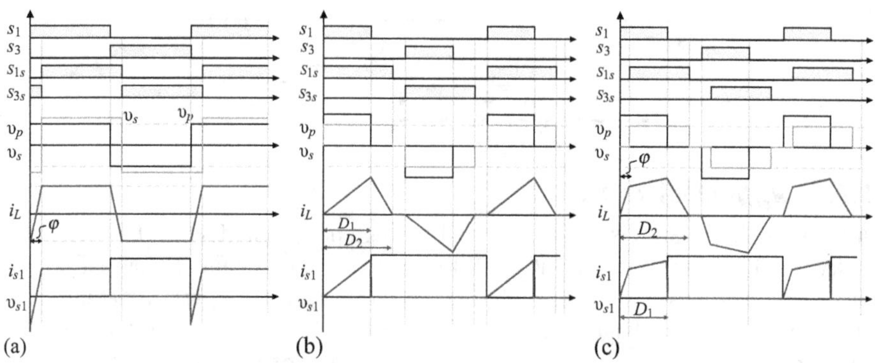

Figure 9.13 Main waveforms of the DAB converter considering different modulation strategies: (a) phase-shift modulation (PSM), (b) triangular current modulation (TCM), (c) trapezoidal modulation (TPM)

investigated for ST applications like in [42], and good results have been presented. However, the DAB converter is analyzed next for the sake of simplicity.

9.7.1 Modulation strategy

To modulate the DAB, many strategies have been presented [43–48], but the phase shift modulation (PSM) is the most adopted one. The main waveforms of the DAB using the PSM is depicted in Figure 9.13(a). In this scheme, rectangular voltages v_p and v_s with 50% duty-cycles and phase shift of φ are applied to the transformer with a constant switching frequency f_s. The power is controlled by the phase difference between the bridges. The PSM is characterized by ZVS turn-on of the semiconductors, low rms current, symmetrical share of the losses in all switches, high power transfer capability besides its simplicity. On the other hand, these advantages are obtained depending on the input–output voltage ratio and also the chosen operation point (nominal phase shift φ).

When the input voltage and/or output voltage of the converter deviate significantly from their nominal values (around 20%, according to [46,49]), the soft-switching operation is normally lost. In this condition, the PSM is not advantageous anymore [46]. To overcome this problem, alternative modulation strategies have been proposed, like the triangular current mode (TCM) and the trapezoidal modulation (TPM), as shown in Figure 9.13(b) and (c), respectively. The TCM imposes a triangular shape current on the transformer, while the TPM generates a trapezoidal current waveform on the transformer, as shown in (b) and (c), respectively. In contrary to the PSM, the TCM and TPM control the duty-cycle for transferring power between the input and output of the converter. These strategies can operate with soft-switching independently from the input and/or output voltage deviation, but with penalty on the peak value of the currents. To transfer the same amount of power, the DAB converter presents higher peak values of currents, when TCM and TPM are adopted in comparison with the PSM, because of the triangular or rectangular shape of the current. It means that the rms values on the semiconductors and transformers are also expected to be higher, implying on possible increase of the conduction losses.

In ST application, the input and output voltage are typically constant without large deviation of the nominal value. In this sense, the PSM is very advantageous, and therefore the DAB converter is analyzed next using this strategy.

9.7.2 Analysis of the DAB using the PSM

For the theoretical analysis of the DAB converter, the equivalent circuit shown Figure 9.12(b) can be used, where the active bridges are replaced by rectangular voltage sources v_p and v_s. The waveforms of v_p and v_s are shown in Figure 9.14. Analyzing the equivalent circuit, the inductor current is described by (9.3).

$$i_L(t) = i_L(t_0) + \frac{1}{L}\int_{t_0}^{t_0+T_s} v_L(t)dt = i_L(t_0) + \frac{1}{L}\int_{t_0}^{t_0+T_s} (v_p - nv_s)dt \tag{9.3}$$

The average power transferred can be defined by the following equation:

$$P = \frac{1}{T_s}\int_{t_0}^{t_0+T_s} p(t)dt = \frac{2}{T_s}\int_{t_0}^{t_0+T_s/2} v_p(t)i_L(t)dt = \frac{2V_i}{T_s}\int_{t_0}^{t_0+T_s/2} i_L(t)dt \tag{9.4}$$

Considering the current waveforms shown in Figure 9.14, the inductor current can be described by (9.5). In this equation, the initial values are defined by (9.6) and (9.7), respectively.

$$i_L = \begin{cases} i_L(t_o) + \dfrac{v_p + nv_s}{L_{DAB}}t, & t_o < t < t_1 \\[2ex] i_L(t_1) + \dfrac{v_p + nv_s}{L_{DAB}}(t - t_1) & t_1 < t < t_0 + \dfrac{T_s}{2} \end{cases} \tag{9.5}$$

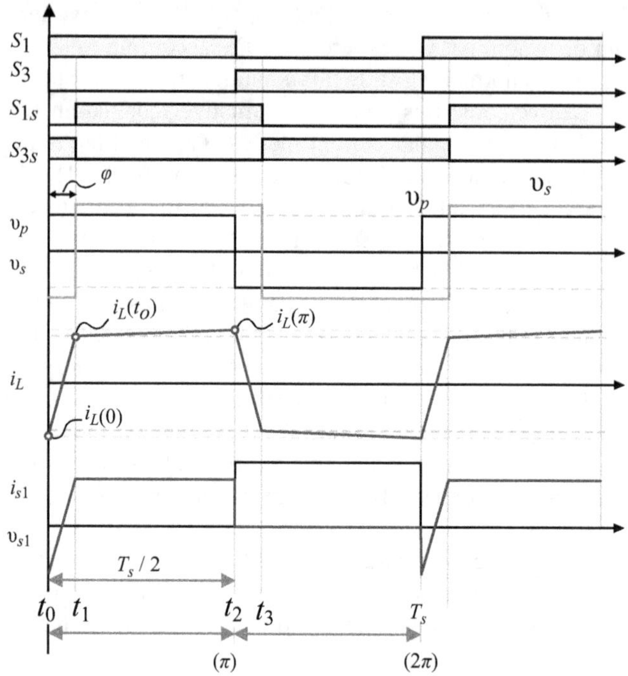

Figure 9.14 Voltage and current waveforms of the DAB converter operating with PSM

$$i_L(t_o) = -\frac{(V_i - nV_o)\pi - 2nV_o\varphi}{4\pi f_s L_{DAB}} \tag{9.6}$$

$$i_L(t_1) = \frac{(nV_o - V_i)\pi - 2V_i\varphi}{4\pi f_s L_{DAB}} \tag{9.7}$$

Note that $i_L(t_o) = -i_L(\pi)$ in steady-state condition. Replacing (9.5) in (9.4) and rearranging it, the equation of the power transference of the DAB is obtained, as shown in (9.8) [49]. The input–output voltage relation of the DAB can be defined as $d = nV_o/V_i$. The output power can be rewritten in terms of d, as shown in the following equation:

$$P = \frac{V_i V_o}{2\pi f_s L_{DAB} n}\varphi\left(1 - \frac{|\varphi|}{\pi}\right) = \frac{V_i^2}{2\pi f_s L_{DAB}} d\varphi\left(1 - \frac{|\varphi|}{\pi}\right) \tag{9.8}$$

The soft-switching is achieved when the devices are conducting at turn-off. This means that the condition $i_L(0) \leq 0$ and $i_L(\varphi) \geq 0$ must be satisfied. The requirement of $i_L(\varphi) \geq 0$ ensures the turn-off under ZVS of the switches s_1 and s_4 (see Figure 9.14), while $i_L(\varphi) \geq 0$ guarantees the soft-switching of the devices s_2 and s_3. Applying these limits in (9.6) and (9.7), the ZVS boundaries conduction is obtained and described in (9.9). Accordingly, the area constrained by (9.9)

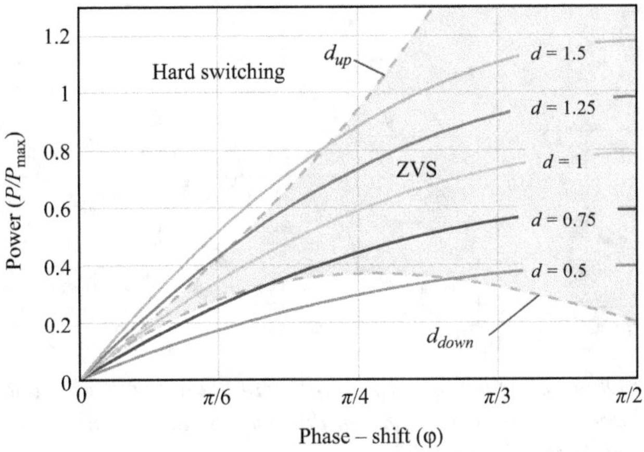

Figure 9.15 Output power of the DAB converter in function of phase-shift angle φ, considering several input–output voltage relation d

indicates the soft-switching region of the DAB converter, which is the desired operation region to reduce losses.

$$\begin{cases} d_{ZVS(up)} = 1 - \dfrac{3\varphi}{2\pi} \\ d_{ZVS(down)} = \dfrac{1}{1 - (3\varphi/2\pi)} \end{cases} \tag{9.9}$$

Figure 9.15 shows the variation of the normalized power as a function of the phase-shift angle for different values of d, in which the soft-switching boundaries are highlighted. Note that the DAB converter operates in soft-switching for the entire range of load if $d = 1$, i.e., $V_i = nV_o$. Consequently, the DAB converter can be properly designed, so that it operates with soft-switching for the whole load range.

9.7.3 Current stresses on the DAB

To calculate the current stresses on the semiconductors and transformer, the peak current on the LV side inductor ($I_{L_{a(pk)}}$) and MV side inductor ($I_{L_{b(pk)}}$) are required, and those can be calculated by (9.10) and (9.11), respectively.

$$I_{L_{PK(LV)}} = \frac{V_L - \sqrt{V_L^2 - 8L_{eq}f_s n V_L I_o}}{4L_{eq}f_s} \tag{9.10}$$

$$I_{L_{PK(MV)}} = \frac{I_{L_{a(pk)}}}{n} \tag{9.11}$$

Considering the current waveforms presented in Figure 9.16 for positive power flow, i.e., from MV to LV side, the current stresses on the semiconductors and transformer are calculated and presented in Tables 9.6–9.8. Table 9.6 shows the

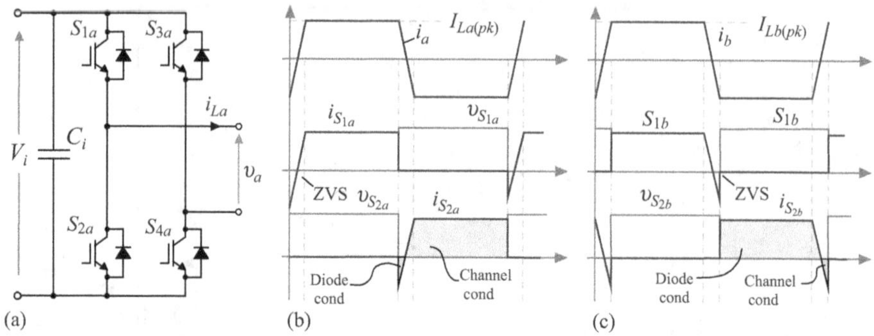

Figure 9.16 Dual active bridge: (a) topology of the active bridge and the current and voltage waveforms on the semiconductors of the (b) MV bridge and (c) LV bridge

Table 9.6 Current stresses on the semiconductor S_{1b} (MV side) of the DAB converter

Component	**AVG**	**RMS**
Channel (S_{1b})	$I_{L_{PK(LV)}}/6n(1-(3\varphi/4\pi))$	$I_{L_{PK(LV)}}/3n\sqrt{1/2(1-(5\varphi/12\pi))}$
Body diode ($i_{D_{1b}}$)	$(I_{L_{PK(LV)}}/3n)(\varphi/8\pi)$	$i_{D_{1b(rms)}} = (I_{L_{PK(LV)}}/3n)\sqrt{(\varphi/12\pi)}$

Table 9.7 Current stresses on the semiconductor S_{1a} (LV side) of the DAB converter

Component	**AVG**	**RMS**
Channel (S_{1a})	$(I_{L_{PK(LV)}}/6n)(1-(3\varphi/4\pi))$	$(I_{L_{PK(LV)}}/3n)\sqrt{1/2(1-(5\varphi/12\pi))}$
Body diode ($i_{D_{1a}}$)	$(I_{L_{PK(LV)}}/3n)(\varphi/8\pi)$	$i_{D_{1b(rms)}} = (I_{L_{PK(LV)}}/3n)\sqrt{(\varphi/12\pi)}$

Table 9.8 Current stresses on the transformer of the DAB converter

MV side i_{Lb}	**MV side i_{La}**
$(I_{L_{PK(LV)}}/3n)\sqrt{1-(2\varphi/3\pi)}$	$I_{L_{PK(LV)}}\sqrt{1-(2\varphi/3\pi)}$

average and rms currents on the channel and body diode of the semiconductor S_1a, while Table 9.6 shows the current effort on the semiconductor of the LV side S_1b of the DAB converter. The current stresses on the inductor regardless it is connected to the LV or MV side are presented in Table 9.8.

As observed in these equations, the current stresses of all components are somehow proportional to defining φ. Furthermore, when the power flows from the MV to the LV, most of the current flows through the channel of the MV side semiconductors, whereas it flows mostly through the body diode on the LV side. Therefore, it is very important to select a power semiconductor for the LV side with low forward drop voltage for the body diode.

9.7.4 Efficiency calculation

High performance is expected for the DAB converter, because of its soft-switching operation, but it depends on the design and the employed semiconductors. For instance, if a large nominal phase-shift is selected, i.e., larger than 60°, the converter operates with high reactive power, deteriorating the efficiency. In this case, even operating under soft-switching, the converter might have low performance. Thus, the design and selected components are of paramount importance.

In [48], a 4 kW DAB prototype based on 1.2 kV Silicon (Si) IGBT has achieved an efficiency of 97.1%, while an efficiency of 97.5% was reported in [50], for a 6.25 kW prototype based on silicon carbide (SiC) MOSFETs. Similarly, an optimized design of the DAB is presented in [51], where 1.2 kV SiC MOSFETs were employed, and an efficiency of 98.2% was obtained.

The quadruple active bridge (QAB) [52], another topology from the active bridge converters family, has been used as modules for the ST architecture in [42]. Differently from the DAB, in which only two bridges are connected to the HFT, the QAB converter has four bridges connected to the HFT, as it is illustrated in Figure 9.17. This converter operates similarly to the QAB when employed to the ST application. Hence, a similar efficiency is expected. In fact, the QAB is more advantageous than the DAB in ST application according to [12], because the QAB can perform similarly to the DAB, but offering a significant cost reduction. The cost reduction is obtained due to the fact that less bridges are used on the LV side (see Figure 9.17), requiring less auxiliary power supplies and gate drivers, which are costly.

In [12,42], an optimum design of the QAB converter is reported for optimizing the efficiency and cost. The converter has presented an efficiency of 97.5% when SiC-MOSFETs are used, which is considerably high. Although the SRC tends to present higher efficiency than the QAB converter, this last converter presents two advantages over the SRC: lower cost and high-control capability. This last feature is of paramount importance, because the QAB converter is able to control the LVDC link and enabling the connection of the dc microgrid directly to the LVDC without an extra converter. It is highly desired in ST applications, and then the QAB converter is highlighted to be a very advantageous choice for ST applications.

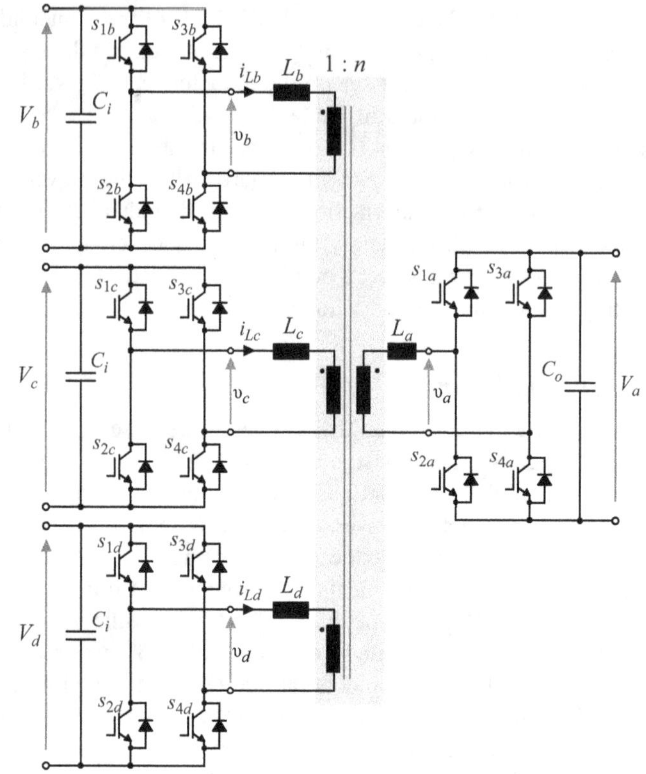

Figure 9.17 The topology of the quadruple active bridge (QAB)

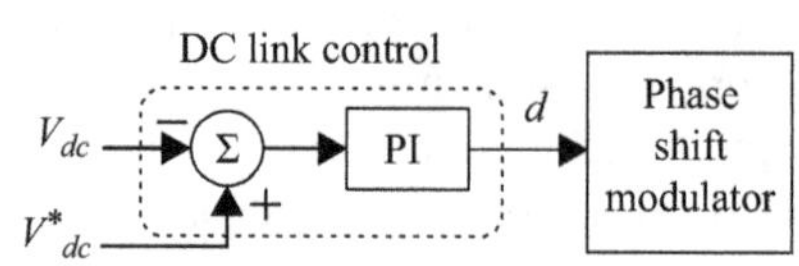

Figure 9.18 General form of a dc voltage control for a DAB

9.8 Active-bridge converter: control description and tuning

The objective of the control is to ensure a stable bus regulation, with a good immunity from the load variations. Figure 9.18 shows a simplified diagram of the control. The normalized phase shift $d = (\varphi/2\pi)$ is the output of a PI regulator that controls the dc voltage.

Considering the equation that regulates the power transfer (9.12), the current that flows into the output port can be obtained by $I_{dc} = P_{dc}/V_{dc}$.

$$P_{dc} = \frac{nV_{in}V_{dc}}{f_{sw}L_{lk}} d(1 - 2|d|) \tag{9.12}$$

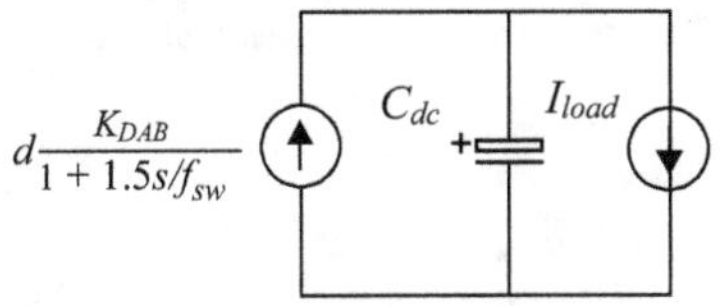

Figure 9.19 Simplified model for the voltage control of the dc bus

By linearizing the relationship, (13) can be written.

$$I_{DC} = \frac{nV_{in}}{f_{sw}L_{lk}}(1 - 4d)|_{d=d_0,V_{in}=V_{in0}}(d - d_0) + \frac{nV_{in}}{f_{sw}L_{lk}}d(1 - 2|d|)|_{d=d_0,V_{in}=V_{in0}}(V_{in} - V_{in0}) \tag{9.13}$$

As can be seen, the small-signal gain depends on the linearization point and is also dependent on the input voltage. The common simplification is to consider the input voltage constant in a switching cycle, neglecting the second part of (9.13). Moreover, since the design of the DAB is normally realized for small phase shift to limit the reactive power, the linearization around the point $d_0 = 0$ is performed. It is important to note that this also constitutes the worst case scenario for the stability point of view (maximum gain).

In order to study the control, a simplified model is shown in Figure 9.19, where $K_{DAB} = 4(V_{in}/f_{sw}L_{lk})$. In order to take into account the switching frequency and the discretized control, a first-order transfer function with a time constant of $T_c = 1.5/f_{sw}$ is added.

Neglecting the load current for the controller tuning, the system presents a pole in the origin due to the capacitor. The tuning of the PI can be performed with the symmetrical optimum criterion that aims at maximizing the phase margin. The target is the crossover frequency of the open-loop transfer function f_{cut}. The PI regulator is chosen so that the ratio between the target crossover frequency and its corner frequency ($1\tau_i$) is the same as the ratio between the high-frequency pole of the plant and the target crossover frequency $R = f_{cut}\tau_i = 1/f_{cut}T_c$.

In this way,

$$K_p = 2\pi f_{cut}\frac{C_{dc}}{K_{DAB}\sqrt{1 + (1/R^2)}} \tag{9.14}$$

$$K_i = \frac{K_p}{\tau_i} \tag{9.15}$$

The control realized with a simple PI regulator has a main drawback: the higher the phase margin of the design, the higher the output impedance becomes. The smaller the output impedance $Z(s) = (V_{dc}(s)/I_{load}(s))$, the more immune the control system is to load variation. As a consequence, fast-changing load will deteriorate the voltage regulation unless a very fast controller is realized.

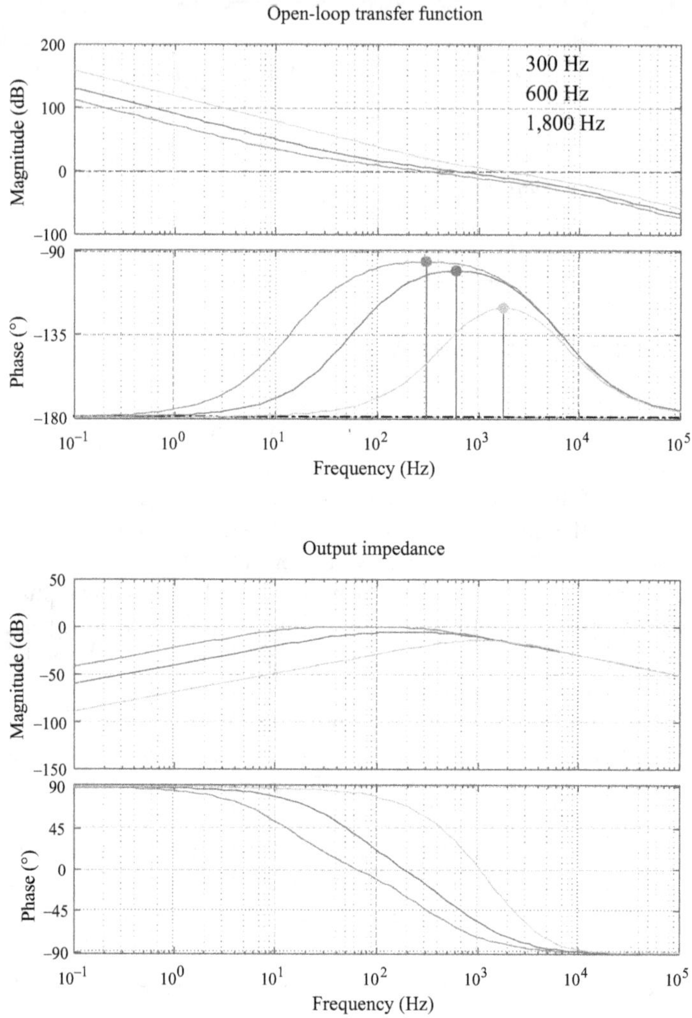

Figure 9.20 Frequency response of the voltage control and of the output impedance

Figure 9.20 shows the frequency responses of the voltage control and of the output impedance with the voltage control. As can be seen, the crossover of the open-loop transfer function always happens at the maximum of the phase. Of course, choosing a higher crossover frequency reduces the phase margin.

9.9 Summary

The SST has several advantages when applied in the modern electric distribution grid. Among them, the availability of the dc link for connection of dc microgrid is mentioned in this chapter. There are many different possibilities to implement the architecture of the SST, and an overview and classification have been presented. For electric distribution, the three power processing stages architecture is cited to be the best choice, because it provides the decoupling between the MV side and LV side and also providing dc link connectivity. Furthermore, the dc–dc stage plays an important role on the SST architecture design, because it is responsible for the major losses of the system, besides to be in charge for controlling the LVDC link. Therefore, the SRC and the DAB have been pointed out as the most promising choice. The SRC can provide a very high efficiency, but it is not able to control the power flow and the regulate the LVDC link properly. The DAB converter, on the other hand, can provide a high efficiency solution and provide full control. The design aspects of the power stage and control stage of the DAB converter focusing on the SST application has also been discussed in this chapter.

References

[1] E. McMurray, "Power converter circuits having a high frequency link," 1968, U.S. Patent US3517300.

[2] D. Dujic, F. Kieferndorf, F. Canales, and U. Drofenik, "Power electronic traction transformer technology," in *Proceedings of The 7th International Power Electronics and Motion Control Conference*, vol. 1, Jun 2012, pp. 636–642.

[3] M. Liserre, G. Buticchi, M. Andresen, G. D. Carne, L. F. Costa, and Z. X. Zou, "The smart transformer: impact on the electric grid and technology challenges," *IEEE Industrial Electronics Magazine*, vol. 10, no. 2, pp. 46–58, Summer 2016.

[4] X. She, A. Q. Huang, and R. Burgos, "Review of solid-state transformer technologies and their application in power distribution systems," *IEEE Journal of Emerging and Selected Topics in Power Electronics*, vol. 1, no. 3, pp. 186–198, Sep 2013.

[5] A. Q. Huang, M. L. Crow, G. T. Heydt, J. P. Zheng, and S. J. Dale, "The future renewable electric energy delivery and management (FREEDM) system: the energy internet," *Proceedings of the IEEE*, vol. 99, no. 1, pp. 133–148, Jan 2011.

[6] A. Q. Huang, "Medium-voltage solid-state transformer: technology for a smarter and resilient grid," *IEEE Industrial Electronics Magazine*, vol. 10, no. 3, pp. 29–42, Sep 2016.

[7] B. Olek and M. Wierzbowski, "Local energy balancing and ancillary services in low-voltage networks with distributed generation, energy storage, and active loads," *IEEE Transactions on Industrial Electronics*, vol. 62, no. 4, pp. 2499–2508, Apr 2015.

[8] J. Wang, A. Q. Huang, W. Sung, Y. Liu, and B. J. Baliga, "Smart grid technologies," *IEEE Industrial Electronics Magazine*, vol. 3, no. 2, pp. 16–23, Jun 2009.

[9] K. Mainali, A. Tripathi, S. Madhusoodhanan, *et al.*, "A transformerless intelligent power substation: a three-phase SST enabled by a 15-kV SiC IGBT," *IEEE Power Electron Magazine*, vol. 2, no. 3, pp. 31–43, Sep 2015.

[10] S. Madhusoodhanan, A. Tripathi, D. Patel, *et al.*, "Solid-state transformer and MV grid tie applications enabled by 15 kV SiC IGBTs and 10 kV SiC MOSFETs based multilevel converters," *IEEE Transactions on Industry Applications*, vol. 51, no. 4, pp. 3343–3360, Jul 2015.

[11] F. Wang, G. Wang, A. Huang, W. Yu, and X. Ni, "Design and operation of a 3.6 kV high performance solid state transformer based on 13 kV SiC MOSFET and JBS diode," in *2014 IEEE Energy Conversion Congress and Exposition (ECCE)*, Sep 2014, pp. 4553–4560.

[12] L. F. Costa, F. Hoffmann, G. Buticchi, and M. Liserre, "Comparative analysis of MAB dc–dc converters configurations in modular smart transformer," in *2017 IEEE 8th International Symposium on Power Electronics for Distributed Generation Systems (PEDG)*, Apr 2017, pp. 1–8.

[13] X. She, X. Yu, F. Wang, and A. Q. Huang, "Design and demonstration of a 3.6-kV 120-V/10-kVA solid-state transformer for smart grid application," *IEEE Transactions on Power Electronics*, vol. 29, no. 8, pp. 3982–3996, Aug 2014.

[14] C. Zhao, D. Dujic, A. Mester, *et al.*, "Power electronic traction transformer: medium voltage prototype," *IEEE Transactions on Industrial Electronics*, vol. 61, no. 7, pp. 3257–3268, Jul 2014.

[15] M. Liserre, M. Andresen, L. Costa, and G. Buticchi, "Power routing in modular smart transformers: active thermal control through uneven loading of cells," *IEEE Power Electronics Magazine*, vol. 10, no. 3, pp. 43–53, Sept 2016.

[16] L. F. Costa, G. Buticchi, and M. Liserre, "Highly efficient and reliable SiC-based dc–dc converter for smart transformer," *IEEE Transactions on Industrial Electronics*, vol. 64, no. 10, pp. 8383–8392, Oct, 2017.

[17] F. Wang, G. Yao, A. Huang, W. Song, and X. Ni, "A 3.6 kV high performance solid state transformer based on 13 kV SiC MOSFET," in *2014 IEEE 5th International Symposium on Power Electronics for Distributed Generation Systems (PEDG)*, Galway, Jun 2014, pp. 1–8.

[18] J. W. Kolar, and J. Huber, "Solid-state transformers (SST) concepts, challenges and opportunities," in *ECPE Workshop on Smart Transformers for Traction and Future Grid Application*, Zurich, Switzerland, Feb 2016.

[19] S. Bifaretti, P. Zanchetta, A. Watson, L. Tarisciotti, and J. C. Clare, "Advanced power electronic conversion and control system for universal and flexible power management," *IEEE Transactions on Smart Grid*, vol. 2, no. 2, pp. 231–243, Jun 2011.

[20] J. Clare, "Advanced power converters for universal and flexible power management in future electricity networks," in *2009 13th European Conference on Power Electronics and Applications*, Sep 2009, pp. 1–29.

[21] H. Q. S. Dang, A. Watson, J. Clare, *et al.*, "Advanced integration of multilevel converters into power system," in *2008 34th Annual Conference of IEEE Industrial Electronics*, Nov 2008, pp. 3188–3194.

[22] M. K. Das, C. Capell, D. E. Grider, *et al.*, "10 kV, 120 a SiC half H-bridge power MOSFET modules suitable for high frequency, medium voltage applications," in *2011 IEEE Energy Conversion Congress and Exposition*, Sep 2011, pp. 2689–2692.

[23] L. Meysenc, P. Stefanutti, P. Noisette, N. Hugo, and A. Akdag, "Multilevel ac/dc converter for traction application," 2009, U.S. Patent US7558087.

[24] N. Hugo, P. Stefanutti, M. Pellerin, and A. Akdag, "Power electronics traction transformer," in *2007 European Conference on Power Electronics and Applications*, Sep 2007, pp. 1–10.

[25] D. Dujic, C. Zhao, A. Mester, *et al.*, "Power electronic traction transformer-low voltage prototype," *IEEE Transactions on Power Electronics*, vol. 28, no. 12, pp. 5522–5534, Dec 2013.

[26] J. Taufiq, "Power electronics technologies for railway vehicles," in *2007 Power Conversion Conference – Nagoya*, Apr 2007, pp. 1388–1393.

[27] J. Martin, P. Ladoux, B. Chauchat, J. Casarin, and S. Nicolau, "Medium frequency transformer for railway traction: soft switching converter with high voltage semi-conductors," in *2008 International Symposium on Power Electronics, Electrical Drives, Automation and Motion*, Jun 2008, pp. 1180–1185.

[28] M. Steiner and H. Reinold, "Medium frequency topology in railway applications," in *2007 European Conference on Power Electronics and Applications*, Sep 2007, pp. 1–10.

[29] M. Glinka and R. Marquardt, "A new ac/ac multilevel converter family," *IEEE Transactions on Industrial Electronics*, vol. 52, no. 3, pp. 662–669, Jun 2005.

[30] M. Glinka, "Prototype of multiphase modular-multilevel-converter with 2 MW power rating and 17-level-output-voltage," in *2004 IEEE 35th Annual Power Electronics Specialists Conference (IEEE Cat. No. 04CH37551)*, vol. 4, 2004, pp. 2572–2576.

[31] M. Liserre, M. Andresen, L. Costa, and G. Buticchi, "Power routing in modular smart transformers: active thermal control through uneven loading of cells," *IEEE Industrial Electronics Magazine*, vol. 10, no. 3, pp. 43–53, Fall 2016.

[32] L. F. Costa, G. Buticchi, and M. Liserre, "Quad-active-bridge as cross-link for medium voltage modular inverters," in *IEEE Energy Conversion Congress and Exposition (ECCE)*, Sep 2015, pp. 645–652.

[33] J. E. Huber and J. W. Kolar, "Solid-state transformers: on the origins and evolution of key concepts," *IEEE Industrial Electronics Magazine*, vol. 10, no. 3, pp. 19–28, Sep 2016.

[34] L. Schrittwieser, M. Leibl, M. Haider, F. Thöny, J. W. Kolar, and T. B. Soeiro, "99.3% Efficient three-phase buck-type all-SiC SWISS rectifier for dc distribution systems," in *2017 IEEE Applied Power Electronics Conference and Exposition (APEC)*, March 2017, pp. 2173–2178.

[35] B. Benkendorff, F. W. Fuchs, and M. Liserre, "Simulated and measured efficiency verification power circulation method of a high power low voltage

NPC converter for wind turbines," in *2016 18th European Conference on Power Electronics and Applications (EPE'16 ECCE Europe)*, Sep 2016, pp. 1–10.

[36] Y. Shi, R. Xie, L. Wang, Y. Shi, and H. Li, "Switching characterization and short-circuit protection of 1200 V SiC MOSFET T-type module in PV inverter application," *IEEE Transactions on Industrial Electronics*, vol. 64, no. 11, pp. 9135–9143, Nov 2017.

[37] J. Colmenares, D. Peftitsis, J. Rabkowski, D. P. Sadik, G. Tolstoy, and H. P. Nee, "High-efficiency 312-kVA three-phase inverter using parallel connection of silicon carbide MOSFET power modules," *IEEE Transactions on Industry Applications*, vol. 51, no. 6, pp. 4664–4676, Nov 2015.

[38] K. Afridi, "Resonant and soft-switching techniques in power electronics," Department of Electrical, Computer and Energy, Colorado University, Colorado, USA, Lectures Note, 2014.

[39] R. L. Steigerwald, "A comparison of half-bridge resonant converter topologies," *IEEE Transactions on Power Electronics*, vol. 3, no. 2, pp. 174–182, Apr 1988.

[40] P. K. Jain, "Power electronic course: resonant dc/dc converters," Department of Electrical and Computer Engineering, Queen's University, Canada, Ontario, Lectures Note, 2016.

[41] S. Inoue and H. Akagi, "A bidirectional isolated dc/dc converter as a core circuit of the next-generation medium-voltage power conversion system," *IEEE Transactions on Power Electronics*, vol. 22, no. 2, pp. 535–542, Mar 2007.

[42] L. F. Costa, G. Buticchi, and M. Liserre, "Quadruple active bridge dc–dc converter as the basic cell of a modular smart transformer," in *2016 IEEE Applied Power Electronics Conference and Exposition (APEC)*, Mar 2016, pp. 2449–2456.

[43] D. J. Costinett, "Analysis and design of high efficiency, high conversion ratio, dc–dc power converters," Ph.D. dissertation, University of Colorado at Boulder, 2010.

[44] S. Han, "High-power bi-directional dc/dc converters with controlled device stresses," Ph.D. dissertation, Georgia Institute of Technology, 2012.

[45] H. van Hoek, "Design and operation considerations of three-phase dual active bridge converters for low-power applications with wide voltage ranges," Ph.D. dissertation, RWTH Aachen University, 2017.

[46] N. Schibli, "Symmetrical multilevel converters with two quadrant dc–dc feeding," EPFL, Lausanne, 2000.

[47] F. Krismer and J. W. Kolar, "Closed form solution for minimum conduction loss modulation of DAB converters," *IEEE Transactions on Power Electronics*, vol. 27, no. 1, pp. 174–188, Jan 2012.

[48] T. Todorčević, R. van Kessel, P. Bauer, and J. A. Ferreira, "A modulation strategy for wide voltage output in DAB-based dc–dc modular multilevel converter for DEAP wave energy conversion," *IEEE Journal of Emerging*

and Selected Topics in Power Electronics, vol. 3, no. 4, pp. 1171–1181, Dec 2015.

[49] R. W. A. A. D. Doncker, D. M. Divan, and M. H. Kheraluwala, "A three-phase soft-switched high-power-density dc/dc converter for high-power applications," *IEEE Transactions on Industry Applications*, vol. 27, no. 1, pp. 63–73, Jan 1991.

[50] C. Gammeter, F. Krismer, and J. W. Kolar, "Comprehensive conceptualization, design, and experimental verification of a weight-optimized all-SiC 2 kV/700 V DAB for an airborne wind turbine," *IEEE Journal of Emerging and Selected Topics in Power Electronics*, vol. 4, no. 2, pp. 638–656, Jun 2016.

[51] R. M. Burkart and J. W. Kolar, "Comparative *eta–rho–sigma* pareto optimization of Si and SiC multilevel dual-active-bridge topologies with wide input voltage range," *IEEE Transactions on Power Electronics*, vol. 32, no. 7, pp. 5258–5270, Jul 2017.

[52] S. Falcones, R. Ayyanar, and X. Mao, "A dc–dc multiport-converter-based solid-state transformer integrating distributed generation and storage," *IEEE Transactions on Power Electronics*, vol. 28, no. 5, pp. 2192–2203, May 2013.

Chapter 10
Bipolar-type DC microgrids for high-quality power distribution

Sebastian Rivera[1,2], Ricardo Lizana[3], Samir Kouro[1] and Bin Wu[4]

10.1 Introduction

The implementation of active distribution networks aims to modernize and improve the sustainability of power systems, in order to include distributed energy generation resources and also to address the growing presence of power electronics in different applications. This will be done through the integration of communication and control technologies for effective control and monitoring of the system, in the form of active and flexible DC distributed networks. The deployment of these active distribution networks improves the performance of the traditional electric system in terms of cost, efficiency, space and reliability. Different factors motivated studies and development of a wide variety of DC distribution systems during the last years [1], and led to a regained popularity in DC energy systems. Some of these reasons are the growing presence of DC-based consumer equipment, the significant increase in the installation of renewable energy conversion systems and their detrimental effects on the power quality in conventional AC grids.

Moreover, the successful introduction of transmission of electrical power using high-voltage DC (HVDC) systems has further validated their enhanced efficiency, higher power capacity and controllability when compared to their AC counterparts, especially in long distance and underground or submarine transmission systems. These features allowed important developments in HVDC transmission configurations over the last years [2–6]. Consequently, these improvements later extended to distribution level, enabled by advances in power electronics, the relative maturity of DC-based consumer appliances and their subsequent widespread growth [7,8]. Currently, DC-based equipment is increasing its penetration in low-voltage kilowatt-scale applications in many sectors: aerospace, automotive, data

[1]Department of Electronic Engineering, Universidad Tecnica Federico Santa Maria, Chile
[2]Faculty of Engineering and Applied Sciences, Universidad de los Andes, Chile
[3]Faculty of Engineering, Universidad Catolica de la Santisima Concepcion, Chile
[4]Department of Electrical and Computer Engineering, Ryerson University, Canada

centers, high-efficiency households or marine, to name a few. It is expected that DC grids will become key players in the large-scale deployment of distributed generators (DGs), allowing increased power capacity, enhanced power quality and resilience against power surges, disturbances and asymmetrical loads [7]. This is also supported by the increasing number of both loads and generation units interfaced by power electronics [8].

In general terms, DC grids represent a much more compact, reliable and simpler approach for power distribution. By minimizing the use of transformers, the size and volume of components of an equivalent DC system are reduced, while they are capable of transferring higher power levels using the same infrastructure. Moreover, when power is delivered at DC it does not exhibit issues with synchronization nor reactive power, and only the regulation of the voltage is required. Additionally, efficiency is enhanced given the DC nature of the majority of appliances, storage elements and DGs, through the elimination of additional power converter stages, required to integrate them to AC grids. In other words, their integration to a DC network is done with a single power conversion stage. More importantly, their capacity of delivering a continuous supply of high-quality power regardless the conditions in the utility system [9,10] facilitates the integration of growing technologies such as renewable generation, electric vehicle charging infrastructure and energy storage solutions.

Nevertheless, several challenges need to be addressed before shifting to DC systems. The main barriers are the considerably challenging design of the protective devices and the lack of standardization, despite several organizations are making efforts promoting and developing standardized practices. The existence of regulation will allow the proper preparation of personnel and equipment, and eventually will create a market for DC networks supporting their proliferation [7,11].

Given the growing adoption of DC distribution architectures, important efforts are being made toward standardization and masiffication of this emerging technology, proposing guidelines regarding DC voltage levels, installation design or maintenance practices to name a few [11]. For example, the EU Low Voltage Directive (LVD 2014/35/EU) defines the boundaries for voltages used in LVDC systems [12]. This directive defines guidelines and requirements for the electrical equipment with voltage ratings within 75 V and 1,500 V to be commercialized and also guarantees the free commercialization of conforming equipment. Additionally, EMerge Alliance, the International Electrotechnical Commission and the Institute of Electrical and Electronics Engineers are other important organizations engaged in developing voltage and performance standards for the wide-scale adoption of DC distribution systems.

As stated earlier in this book, there are many ways for implementing a DC distribution system [11]. However, two of them are the dominant methods: using a unipolar or monopolar architecture, where two conductors are employed for generating a single voltage level; and a bipolar one, that resembles the three-phase AC system and uses three wires to generate two different voltage levels [8,13]. The first option features a simpler structure, with a reduced supply reliability, while the latter one represents a more reliable and flexible alternative, given the possibility of power transfer when a line fails or that allows the connection of the loads to two regulated voltages [13]. In addition, the converters in bipolar systems can either operate with relatively low-rated voltage, thereby securing converter cost competitiveness, or be connected to higher AC voltages while maintaining the step-down effort of the DC–DC stages. Efficiency is also enhanced in three-wire systems, as the currents are reduced for a system with the same power rating [14].

On the other hand, grounding of DC systems is a complex issue and its design involves several factors [11]. One of the features of a bipolar network is that it has higher safety levels since the grounding connection is easier and simpler using the neutral conductor [1]. Unlike unipolar systems, the lack of uncertainty in the pole-to-ground voltages allows to clear faults easily and quickly [1,8,15]. However, if bipolar systems present differences in the generation or power consumption of the loads connected to each pole, voltage imbalance can occur, which can adversely affect the receiving voltage end or the loads connected to it, if not addressed properly [8,16,17]. The current circulating through the neutral conductor affects the voltage balance of the system and also its efficiency, as during such scenarios the losses in the feeder are increased [7].

For these reasons, balance control becomes essential in bipolar systems [10], as there is no way to guarantee the loads will be identical in both of the buses of the distributed system. Aiming in this direction, different balancing methods have been proposed in the literature, which will be discussed in detail in the upcoming sections. Considering the tremendous potential that bipolar DC microgrids hold for the development of DC distribution systems, the remainder of this chapter will provide an overall view in the efforts being made in this architecture and the current technologies existing in the literature.

10.2 Bipolar-type DC distribution systems

The bipolar architecture for DC distribution networks offers interesting features over the conventional unipolar counterpart. This structure is exhibited in Figure 10.1, where it can be seen that the utility AC voltage is converted to DC with the use of a distribution transformer and an active rectifier, also called distribution converter. Then, at DC level the system adopts a three-wire system that consists of the positive conductor (*p*), the negative conductor (*n*) and the neutral conductor (*z*). As it has been demonstrated at transmission level for HVDC, and despite being a more complex technical solution, the aforementioned configuration presents clear advantages in terms of efficiency, reliability, safety and transmission capacity when compared to conventional two-wire systems. Advances in power electronics and their decreasing costs have allowed an increased penetration of power converters in different applications, thus enabling the consequent expansion of DC systems at distribution level [1].

Overall, the system exhibited in Figure 10.1 features several benefits. First, the three-wire configuration resembles the traditional AC system in the sense that it provides the connection to two different regulated voltages. The voltage between the positive and negative poles is analogous to the line-to-line voltage, while the

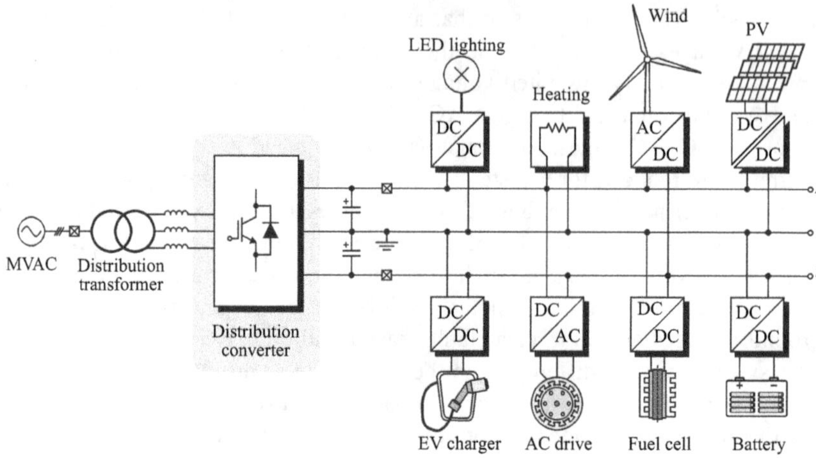

Figure 10.1 Concept of low-voltage bipolar-type DC network

voltage of a pole with respect to the neutral connector is analogous to the phase-to-neutral one [8]. This allows to accommodate a wide set of DGs and loads with different voltage and power ratings combinations in a single DC network [18–20].

From an economical point of view, bipolar systems secure the cost competitiveness of the converters interfacing the different stages of the network. This is related with the reduction of the voltage ratings of the power electronic components, the efficiency improvement due the reduction of the rated current, besides the elimination of unnecessary DC–AC stages [14]. The latter is because most of the stages involved in DGs are either DC-based, e.g., photovoltaic (PV) panels, fuel cells and batteries or generate outputs with variable voltage/frequency, e.g., wind, small hydro, wave/tidal energy conversion systems, hence require power electronics devices to accommodate their output to network conditions [14]. In addition, the presence of batteries and storage stages further enhances the benefits of DC distribution, as it yields to greater efficiency improvements [21,22]. The result is a substantial enhancement in the quality of the electric system and at the same time, reduction of the costs when compared to conventional AC solutions [13].

The aforementioned resemblance with three-phase AC systems is also beneficial for the migration process to bipolar DC networks. Considering a conventional grounded three-phase system, which requires five conductors for its realization (three-phase conductors, one neutral and the protective earth or grounding conductor). These installed cables can serve in a retrofitted DC system that will be superior in terms of power ratings capability and efficiency than the original one [14]. Typically, low-voltage power cables can be found in different sizes and ratings. As an example, multi-wire 230 V AC installations use 300 V/500 V 10 A cables, following the structure in Figure 10.2(a). The system is then rated for 6.21 kW (assuming a power factor of 0.9). If such system would be replaced by (a) a 325 V DC network or (b) a ±325 V DC network, using the configuration

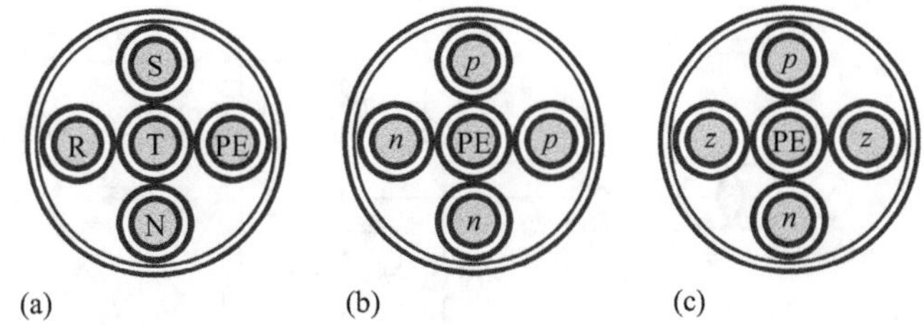

Figure 10.2 Use of cables in retrofitted systems or new DC systems: (a) AC cable; (b) unipolar DC cable; and (c) bipolar DC cable

proposed in Figure 10.2(b) and (c), the power transfer capability will be enhanced to 6.5 kW in both cases, with no considerations for reactive power and using the same cables. However, bipolar DC system cables only have to withstand 10 A while the unipolar one should be rated for 20 A. This illustrates some of the benefits of implementing a bipolar DC microgrid from an existing AC installation.

10.3 Topologies and operational aspects of bipolar LVDC grids

Power electronics are an essential part in this modernization of the electrical system. The large presence of DC-based equipment, along with variable voltage/frequency DGs require power converters for the connection to DC grids. For these reasons, the following sections will provide a topological analysis of the different stages involved in this promising DC network architecture. Both, the distribution converters, that enable the split-DC bus network and balancing converters, which perform the power consumption relocation to keep the system balanced will be covered and briefly discussed.

10.3.1 Distribution converter topologies

The grid-tied or distribution converter stage plays a fundamental role in the DC network of Figure 10.1, as it acts as the power hub on the LVDC grid [8,23,24]. For this reason, the bidirectional power flow capability is mandatory, besides a wide set of additional features given the challenges associated to distributed generation. High power quality, adjustable power factor, low filtering requirements, reliability, while simultaneously featuring a reduced number in both active and passive components are desirable features in this stage [25,26].

A brief review of the most common converters used as the grid interface of DC microgrids can be found below. Despite several topologies are able to act as the rectifier stage of the system, only two alternatives will be analyzed in detail: the two-level voltage source converter (2L-VSC) and the three-level neutral point clamped (3L-NPC) converter. This lies in their proven cost effectiveness, efficiency and reliability, besides being the two most widespread three-phase converters employed in the industry [23].

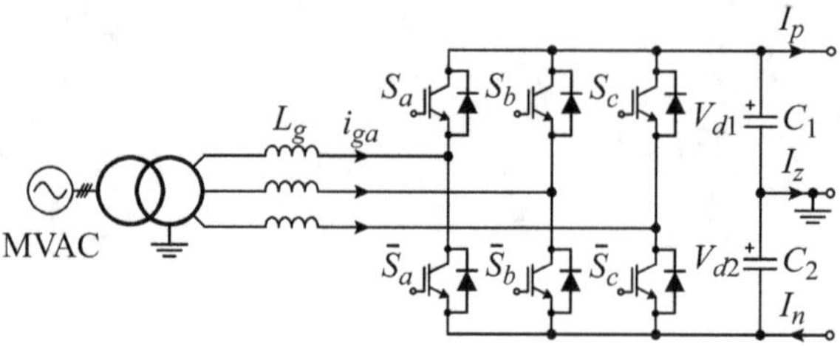

Figure 10.3 2L-VSC with neutral line connected to DC midpoint

10.3.1.1 Two-level voltage source converter

The 2L-VSC displayed in Figure 10.3 has a strong presence in the industry, acting as the grid interface for a wide series of applications. As exhibited in the figure, the circuit consists of an array of six switching devices, typically insulated-gate bipolar transistors (IGBTs) and a DC-link capacitor. The presence of these active switches, together with a proper control scheme, generates sinusoidal currents at the input side, along with adjustable power factor and bidirectional power flow [27]. This converter also steps up the voltage to higher values than the input grid voltages.

The generation of the gating signals can be done using carrier-based pulse-width modulation (PWM), space vector modulation (SVM), besides variable switching frequency methods, such as table or prediction-based methods [28–30]. The resulting binary pulse train in the phase voltages, which alternate between 0 and V_d originates the name of the converter. The topology presents a line-to-line voltage with a three-level waveform. This impacts the total harmonic distortion (THD) of the input current, as larger active/passive filters or higher switching frequencies are needed in order to meet the requirements imposed by the grid code.

This topology has been widely applied to different types of bipolar DC networks [7,8,16,23,31,32]. Among the options explored, the simplest approach for implementing a bipolar system is through the use of two cascaded 2L-VSC and a multi-winding transformer, as exhibited in Figure 10.4. In this configuration, each 2L-VSC is employed to enable independent DC voltage sources. Consequently, this leads to independent operation of the positive and negative poles, hence there are no issues during asymmetrical operation. However, this alternative is not the preferred method for operating as the distribution converter. Despite the simplicity of the operation offered by the cascaded connection, this implementation requires additional hardware as two separate converters along with the distribution transformer based in two isolated windings. This increases the cost and the size of the total system. Additionally, the topology in Figure 10.4 requires its secondary windings to be designed to withstand the DC offset caused by the series connection of the converters and the asymmetries that could arise in loads being fed [32].

A single 2L-VSC can also generate a bipolar DC output with additional circuitry. These configurations are all based in the generation of alternative paths for the returning currents, which otherwise flow through the DC bus midpoint during asymmetrical operation and therefore unbalance the DC voltages. The system

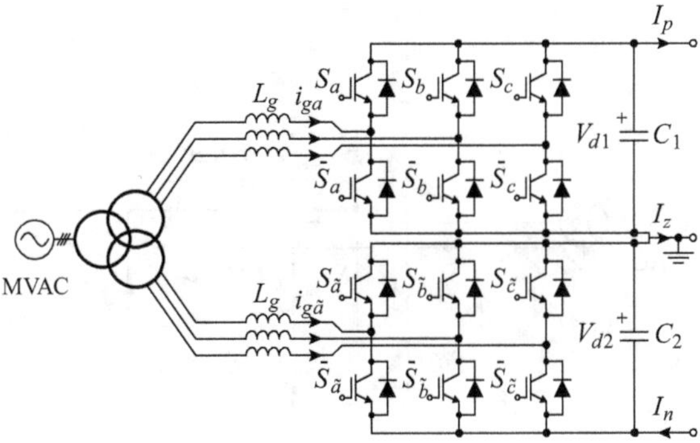

Figure 10.4 Cascaded 2L-VSC with bipolar DC output

exhibited in Figure 10.3 does it by connecting to the neutral point of the transformer with the DC bus midpoint. This additional current path allows the regulation of the midpoint current, in order to prevent DC bus voltages from drifting and maintain their proper regulation. However, the generation of non-zero DC currents at the AC side can lead to transformer saturation and thereby it should be strictly limited. To avoid this issue, different approaches have been proposed in the literature [7,8,16,23,31,32], both based in passive or active methods, the latter ones enabled by power electronics. The main solutions will be covered in the upcoming sections.

10.3.1.2 Three-level neutral point clamped converter

Another topology with a significant acceptance in industry applications is the 3L-NPC converter. Despite being originally proposed for MV drives, the 3L-NPC has found his way to lower power applications during the last years, mainly in PV applications. This inherent versatility in a wide voltage-power combination makes it a natural candidate to act as the distribution converter. At present, commercial NPCs reach voltages in the range of 2.2–6.6 kV, within a wide power rating range (3 kW–50 MW). This converter is used either as an inverter that feeds AC loads or as an active front-end converter that interfaces with the utility grid [33].

From the power circuit displayed in Figure 10.5(a), each phase of the 3L-NPC converter is composed by four switching devices, which can either be IGBTs or integrated gate-commutated thyristors, and two diodes connected to the neutral point. The presence of these diodes allows generating a third voltage level in the synthesized voltage, through connection of the phase with this neutral point when the inner devices are turned on. Thereby, the converter is able to generate an output voltage equal to $-V_{d_2}$, 0 and V_{d_1}. The proper regulation of this floating voltage represents a critical aspect for the 3L-NPC. Otherwise, the generated voltage becomes distorted, worsening the power quality and also potentially damaging the switching devices. These reasons have motivated important efforts in the literature to solve this matter [34–36].

The selection of this topology as the distribution converter of the DC network in Figure 10.1 is because of its higher grid power quality (three levels in the voltage

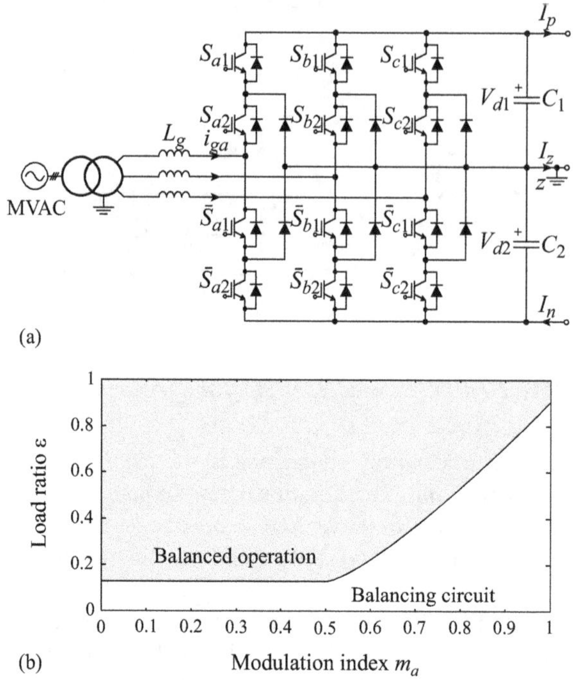

Figure 10.5 Grid-connected three-level neutral point clamped converter: (a) power circuit and (b) unbalanced load limitation

waveform), superior harmonic performance (reduced THD), lower switching frequency, lower transformer ratio and the possibility of providing certain room for unbalanced bipolar dc bus operation [17,23,34,37]. When feeding a bipolar DC network, this topology enables a limited asymmetrical load operation.

By injecting a DC bias in the phase-to-neutral voltage, the 3L-NPC is able to absorb current through its neutral point, hence extending the operation of the active distribution network to a certain range of asymmetries. However, this midpoint current capability is limited to a fraction of the DC current being fed, i.e., the asymmetrical scenarios that the system can handle while keeping the midpoint voltage controlled are limited and dependent of the amplitude modulation index [17,23].

The unbalanced operation limitation is well defined for this converter as presented in Figure 10.5(b), and it can be used as a foundation for the balancing scheme [17]. Unlike the previous converter, this topology can contribute to the balancing of the buses, leading to two different kinds of balancing strategies that extend the operation of the network. Despite both approaches require additional circuitry they differ in the balancing efforts made by the stages. Following the balancing schemes for the 2L-VSC, one possibility is to make all the balancing efforts with the added circuits, through a virtual disconnection of the neutral point from the unbalanced loads. In this way, the 3L-NPC focuses on regulating the input currents and power quality while the DC voltages remain unaltered. This leads to higher current stresses on the balancing stages, so interleaving channels can be promoted in order to reduce their ratings [37–39].

A different concept is to implement a coordinated balancing scheme, where both the central converter and the balancing stages share the current redistribution efforts [17,20]. The balancing principle is the same as with the 2L-VSC, and the exceeding current should be redistributed by the means of a power converter stage. The DC-link structure opens the possibility to three-level DC–DC stages as it will be covered in the upcoming section.

10.4 Balancing topologies

As depicted in the previous sections, the realization of a bipolar system using a single power converter is not sufficient to handle asymmetrical generation or unbalanced loads [7], hence additional circuitry is mandatory. The presence of these circuits that balance the currents at DC side is beneficial for several reasons. Depending on the way these converters interact with the DC systems can be denoted as voltage balancers or current redistributors. Voltage balancers are DC–DC converters that generate a split-DC bus based on the total DC voltage supplied by the rectifier V_d, and are also able to perform the regulation of the center line. Thus the voltages of the bipolar system (V_{d_1} and V_{d_2}) are balanced to $V_d/2$.

On the other hand, current redistributors are used to balance the poles in bipolar systems. This redistribution reduces the current flowing through the neutral conductor to virtually zero, thereby the efficiency and stability of the feeder are enhanced [7]. Moreover, balanced operation also eases the implementation and operational requirements of the rectifier stage. This is related with the possibility of using off-the-shelf products for the rectifier stage, as there are no special requirements on its operation. By adding a balancing stage, the rectifier can be used exclusively for the grid-connection related tasks adjusting the power factor, regulating V_d and shaping the input currents [7,19].

The presence of these circuits also allows to isolate the neutral line in the system and facilitate its connection to ground, thus improving the safety levels in case of remote energy supply.

Taking into consideration the structure of the DC bus enabled by these power converter stages, the balancing circuitry will be denoted as voltage balancer or current redistributor. The basic operational principle of these complementary circuits is the possibility of relocating the energy consumption or generation, in such a way that voltage balancing is achieved regardless the power flow between the DC buses. To perform such corrections, these stages should have access to both of the DC buses [37].

This section will provide an overview of the main power converter topologies, as well as other non-power-electronics-based approaches for guaranteeing the independent operation of the DC buses, regardless the load/generation condition.

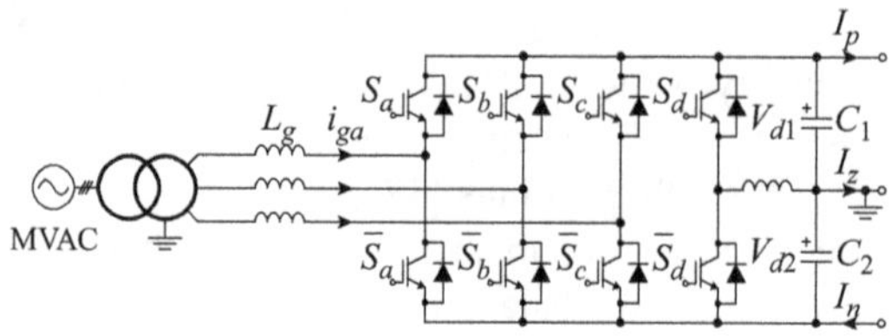

Figure 10.6 2L-VSC with voltage balancer enabling a bipolar DC output

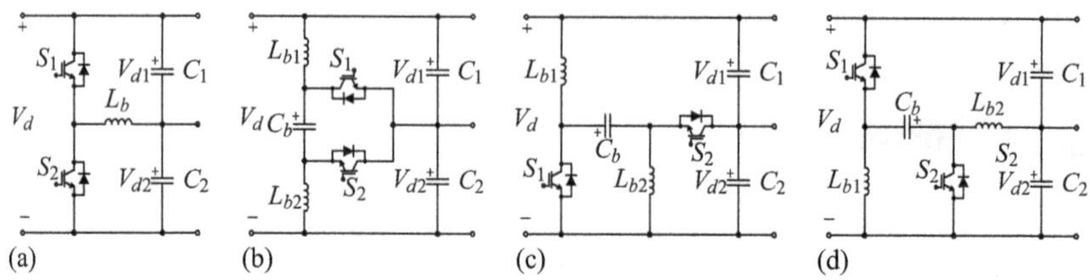

Figure 10.7 Bipolar balancer topologies derived from the bidirectional buck/boost concept: (a) buck/boost; (b) Cuk; (c) super-Sepic; and (d) super-Zeta

10.4.1 Bidirectional buck-boost topologies

The most elemental topology to generate a regulated bipolar DC bus with a single 2L-VSC is shown in Figure 10.6, where a buck-type voltage balancer has been added. The presence of this additional half-bridge leg permits splitting the DC bus voltage V_d into V_{d_1} and V_{d_2}. Additionally, these balancers allow the active redistribution of the current during unbalanced scenarios [7,8]. This bidirectional buck/boost structure is beneficial for the distribution transformer as well, because there are no DC currents flowing through the AC neutral in contrast to the circuit of Figure 10.3. The idea of this configuration is to reduce the dependency between the operation of the poles, by properly modeling the inter-pole dynamics at the DC side in order to adjust the currents accordingly. The result of this relocation of the power consumption allows to keep the floating midpoint voltage tightly regulated.

Following the buck/boost concept, many balancing converters can be derived using conventional DC–DC stages as building blocks, however it is important to note that the bidirectional power flow is mandatory for modern distribution networks [7]. Figure 10.7 presents different balancing stages based on this structure, which is displayed in Figure 10.7(a).

The circuit displayed in Figure 10.7(b) is a Cuk-type voltage balancer, which besides providing the regulation of the bipolar bus eliminates the shoot-through possibility of the switching devices. The reason for this feature is the presence of an impedance instead of a purely capacitive element as the DC-link, which is typical in voltage source converters. Following the same concept, super-Sepic- and super-Zeta-type voltage balancers have also been developed in the literature, and are displayed in Figure 10.7(c) and 10.7(d), respectively [20]. Another way of enhancing the benefits of

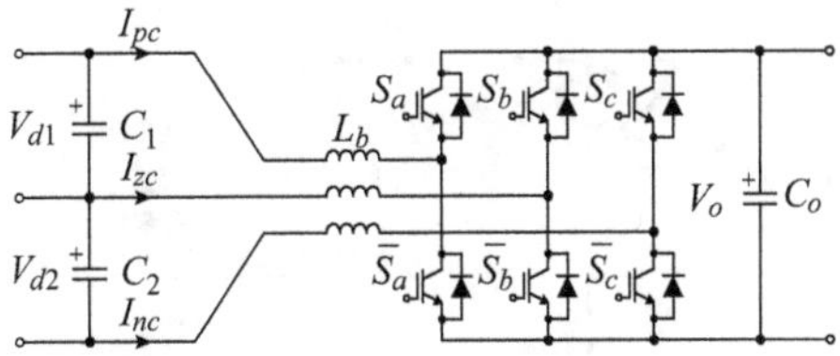

Figure 10.8 2L-VSC-based current redistributor

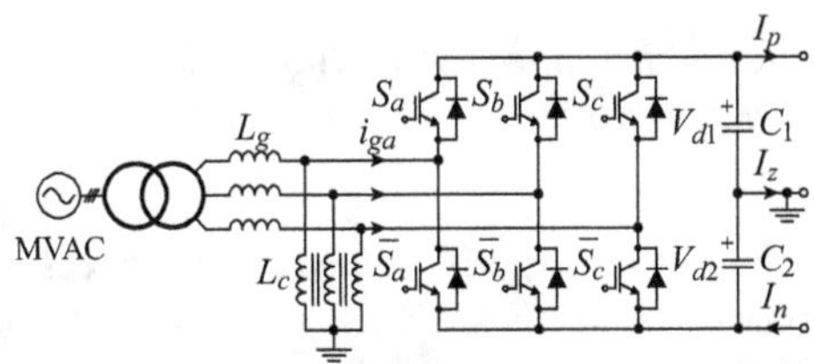

Figure 10.9 2L-VSC with a grounding inductor balancer

the voltage balancing stage is to use additional channels and operate them interleaved. This connection will reduce the high current ripple exhibited in the DC inductor L_b of the buck/boost stage shown in Figure 10.6, as the current will be equally shared by the additional channels. By means of this configuration it is possible to reduce the inductor size, as the interleaved operation increases the output equivalent frequency. This leads to a smooth output current, which allows the reduction in the output inductor by a factor of $1/n$, for an n-channel converter [24].

The circuit in Figure 10.8 can also serve towards the power balance of the buses. The topology proposed in [7] regulates the positive and negative bar currents in such a way that the neutral current I_{zc} is almost zero. However, this is not sufficient to guarantee a stable operation, as the exceeding power must be relocated to the opposite pole and returned to the grid. The result is a control scheme based on the sum–difference domain, that ensures balance for asymmetrical operation in split-DC networks. As it will be shown in the next section, the resemblance with a three-phase system motivated a control scheme based on the symmetrical component method applied for grid-connected converters.

10.4.2 Coupled inductor current redistributor

Following the same concept of Figure 10.3, the topology proposed in Figure 10.9 uses a grounding reactor to provide the current injection path between the AC side terminals and the DC neutral conductor [32]. The main idea is to regulate the zero-sequence DC current during unbalanced scenarios while limiting the AC currents, in order to maintain the system's efficiency. This is achieved through the proper design of the grounding inductor, which uses a coupled magnetic core that reduces the chances of saturation due the presence of DC flux.

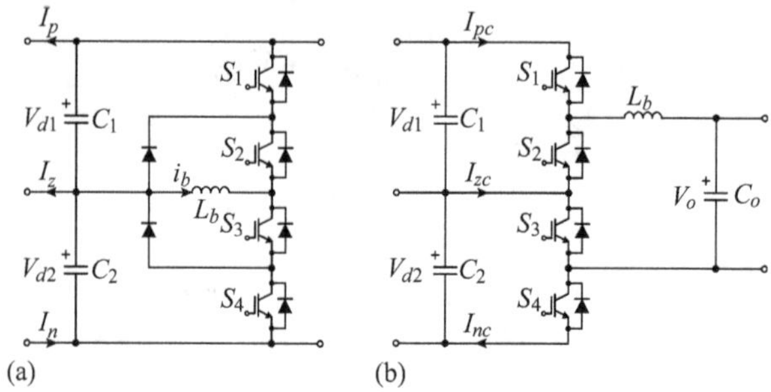

Figure 10.10 Three-level current redistributing topologies: (a) 3L-NPC balancer and (b) three-level buck/boost

The design of the core leads to zero-sequence inductances that are substantially lower than the stationary frame ones, which are desired to be large in order to reduce sustained AC currents given the connection. The structure of the inductor enables a viable path for the returning DC currents, enhancing the controllability of the converter and regulating the DC bus voltages regardless the load condition. This approach offers a trade-off between having less power electronics and increasing the losses in the coupled inductor showing promising results.

10.4.3 Three-level DC–DC current redistributors

Given the 3L-NPC structure, which already includes a split DC bus, makes it a natural candidate for bipolar DC networks. Nevertheless, despite this converter is able to balance the midpoint voltage within a certain imbalance range, it cannot guarantee the stability for any operation mode. This limitation motivated the proposal in [17], where an additional 3L-NPC leg has been added and connected as exhibited in Figure 10.10(a). The balancing leg can be seen as a bidirectional buck, whose input is the full DC voltage and the output is the midpoint voltage, i.e., the duty cycle will tend to 0.5 when the balance is reached. This is a crucial feature, as it allows to determine a priori the current flowing through the DC inductor needed to achieve balance. Consequently, this fourth leg performs the complementary balancing functions, extending the range achieved by the 3L-NPC and allowing balanced DC voltages even under severe unbalanced load conditions (rated load/generation in one pole while the other one has no load/generation).

Please note that, if the DC inductor is properly sized, the aforementioned approach is also able to exclusively handling the asymmetries at the DC side, performing the virtual disconnection of the midpoint while the distribution converter focuses on the regulation of the grid-side quantities. The main drawback will be that L_b should handle the rated DC current, instead of a fraction of it as the approach in [17].

A different approach is proposed in [19], where this 3L-NPC leg is modulated differently, in order to include free-wheeling times in the switching sequence in such a way that the exceeding current is rearranged and keeps the DC voltages controlled. The idea is the following, when asymmetrical power consumption exists, the modulator takes the exceeding current from the bus lightly loaded and supplies it to the other bus through the DC inductor.

The works in [38,39] use a different three-level DC–DC converter for feeding high-power loads and also contribute to minimize power fluctuations in bipolar DC systems. The power circuit is presented in Figure 10.10(b), where it can be seen that the basic unit is composed by four switching devices and their corresponding free-wheeling diodes, along with an output inductor and the common output filter capacitor. Depending on the power level, more units can be connected in interleaved configuration to minimize current stress in the switches and inductors [38].

One bipolar application of the aforementioned topology is interfacing an ESS [39]. By replacing the additional balancing leg, this converter allows to accomplish two simultaneous objectives: the main one is to perform the charge and discharge of the storage element according to the energy management supervisor controller of the DC network, but at the same time uses this process to relocate the power consumption at DC level and complement the balancing task performed by the distribution converter. Given the nature of the topology, the system is able to guarantee stable operation with an ESS with reduced power ratings.

Alternatively, in [38], two interleaved channels of the converter in Figure 10.10(b) are used to feed high-power loads. Besides enhancing the output power quality, reduce the current stress and the filter size and its modularity, this three-level topology can also assist the distribution converter to perform DC power balancing, allowing to improve the overall system efficiency and the quality of the AC input currents. From the controllability point of view, this stage can be either complementing the balancing efforts carried by the central converter, or exclusively perform the balance in a distributed manner between the remaining stages as suggested by the authors.

10.5 Control schemes

Following the topological analysis of bipolar DC distribution networks, a brief survey on control techniques used for enabling these configurations is presented. First, the main schemes used for the balancing the currents at the DC side, followed by grid-tied AC–DC stages, both single and three-phase approaches will be reviewed.

Given the complexity in the dynamics involved in the aforementioned topic, leads to the necessity of implementing sophisticated control schemes that share the same structure. This architecture corresponds to a two degree-of-freedom configuration for improving the dynamics while dealing with disturbances. Such structure is known as *Cascade Control*. The main idea is based on the feedback of intermediate variables, which have a direct relation with the main output of the system, in order to improve the dynamic response of the loop and the robustness of the closed loop.

10.5.1 Cascade control

Figure 10.11 shows the diagram of the cascade control structure. In the figure, two control loops are shown: the primary (or outer) loop with its controller $C_1(s)$ and a secondary loop with its controller $C_2(s)$. The design of the secondary loop is made

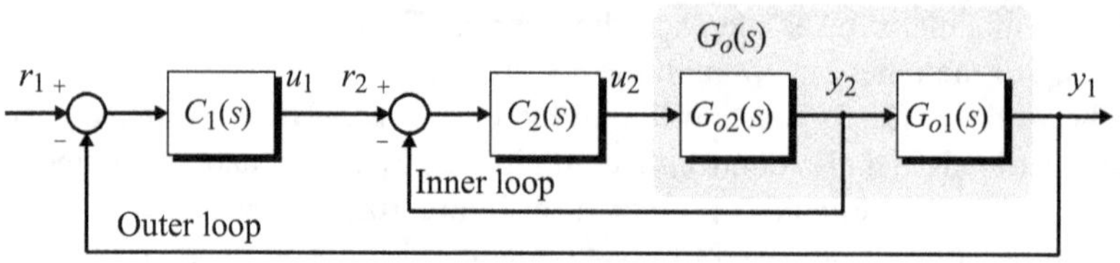

Figure 10.11 Cascade control structure

in such a way that it mitigates the effects of disturbances before they significantly affect the system output y_1 [40].

The main benefits of the cascaded control are obtained when when $G_{o_2}(s)$ presents significant non-linearities that limit loop performance or when $G_{o_1}(s)$ limits the bandwidth in a basic control loop (i.e., the presence of non-minimum phase zeros and/or pure time delays). The first benefit, is explained because the inclusion of $C_2(s)$ allows a high gain control, which tends to reduce the effect of nonlinearities. The latter one, is because this inner controller pre-compensates the effect of the disturbances that $C_1(s)$ has to deal with [40].

10.5.2 AC–DC converter control

The main goal of a grid-tied converter is enabling a regulated DC voltage, while controlling the input currents at the grid side. However, these are not the only features desired for a grid-tied converter, as requirements for these converters have increased through time. The significant presence of power converters connected to the utility grid, has imposed harder requirements through time. Thereby, in addition to their main functions they are also required to reduce their electromagnetic interference, harmonic content, reactive power generation, etc. The control schemes have updated accordingly in order to maintain their performance and cope with the additional objectives imposed by more stringent grid codes. Besides controlling the aforementioned quantities, ideally an active rectifier stage should provide additional functionalities such as:

- The reduction of the harmonic content because of its negative effect on the electric system (e.g., voltage distortion, electromagnetic interference, increased power ratings of power system equipment and so on).
- Adjustable power factor
- Bidirectional power flow

The appearance of these requirements has resulted in new challenges for the control schemes and has prompted researchers to improve them to meet the control objectives and at the same time maintain high performance.

10.5.2.1 Single-phase AC–DC converter control

For the regulation of single-phase rectifiers the control scheme presented in Figure 10.12. From the figure, it can be seen that the scheme uses a cascade control structure, where the grid-current is regulated through a proportional resonant

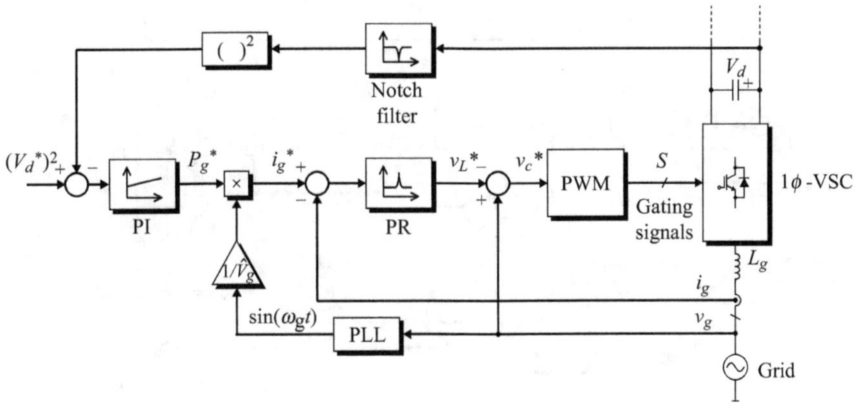

Figure 10.12 Single-phase AC–DC converter stage control scheme

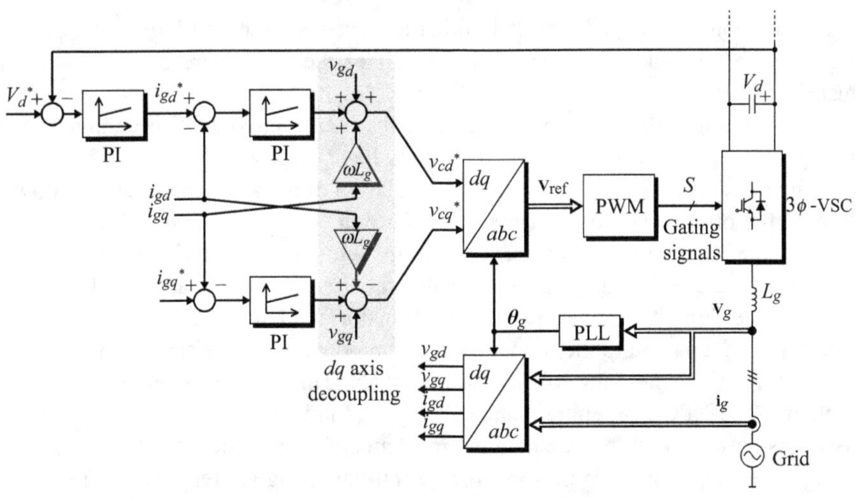

Figure 10.13 Three-phase AC–DC converter stage VOC control scheme

controller in order to enhance the DC-link voltage dynamics, which are regulated by an outer proportional integral (PI) controller. Note that the feedback of the DC voltage is filtered with a notch in 2 ω_g to minimize the impact of the rectification ripple in the grid current. For the synchronization of the scheme, a phase-locked loop (PLL) is employed to guarantee the unity power factor operation.

10.5.2.2 Voltage-oriented control

Figure 10.13 displays the three-phase counter part of the previous control method. This control scheme is called voltage-oriented control (VOC). Derived from the flux-oriented control (FOC) of an induction motor [41], VOC is based on a cascade control structure and a coordinate transformation between the $\alpha\beta$ stationary frame and the

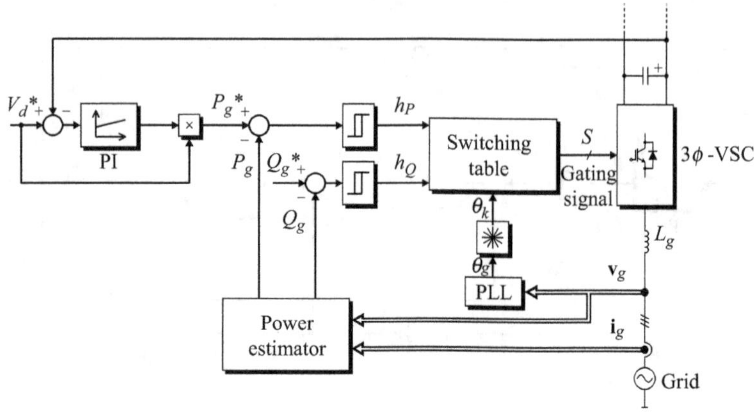

Figure 10.14 Three-phase AC–DC converter stage DPC control scheme

dq synchronous frame. The result is fast transient response and high static performance, with no steady-state error, as a result of the DC nature of the controlled quantities [29].

From Figure 10.13, it can be seen that the measured signals are converted to the *dq* frame, which is aligned with the grid voltage using a PLL, hence the name of the scheme. This allows to decouple the regulation of the active and reactive power through the orthogonal components of the current, i_{gd} and i_{gq}, respectively.

For the regulation of the current components and the DC-link voltage, PI controllers are used, leading to zero steady-state tracking errors and good dynamic performance. Finally, their actuations are transformed back to a three-phase reference frame and passed to the modulator. This modulation stage can be either PWM or SVM, for the generation of the gating signals. This control scheme can also be applied to the NPC converter, by modifying the modulation stage accordingly and also considering an midpoint balancing mechanism. Conventionally, a PI controller is implemented in order to perform the partial balancing strategy, which regulates the midpoint deviations depending on the selected modulation strategy [17].

10.5.2.3 Direct power control

Another mainstream method for regulating three-phase converters is the direct power control (DPC), presented in Figure 10.14. The scheme is also based on the cascade control structure, except that the inner loop controllers are nonlinear.

The DC-link voltage V_d is controlled by an outer linear PI controller, which provides the reference for the active power P_g^*, whereas the reactive power reference Q_g^* can be set arbitrarily depending on the selected V_d reference. Both powers are estimated from the grid voltage and current measurements feedback and controlled with nonlinear hysteresis comparators. Their outputs h_P and h_Q along with the voltage angle θ_s, are used to access a voltage vector look-up table. Finally, the table delivers the gating signals to generate the selected voltage vector. Although this method is designed for the conventional 2L-VSC, various adaptations for the

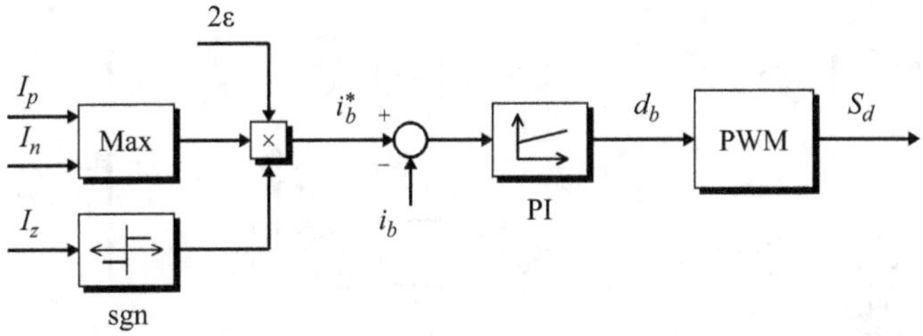

Figure 10.15 Voltage balancing circuit controller for the additional NPC leg

3L-NPC converter exist, which add an additional hysteresis comparator to regulate the midpoint corrections [42–44].

10.5.2.4 Balancing leg control

The idea behind this approach is to extend the bipolar NPC operation through the entire load range, either using the maximum balancing capabilities offered by the central converter modulator or to completely perform the balancing of the system. To do so for the first case, the balancing leg must be able to act as a virtual impedance, such that the minimal load condition is met, and the modulator is able to keep the voltages balanced. For the latter case, the balancing leg should absorb the full difference between the currents flowing through the poles, which in other words can be seen as the virtual disconnection of the midpoint.

Given the connection of the balancing leg, the duty cycle required to balance the system is known a priori as the midpoint voltage should be equal to $V_d/2$, which allows to exactly determine the current required to be handled by the balancing converter in order to balance the bipolar system, and a simple PI controller will suffice to perform the corrective actions and provide a good dynamic performance. The control scheme for this converter is exhibited in Figure 10.15, considering the complementary balance case and operates whenever the load ratio of the dc buses is lower than the critical ratio ε allowed by the modulation scheme [17]. If the current distributor is performing exclusively the balancing tasks, then the balancing current reference i_b^* is given by the difference $I_p - I_n$. This will result in the virtual disconnection of the midpoint, hence keeping the DC voltages balanced in the presence of any imbalance.

10.5.2.5 Symmetrical component control-based methods

The similarity of the analyzed DC network with three-phase AC systems is also beneficial from the control scheme perspective. The symmetrical component method is a powerful and useful tool for analyzing and controlling unbalanced AC systems [45–47]. Accordingly, this technique can also be applied in bipolar DC networks, so a reference frame transformation **T** can be derived from AC theory that leads to:

$$\begin{bmatrix} x_\Sigma \\ x_\Delta \end{bmatrix} = \underbrace{0.5\begin{bmatrix} 1 & 1 \\ 1 & -1 \end{bmatrix}}_{\mathbf{T}} \begin{bmatrix} x_p \\ x_n \end{bmatrix}. \tag{10.1}$$

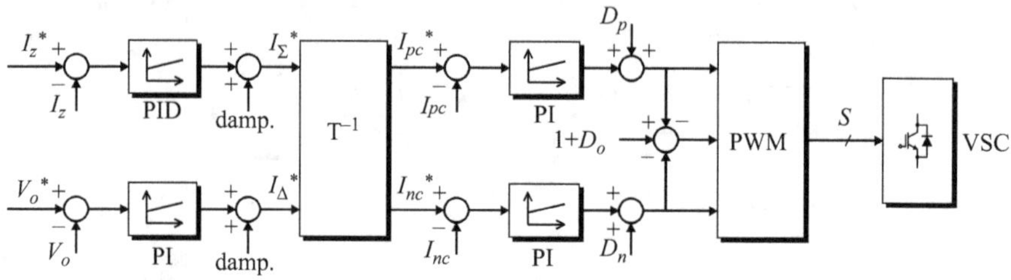

Figure 10.16 DC symmetrical component control scheme

The importance of this transformation is that decouples the collective power transfer of the two poles and the inter-pole interaction to different components [7,8], namely the differential mode x_Δ and the common mode x_Σ. This is relevant for the analysis and control of the system, as it allows to develop decoupled circuits and therefore allows to precisely model the inter-pole dynamics. Based on the control scheme in Figure 10.13, the symmetrical component control scheme is presented in Figure 10.16. From the figure, it becomes clear how the balancing current I_z holds a relation with the common mode current I_Σ while the output voltage V_o is related with the differential current I_Δ. Both, the output voltage and the neutral current are regulated with linear controllers and set the references for the sequence components. These references are transformed back to the DC side and the references for the positive and negative bar currents in the redistributor, for example the circuit in Figure 10.8. Finally, these currents I_{pc} and I_{nc} are regulated using PI controllers to generate the duty cycles for each leg in the converter.

10.6 Summary

During the last decades, AC systems dominated the power transmission and distribution applications almost exclusively. However, a recent convergence of needs originated in different sectors (renewable energy conversion, information technology and transportation) have accelerated the development of DC systems. Nowadays, DC systems are present at both transmission and distribution levels, offering high-performance solutions with enhanced efficiency and reliability, besides reducing the number of power conversion stages involved and uninterrupted power delivery.

For LVDC active networks, two kinds of architectures are possible: unipolar and bipolar. Despite being a more sophisticated and technically complex solution, bipolar structure provide several advantages over conventional unipolar ones. Higher availability, efficiency and flexibility are just a few advantages featured by bipolar systems.

A typical bipolar grid has certain similarities with a three-phase AC system, where the phase voltage in an AC system is the analog to the voltage between one pole and the neutral point in the bipolar LVDC grid, and the line-to-line AC voltage can be considered as the voltage between the positive and negative poles of the bipolar LVDC grid. This resemblance with AC systems is beneficial for the operation, retrofitting and also for developing the control schemes.

However, if the loads connected to each pole of this LVDC grid are unbalanced, the system operates asymmetrically. This uneven power distribution along with the dependency between the poles can be problematic if not taken care properly. The main issues are the imbalances in the DC voltages, higher losses and a decreased power quality of the system.

In order to solve these imbalance problems additional circuitry is required. Voltage balancers and/or current redistributors are the key players that allow to deal with asymmetrical operation and keep the system in a balanced and stable operation. As it was discussed in the chapter, the converter topologies employed in bipolar LVDC systems need to be bidirectional, in order to allow the reversible power flow given the flexible structure of distributed networks.

This chapter presented a brief overview covering the different aspects of bipolar LVDC networks. Distribution converter topologies, balancing stages and also their control schemes are discussed in order to highlight the efforts being made in this growing architecture.

Acknowledgement

The authors acknowledge the support provided by the CONICYT projects FONDECYT 11170774, AC3E (Basal/FB0008) and SERC Chile (FONDAP/15110019).

References

[1] D. Salomonsson. *Modeling, Control and Protection of Low-Voltage DC Microgrids*. PhD thesis, Royal Institute of Technology, Sweden, Apr. 2009.

[2] C. W. Taylor and S. Lefebvre. HVDC Controls for System Dynamic Performance. *IEEE Trans. Power Syst.*, 6(2):743–752, May 1991.

[3] B. H. Bakken and H. H. Faanes. Technical and Economic Aspects of Using a Long Submarine HVDC Connection for Frequency Control. *IEEE Trans. Power Syst.*, 12(3):1252–1258, Aug. 1997.

[4] L. Zhang, L. Harnefors, and H. P. Nee. Interconnection of Two Very Weak AC Systems by VSC-HVDC Links Using Power-Synchronization Control. *IEEE Trans. Power Syst.*, 26(1):344–355, Feb. 2011.

[5] L. Zhang, L. Harnefors, and H. P. Nee. Modeling and Control of VSC-HVDC Links Connected to Island Systems. *IEEE Trans. Power Syst.*, 26(2): 783–793, May 2011.

[6] G. P. Adam, K. H. Ahmed, S. J. Finney, K. Bell, and B. W. Williams. New Breed of Network Fault-Tolerant Voltage-Source-Converter HVDC Transmission System. *IEEE Trans. Power Syst.*, 28(1):335–346, Feb. 2013.

[7] J. Lago and M. L. Heldwein. Operation and Control-Oriented Modeling of a Power Converter for Current Balancing and Stability Improvement of DC Active Distribution Networks. *IEEE Trans. Power Electron.*, 26(3):877–885, Mar. 2011.

[8] Y. Gu, W. Li, and X. He. Analysis and Control of Bipolar LVDC Grid With DC Symmetrical Component Method. *IEEE Trans. Power Syst.*, 31(1): 685–694, Jan. 2016.

[9] H. Kakigano, Y. Miura, and T. Ise. Distribution Voltage Control for DC Microgrids Using Fuzzy Control and Gain-Scheduling Technique. *IEEE Trans. Power Electron.*, 28(5):2246–2258, May 2013.

[10] H. Kakigano, Y. Miura, and T. Ise. Low-Voltage Bipolar-Type DC Microgrid for Super High Quality Distribution. *IEEE Trans. Power Electron.*, 25(12):3066–3075, Dec. 2010.

[11] T. Dragicevic, X. Lu, J. C. Vasquez, and J. M. Guerrero. DC Microgrids – Part II: A Review of Power Architectures, Applications, and Standardization Issues. *IEEE Trans. Power Electron.*, 31(5):3528–3549, May 2016.

[12] Directive 2014/35/EU of the European Parliament and of the Council of 26 February 2014 on the Harmonisation of the Laws of the Member States Relating to the Making Available on the Market of Electrical Equipment Designed for Use within Certain Voltage Limits (recast) (Text with EEA Relevance). *Official Journal of the European Union*, L 96:357–374, Mar. 2014.

[13] T. Kaipia, P. Salonen, J. Lassila, and J. Partanen. Possibilities of the Low Voltage DC Distribution Systems. In *Nordic Distribu. and Asset Manag. Conference (NORDAC)*, Stockholm, Sweden, Aug. 2006.

[14] D. Salomonsson and A. Sannino. Low-Voltage DC Distribution System for Commercial Power Systems With Sensitive Electronic Loads. *IEEE Trans. Power Del.*, 22(3):1620–1627, Jul. 2007.

[15] H. Kakigano, Y. Miura, T. Ise, and R. Uchida. DC Voltage Control of the DC Micro-grid for Super High Quality Distribution. In *Power Conversion Conference – Nagoya, 2007. PCC '07*, pp. 518–525, Apr. 2007.

[16] Byung-Moon Han. A Half-Bridge Voltage Balancer with New Controller for Bipolar DC Distribution Systems. *Energies*, 9(3): 182, Mar. 2016.

[17] S. Rivera, B. Wu, S. Kouro, V. Yaramasu, and J. Wang. Electric Vehicle Charging Station Using a Neutral Point Clamped Converter with Bipolar DC Bus. *IEEE Trans. Ind. Electron.*, 62(4):1999–2009, Apr. 2015.

[18] J. Lago, J. Moia, and M. L. Heldwein. Evaluation of Power Converters to Implement Bipolar DC Active Distribution Networks – DC-DC converters. In *2011 IEEE Energy Conversion Congr. and Expo. (ECCE)*, pp. 985–990, Sep. 2011.

[19] X. Zhang, C. Gong, and Z. Yao. Three-Level DC Converter for Balancing DC 800-V Voltage. *IEEE Trans. Power Electron.*, 30(7):3499–3507, Jul. 2015.

[20] F. Wang, Z. Lei, X. Xu, and X. Shu. Topology Deduction and Analysis of Voltage Balancers for DC Microgrid. *IEEE Trans. Emerg. Sel. Topics Power Electron.*, 5(2):672–680, Jun. 2017.

[21] E. Rodriguez-Diaz, M. Savaghebi, J. C. Vasquez, and J. M. Guerrero. An Overview of Low Voltage DC Distribution Systems for Residential Applications. In *2015 IEEE 5th International Conference on Consumer Electronics – Berlin (ICCE-Berlin)*, pp. 318–322, Sep. 2015.

[22] E. Vossos. *Optimizing Energy Savings from Direct-DC in U.S. Residential Buildings*. MSc thesis, San Jose State University, San Jose, California, USA, Aug. 2011.

[23] J. Moia, J. Lago, A. J. Perin, and M. L. Heldwein. Comparison of Three-Phase PWM Rectifiers to Interface AC Grids and Bipolar DC Active Distribution Networks. In *Power Electron. for Distributed Generation Syst. (PEDG), 2012 3rd IEEE Int. Symp. on,* pp. 221–228, Jun. 2012.

[24] S. Rivera, S. Kouro, and B. Wu. *Charging Architectures for Electric and Plug-In Hybrid Electric Vehicles*, pp. 111–149. Springer International Publishing, Basel, Switzerland, 2017.

[25] S. Dusmez, A. Cook, and A. Khaligh. Comprehensive Analysis of High Quality Power Converters for Level 3 Off-Board Chargers. In *2011 IEEE Vehicle Power and Propulsion Conference*, pp. 1–10, Sep. 2011.

[26] A. Khaligh and S. Dusmez. Comprehensive Topological Analysis of Conductive and Inductive Charging Solutions for Plug-In Electric Vehicles. *IEEE Trans. Veh. Technol.*, 61(8):3475–3489, Oct. 2012.

[27] James W. A. Wilson. The Forced-Commutated Inverter as a Regenerative Rectifier. *IEEE Trans. Ind. Appl.*, IA-14(4):335–340, Jul. 1978.

[28] B. Wu. *High-Power Converters and AC Drives*. Wiley-IEEE Press, Hoboken, New Jersey, USA, Apr. 2006.

[29] M. P. Kazmierkowski, R. Krishnan, and F. Blaabjerg (editors). *Control in Power Electronics: Selected Problems*. Academic Press Series in Engineering, Cambridge, Massachusetts, USA, Sep. 2002.

[30] J. Rodriguez and P. Cortes. *Predictive Control of Power Converters and Electrical Drives*. Wiley-IEEE Press, Chichester West Sussex, UK, Apr. 2012.

[31] T. H. Jung, G. H. Gwon, C. H. Kim, J. Han, Y. S. Oh, and C. H. Noh. Voltage Regulation Method for Voltage Drop Compensation and Unbalance Reduction in Bipolar Low-Voltage DC Distribution System. *IEEE Trans. Power Del.*, 33(1):141–149, Feb. 2018.

[32] Y. Li, A. Junyent-Ferré, and J. M. Rodriguez-Bernuz. A Three-Phase Active Rectifier Topology for Bipolar DC Distribution. *IEEE Trans. Power Electron.*, 33(2):1063–1074, Feb. 2018.

[33] S. Kouro, M. Malinowski, K. Gopakumar, *et al.* Recent Advances and Industrial Applications of Multilevel Converters. *IEEE Trans. Ind. Electron.*, 57(8):2553–2580, Aug. 2010.

[34] N. Celanovic and D. Boroyevich. A Comprehensive Study of Neutral-Point Voltage Balancing Problem in Three-Level Neutral-Point-Clamped Voltage Source PWM Inverters. *IEEE Trans. Power Electron.*, 15(2):242–249, Mar. 2000.

[35] A. Yazdani and R. Iravani. A Generalized State-Space Averaged Model of the Three-Level NPC Converter for Systematic DC-Voltage-Balancer and Current-Controller Design. *IEEE Trans. Power Del.*, 20(2):1105–1114, Apr. 2005.

[36] Z. Mohzani, B. P. McGrath, and D. G. Holmes. The Balancing Properties of DC Link Compensation for 3-Phase Neutral Point Clamped (NPC)

Converter. In *Proceedings of The 7th International Power Electronics and Motion Control Conference*, volume 1, pp. 574–579, Jun. 2012.

[37] S. Rivera and B. Wu. Electric Vehicle Charging Station With an Energy Storage Stage for Split-DC Bus Voltage Balancing. *IEEE Trans. Power Electron.*, 32(3):2376–2386, Mar. 2017.

[38] L. Tan, B. Wu, S. Rivera, and V. Yaramasu. Comprehensive DC Power Balance Management in High-Power Three-Level DC-DC Converter for Electric Vehicle Fast Charging. *IEEE Trans. Power Electron.*, 31(1):89–100, Jan. 2016.

[39] L. Tan, B. Wu, V. Yaramasu, S. Rivera, and X. Guo. Effective Voltage Balance Control for Bipolar-DC-Bus-Fed EV Charging Station With Three-Level DC-DC Fast Charger. *IEEE Trans. Ind. Electron.*, 63(7):4031–4041, Jul. 2016.

[40] G. C. Goodwin, S. F. Graebe, and M. E. Salgado. *Control System Design*. Prentice Hall, Upper Saddle River, New Jersey, USA, Oct. 2000.

[41] F. Blaschke. The Principle of Field Orientation as Applied to the New Transvector Closed-Loop System for Rotating-Field Machines. *Siemens Review*, 34:217–220, May 1972.

[42] L. A. Serpa and J. W. Kolar. Virtual-Flux Direct Power Control for Mains Connected Three-Level NPC Inverter Systems. In *Power Conversion Conference - Nagoya, 2007. PCC '07*, pp. 130–136, Apr. 2007.

[43] J. Eloy-Garcia, S. Arnaltes, and J. L. Rodriguez-Amenedo. Extended Direct Power Control for Multilevel Inverters Including DC Link Middle Point Voltage Control. *IET Electronic Power Applications*, 1(4):571–580, Jul. 2007.

[44] S. Rivera, S. Kouro, B. Wu, *et al.* Multilevel Direct Power Control – A Generalized Approach for Grid-Tied Multilevel Converter Applications. *IEEE Trans. Power Electron.*, 29(10):5592–5604, Oct. 2014.

[45] G. C. Paap. Symmetrical Components in the Time Domain and Their Application to Power Network Calculations. *IEEE Trans. Power Syst.*, 15(2):522–528, May 2000.

[46] F. J. Alcantara and P. Salmeron. A New Technique for Unbalance Current and Voltage Estimation with Neural Networks. *IEEE Trans. Power Syst.*, 20(2):852–858, May 2005.

[47] M. Karimi-Ghartemani and H. Karimi. Processing of Symmetrical Components in Time-Domain. *IEEE Trans. Power Syst.*, 22(2):572–579, May 2007.

Chapter 11
Aircraft DC microgrids

Fei Gao[1], Tao Yang[2], Serhiy Bozhko[2], and Pat Wheeler[2]

11.1 Introduction

Similar to utility grid and other terrestrial microgrids, aircraft electrical power system (EPS) has its own power generation, distribution, utilization and energy storage. The rapid development of power electronics technology has allowed the converters to operate at DC voltage levels required for transmission, distribution and consumption. Today there is a clear tendency that high voltage (HV) and low voltage have seen the proliferation of DC systems and its implementation for transmission and distribution of electricity in aircraft electrical systems.

This chapter will provide an overview of aircraft DC microgrids. Section 11.2 introduces the aircraft EPS, covering the topics from power generation, distribution and utilization. Section 11.3 reviews aircraft electrical system standards and highlights the power quality and power factor requirement for aircraft applications. Section 11.4 presents the concept of aircraft electric starter/generator system. Control and stability analysis of aircraft DC microgrids are performed in Sections 11.5 and 11.6, respectively.

11.2 Aircraft electrical power system

Traditionally, there are two energy sources on conventional aircrafts. The primary source is the engines on the aircraft. The secondary source includes electric, hydraulic, pneumatic and mechanical systems powered by the primary source. The power source distribution in conventional aircrafts is illustrated in Figure 11.1.

The more electric aircraft (MEA) concept is one of the major trends in modern aerospace engineering aiming for reduction of the overall aircraft weight, operation cost and environmental impact. As shown in Figure 11.2, electrical systems are employed to replace existing hydraulic, pneumatic and mechanical actuators. As a consequence, the onboard-installed electrical power increases significantly, and this results in challenges in the design of the aircraft EPSs. In alignment with the tendency

[1]Department of Engineering Science, University of Oxford, UK
[2]Department of Electrical and Electronic Engineering, The University of Nottingham, UK

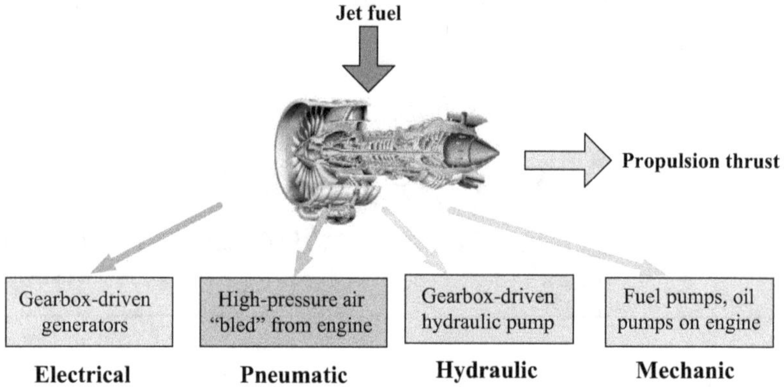

Figure 11.1 Power sources distribution in conventional aircraft [1]

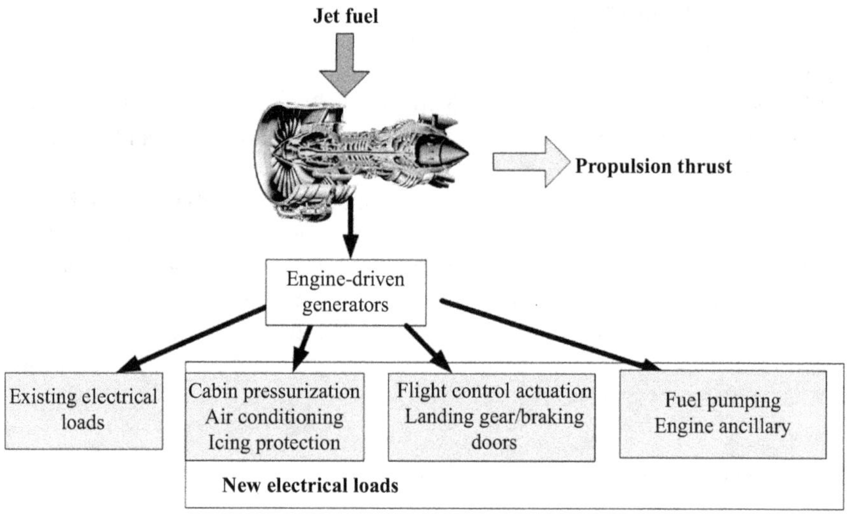

Figure 11.2 Power sources distribution in more electric aircraft [1]

in the utility grid, the EPS architecture with DC distribution is considered as one of the most promising candidates for MEA due to the potential advantages such as higher efficiency compared to AC, less weight of harness, absence of reactive power related issues and also reduced costs. Compared with existing DC distribution systems in aircraft, the future architectures will move towards HV distribution.

11.2.1 Power generation

In large commercial aircrafts, one generator per engine typically performs electrical generation. Depending on the aircraft type, there could be more than one generator connected to each engine, such as back-up generators, in order to meet redundancy and extended range twin-engine operations performance standard [2]. An additional source of electric power on an aircraft is an auxiliary power unit (APU), which is essentially a small gas turbine-driven generator that can be used as an additional power source. APU typically provides power when the aircraft is on the ground and can also provide power, while in the air under emergency conditions. Various batteries exist in the aircraft to start the APU and to provide back-up power for critical equipment in the cockpit as well as other important functions such as the emergency lighting for the aisles [3].

The 115 VAC at constant frequency (CF) aircraft system started in the 1960s. This generation of electrical networks was equipped with fixed frequency (115/200 VAC) integrated drive generators as on the Airbus A320, A330 and A340. With this architecture, the generator is connected to the main engine via a mechanical drive, which keeps the mechanical speed, and hence the electrical frequency, constant on the aircraft's electric bus. The variable-frequency (VF) generations have appeared since the 1990s. In order to simplify the constant-speed mechanical gearbox, the VF generators are preferred, as on the Airbus A380 and on the next A350, for which the voltage standard is doubled (230/400 VAC) with respect to the increase in electrical power needs. Variable-speed CF (VSCF) aircraft appeared in the 2000s. Due to the introduction of VF generation, several power loads are supplied at fixed frequency, such as induction motor driven pumps, have been replaced with power electronics controlled loads compatible with VF standards. For example, this was the case for flight control actuation systems in which electro-hydrostatic actuators were embedded on the A380 for hydraulic network backup.

As discussed above, the power generation system gradually evolves from 115 VAC CF, VSCF, to variable speed variable-frequency (VSVF) Table 11.1 lists the power generation types of recent civil and military aircrafts.

11.2.2 Power distribution

EPS in conventional aircrafts often consists of two or more engine-driven generators to power AC loads. The engine-driven generators are connected to the distribution buses in some civil aircraft configurations (i.e. each generator is responsible for a specific numbers of buses). In the event that one generator fails, it is automatically isolated from its respective busbar, and all busbar loads are then taken over by the alternative generator.

If the electrical system and associated loads can be adapted to operate with a VF, then it is possible to link a generator directly to the engine shaft, i.e., the electrical output of the generator provides a VF supply with the frequency related to the speed of the gas turbine, typically in the range of 360–800 Hz. EPS architectures with such VF distributions are known as "frequency-wild" systems. The key advantage of these is the direct connection between the generator output and the EPS, giving a simple and potentially very reliable configuration. The drawback is that nearly all aircraft loads will require power converters for control, as the VF supply cannot be used directly by some onboard loads. However, many loads, for example, actuators, require this power conversion stage for control even when

Table 11.1 Power generation type for recent civil and military aircrafts [4]

Generation type	Civil aircrafts		Military aircrafts	
Constant frequency	B777	2 × 120 kV A	Eurofighter Typhoon	
	A340	4 × 90 kV A		
	MD-12	4 × 120 kV A		
	B747-X	4 × 120 kV A		
	B717	2 × 40 kV A		
	B737NG	2 × 90 kV A		
	B767-400	2 × 120 kV A		
Variable speed constant frequency (cycloconverter) 115 VAC			F-18C/D	2 × 40/45 kV A
			F-18E/F	2 × 60/65 kV A
Variable speed constant frequency (DC link) 115 VAC	B777 (Backup)	2 × 20 kV A		
	MD-90	2 × 75 kV A		
Variable frequency (115 VAC/380-760 Hz)	Horizon	2 × 20/25 kV A	Boeing JSF	2 × 50 kV A
	A380	4 × 150 kV A		
Variable frequency 230 VAC	B787	4 × 250 kV A		
	A350	4 × 100 kV A		

operated from a fixed frequency supply. Having many distributed power converters gives more options for a safe aircraft system design as redundancy can be built-in at the systems level, avoiding any single points of failure within the design [1]. In new aircrafts such as the Boeing B787, Airbus A380 and A350, this VSVF architecture has been employed [5]. In these aircrafts, while the voltage is regulated at either 115 V (A380) or 230 VAC (B787), the frequency changes with the engine speed and typically varies between 360 and 800 Hz [6].

Figure 11.3 shows the distribution system in B787. The 230 VAC bus, generated by the variable speed synchronous generator, is then transformed and rectified by an auto-transformer rectifier unit (ATRU) to supply ±270 VDC bus. The distribution buses also supply 115 VAC bus bars via auto transformer units, and 28 VDC buses via transformer rectifier units (TRUs). This permits more conventional consumers such as window heat, APU fuel pump and avionics components, to be powered from less conventional 230 VAC supply. The availability of the 115 VAC and 28 VDC buses therefore avoids the need to develop specialized power conversion for each of the client loads. The distribution architecture also furnishes 230 VAC direct to certain users for which no transformation is necessary. These loads include the wing anti-ice heater mats, environment control system (ECS) recirculation fans and cargo heaters. To provide a temporary source of emergency power for essential loads, and to provide a source of energy for starting the APU, the B787 has two 28 V batteries.

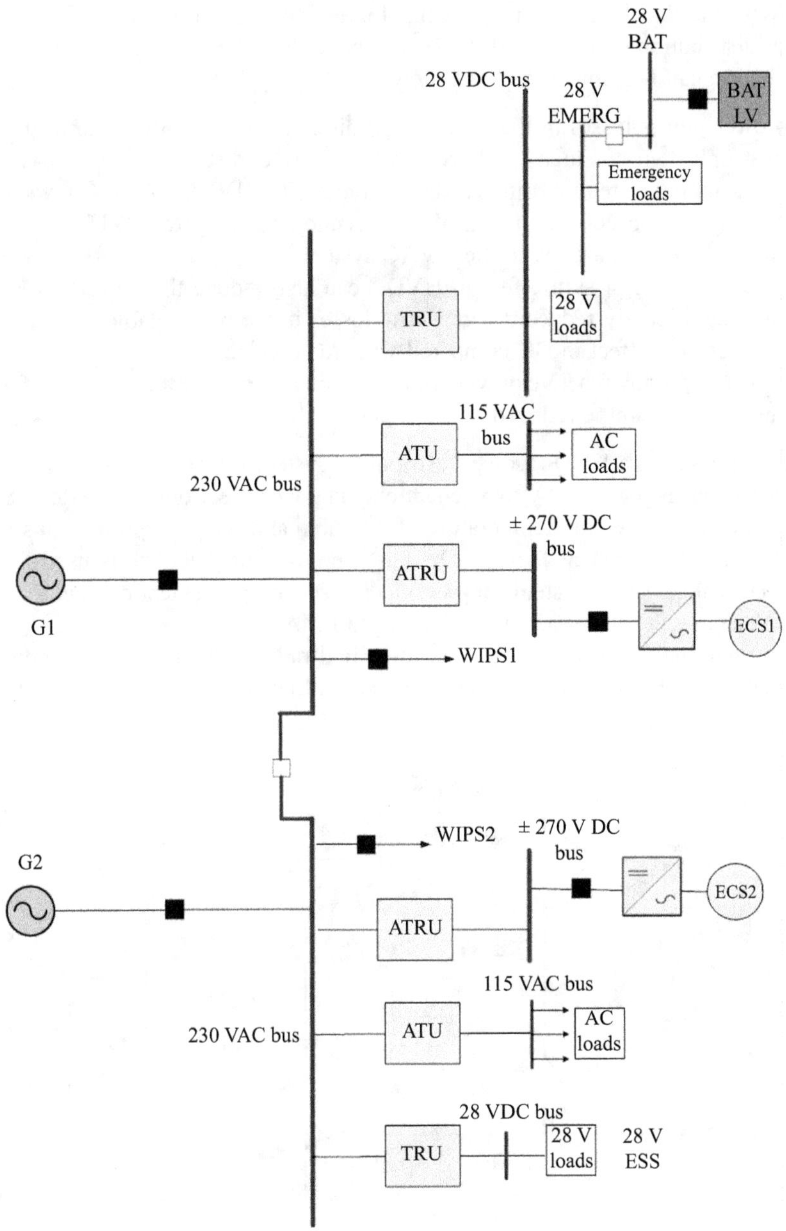

Figure 11.3 Power generation and distribution system for Boeing 787 having twin generators

With the development of HV technologies, there is a trend for using HVDC power distribution system in MEA [7]. The benefits of VF generation with HVDC distribution can be summarized as follows:

- Lower power losses in the power transmission cables. This is due to the fact that only two conductors are required in HVDC distribution, whereas three conductors are required in AC distributions. The HVDC system shows about 67% reduction compared with the conventional AC system (115 V CF) and about 33% compared with the HVAC system.
- The reduction of skin effect in HVDC can also reduce the power losses and also significantly reduce the dielectric losses in the power cables.
- Less corona effect in DC compared with AC conductors.
- No need for any reactive power compensation equipment: the capacity of wires and devices can be reduced.

Figure 11.4 shows a HVDC power distribution system for future modern aircrafts. 270 V DC bus is powered by two generators via power electronic converters. ECSs that maintain the pressure/temperature of the cabin and wing ice protection system (WIPS) are also connected to the DC bus. The essential DC bus is used for the flight critical actuation systems, represented by several permanent magnet machine driven electromechanical actuators (EMA1 to EMA6).

Further analysis leads to the next identified radical change: one can consider parallel operation of sources that was not allowed in previous EPS design

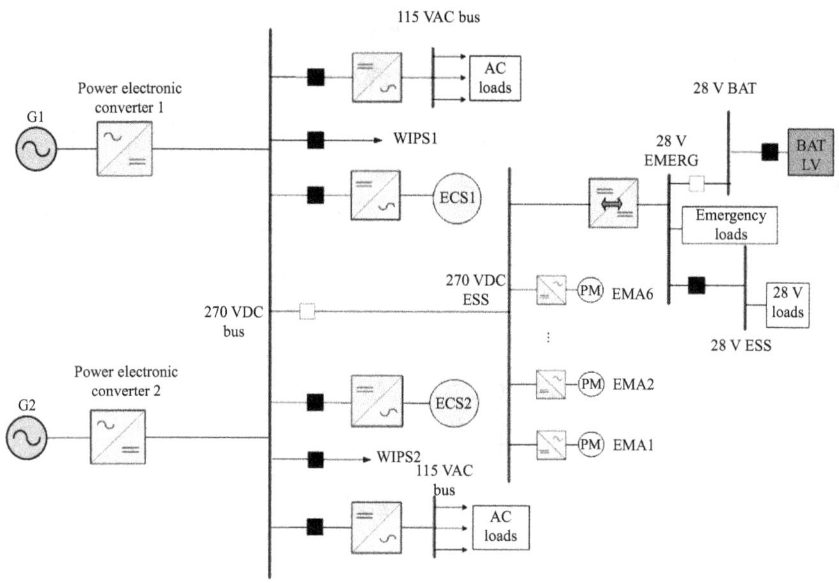

Figure 11.4 A single line diagram of an aircraft DC microgrid

approaches mainly for safety reasons. However, if the approach is properly established, one can assume that the appropriate safety requirements can be met in future. In return, sources paralleling idea will lead to (1) possibility of power sharing among different sources with minimization of total weight of generation systems, (2) ease of energy management and power flow control, (3) convenient integration with power sources of different nature (fuel cells, energy storage devices, batteries, etc.), (4) improved availability of electrical energy onboard.

11.2.3 Power utilization

After the electrical power has been generated and distributed throughout the aircraft, it is wide to the aircraft services, i.e., power utilization. These electrical services/loads cover a range of functions: lighting services, heating devices, motors and actuation, subsystem controllers and avionics systems. Figure 11.5 shows the power consumption of onboard loads in an aircraft.

As it is known to all, the availability of adequate lighting is of great importance to the safe operation of the aircraft. Lighting systems consist of internal lighting and external lighting. Internal lighting includes cockpit/flight deck, passenger service and emergency/evacuation lighting. On the other hand, external lighting includes navigation, strobe, landing/taxi, formation, inspection, emergency evacuation and logo lights. The heating load represents the power needed for WIPS. Motors used onboard provide the linear actuation (engine control); rotary actuation (position actuations for flap/slat operation); valve operation (electrical operation of fuel, hydraulic, air and ancillary systems control valves); engine starter: starting of the engine, APU and others to reach self-sustaining operation; gyroscope devices: provision of power to run gyroscopes for flight instruments and auto-pilots.

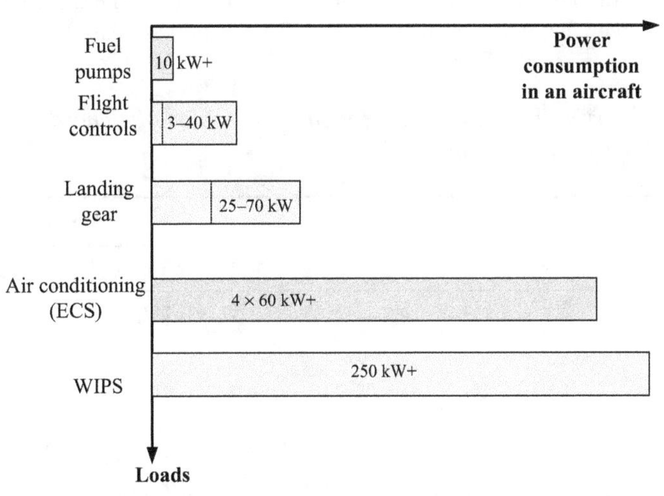

Figure 11.5 Power consumption of loads onboard in an aircraft

11.2.4 Energy storage system

With the improvement of the system reliability and the potential to reduce the size of the main generators, the installation of energy storage system (ESS) is essential in the MEA [8]. ESS can balance the power between sources and loads [9], improve the transient performance and stability [10] and provide the backup electrical power in the case of a main power fault [11]. A corresponding coordinated control strategy needs to be developed dealing with the power balance between main generators and ESS. When the load is within the capacity of the main generators, they will supply power to feed the load, while the ESS can work in charging mode or "offline" mode. When the load exceeds, the generator capacity under certain scenarios, together with main generators ESS, works in discharging mode and provides power to the load. If more than one set of ESS is installed onboard, the ESS with higher state of charge should provide more power to the load. In this sense, optimal-coordinated control methods using dynamic consensus algorithm, fuzzy logic or droop principle can be developed.

11.3 Power quality requirements in aircrafts

There are several stands to specify the electrical system in aircrafts, such as ISO-1540 Aerospace—Characteristics of aircraft electrical systems [12], DO-160F Environmental conditions and test procedures for airborne equipment [13], MIL-STD-704 Aircraft Electrical Power Characteristics [14], etc.

According to DO-160F, current harmonic limits for single phase and balanced three phase systems are shown in Table 11.2. I_1 is the maximum fundamental current.

To deal with the AC/DC power conversion, conventional TRU [15] is suffering from poor power factor and power quality. The 12- or 18-pulse ATRU is used in

Table 11.2 Current harmonic limits for single phase and balanced three phase system in aircrafts [13]

Harmonic order	**Limits**	
	Single phase	**Three phase**
$h=3$	$I_h=0.15I_1/h$	$I_h=0.02I_1$
$h=5, 7$		$I_h=0.02I_1$
$h=11$		$I_h=0.1I_1$
$h=13$	$I_h=0.3I_1/h$	$I_h=0.08I_1$
$h=17, 19$		$I_h=0.04I_1$
$h=23, 25$		$I_h=0.03I_1$
$h=29, 31, 35, 37$		$I_h=0.3I_1/h$
Odd triplen harmonics ($h=9, 15, 21, \ldots, 39$)	$I_h=0.15I_1/h$	$I_h=0.1I_1/h$
$h=2, 4$	$I_h=0.01I_1/h$	$I_h=0.01I_1/h$
Even harmonics $h>4$ ($h=6, 8, 10, \ldots, 40$)	$I_h=0.0025I_1$	$I_h=0.0025I_1$

order to meet the power quality requirement. Recent research showed that, to meet the low input harmonic current requirements, the converters with active front-end are attractive solutions for interfacing with the AC bus [16]. However, to effectively control the input current harmonics to meet stringent specifications remains a major challenge. It would be more difficult to meet the entire requirement simultaneously without reducing the other performances, such as converter's voltage transfer ratio, input power factor and robustness against unbalanced and/or distorted voltage input. On the other hand, the requirements also present a major design constraint for the filter design of the converters. A relatively small second-order input filter is typically employed in the converters to smooth out the switching ripple current at the power input lines. Alternative topologies for active front-side control and input power filter designs to further reduce the weight and size are also a big challenge.

MIL-STD-407F, US Military Standard, defines a standardized power interface between a military aircraft and its equipment and carriage stores, covering such topics as voltage, frequency, power factor, electrical noise and abnormal conditions (overvoltage and undervoltage), for both AC and DC systems. Figure 11.6 shows the limit for 115 V AC, 400 Hz and 270 V DC, 28 V DC. The envelope of the nominal voltage transients is shown in Figure 11.6. It can be seen that in 270 V DC system, the tolerable steady-state range is between 250 and 280 V, which indicates the possibility of droop control (will be discussed in the following section). The voltage at the transient can go up to 330 V and down to 200 V. Similarly, the steady-state limit of 28 VDC system is 22–29 V. The overvoltage and under voltage for 270 and 28 V DC system are illustrated in Table 11.3.

The high fundamental frequency (400 Hz in CF EPS and 360–800 Hz in VF EPS) is also a challenge for aircrafts. Achieving low input current distortion and unity input power factor at such high line frequencies require much wider control bandwidth compared to the industry 50/60 Hz systems. Moreover, VF EPS will

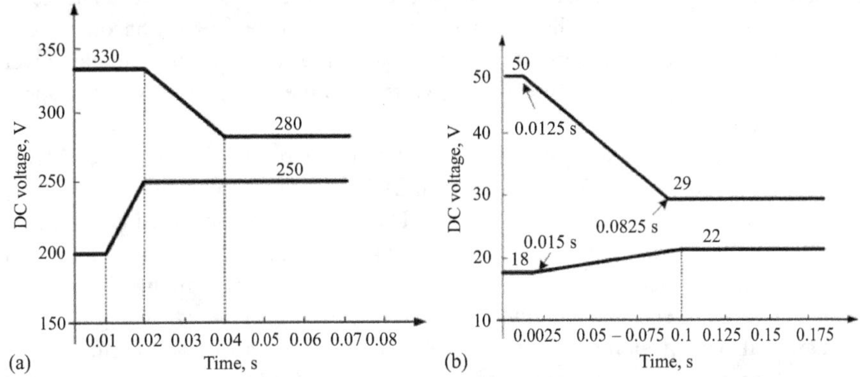

Figure 11.6 Envelope of normal voltage transients for (a) 270 V DC system and (b) 28 V DC system

Table 11.3 Voltage limits for 270 and 28 V DC system in aircraft DC MGs

	Steady-state limit	Limit for under voltage	Limit for overvoltage	Transient high voltage level
270 V	[250,280]	240	290	350 ($t < 0.05$ s)
28 V	[22,29]	20	31.5	50 ($t < 0.05$ s)

have more challenges for power factor correction (PFC). It is because that, the cable impedance will vary with the AC frequency in VF EPS. As a result, the PFC needs to provide reactive power varying with the frequency. Another reason is that, because the frequency range of VF EPS is typically 360–800 Hz, most of which is higher than CF, the voltage drop of cables in VF EPS is higher than in the system of CF EPS.

11.4 Aircraft starter/generator control

Electric starter/generator is beneficial to facilitate electric engine crank, assist to the engine in reaching the desired idle speed, generate power to drive the onboard auxiliaries and improve the overall system efficiency [17]. To ensure that an adequate supply of electrical power is available under all operational conditions, including those conducted in the presence of failures, the B787 is equipped with four VF starter generators of 250 kV A capacity each. Figure 11.7 shows the architecture of the electric starter/generator system. In the motoring mode, the machine is driven by a bi-directional power converter and uses the power coming from the supply bus to accelerate the aircraft engine. Since the control target in the motoring mode is the machine speed, as it can be seen, the speed signal is sensored and the control system regulates the machine speed to its reference value. This mechanical energy can be harnessed to the engine shaft and cause the engine to spool up to a speed where the combustion ignites and the engine becomes self-sustaining. In the generating mode, the engine acts as a source of mechanical power that is converted by the machine and converter into the electrical power to supply the onboard loads.

The University of Nottingham contributes to "Aircraft Electrical Generation with Active Rectification Technology" (AEGART) project, a part of the Clean Sky JTI ITD EU FP7 framework program [18]. The project has conducted an intensive trade-off study of potential S/G topology candidates and, due to the advantages of the permanent magnet machine (PMM), the topology of surface-mounted PMM fed by an active front end (AFE) rectifier has been chosen. As shown in Figure 11.8, fundamental vector control is used. After transforming measured currents to rotating frame, the linear control (conventional PI controller) adjusts current in the synchronously rotating reference frame (*dq* domain) and outputs *dq* voltage demands. Voltage demands are inversely transformed into three-phase demand

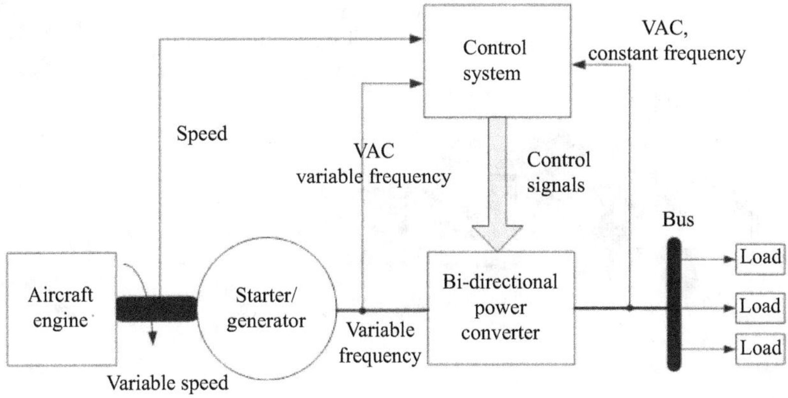

Figure 11.7 Electric starter/generator system including loads and control

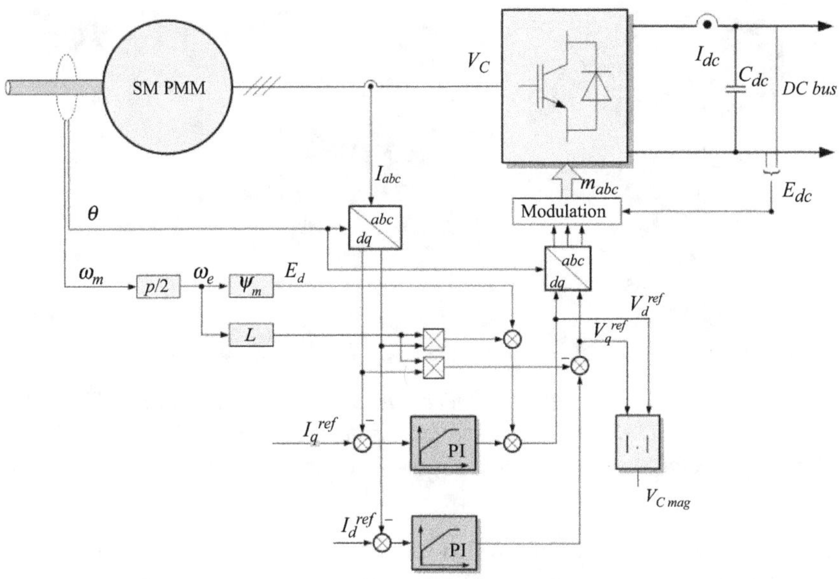

Figure 11.8 Inner current loop control of the PMM-based S/G [19]

modulation indexes for pulse width modulation. Then the core system can be fully controlled by using both *dq* current demands.

Figure 11.9 shows the outer loop of the S/G. Since the designed PMM is for high speed operations, the flux weakening controller needs to be taken into account when the machine speed exceeds the base speed. For the starter mode, speed controller and flux weakening controllers are employed to generate the *q*-axis current reference and *d*-axis current reference. When the stator voltage is less than its

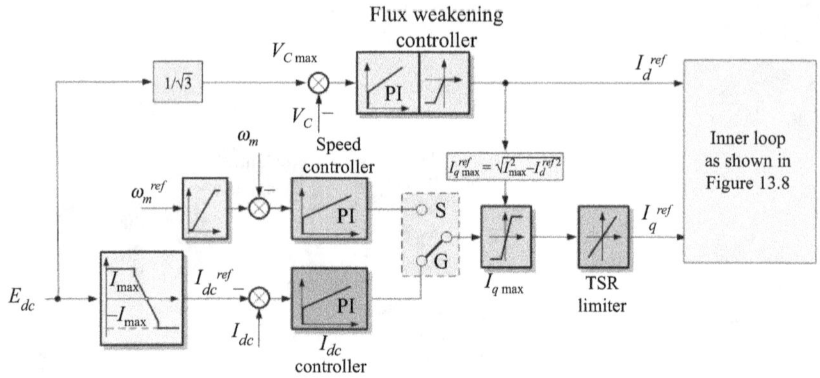

Figure 11.9 Outer loop control for Aircraft Electrical Generation with Active Rectification Technology (AEGART) S/G system

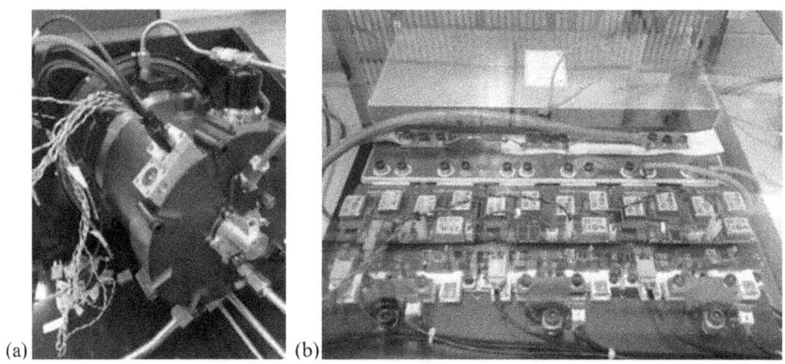

Figure 11.10 Aircraft Electrical Generation with Active Rectification Technology (AEGART) starter/generator system: (a) 45 kW 32 krpm PMM starter/generator and (b) bi-directional power converter

maximum value, the output of the flux weakening controller is saturated to zero. The controller starts to work when the stator voltage exceeds the maximum voltage limit. With the outer voltage regulation loop (V_c loop), the flux-weakening control is fulfilled by injecting the negative d-axis current. For the generator mode, the conventional PI controllers are used to deflux the machine (d-axis) at high speed and control the output DC current (q-axis). The stator current references in the d and q axes are obtained from the output of the flux weakening controller and I_{DC} controller respectively. The reference of the AC voltage (V_c) is dependent on the DC voltage. The DC current reference (I_{DC}^{ref}) is determined by the desired droop characteristic.

Figure 11.10 shows a lab prototype for the electric S/G developed in Nottingham [20].

11.5 Control strategies in aircraft DC microgrids

A "single-bus" EPS architecture with multiple generators feeding to the same bus is illustrated in Figure 11.11. The expected benefits of using parallel generation with single bus include improvements in power system redundancy, i.e., the parallel system can provide 'no break' power in the event of a generator failure. As shown in Figure 11.11, two generators are connected directly to the turbine shafts through active rectifiers and can operate in variable speed mode. The two generators take power from the main engine through the high pressure and low pressure shafts, guaranteeing an efficient exploitation of the power generated by the engine, and operate in parallel to transfer the power to the DC bus. Normally, this 270 V DC power can be obtained from the main engine-driven generators or through active rectification. Alternatively for ground starting of the engine, the 270 V could also be obtained from a ground-based APU.

In an example aircraft DC MG shown in Figure 11.11, the sources in a distributed control scheme operate cooperatively to regulate the bus voltage, but a load sharing problem arises, where each source must provide power to the load proportional to its power capacity. In such context, load sharing is critical; to avoid that, some sources become overloaded, losing the reliability of distributed power systems. Appropriate power sharing among the sources is of importance in multi-source configuration.

Control strategies applied in utility DC MGs can be employed in aircraft DC MG as well. Following the definition of hierarchical control in [21], the level of control can be composed of primary control (inner control loop and primary power sharing control), secondary control (voltage and frequency restoration) and tertiary control (optimal load dispatch and system operation) which are explained in the followings.

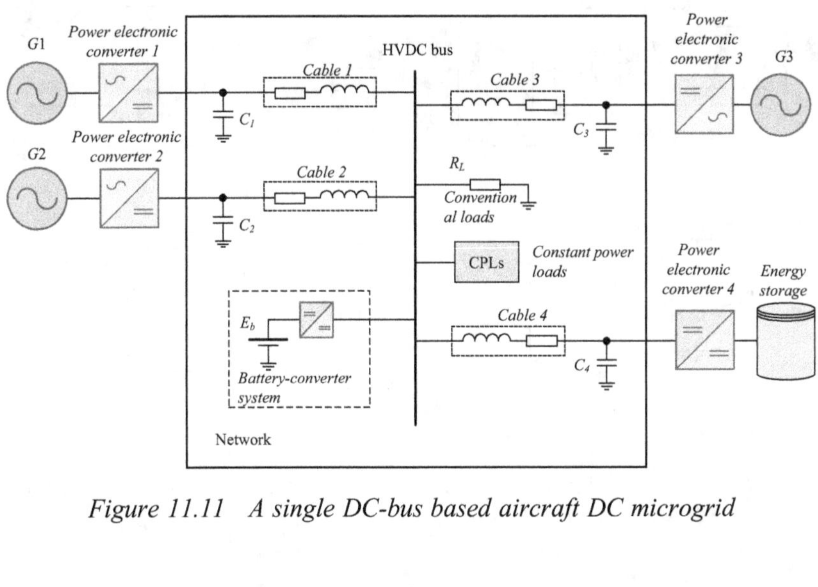

Figure 11.11 A single DC-bus based aircraft DC microgrid

11.5.1 Primary control

For the voltage-source-converter based DC MGs, the basic droop control concept can be implemented either as current/power mode droop, including current–voltage (*I–V*) and power–voltage (*P–V*) strategies or as voltage mode droop, including *V–I* and *V–P* strategies [18].

The basic concept of droop control, as discussed above, is implemented by adding a "virtual resistance" into the existing local voltage controllers for the sources, enabling their parallel operation on common load(s). Due to the modularity and reliability, droop control has been successfully applied in DC systems.

As discussed in previous sections, AFEs will be widely used in aircrafts rather than TRU or ATRU due to the power quality requirements. In the AFEs-based DC EPS, the basic droop control concept can be implemented either as current/power mode droop, including *I–V* and *P–V* strategies or as voltage mode droop, including *V–I* and *V–P* strategies [22]. The *I–V* and *P–V* droop methods are shown in Figure 11.12(a) and (b) [23–25]. In the implementation of these control methods, the DC voltage is measured and the injected current/power is controlled according to the droop characteristic. Alternatively, for the *V–I* and *V–P* methods shown in Figure 11.12(c) and (d), current or power is measured and the DC voltage is regulated accordingly [25].

Figure 11.13 shows the voltage current/power characteristics for both power and current controlled terminals. It is obvious that both models are almost equivalent with little difference when small voltage errors are considered. In [26], *I–V* and *P–V* demonstrated similar performance, and for this reason only *I–V* droop is considered in this chapter as a representative of the current/power mode strategies. Similarly, only *V–I* droop is investigated within the voltage mode methods.

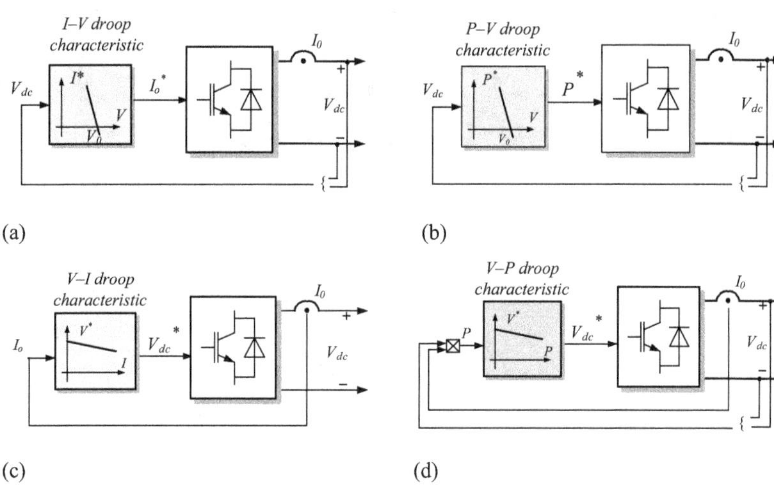

Figure 11.12 Droop characteristic employed in active front ends (AFEs). (a) I–V droop, (b) P–V droop, (c) V–I droop, (d) V–P droop [19]

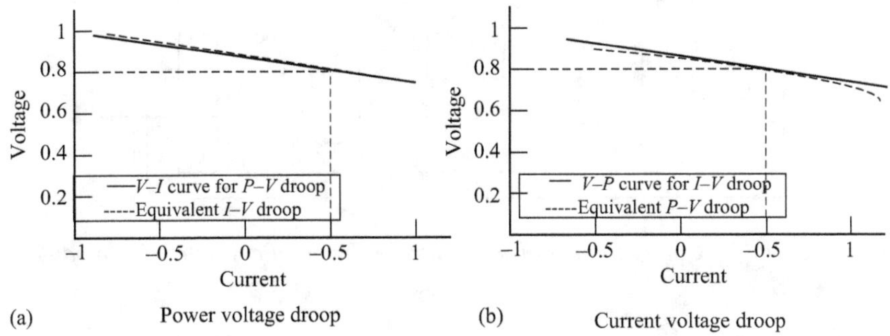

Figure 11.13 Similarity validation of P–V droop and I–V droop [26] in DC MGs. (a) Power voltage droop; (b) Current voltage droop

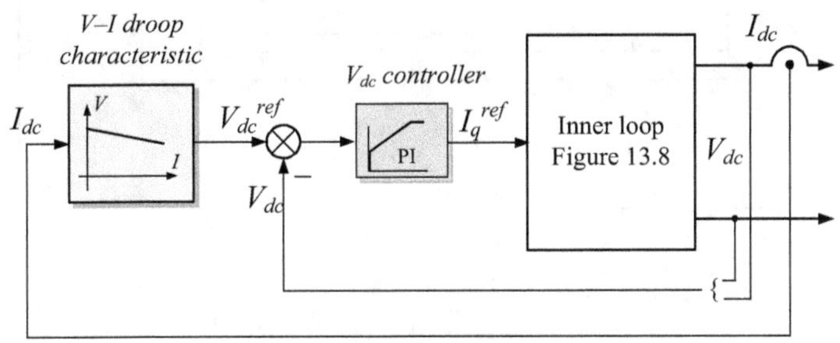

Figure 11.14 Voltage-mode droop control scheme of a single AFE module

The voltage-mode approach employs the *V–I* droop characteristic which uses the measured branch current to generate the terminal voltage reference. The control scheme of the voltage-mode droop-controlled AFE is shown in Figure 11.14. As expressed in (1), the DC voltage reference is generated according to the branch output DC current using the *V–I* droop characteristic.

$$V_{DC}^{*} = V_o - kI_{DC} \tag{11.1}$$

The current-mode approach uses the measured voltage to calculate the desired injecting DC current. The current-mode droop control scheme is shown in Figure 11.15, with the current reference derived from the *I–V* droop characteristic, based on the DC voltage measurement:

$$I_{DC}^{*} = \frac{V_o - V_{DC}}{k} \tag{11.2}$$

where V_o is the nominal bus voltage; k is the droop gain; V_{DC} is the DC voltage measurement; I_{DC}^{*} is the generated DC current reference.

Figure 11.16(a) and (b) shows the primary control for current-mode and voltage-mode droop controlled systems, respectively.

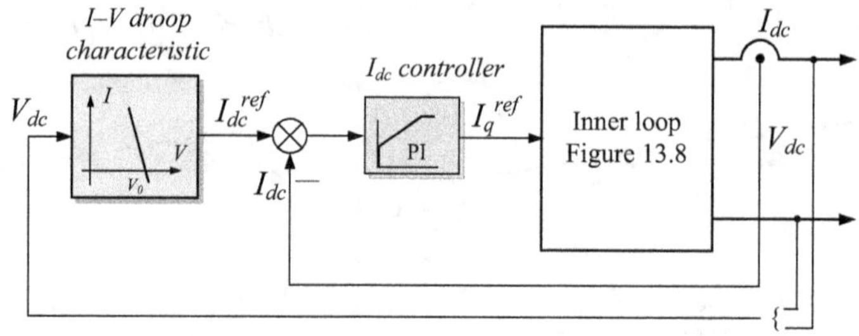

Figure 11.15 Current-mode droop control scheme of a single AFE module

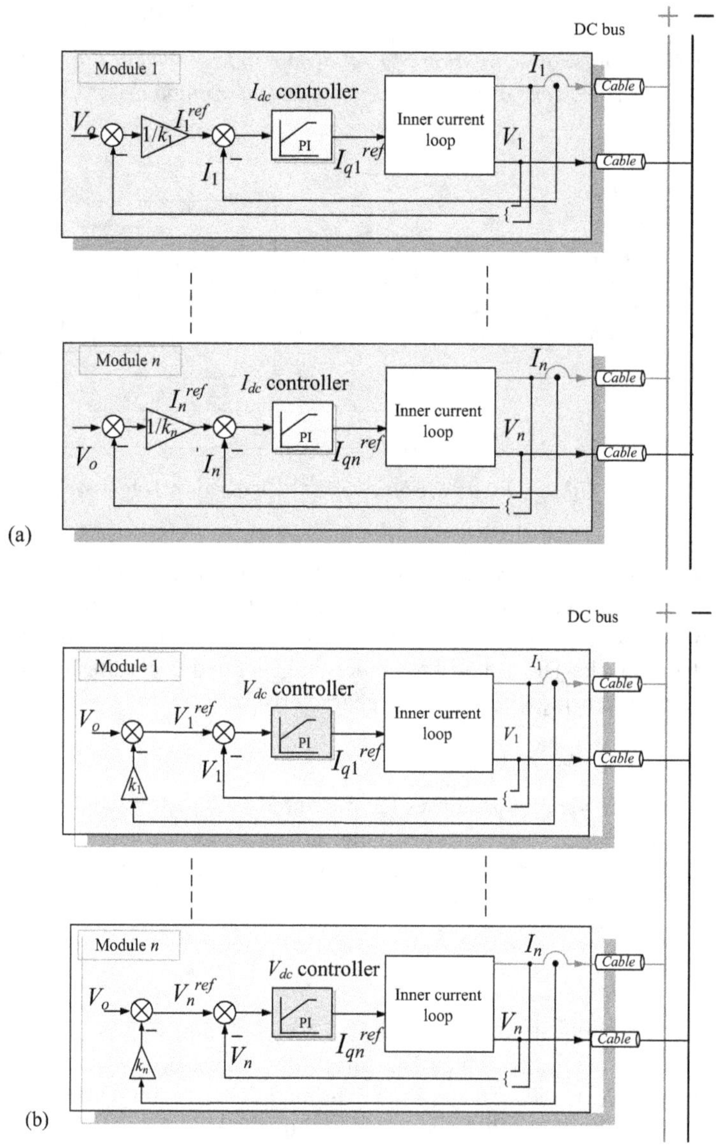

Figure 11.16 Primary control for AFEs-based aircraft DC MGs (a) current-mode droop control and (b) voltage-mode droop control

11.5.2 Secondary control

As discussed in [27], there is a trade-off between droop coefficients and voltage regulation. High droop gain can guarantee the precise power sharing among the sources, while the voltage regulation performance is poor, i.e., voltage drop is high under high droop gains.

As abovementioned voltage profile in 270 V DC system (see Section 11.4), the rated voltage of the main bus is 270 V, but the acceptable steady-state range is between 250 and 280 V as depicted in the standard MIL-STD-704F.

In order to maintain the bus voltage for a droop-controlled aircraft DC MG at the nominal value, a conventional method is to employ a secondary control to compensate the voltage drop. The principle of the secondary control in DC MGs is to restore the voltage drop due to the droop characteristic, as shown in Figure 11.17. When the primary control is implemented, the system operating point will move from V_o (no load voltage) to OP1 under i_{DC1} load condition and from V_o to OP2 under i_{DC2} load condition, respectively. After activating the secondary control, the operating point will move from OP1 to OP1_new and OP2 to OP2_new, i.e., the system always work at nominal voltage level.

Similar to the droop characteristic of each individual module at the DC terminal, the main bus of the multi-source system also has the droop characteristic for the DC bus voltage and total load current. The global droop gain, which defines the main bus *V–I* characteristic, will be derived as given and shown in Figure 11.18.

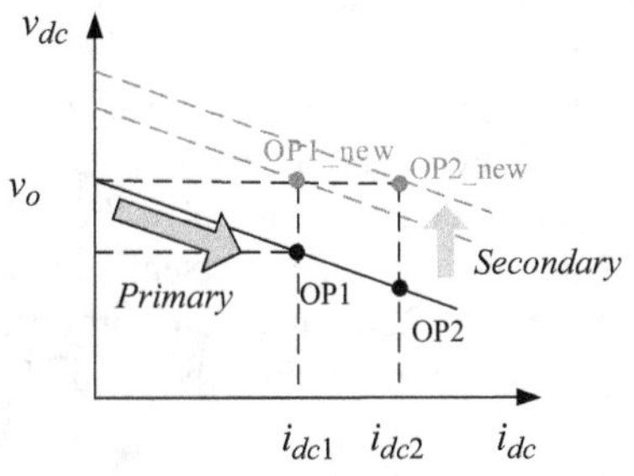

Figure 11.17 Principle of secondary control on DC voltage restoration in DC MGs

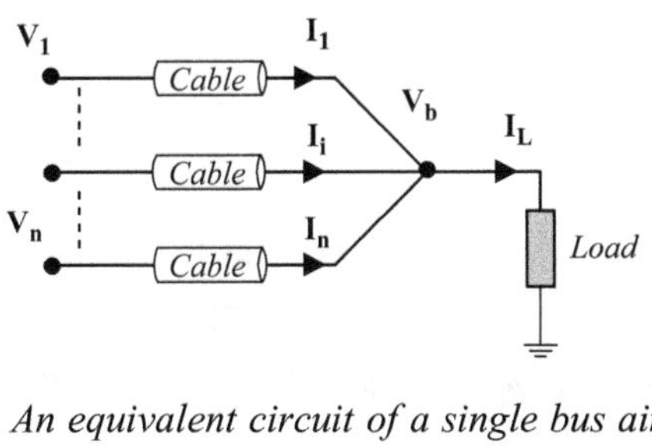

Figure 11.18 An equivalent circuit of a single bus aircraft DC MG

As one can see from Figure 11.18, at steady state, the bus voltage can be expressed as follows:

$$V_b = V_o - I_1(k_1 + R_1) = V_o - I_2(k_2 + R_2) = \cdots = V_o - I_n(k_n + R_n) \tag{11.3}$$

The total load current can be written as

$$I_L = I_1 + I_2 + \cdots + I_n = (V_o - V_b)\sum_{i=1}^{n}\frac{1}{k_i + R_i} \tag{11.4}$$

Then (11.3) can be reformatted to

$$V_b = V_o - I_L\frac{1}{\sum_{i=1}^{n}(1/(k_i + R_i))} \tag{11.5}$$

and the main bus voltage-current characteristic can be formulated as a linear line characterized by the droop gain as

$$k_t = \frac{1}{\sum_{i=1}^{n}(1/(k_i + R_i))} \tag{11.6}$$

The relationship between global droop gain and individual droop gain is depicted in Figure 11.19. It can be seen that the global droop slope k_t is stiffer than individual droop gain k_i. In other words, under the same load power, the voltage drop at the main bus with multi-source operation is smaller than the voltage deviation under the single source operation scenario.

In [27], a three-level control structure is proposed so that the load sharing and voltage regulation can be achieved simultaneously. Secondary control is implemented via a centralized voltage controller to solve the voltage deviation owing to primary droop control. References [27–33] present several distributed secondary control approaches in DC MGs. Reference [34] proposed a secondary control method for DC bus voltage restoration without the need of communication line and additional controllers.

According to the droop control approach, the DC bus voltage will reduce according to the increase of the load power/current. The method restores the terminal voltage to its nominal value autonomously. A feedforward term ΔV is

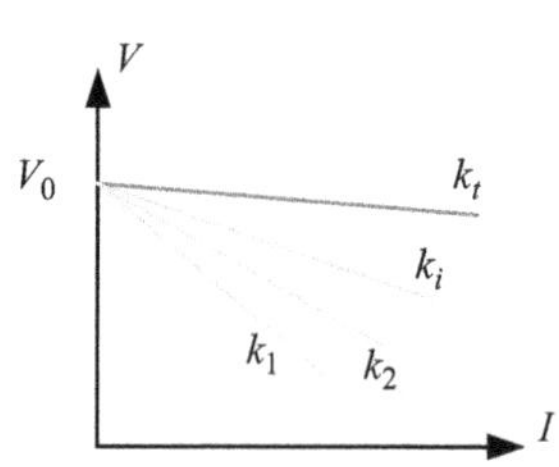

Figure 11.19 Relationship between global droop gain and individual droop gain in DC MGs

added to the terminal voltage reference for each module which can be expressed as follows

$$\Delta V = I_{Lt}\, k_t \tag{11.7}$$

where I_{Lt} is the total load current and k_t is the global droop gain. It is easily implemented since only the load current needs to be measured and no DC voltage controller is in need. If multiple loads are in presence and distributed among the EPS, the total load current measurement can be obtained at the main feeders (if this exists) which supplies the power to all loads.

However, the application is limited to small-scale DC MGs, i.e., the converters need to be close to the main bus and an additional current sensor need to be installed at the main bus to measure the total load current. It is not applicable for large-scale MGs, where the converters and loads are distributed among the network, and it is not practical to install a current sensor at the main bus for the total load measurement. In this sense, a communication line is still required for the information exchange, as shown in Figure 11.20.

11.6 Stability analysis

Control and stability are two important topics that interest researchers in aerospace engineering. Typical loads within aircraft DC microgrids are tightly controlled by power electronic converters and can often behave like constant-power loads (CPLs).

Figure 11.21 shows an example of a tightly controlled motor drive system, which behaves as a CPL. A DC/AC inverter drives an electric machine and tightly regulates the rotating speed. Therefore the speed (ω) is almost constant under this rigorous regulation. Since the rotating load has a one-to-one torque–speed characteristic, for every speed, there is only one corresponding torque. Hence, for a given constant speed (ω), torque (T) is constant and hence power (the multiplication of speed and torque) is also constant. Thus, the tightly controlled DC/AC inverter and its associated motor drive present a CPL characteristic.

CPLs have negative incremental impedance characteristics which can lead to interaction problems among subsystems, deteriorating system stability and also power quality [35,36]. Thus, this undesirable destabilizing effect induced by CPLs is a great concern for the MEA EPS. The candidate architecture should be carefully examined for stability in order to guarantee safe EPS operation for a wide range of operation scenarios.

Two types of stability analysis can be remarked here: large signal and small signal analysis. Large signal stability analysis investigates the ability of a power system to maintain stable operation and reach acceptable steady-state operating points when subjected to large disturbances. The Lyapunov-based method has been popular in terms of large signal stability analysis, where non-linear mathematical modelling is used instead of small signal linearization [37,38]. The main advantage of Lyapunov-based approaches is that a developed Lyapunov function allows the

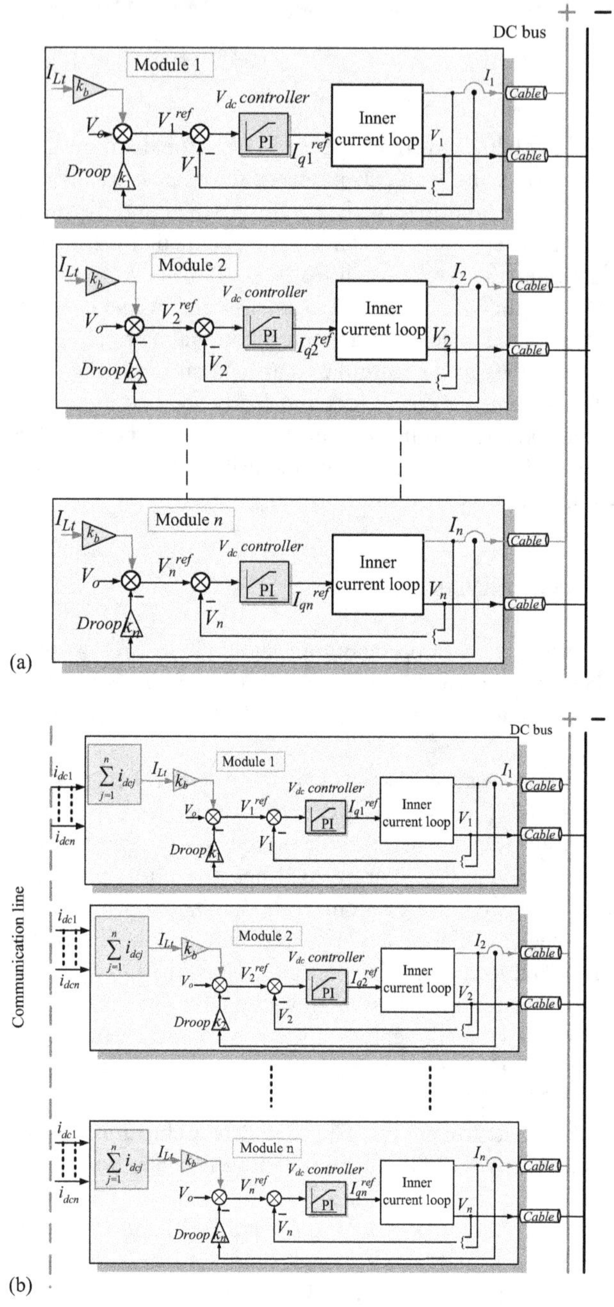

Figure 11.20 Secondary control approach in aircraft DC MGs with multiple controllers (a) without communication link and (b) with communication link

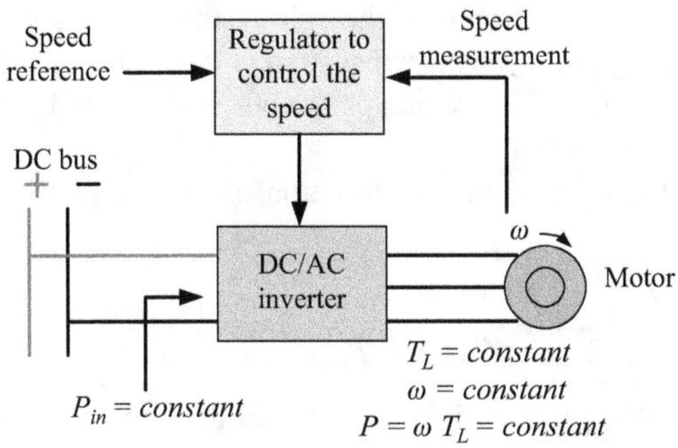

Figure 11.21 A typical presentation of CPLs in aircraft MGs (motor drives)

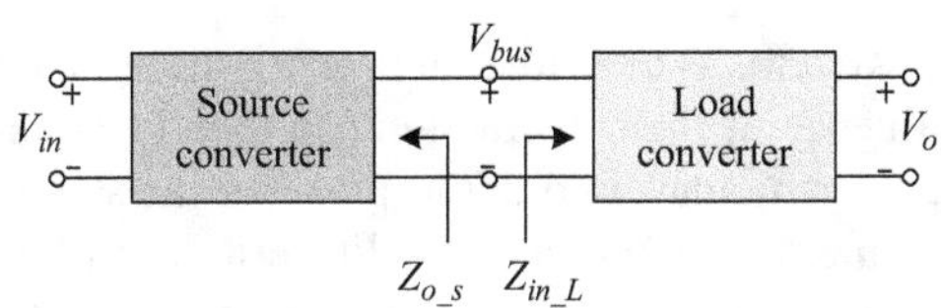

Figure 11.22 Impedance-based method to assess system stability

estimation of the attraction region of a stable operating point, thus providing an estimate of the size of disturbance that can be tolerated [39–42]. On the other hand, small signal stability ensures that the system can revert to its original equilibrium point when subject to a small signal disturbance. Thus, a linearized model around the system equilibrium point is used to study the stability. Linear analysis tools (Bode diagram, Nyquist diagram and Routh–Hurwitz criterion) can then be employed to derive local stability conditions. These approaches are relatively easy and familiar to applied engineers. Two dominant approaches for small signal stability analysis can be remarked on here. The first option is to consider the stability problem with a general state space model, design the control and evaluate the impact of different parameters by studying eigenvalues.

Alternatively, the impedance method, which is based on an analysis of the output impedance of the source and equivalent load impedance, has also been widely used. The contribution of each component to the system's stability can be assessed in the frequency domain. In addition, the output impedance of the components or subsystems can be reshaped to stabilize the overall power system using an active damping method, an impact of connection of new converters that can be easily assessed. For these reasons, the impedance-based method has been successfully applied in power system stability analysis [43,44]. Models of individual components are assembled into a full system model which is then typically broken down into two subsystems at a DC point of interest, i.e. a load subsystem and a source subsystem, as shown in Figure 11.22. Consequently, analytical expressions are derived for input impedance of the load Z_{in_L} and output impedance of the source Z_{o_s} subsystems. The input of the full system is V_{in} and output is V_o. The voltage at the breaking point for the subsystem is V_{bus}. The definition of the source converter is the converter which regulates the bus voltage, and the load converter is responsible of current control.

The total input-to-output transfer function of the cascaded subsystems can be written by

$$V_o = V_{in}\frac{Z_{in\_L}}{Z_{in\_L} + Z_{o\_s}} = \frac{V_{in}}{Z_{in\_L}}\frac{1}{1 + T_{MLG}} \tag{11.8}$$

where the ratio Z_{o_s}/Z_{in_L}, which is often referred to as the minor loop gain, TMLG can be given by

$$T_{MLG} = \frac{Z_{o\_s}}{Z_{in\_L}} \tag{11.9}$$

If each of the two subsystems is individually properly designed with good dynamic performance, the influence of their interaction can then be studied by looking into the minor loop gain. In particular, in order to preserve the stability, it is mandatory that minor loop gain meets the Nyquist stability criterion [45]. It should also be noted that if the detailed information about source and load systems is not available and the respective impedances cannot be analytically constructed, then they should be measured online. Since the source and load converter are assumed to be stable standalone, the minor loop gain term in (11.9) is the one responsible for stability.

In the specific context of aircraft power systems, so far, several publications have discussed the system stability in the MEA electric power systems. The stability of a switched reluctance motor-based 270 V DC power system has been analysed in [46]. A permanent magnet synchronous generator-based aircraft DC MG is investigated in [47–49], and the influence of parameter variation on system stability is presented. Components (R, L, C parameters), operating points (voltage, power) and control parameters (controller gains, droop gains, etc.) will influence the system stability. Take the droop gain as an example, the Bode plot of the source and load impedance in the voltage-mode droop-controlled system is shown in Figure 11.23. The source impedance magnitude in the low-frequency range increases with an increase in the droop gain, whilst the load impedance magnitude decreases. Therefore, the droop gain not only affects the power sharing ratio but also the system stability.

11.7 Chapter summary

This chapter provides an overview of architecture, power quality, control and stability in aircraft DC microgrids. First the evolution of aircraft DC microgrids including power generation, distribution, utilization and energy storage is reviewed. Then the chapter presents the power quality requirement in aircraft DC microgrid including the AC current harmonics and DC voltage envelope. As an

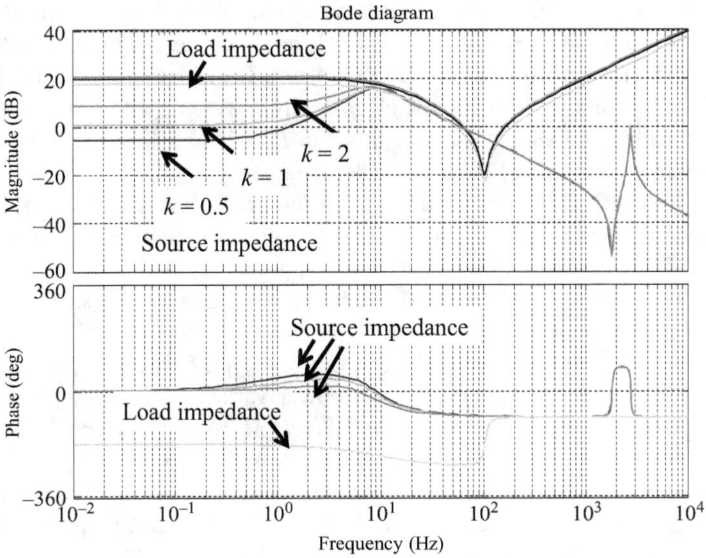

Figure 11.23 Bode diagram of source/load impedances with different values of droop gain using the voltage-mode droop control approach

important part in aircraft power system, electric starter/generator system is introduced and corresponding control in starter/generator mode is presented. In addition, hierarchical control in the aircraft DC MG is presented. Since the bus voltage envelop allows some voltage deviations, it provides the possibility of the voltage droop control in a multiple source-based aircraft DC MGs. Primary control deals with the inner current/voltage regulation and preliminary droop control. To overcome the potential issue in the conventional droop control, distributed/decentralized secondary control deals with the DC voltage restoration and improves the power sharing performance in the meantime. At last, due to the presence of the CPLs, stability is a concern in the aircraft DC MGs. To design a stable aircraft DC MGs over a wide range of operation scenarios, eigenvalue-based and impedance-based stability analysis methods have been discussed.

Acknowledgement

The authors acknowledge the support from CleanSky JTI Projects, FP7 European Integrated Projects-http://www.cleansky.eu.

References

[1] P. Wheeler and S. Bozhko, "The more electric aircraft: technology and challenge," *IEEE Electrif. Mag.*, vol. 2, no. 4, pp. 6–12, Dec. 2014.

[2] P. E. Gartz, "Commercial systems development in a changed world," *IEEE Trans. Aerosp. Electron. Syst.*, vol. 33, no. 2, pp. 632–636, Apr. 1997.

[3] A. Eid, H. El-Kishky, M. Abdel-Salam, and T. El-Mohandes, "Modeling and characterization of an aircraft electric power system with a fuel cell equipped APU paralleled at main ac bus," in *Proc. IEEE Conf. Int. Power Modulator High Voltage*, 2010, pp. 229–232.

[4] X. Roboam, B. Sareni, and A. D. Andrade, "More electricity in the air: toward optimized electrical networks embedded in more-electrical aircraft," *IEEE Ind. Electron. Mag.*, vol. 6, no. 4, pp. 6–17, Dec. 2012.

[5] R. Abdel-Fadil, A. Eid, and M. Abdel-Salam, "Electrical distribution power systems of modern civil aircrafts," in *Proc. 2nd International Conference on Energy Systems and Technologies*, Cairo, Egypt, Feb. 2013, pp. 1–10.

[6] M. Olaiya and N. Buchan, "High power variable frequency generator for large civil aircraft," in *Proc. IET Elect. Mach. Syst. More Elect. Aircr.*, 1999, pp. 1–4.

[7] B. H. Nya, J. Brombach, and D. Schulz, "Benefits of higher voltage levels in aircraft electrical power systems," in *Proc. Electrical Systems for Aircraft, Railway and Ship Propulsion (ESARS)*, Bologna, Oct. 2012, pp. 1–5.

[8] F. Gao, S. Yeoh, S. Bozhko, and G. Asher, "Coordinated control of a DC electrical power system in the more electric aircraft integrated with energy storage," in *ECCE 2015, IEEE Energy Conversion Congress & Expo, Montreal, Canada*, Sep. 2015, pp. 5431–5438.

[9] H. Zhang, F. Mollet, C. Saudemont, and B. Robyns, "Experimental validation of energy storage system management strategies for a local DC distribution system of more electric aircraft," *IEEE Trans. Ind. Electron.*, vol. 57, no. 12, pp. 3905–3916, Dec. 2010.

[10] K. Jong-Yul, J. Jin-Hong, K. Seul-Ki, *et al.*, "Cooperative control strategy of energy storage system and microsources for stabilizing the microgrid during islanded operation," *IEEE Trans. Power Electron.*, vol. 25, no. 12, pp. 3037–3048, Dec. 2010.

[11] Y. Guan, L. Meng, C. Li, J. Vasquez, and J. Guerrero, "A dynamic consensus algorithm to adjust virtual impedance loops for discharge rate balancing of AC microgrid energy storage units," *IEEE Trans. Smart Grid,* To be published.

[12] Characteristics of aircraft electrical systems. ISO-1540, 2006.

[13] Environmental conditions and test procedures for airborne equipment. DO-160F, 2010.

[14] *Aircraft Electric Power Characteristics*, American Military St. MIL-STD-704F, Mar. 2004.

[15] T. Yang, S. Bozhko, and G. Asher, "Functional modeling of symmetrical multipulse autotransformer rectifier units for aerospace applications," *IEEE Trans. Power Electron.*, vol. 30, no. 9, pp. 4704–4713, Sep. 2015.

[16] X. Zheng, F. Gao, H. Ali, and H. Liu, "A droop control based three phase bidirectional AC-DC converter for more electric aircraft applications," *Energies*, vol. 10, no. 3, pp. 1–17, 2017.

[17] B. S. Bhangu and K. Rajashekara, "Electric starter generators: the integration into gas turbine engines," *IEEE Ind. Appl. Mag.*, vol. 20, no. 2, pp. 14–22, Apr. 2014.

[18] Cleansky. AEGART Project Summary. 2015. Available: http://cordis.europa.eu/result/rcn/143557_en.html. Accessed: 11/11/2015.

[19] F. Gao, S. Bozhko, A. Costabeber, *et al.*, "Comparative stability analysis of droop control approaches in voltage source converters-based DC microgrids," *IEEE Trans. Power Electron.*, vol. 32, no. 3, pp. 2395–2415, Mar. 2017.

[20] F. Gao, "Decentralized control and stability analysis of a multi-generator based electrical power system for more electric aircraft," PhD dissertation, University of Nottingham, 2016.

[21] J. M. Guerrero, M. Chandorkar, T.-L. Lee, and P. C. Loh, "Advanced control architectures for intelligent microgrids—Part I: Decentralized and hierarchical control," *IEEE Trans. Ind. Electron.*, vol. 60, no. 4, pp. 1254–1262, Apr. 2013.

[22] F. Gao, Y. Gu, S. Bozhko, G. Asher, and P. Wheeler, "Analysis of droop control methods in DC microgrids," in *Proc. 16th IEEE European Conference on Power Electronics and Applications (EPE-ECCE' Europe)*, Lappeenranta, Finland, Aug. 2014, pp. 1–9.

[23] K. Rouzbehi, A. Miranian, A. Luna, and P. Rodriguez, "DC voltage control and power sharing in Multiterminal DC grids based on optimal DC power flow and voltage-droop strategy," *IEEE J. Emerg. Sel. Topics Power Electron.*, vol. 2, no. 4, pp. 1171–1180, Dec. 2014.

[24] W. Wang, M. Barnes, O. Marjanovic, and O. Cwikowski, "Impact of DC breaker systems on multiterminal VSC-HVDC stability," *IEEE Trans. Power Del.*, vol. 31, no. 2, pp. 769–779, Apr. 2016.

[25] X. Lu, J. M. Guerrero, K. Sun, J. C. Vasquez, R. Teodorescu, and L. Huang, "Hierarchical control of parallel AC–DC converter interfaces for hybrid microgrids," *IEEE Trans. Smart Grid*, vol. 5, no. 2, pp. 683–692, Mar. 2014.

[26] J. Beerten and R. Belmans, "Analysis of power sharing and voltage deviations in droop-controlled DC grids," *IEEE Trans. Power Syst.*, vol. 28, no. 4, pp. 4588–4597, Nov. 2013.

[27] J. M. Guerrero, J. C. Vasquez, and L. G. D. Vicuna, "Hierarchical control of droop-controlled ac and dc microgrids—a general approach toward standardization," *IEEE Trans. Ind. Electron.*, vol. 58, no. 1, pp. 158–172, Jan. 2011.

[28] X. Lu, J. M. Guerrero, K. Sun, and J. C. Vasquez, "An improved droop control method for DC microgrids based on low bandwidth communication with DC bus voltage restoration and enhanced current sharing accuracy," *IEEE Trans. Power Electron.*, vol. 29, no. 4, pp. 1800–1812, Apr. 2014.

[29] S. Anand, B. G. Fernandes, and J. M. Guerrero, "Distributed control to ensure proportional load sharing and improve voltage regulation in low voltage DC microgrids," *IEEE Trans. Power Electron.*, vol. 28, no. 4, pp. 1900–1913, Apr. 2013.

[30] N. Yang, D. Paire, F. Gao, A. Miraoui, and W. Liu, "Compensation of droop control using common load condition in dc microgrids to improve voltage regulation and load sharing," *Int. J. Elect. Power Energy Syst.*, vol. 64, pp. 752–760, Jan. 2015.

[31] P. Wang, X. Lu, X. Yang, W. Wang, and D. Xu, "An improved distributed secondary control method for DC microgrids with enhanced dynamic current sharing performance," *IEEE Trans. Power Electron.*, vol. 31, no. 9, pp. 6658–6673, Sep. 2016.

[32] T. Dragicevic, J. C. Vasquez, J. M. Guerrero, and D. Skrlec, "Advanced LVDC electrical power architectures and microgrids: a step toward a new generation of power distribution networks," *IEEE Electrifi. Mag.* vol. 2, no. 1, pp. 54–65, Mar. 2014.

[33] X. Lu, J. M. Guerrero, K. Sun, and J. C. Vasquez, "An improved droop control method for dc microgrids based on low bandwidth communication with dc bus voltage restoration and enhanced current sharing accuracy," *IEEE Trans. Power Electron.*, vol. 29, no. 4, pp. 1800–1812, Apr. 2014.

[34] F. Gao, S. Bozhko, G. Asher, P. Wheeler, and C. Patel, "An improved voltage compensation approach in a droop-controlled DC power system for the more electric aircraft," *IEEE Trans. Power Electron.*, vol. 31, no. 10, pp. 7369–7383, Oct. 2016.

[35] A. Emadi, A. Khaligh, C. Rivetta, and G. Williamson, "Constant power loads and negative impedance instability in automotive systems: definition, modeling, stability, and control of power electronic converters and motor drives," *IEEE Trans. Veh. Technol.*, vol. 55, no. 4, pp. 1112–1125, Jul. 2006.

[36] A. Rahimi and A. Emadi, "Active damping in DC/DC power electronic converters: a novel method to overcome the problems of constant power loads," *IEEE Trans. Ind. Electron.*, vol. 56, no. 5, pp. 1428–1439, May 2009.

[37] M. Kabalan, P. Singh, and D. Niebur, "Large signal Lyapunov-based stability studies in microgrids: a review," *IEEE Trans. Smart Grid*, vol. 8, no. 5, pp. 2287–2295, Sep. 2017.

[38] P. Kundur, J. Paserba, V. Ajjarapu, *et al.*, "Definition and classification of power system stability IEEE/CIGRE joint task force on stability terms and definitions," *IEEE Trans. Power Syst.*, vol. 19, no. 3, pp. 1387–1401, Aug. 2004.

[39] P. W. Sauer and M. A. Pai, *Power System Dynamics and Stability*. Champaign, IL, USA: Stipes, 2006.

[40] W. Du, J. Zhang, Y. Zhang, and Z. Qian, "Stability criterion for cascaded system with constant power load," *IEEE Trans. Power Electron.*, vol. 28, no. 4, pp. 1843–1851, Apr. 2013.

[41] A. Griffo and J. Wang, "Large signal stability analysis of 'more electric' aircraft power systems with constant power loads," *IEEE Aerosp. Electron. Syst. Mag.*, vol. 48, no. 1, pp. 477–489, Jan. 2012.

[42] D. Marx, P. Magne, B. Nahid-Mobarakeh, S. Pierfederici, and B. Davat, "Large signal stability analysis tools in DC power systems with constant

power loads and variable power loads—a review," *IEEE Trans. Power Electron.*, vol. 27, no. 4, pp. 1773–1787, Apr. 2012.

[43] M. Cespedes and J. Sun, "Impedance modeling and analysis of grid-connected voltage-source converters," *IEEE Trans. Power Electron.*, vol. 29, no. 3, pp. 1254–1261, Mar. 2014.

[44] A. Radwan and Y. Mohamed, "Assessment and mitigation of interaction dynamics in hybrid AC/DC distribution generation systems," *IEEE Trans. Smart Grid*, vol. 3, no. 3, pp. 1382–1393, Sep. 2012.

[45] R. D. Middlebrook, "Input filer consideration in design and application of switching regulators," in *Proc. IEEE IAS Annu. Meet.*, Chicago, USA, Oct. 1976, pp. 336–382.

[46] L. Han, J. Wang, and D. Howe, "Small-signal stability studies of a 270V DC more-electric aircraft power system," in *Proc. Power Electronics Machines and Drives 2006. The 3rd IET International Conference (PEMD)*, Apr. 2006, pp. 197–201.

[47] K.-N. Areerak, T. Wu, S. V. Bozhko, G. M. Asher, and D. W. P. Thomas, "Aircraft power system stability study including effect of voltage control and actuators dynamic," *IEEE Trans. Aerosp. Electron. Syst.*, vol. 47, no. 4, pp. 2574–2589, Oct. 2011.

[48] K.-N. Areerak, S. V. Bozhko, G. M. Asher, L. De Lillo, and D. W. P. Thomas, "Stability study for a hybrid AC–DC more-electric aircraft power system," *IEEE Trans. Aerosp. Electron. Syst.*, vol. 48, no. 1, pp. 329–347, Jan. 2012.

[49] F. Gao and S. Bozhko, "Modeling and impedance analysis of a single DC bus based multiple-source multiple-load electrical power system," *IEEE Trans. Transp. Electrif.*, vol. 2, no. 3, pp. 335–346, Sep. 2016.

Chapter 12
Shipboard MVDC microgrids

Dong-Choon Lee[1]*, Yoon-Cheul Jeung*[1]*, Dinh Du To*[1]*, and Duc Dung Le*[1]

12.1 Introduction

Electric power was successfully applied to marine vessels in 1880s. Due to the advance of technology in batteries, motors, and engines, the primitive electric devices were installed on the ship, and in the early twentieth century the electric propulsion system was developed [1]. The electric propulsion has a lot of advantages such as better boarding comport, design flexibility, low vibration and noises, low fuel consumption and emission, etc.

In electric propulsion systems, the propulsion motors are supplied by the dedicated generators and the other loads are supplied by the auxiliary generators. To avoid the inefficient structure of system, multiple gen-sets are connected to one common bus, which supply propulsion and service loads together. This system is called as *integrated propulsion system* (IPS) [2]. When the load power demand in the IPS is low, some of the prime movers are operated at high output power region and the others are deactivated, which increases the overall system efficiency. *Queen Elizabeth II* launched in 1987 was the first diesel-electric ship that embeds the IPS [3].

Due to the evolution of the power electronics technology, the concept of *all-electric ship* (AES) has been established, where, needless to say about propulsion and service loads, even the loads using hydraulic and compressed air power are replaced by electric devices. In 2015, the world's first purely battery-driven car and passenger ferry *Ampere* was launched in Norway [4].

So far, the AC power distribution system has been commonly used in shipboard power systems since the AC motors are mostly employed for electric propulsion. Recently, however, the shipboard medium voltage DC (MVDC) power system is attracting a great attention since it has lots of advantages such as smaller high-frequency transformer, easy connection and disconnection, no harmonic and unbalance problems, no need of synchronization, no reactive power, high efficiency, flexible design in space, easy monitoring of power quality, fuel saving by variable speed operation, and so on [5].

[1]Department of Electrical Engineering, Yeungnam University, South Korea

In this chapter, the shipboard MVDC microgrid system is introduced. First, the architecture and components of the system are described, and then the modeling and control are dealt with. Next, power management based on hierarchical control is described and then fault and protection schemes are discussed. Some simulation results for the case study are provided.

12.2 Architecture of shipboard MVDC power systems

The shipboard DC power system consists of generation, distribution, power conversion, and ship loads. In the IPS, the gen-sets and ship loads are connected to the common MVDC bus through the power converters. Electric generators are driven by gas turbines (GTs) or diesel engines (DEs), which supply power to the bus. In large ships, multiple generators are employed to supply the high power demand. In addition, the ESS (energy storage system), which can compensate for the fluctuations in the DC bus voltage due to load variations, is integrated. The ship loads are usually composed of propulsion, services, and pulsed power loads (PPLs). The optional configuration of the shipboard MVDC power system is shown in Figure 12.1.

12.2.1 Structure of DC bus

The architecture of the MVDC distribution in ships is mostly of a ring or radial type. The radial bus is an economic structure since it can supply power to various loads through independent power lines. At fault conditions, however, the survivability of the ship is low. To overcome this weakness, the additional energy sources can be installed for service loads, but it is not easy in ships due to the limitation of the space.

Meanwhile, the ring bus fed by multiple power sources can supply the power to the loads through the common power lines. Figure 12.2 shows the ring bus structure of the MVDC distribution system with several zones [5]. This structure has not only an intrinsic advantage of DC distribution but also an ability to supply power continuously to the load under a certain fault condition. Furthermore, it is

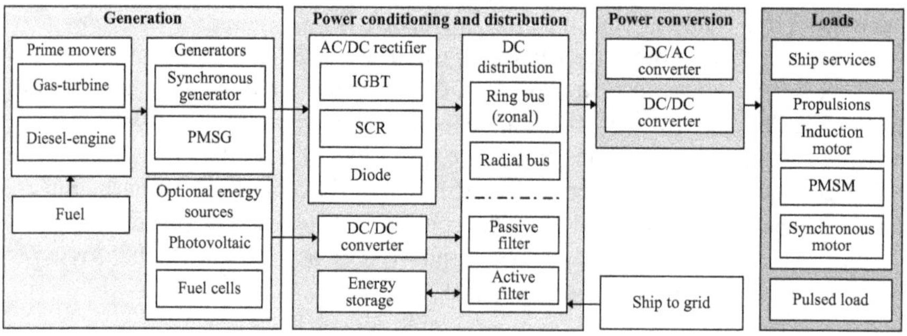

Figure 12.1 Configuration of shipboard MVDC power system

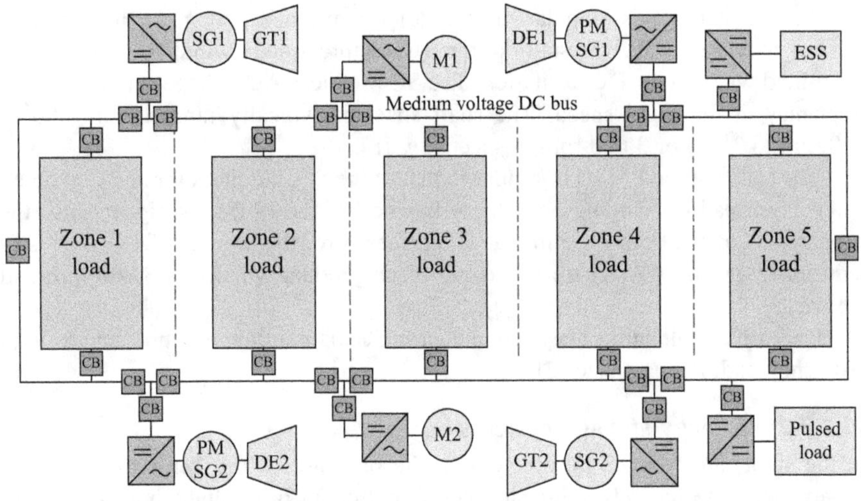

Figure 12.2 Ring bus structure of shipboard MVDC power system with zonal distribution [5]

expected that fuel consumption and emission can be reduced by up to 20% and that footprints and weight of electrical equipment can be reduced by up to 30% compared with the AC distribution systems [6].

12.2.2 Prime movers

In ships, the main electric power sources are mostly synchronous generators (SGs) which are driven by prime movers. The GT is preferred for the high power level of ships, but the DE is often used below a medium power level. In certain cases, both types of prime movers can be used at the same time.

Conventionally, the prime mover in ships is operated at a fixed speed for AC distribution. In this case, the fuel efficiency of the prime movers is low at light loads. Even though higher fuel efficiency can be achieved by a variable speed operation, it is difficult to apply to the shipboard AC power system since the output voltage frequency of the generators should be limited by the standards of AC distribution.

In the DC power system, however, there is no critical hurdle to apply the variable speed operation scheme since the AC voltage with a variable frequency is rectified to the DC bus voltage. It has been reported that up to 15% of the fuel is saved in the 5.17 MW GT system by the variable speed operation [7].

12.2.3 Components of shipboard MVDC power systems

12.2.3.1 AC generators

For high power ships, a wound-field SG is usually used, whereas a permanent magnet SG (PMSG) is applied to the low-to-medium power capacity ships.

The SG is the most popular machine of which power rating is ranged from a few kW to tens of MW. The stator of the SG has three-phase windings and the rotor has a field winding for DC excitation. The SG produces AC voltages at 50 or 60 Hz frequency at the rated speed. The shaft speed is typically ranged from 600 or 720 rpm to 3,000 or 3,600 rpm, respectively [8].

The stator of the PMSG is similar to that of the SG, but the rotor has permanent magnets instead of windings. So, the efficiency of the PMSG is higher since the copper loss of the rotor is eliminated. Furthermore, the weight of the PMSG is reduced by up to 30%. There is a commercial product which has 3,000 rpm and 8 MW [9].

Recently, application of superconductor SGs is gaining attention due to high efficiency and compact size [2].

12.2.3.2 Energy storage systems

In the shipboard MVDC power system, the response of the gen-sets as a main power source is relatively sluggish. Therefore, the DC bus voltage is vulnerable to variations of the load, which will drop much in the case of abrupt application of high power loads. Thus, the ESS is indispensable to supply power quickly since the voltage drop in the DC bus causes harmful influences on propulsion and service loads. According to the IEEE Standard 1709–2010, the bus voltage tolerances should be ±10% in steady states [5].

Warships or other electric ships have PPLs like rail guns, high power radars, cranes, etc., which demand a high power in a short interval. The ESS needs to cover this requirement as well as supply loads for survivability in certain cases. When determining the power capacity of the ESS, the initial installment cost, payback time, weight, reserve margin of single failure, optimal efficiency operation of turbines, and power supplying capability to PPLs should be considered [10,11].

12.2.3.3 Ship loads

Ship loads can be classified into four categories: propulsion, dedicated high-power load, ship services, and PPL [5]. The electric propulsion forms 80% of the total ship load power. The electric propulsion needs no speed reduction gears, which uses a fixed pitch of propeller blades. Also, it has a lot of advantages such as small volume, low maintenance, low fuel consumption, longer life, and flexibility of design, low emission, and better boarding comport. The dedicated high-power load includes military radars and cranes, which consume a few MW of power. The service load includes small pumps and fan motor drives and hotel load. The rail gun is a kind of PPLs which consumes a few MW of power in several seconds [5].

The load power of propulsion depends on speed and sailing mode of the ship. For instance, the load profile of liquefied natural gas (LNG) carrier consists of unloading, loading, loaded voyage, and ballast voyage modes [12].

12.2.3.4 Power converters

In the shipboard MVDC power system, different types of power converters are employed such as AC/DC converters, DC/DC converters, and DC/AC inverters.

- AC/DC converters

In shipboard MVDC power systems, the output voltage of the main generators is connected to the DC bus through the AC/DC converters that are diode rectifiers or phase-controlled converters with silicon-controlled rectifier (SCR). In the case of diode rectifiers, if the SG is used, its output voltage is controlled by an exciter, as shown in Figure 12.3(a). If the PMSG is used, however, the back-end DC/DC converter is required to match the MVDC bus voltage, as shown in Figure 12.3(b). If the SCR rectifier instead of diode rectifier is employed, the back-end converter is not necessary.

On the other hand, multilevel converters have been so far used for medium voltage applications, in which neutral-point clamped (NPC), flying capacitor, cascade-H bridge converters are included. Recently, the modular multilevel converter (MMC) topology has been applied to the high voltage DC (HVDC) and MVDC systems, as shown in Figure 12.4. Due to the modular structure, the voltage rating of MMC can be expanded easily.

- DC/DC converters

The DC/DC converter is required for connecting the ESS to the MVDC bus. A dual active bridge (DAB) DC/DC converter is popularly used since it has advantages of bidirectional power flow, high efficiency, modular structure, isolation, and soft switching. By a multiple connection of the single DAB, the power capacity of the converter can be expanded easily, which is shown in Figure 12.5 [13].

Similarly, modular multilevel DC/DC converters are also a candidate for high power applications [14].

- DC/AC converters

In electric propulsion ships, DC/AC converters, that is, inverters are employed to feed the propulsion motors and other AC loads. The two-level inverter is sufficient for small pump and fan motor drives and hotel loads. For propulsion drives,

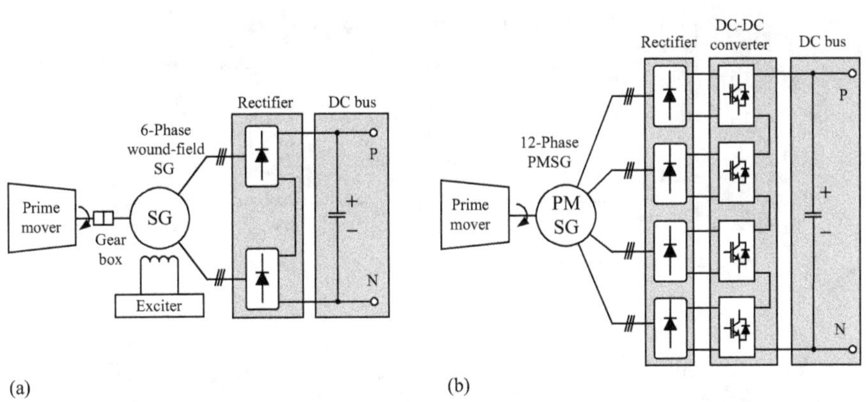

Figure 12.3 AC/DC converter for MVDC shipboard system: (a) Wound-field SG and (b) PMSG

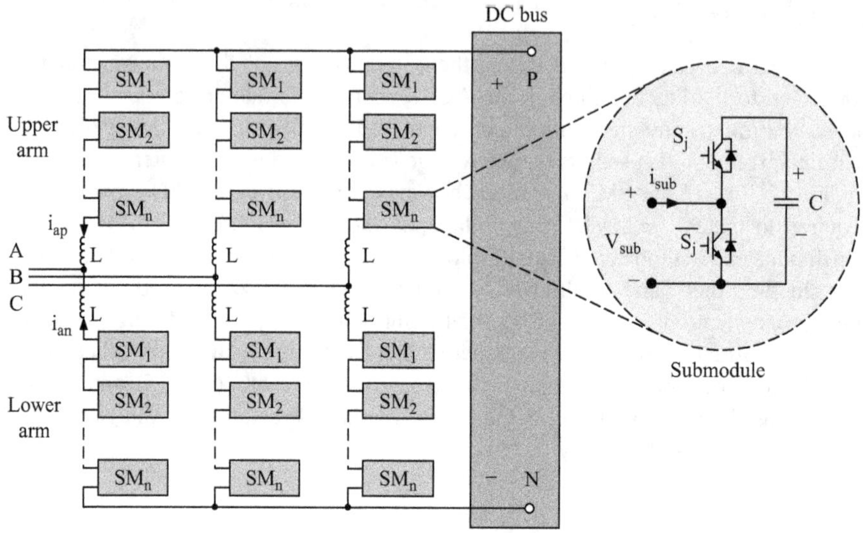

Figure 12.4 Topology of MMC

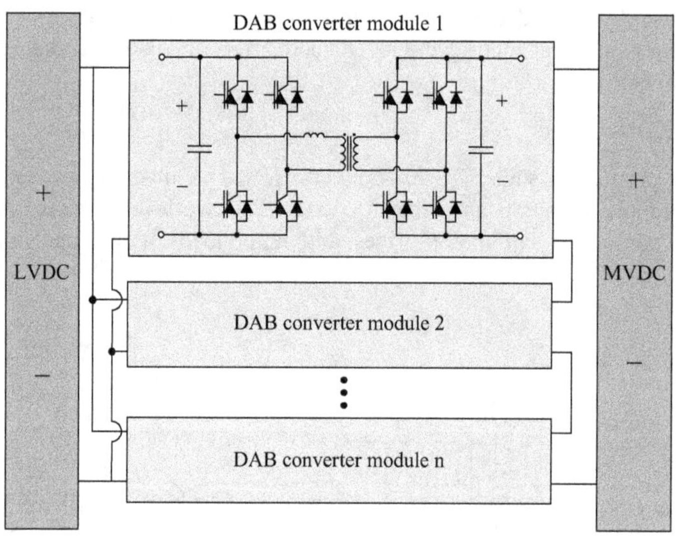

Figure 12.5 Modular dual active bridge DC–DC converter

cycloconverters and load-commutated inverters have been conventionally adopted for high power AC motor drives. For the shipboard MVDC power system, however, voltage-source inverters are appropriate since the inverter is supplied from the DC bus. For high power AC drives, multilevel inverters can be applied. It has been reported that the NPC inverter available commercially can be ranged up to 33 MW [15]. In the future, the MMC topology is a good candidate since it has a lot of advantages such as modularity, fault tolerant capability, low-voltage device, high efficiency, and better harmonic performance [16].

- DC circuit breakers (CBs)

In AC power systems, the alternating current naturally crosses zero at every half cycle, which creates a condition for arc self-extinction to isolate the fault. However, due to no zero-crossing of the direct current, it is more challenging to design the DC CB in the DC power system.

The DC CB topologies are classified into the three categories: mechanical CBs, solid-state CBs (SSCBs) and hybrid CBs [17]. The mechanical CB has an advantage of low-contact resistance, but a long opening time (5–100 ms). In the SSCB, the insulated-gate bipolar transistors (IGBTs) and integrated-gate commutated thyristors (IGCTs) are the key components to break the fault current with a shorter opening time (100 μs). The SSCB has some critical issues such as the high conduction loss of power switches, the expensive cooling system, and a lot of power switches connected in series and/or parallel for high current and voltage ratings. A combination of the mechanical CB and the power switches results in a hybrid CB, which is featured with low conduction loss and short opening time. When a fault occurs, the fault current is commutated from the mechanical CB to the power switches during the fault. In the near future, wide band-gap devices (SiC or GaN) will be used to achieve the lower conduction loss, the higher speed fault clearance as well as the cost reduction of cooling system due to the wider range of operating temperature.

12.3 Control of shipboard MVDC power systems

12.3.1 Gen-sets

The gen-set consists of a prime mover and a SG. A classical gen-set is operated at a constant speed to generate a fixed electrical frequency and output voltage. Although such a gen-set is of low cost, its fuel efficiency is low at light load conditions, and the output voltage and frequency are fluctuated when the load varies. If a variable speed operation scheme is applied, a significant amount of fuel saving is achieved. The optimal speeds of DEs can be derived from the specific fuel consumption curve [18].

Figure 12.6 shows the control block diagram of the gen-set system, which consists of a GT, a wound-field SG with an exciter, and a diode rectifier. The governor can utilize conventional PI or lead-lag controllers to regulate the speed of the prime mover. Recently, some advanced control techniques like sliding mode control, fuzzy logic control, and model predictive control have been applied [19]. In the wound-field SG, the magnitude of output voltage is controlled by an exciter, so that the DC output voltage rectified by the diode rectifier is regulated indirectly. In the case of PMSG, the thyristor phase-controlled rectifier or the diode rectifier with back-end DC/DC converter are usually used.

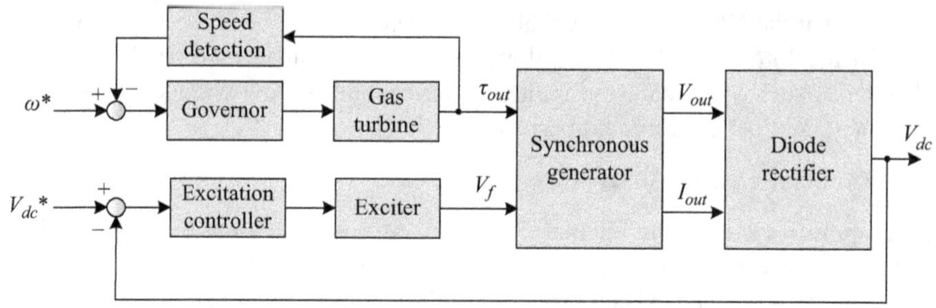

Figure 12.6 Control structure of gen-set system

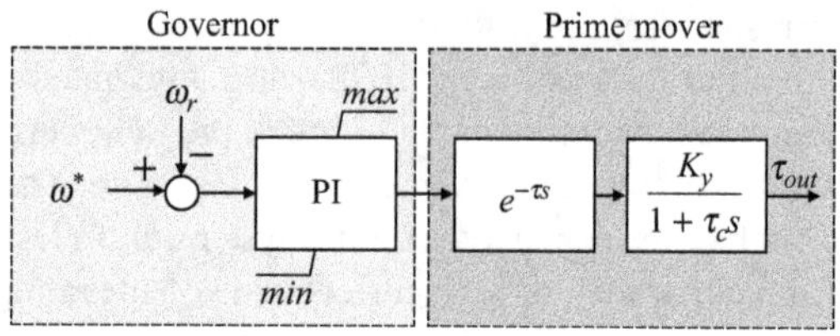

Figure 12.7 Speed control of prime mover

To simplify the design of the speed control loop, the dynamics of prime movers are often modeled as a first-order or second-order delay. Figure 12.7 shows the speed control block diagram of the prime mover with PI regulators, where τ is a time delay, τ_c is the time constant, and K_y is the torque constant. For better control performance, the characteristics of the engine and turbine, effects of temperature, shaft speed, combustion, and cooling air flow need to be included in the detailed modeling [20].

For the long-term system analysis, the generator model is effectively simplified in a series circuit of variable AC voltage source, resistance, and inductance [21]. The exciter system is often modeled as a first-order or second-order time delay, which is usually controlled by thyristor phase-controlled converters [22].

12.3.2 Energy storage systems

In the shipboard MVDC power system, the ESS can be used to compensate for the fluctuations of the DC bus voltage and to manage the number of generators being operated for fuel savings.

For the operation of the ESS, the following control strategies are considered. The ESS controller regulates the voltage and current of the ESS during charging and discharging operations. If multiple ESSs are used, the output powers and state of charge (SOC) values among the ESSs can be balanced by power sharing. If the droop control is applied for power sharing, or if the power line impedance is high, the DC bus voltage becomes lower than the nominal voltage. Then, the control scheme which increases the output voltage of the ESS should be applied. Finally, the power management controller produces a power reference for the ESS [23].

12.3.3 Power management control

The power management controller should be able to ensure the energy balance between generated powers and load power consumptions both at steady-state and transient-state conditions. For the shipboard power system, the same hierarchical control structure as in the conventional power grid can be applied for power management [24].

Figure 12.8 shows a hierarchical control structure of shipboard MVDC power system. The primary controllers regulate voltages and currents of the local power converters. The secondary controller performs the DC bus voltage control. The tertiary controller produces a command for power management of the whole ship with information of generated power and load power demand.

12.3.3.1 Local control

Local (primary) control requires the fastest responses among three hierarchical control levels. The primary controller regulates the output voltage at the coupling point of the DC bus where the power converter is connected. The voltage controller is of PI type, which usually has an inner current control loop and selectively a feedforward control term. Also, the power sharing can be included in this control level.

The droop control is the most popular scheme for power sharing since it needs no communication links, which results in a simple circuit and low cost. With this scheme, each output current of the converters can be balanced by adjusting the DC voltage reference, which can be expressed as

$$v_i^{ref} = v_{nom} - R_{ii} \cdot i_i \tag{12.1}$$

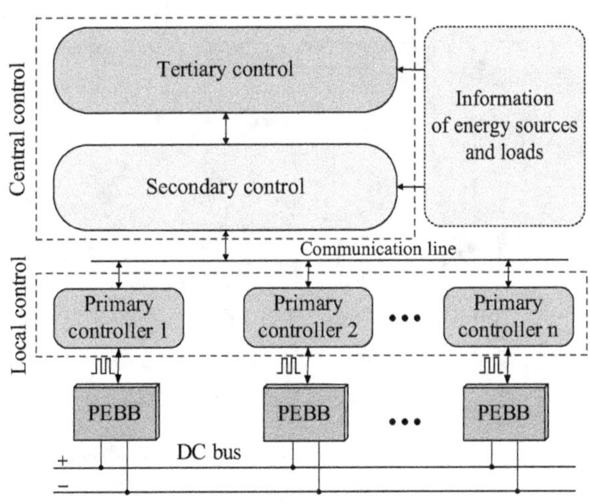

Figure 12.8 Hierarchical control structure of shipboard power system

where v_i^{ref} is the voltage reference for ith unit, v_{nom} is the nominal voltage, i_i is the output current, and R_{ii} is the droop coefficient.

The droop gain can be obtained from the maximum output current (i_{dc_max}) and the allowable voltage variation (Δv_{dc}), as [25]

$$R_{ii} = \frac{\Delta v_{dc}}{i_{dc\_max}}. \tag{12.2}$$

When the gen-sets and ESSs supply power to the DC grid, each local controller regulates its output voltage and performs the power sharing. In the case of battery ESS, the management of the SOC can be included. If only one generator is operated or if the generator is operated at rated power, the droop control is not applied. When switching the control modes, the higher-level controller sends a control mode selection signal to the primary one via communication as shown in Figure 12.9.

12.3.3.2 Central control

The secondary control can be designed to meet the power quality requirements. Figure 12.10 demonstrates the droop characteristic of the secondary control. If the

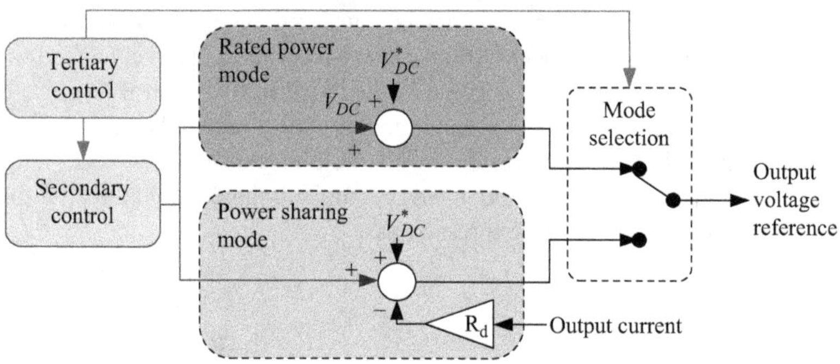

Figure 12.9 Hierarchical control block diagram

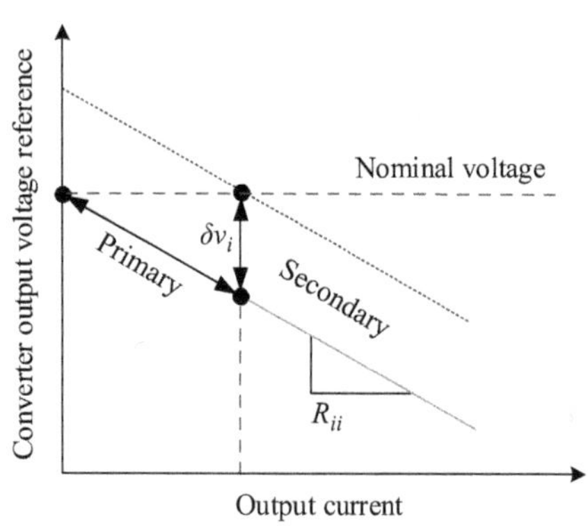

Figure 12.10 Droop characteristic of the secondary control

output current in the primary level is increased according to the load, the output voltage is decreased due to droop control. Then, the secondary control yields a DC-bus voltage compensation term, δv_i, by which the modified voltage reference is obtained as [26]

$$v_i^{ref} = v_{nom} - R_{ii} \cdot i_i + \delta v_i. \tag{12.3}$$

The tertiary control, which is the highest-level control, decides whether the gen-sets and ESSs are operated or not, depending on the load power demand. Furthermore, the tertiary controller can perform the economic power dispatch control and energy balancing control.

12.3.3.3 Case study

In this subsection, the simulation study of MVDC shipboard power system with the hierarchical control is described. The bus structure and system power capacity are selected referred to the IEEE Standard 1709–2010 [5]. The parameters of the simulation model are listed in Table 12.1.

- Model and control of MVDC IPS

In simulation models, the gen-sets consist of two GTs with SGs and two DEs with PMSGs. The prime movers are modeled as shown in Figure 12.7, where $\tau = 0.02$, $\tau_c = 0.0025$, and $K_y = 1$ in p.u. For fuel savings, the optimal speed control of the prime movers is applied [18]. In addition, the generators are modeled as a variable voltage source with internal impedance.

The output of the SG is connected to the MVDC bus through diode rectifiers, but that of the PMSG is connected through diode rectifiers and back-end DC/DC converters. For speed and excitation controls of the gen-sets, the PI controllers are used. The ESS is modeled as a current source with a slew rate of 100 MW/s, which is operated in a bidirectional power flow mode. The operating mode and power reference for the ESS are commanded by the tertiary control.

There are three kinds of loads applied, where the zonal load is modeled as a resistance, the propulsion drive as a constant power load, and the PPL as a pulse charging circuit with a large capacitance.

Table 12.1 Parameters of MVDC shipboard power system

Parameters	Values
MVDC-bus voltage	6 kV
GT-SG set	36 MW × 2
DE-PMSG set	4 MW × 2
Propulsion (M1, M2)	36 MW × 2
Zonal loads	5 MW
Pulsed power load	4 MW for 1 s
Energy storage system	4 MW h

The secondary controller regulates the MVDC bus voltage with information of gen-sets. The output voltage references of each gen-set system are calculated in (12.3), which will compensate for the variations of the MVDC bus voltage. Finally, the tertiary controller selects the operating mode of the gen-sets according to the load profile.

- Load profile

The load profile has four operating modes, as shown in Figure 12.11, which are classified by the power demand of ship loads.

Mode I: The consumption of the load power is lower than 5 MW. The generated power is supplied only to services loads since the ship is in port. In this mode, only DEs except GTs are operated.

Mode II: The consumption of the load power is ranged between 5 and 36 MW. In this mode, only one GT can support the whole power demand for propulsion and service loads.

Mode III: The consumption of the load power is ranged between 36 and 72 MW. In this mode, both GTs are operated and DEs are not operated. The PPL is applied in this mode.

Mode IV: The consumption of the load power is higher than 72 MW. The four gen-sets should be operated to satisfy the power demand while the ship is on the voyage at the maximum speed.

The operating modes of the four gen-sets are summarized in Table 12.2.

- Simulation results

Figure 12.12 shows the system performance in operating modes I and II. Initially, the system is operating in mode I, where two DEs supply power to loads. At 10 s, the propulsion drives are activated. Then, the load power is increased and the operating mode is changed to mode II. The central control sends a start command to the GT1. At the same time, the governor adjusts the mechanical torque to increase

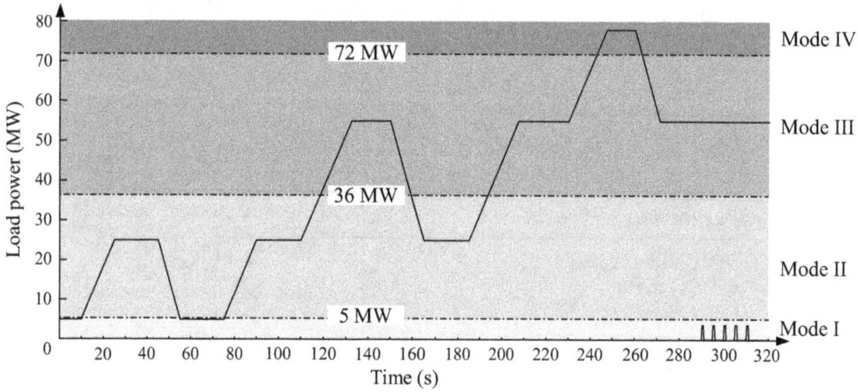

Figure 12.11 Load profile

Table 12.2 Control mode of shipboard power system

Operating mode	Control mode of gen-sets				Load power
	*DE1	*DE2	**GT1	**GT2	
Mode I	Droop control	Droop control	Off	Off	Lower than 5 MW
Mode II	Off	Off	Voltage control	Off	5–36 MW
Mode III	Off	Off	Droop control	Droop control	36–72 MW
Mode IV	Rated power control	Rated power control	Droop control	Droop control	Higher than 72 MW

**DE, diesel engine; **GT, gas turbine.*

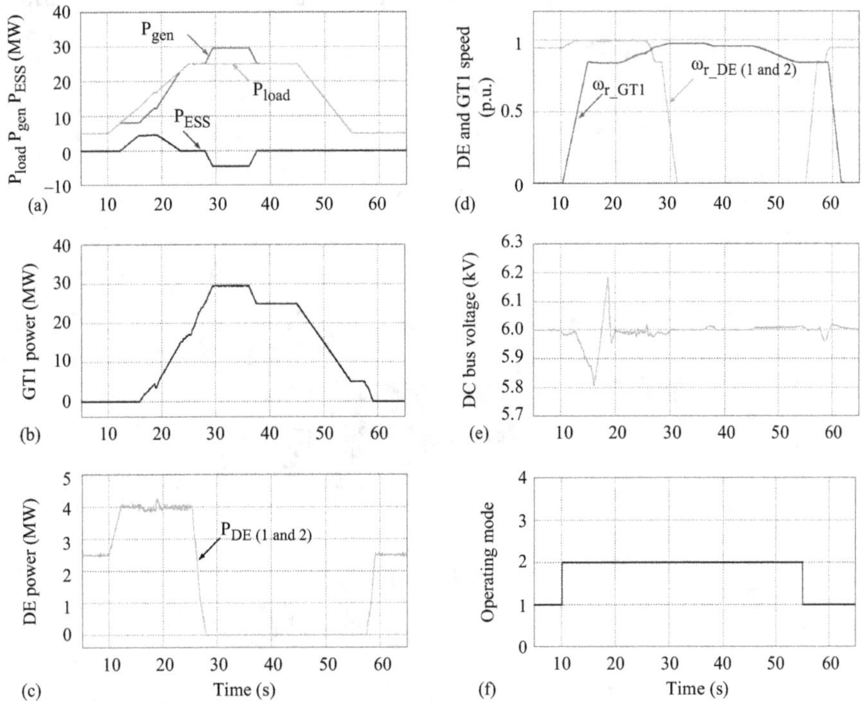

Figure 12.12 System performance in operating modes I and II: (a) generator, ESS, and load powers, (b) output power of GT1, (c) output power of DE1 and DE2, (d) speed of DE1, DE2, and GT1, (e) DC bus voltage, (f) operating mode

the speed. During this interval, the load power exceeds the maximum power of two DEs (DE1 and DE2). The central controller requests the ESS to support the demanded power at 13 s. When the GT reaches to the optimal speed at 15 s, the output voltage is regulated by the exciter system and the GT begins to support the power. The three generators are still operated together until the load power stops increasing at 25 s. After that, the central controller sends a stop command to the DE1 and DE2 at 27 s. At 28 s, the charging operation of the ESS is activated.

At 45 s, the propulsion load begins to decrease (not shown). As it reaches 5 MW at 55 s, the operating mode is changed back to mode I. The GT1 continues supplying power to the load. If the propulsion stops completely, then, only zonal loads consume the power. The DE1 and DE2 are requested to supply the power. When the DE1 and DE2 can afford to supply the load demand, the central controller gives a stop command to the GT1 at 57 s.

Figure 12.13 shows the system performance in operating modes II and III. Initially, the system is operating in mode II, where one GT1 supplies the power of 25 MW to the loads. At 110 s, the propulsion loads begin to increase to accelerate the ship. At 118 s, the load power demand becomes higher than 36 MW, from

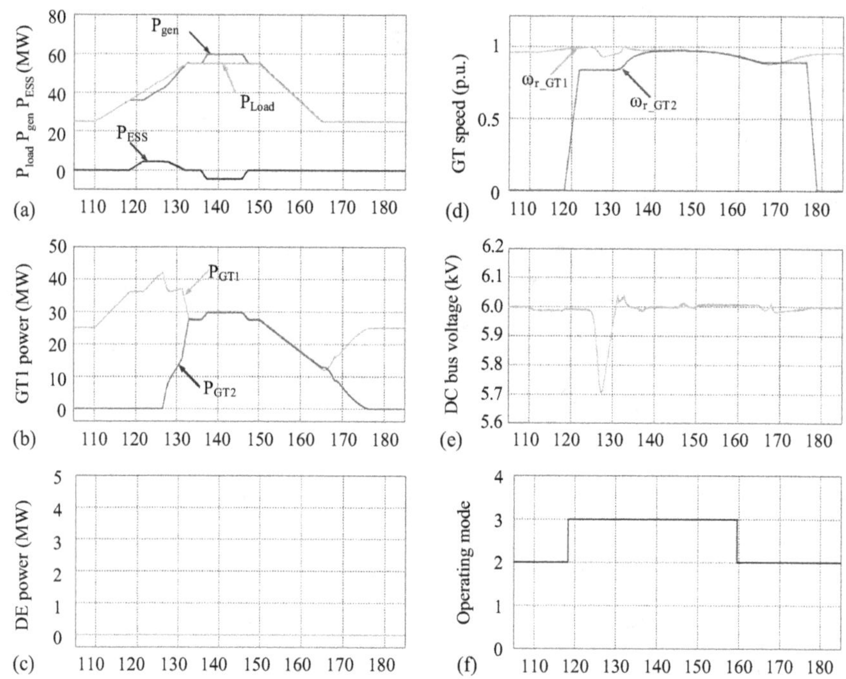

Figure 12.13 System performance in operating modes II and III: (a) generator, ESS, and load powers, (b) output power of GT1 and GT2, (c) output power of DE1 and DE2, (d) speed of GT1 and GT2, (e) DC bus voltage, (f) operating mode

which the operating mode is changed to mode III. The central controller requests the ESS to support the demanded power at 118 s and at the same time sends a start command to the GT2. After some delay, the output power of GT2 begins to increase and the support of the ESS decreases. At 133 s, two GTs begin to operate at a balanced power sharing mode without the support of the ESS. While the system is operating in steady states, the ESS energy is recovered by a charging operation.

At 150 s, the propulsion load begins to decrease (not shown). As it reaches 36 MW at 159 s, the operating mode is changed back to mode II. Two GTs are still operating in parallel until the load demand stops decreasing at 165 s. Then, the central controller sends a stop command to the GT2. At 175 s, the GT2 is deactivated completely and the GT1 alone supplies the power.

Figure 12.14 shows the system performance in operating modes III and IV. Initially, the system is operating in mode III, where two GTs supply the power of 55 MW to the loads. At 230 s, the propulsion loads begins to increase for accelerating the ship. At 242 s, the load power demand becomes higher than 72 MW,

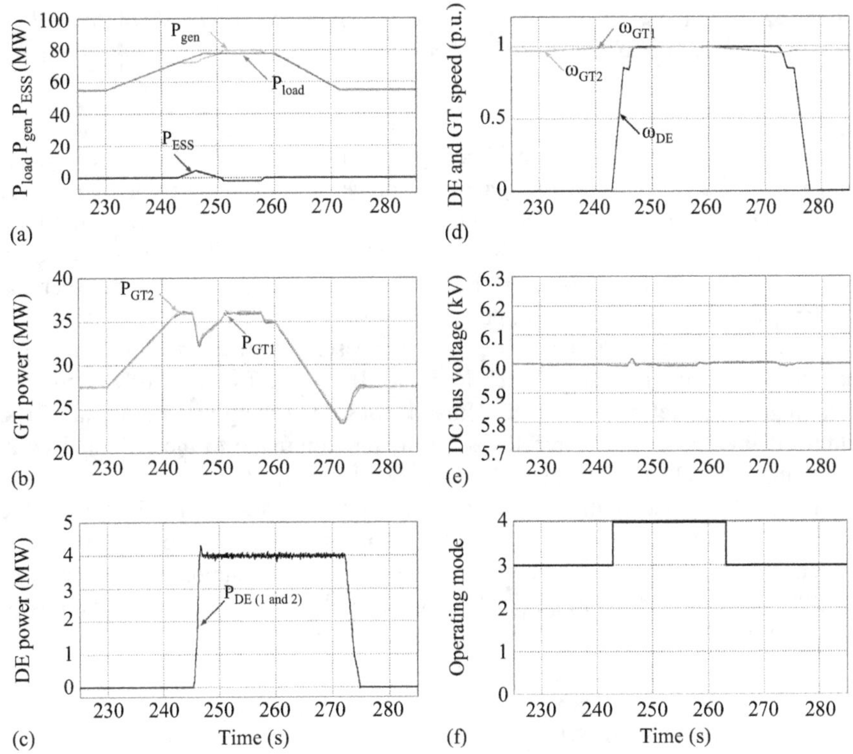

Figure 12.14 System performance in operating modes III and IV: (a) generator, ESS, and load powers, (b) output power of GT1 and GT2, (c) output power of DE1 and DE2, (d) speed of DE1 and 2 and GT1 and 2, (e) DC bus voltage, (f) operating mode

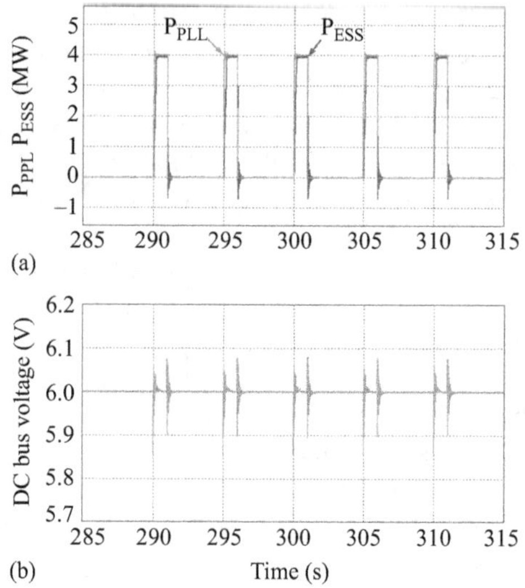

Figure 12.15 System responses for pulsed power load: (a) pulsed power load and ESS power, (b) DC bus voltage

from which the operating mode is changed to mode IV. The central controller requests the ESS to support the demanded power at 243 s and at the same time sends a start command to the DE1 and DE2. At 247 s, both of the DE1 and DE2 generate their rated powers, from which the output power of the ESS is decreased. During this interval, the output power of GTs varies to make a balance between the total supplied power and the load power demand.

At 260 s, the propulsion load begins to decrease. As it reaches 72 MW at 263 s, the operating mode is changed back to mode III. Two GTs and two DEs are still operating in parallel until the load demand stops decreasing at 55 MW, at 271 s. After a delay, the central controller sends a stop command to the DE1 and DE2. Then, only the GT1 and GT2 supply the power to the loads.

Figure 12.15 shows the responses in the case of applying the PPLs while the system is operating in mode III. The PPL consumes a high power of 4 MW for 1 s. The central controller sends a command to the ESS, which supplies the demand of the PPL quickly. It is seen that the MVDC bus voltage has a little fluctuations of about 150 V, which is within the allowable range of $\pm 10\%$.

12.4 Stability analysis

In shipboard MVDC power systems, the stability analysis is needed to ensure reliable operation and power quality [5]. The shipboard MVDC power systems have multiple power sources and a lot of linear and nonlinear loads which are complicatedly connected. Even though the operation of each unit is designed to be stable, the system control performance may be unstable when they are integrated, due to various operating modes, interferences between subsystems, and different load power characteristics [27].

For system stability analysis, the simulation based on a precise model can be performed in a time domain. However, it takes a long time to confirm the stable operation over the wide operating range. To reduce the simulation time, the concept of average model may be utilized [28].

On the other hand, there are analytic methods, which are based on large-signal analysis and small-signal analysis. The large-signal stability analysis is based on Lyapunov theory which gives the region of the asymptotic stability, where the effects of large perturbations and nonlinearity are involved. Meanwhile, the stability analysis based on small-signal analysis is used to investigate the system stability for small variations around the operating points.

For stability analysis of the shipboard MVDC power system, the Lyapunov method is difficult to apply since the MVDC IPS is a complicated system. On the other hand, the time-domain simulation can be employed, but it usually requires a lot of repetitive tasks to find stable operating conditions. So, the small-signal-based stability analysis is preferred, of which analysis procedure is described as the following [27–31]:

1. Modeling of shipboard MVDC power system

   At first, it is necessary to derive the mathematical model of the power system network, which is usually modeled in π-equivalent circuit. In the model, the power line impedance can be neglected since the cable length is sufficiently short in ships.
2. Simplification of system

   For the analysis, each subsystem is classified into two categories as source subsystem and load subsystem, from which a two-port system is obtained.
3. Determination of stability criteria

   In general, the stability analysis of a two-port system is performed by applying the Middlebrook criteria based on Nyquist plot. However, this method is difficult to apply where the power flow direction changes after system reconfiguration. However, the passivity-based stability analysis can be applied to solve the problem. That is, the stability can be discerned from the passivity of the bus impedance of one-port system which is transformed from the two-port system.
4. Identification of bus impedance

   The DC bus impedance should be known before applying the stability criteria. The input and output impedances of the power converters which are connected to the DC bus can be directly measured by digital network analyzers. Otherwise, they can be estimated from the transfer function of each subsystem. For this, the closed-loop transfer functions including the controllers of each subsystem are derived. From these transfer functions, the DC bus impedance can be determined. An MVDC bus where m power sources and n 7 loads connected in parallel can be simplified by one-port system, which can be expressed as

$$Z_{bus}(s) = \frac{V_{bus}(s)}{I_{inj}(s)} = Z_1//Z_2//\ldots Z_n//\ldots Z_{n+m} \tag{12.4}$$

where Z_{bus} is the overall bus impedance, Z_1 to Z_{n+m} are the impedances of the subsystems.

5. Passivity-based stability analysis [30]

   If the one-port system with Z_{bus} can only absorb the energy, it is said that the system is passive. This condition can be expressed as

$$\int_{-\infty}^{T} v_{bus}(t)i_{inj}(t)dt \geq 0, \quad \text{for all } T \tag{12.5}$$

   For the linear time-invariant (LTI) system, the system is stable if and only if the following conditions are satisfied:
   (i) $Z_{bus}(s)$ includes no right-half plane poles.
   (ii) $\mathrm{Re}\{Z_{bus}(j\omega)\} \geq 0, \forall\omega$

12.5 Faults and protection

According to the IEEE Std. 1709–2010, the ring bus structure with zonal distribution is recommended for the shipboard MVDC power system. This bus structure provides a good survivability and reliability of ships since the faulty zone can be isolated easily. In addition, it has an advantage that the load can be supplied from the two longitudinal MVDC buses, resulting in redundancy. Therefore, the system can be autonomously reconfigured to supply the power to the load at healthy zones.

The most frequent faults in the shipboard MVDC power system are line-to-line faults and line-to-ground faults. For the shipboard DC distribution system, the ungrounded system is preferred since it has high survivability and reliability when the line-to-ground fault occurs. Moreover, it is simple and economic [32].

To ensure the continuous operation of the critical loads at fault conditions, the appropriate fault protection scheme is needed to isolate the fault location quickly. Since there are no zero-crossing points in the DC system, it is more challenging to develop the protection scheme including fault detection and fault location [33,34].

12.5.1 Architecture of protection schemes

There are two structures of protection schemes for the shipboard DC power system. One is the circuit-breaker-based scheme and the other is the converter-based method. The former scheme mostly employs the SSCB, which has been described in Section 12.2.3 [32]. The latter scheme utilizes a current limit algorithm of the power converters which can be regarded as a protective device to isolate faults. This can reduce the weight and volume of the system significantly without any

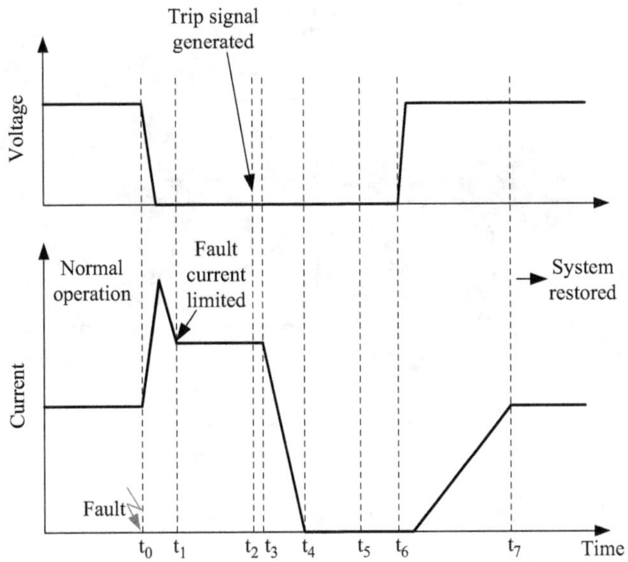

Figure 12.16 Time sequence for converter-based protection scheme [35]

additional devices. However, the reliability and survivability of the converter-based scheme is not so high.

A time sequence for the converter-based protection scheme in the shipboard MVDC power system is shown in Figure 12.16 [35]. When a fault occurs, the protection schemes of all the power converters connected to the faulty bus are triggered. At t_0, a fault occurs and the current increases rapidly. At t_1, a current limiting mode is operated by limiting the currents of the converters which feed the faulty bus. During this interval, the central controller detects the fault and identifies the location.

After the fault is identified, a tripping signal is generated at t_2 to block the operation of power converters. After some delay time, the faulty bus is de-energized by reducing the converter current at t_3. When the currents of the power converters are decreased to zero, at t_4, the magnetic contactor opens to isolate the faulty branch. At t_5, healthy sections are reconfigured by closing relevant magnetic contactors. At t_6, the remaining systems are reenergized and resume the normal operation at t_7.

12.5.2 Fault detection and location

In this section, the fault detection and location are introduced, where the single line-to-ground fault is investigated for a simplified DC power system of the ungrounded port bus, as shown in Figure 12.17. An AC/DC converter is used for the power conversion which feeds the 6 kV MVDC bus. The two DC/DC converters and a DC/AC converter are connected to the MVDC bus, nominal voltages

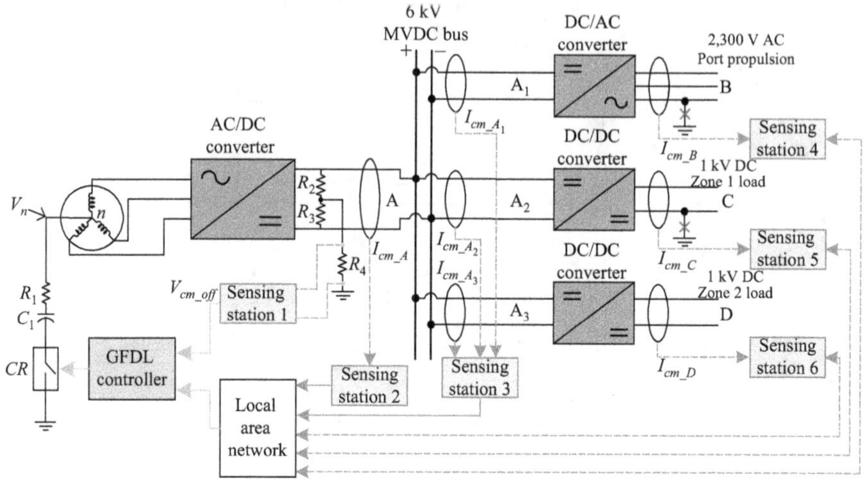

($R_1 = 25\ \Omega$, $C_1 = 0.75\ \mu F$, $R_2 = R_3 = 1\ M\Omega$ and $R_4 = 10\ M\Omega$)

Figure 12.17 Configuration of ground-fault detection and location scheme

of which are 1 kV DC for zone 1 and 2 loads and 2,300 V AC for port propulsion drives, respectively.

Depending on the location of the ground fault, the voltages at all interfaces with respect to the ground shift by different levels [36]. For instance, if a ground fault occurs at positive or negative DC buses, the voltages at all the interfaces shift negatively or positively by a half of DC bus voltage, respectively. However, if the ground fault occurs at AC bus, the voltages at all the interfaces shift positively and negatively by a peak value of phase voltages in AC bus.

Figure 12.18 shows the line-to-ground voltages at each interface when the ground fault occurs in the system of Figure 12.18. If the ground fault occurs at the negative bus of 1 kV DC in zone 1, all the output voltages of power converters are shifted by 500 V, as shown in Figure 12.18(a). For example, the positive MVDC bus voltage shifts from 3 to 3.5 kV DC and the negative bus voltage shifts from −3 to −2.5 kV DC. In the case of ground fault at one phase of 2,300 V AC bus, the voltages at all the interfaces are shifted by a peak phase voltage, ±1,878 V, shown in Figure 12.18(b).

To protect the equipment, a strategy for detecting and locating the ground fault is needed, which is discussed as follows:

- Ground-fault detection

A ground fault condition is checked by measuring a common-mode offset voltage, V_{cm_off}, at the output terminal of the AC/DC converter. V_{cm_off} is the voltage across a resistor, R_4, which is connected from the mid-point of two identical resistors, R_2 and R_3, to the ground, as shown in Figure 12.17.

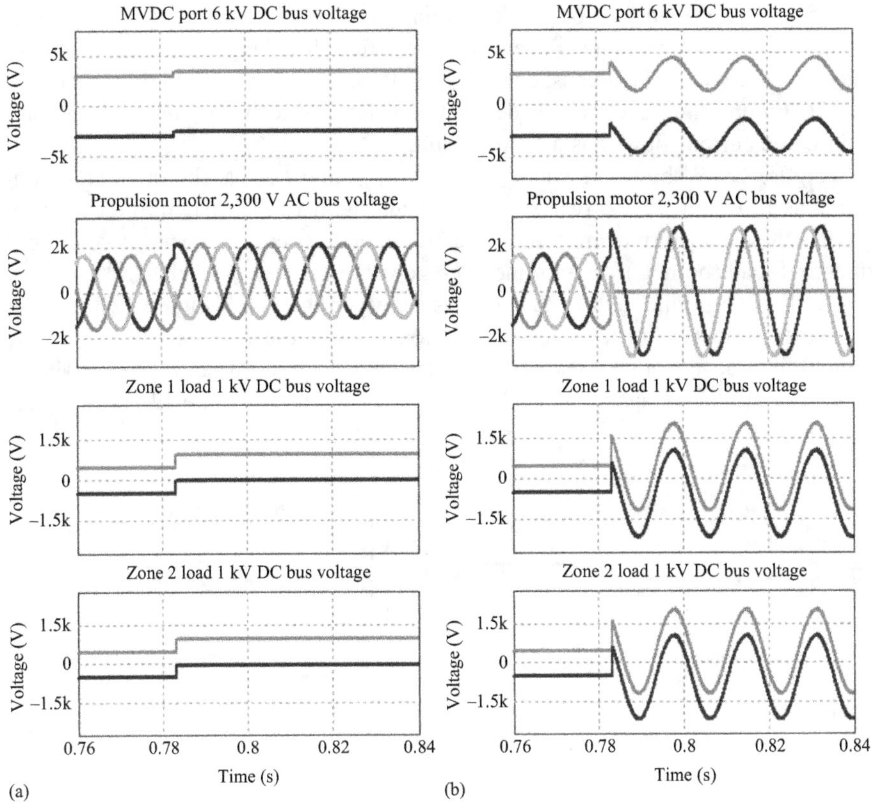

Figure 12.18 Line-to-ground voltages when a ground fault occurs at (a) negative bus of the 1 kV DC zone 1 load, (b) one phase of 2,300 V AC bus

In the ground-fault detection and location (GFDL) controller, the rms and average values of V_{cm_off} are calculated. If both of them are higher than the thresholds for an interval long enough not to be considered as a transient state, it is concluded that a DC ground fault has occurred. On the other hand, if only its rms value is higher than the threshold, it is known that an AC ground fault occurs.

- Ground-fault location

To locate the ground fault, the series R_1C_1 circuit which is connected to the neutral point of the three-phase power source is utilized. If the line-to-ground fault is detected, the controller gives a command to close the contactor relay (CR). Then, a third-order harmonic current is conducted from the chassis ground to the fault point through the CR. At all the interfaces of power converters, the common-mode currents, I_{cm_x}, which are the sum of currents through positive and negative buses are measured. Then, each sensing station calculates the rms value of its I_{cm_x} and sends to the GFDL controller through communication links in a delay time.

If the DC ground fault occurs, the common-mode currents of AC/DC and DC/DC converters are examined. Then, it is identified that the ground fault occurs in the converter where the highest common-mode current flows. On the other hand, if the AC ground fault occurs, it is decided that the inverter where the highest common-mode current flows has a ground fault.

Figure 12.19 shows the waveforms of V_{cm_off} and I_{cm_x} at the same fault condition as in Figure 12.19. If there is no ground fault in the system, the common-mode voltage and current are zero. If the ground fault occurs at the negative DC bus of zone 1 load (point C), the rms and average values of V_{cm_off} reach to 500 V, from which the DC ground fault is detected, as shown in the top of Figure 12.19(a).

After a delay, the fault location is found from the I_{cm_A} and I_{cm_C}. If the ground fault occurs at one phase of inverter (point B), the rms and value of V_{cm_off} is 500 V, but its average value is almost zero, as shown in Figure 12.19(b). In this

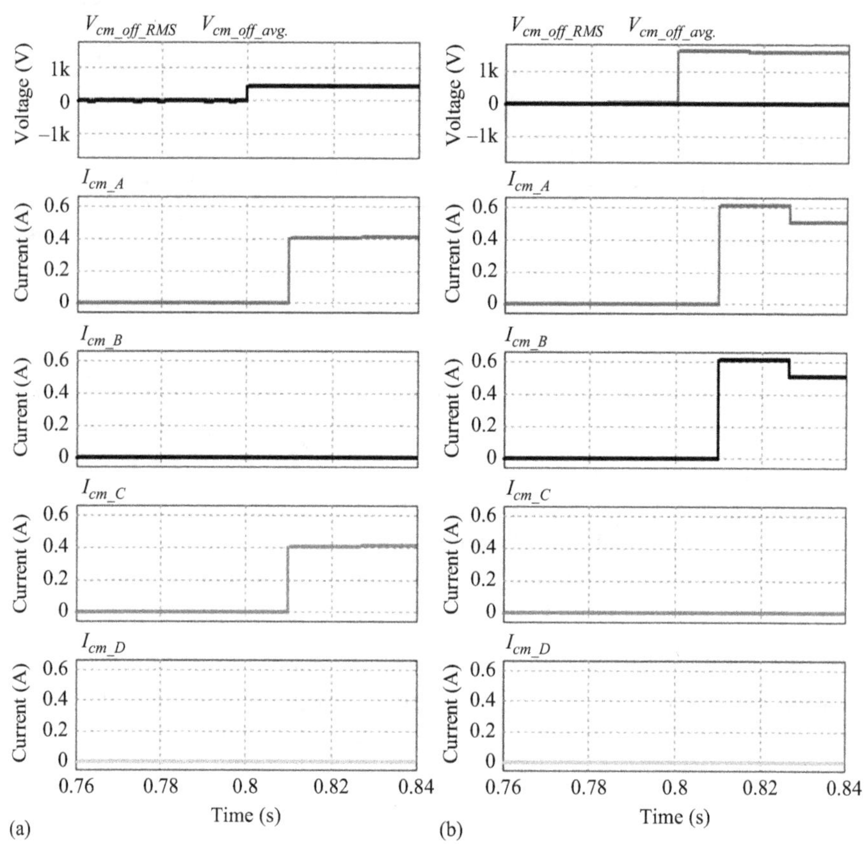

Figure 12.19 Common-mode offset voltages (in rms) and currents (in rms) when a ground fault occurs at (a) negative DC bus of zone 1 (b) one phase of 2,300 VAC bus at B

case, it is diagnosed that the AC ground fault has occurred. It is seen that the common-mode currents of I_{cm_A} and I_{cm_B} have a significant value.

12.6 Summary

As the environmental regulations have become stricter and stricter recently, the demand on higher fuel efficiency and lower emission for the ship is more pressing. Therefore, the AES, which has high survivability and reliability, is predicted to be a future trend of the ship. Since the MVDC IPS well fits with the AES, more research efforts should be devoted to the development of this technology up to the maturity stage.

References

[1] E. Skjong, R. Volden, E. Rødskar, M. Molinas, T. A. Johansen, and J. Cunningham, "Past present and future challenges of the marine vessel's electrical power system," *IEEE Trans. Transport. Electrific.*, vol. 2, no. 4, pp. 522–537, 2016.

[2] M. Patel, *Shipboard propulsion and power electronics and ocean energy*, USA, FL, Boca Raton, CRC Press, 2012.

[3] "Queen Elizabeth 2," 2002. Available from: https://en.wikipedia.org/w/index.php?title=Queen_Elizabeth_2&offset=20041114142442&limit=500&action=history.

[4] O. Alnes, S. Eriksen, and B.-J. Vartdal, "Battery-powered ships: A class society perspective," *IEEE Electrific. Mag.*, vol. 5, no. 3, pp. 10–21, 2017.

[5] IEEE Std. 1709-2010, "IEEE recommended practice for 1 kV to 35 kV medium voltage DC power systems on ships," *in IEEE Std. 1709-2010*, Nov. 2010, pp. 1–54.

[6] J.-F. Hansen, J. Lindtjoern, T. Myklebust, and K. Vanska, "Onboard DC grid: The newest design for marine power and propulsion systems," *ABB Rev.*, no. 2, 2012. Available from: https://library.e.abb.com/public/b4f3f099e9d21360c1257a8a003beac2/ABB%20Generations_20%20Onboard%20DC%20grid.pdf.

[7] D. Li, R. A. Dougal, E. Thirunavukarasu, and A. Ouroua, "Variable speed operation of turbogenerators to improve part-load efficiency," *in Proc. IEEE Electr. Ship Technol. Symp.*, 2013, pp. 353–359.

[8] Energy and Environmental Analysis, Inc. an ICF Company, *Technology characterization: Reciprocating engines*, Environmental Protection Agency, Combined Heat and Power Partnership Program, 2008.

[9] ABB, *Generators for wind power, proven generators – reliable power*, ABB Motors and Generators, 2010.

[10] Y. Tang and A. Khaligh, "Bidirectional hybrid battery/ultracapacitor energy storage systems for next generation MVDC shipboard power systems," *in Proc. IEEE VPPC*, 2011, pp. 1–6.

[11] S. Kim, S. Choe, S. Ko, and S. Sul, "A naval integrated power system with a battery energy storage system: Fuel efficiency reliability and quality of power," *IEEE Electrific. Mag.*, vol. 3, no. 2, pp. 22–33, 2015.

[12] J. F. Hansen and R. Lysebo, *Comparison of electric power and propulsion plants for LNG carriers with different propulsion systems*, ABB AS, 2007.

[13] M. Stieneker and R. W. D. Doncker, "Dual-active bridge dc-dc converter systems for medium-voltage DC distribution grids," *in Proc. IEEE Power Electronics Conference and 1st Southern Power Electronics Conference (COBEP/SPEC)*, 2015, pp. 1–6.

[14] R. Mo and H. Li, "Hybrid energy storage system with active filter function for shipboard MVDC system applications based on isolated modular multi-level DC/DC converter," *IEEE J. Emerg. Sel. Top. Power Electron.*, vol. 5, no. 1, pp. 79–87, 2017.

[15] P. Manuelle, B. Singam, and S. Siala, "Induction motors fed by PWM MV7000 converters enhance electric propulsion performance," *in Proc. IEEE Power Electronics and Applications, 2009. EPE'09. 13th European Conference on*, 2009, pp. 1–9.

[16] M. Spichartz, V. Staudt, and A. Steimel, "Modular multilevel converter for propulsion system of electric ships," *in Proc. IEEE Electric Ship Technol. Symp*, 2013, pp. 237–242.

[17] X. Pei, O. Cwikowski, M. Barnes, A. C. Smith, and R. Shuttleworth, "A review of technologies for MVDC circuit breakers," *in Proc. Industrial Electronics Society, IECON 2016—42nd Annual Conference of the IEEE*, 2016, pp. 3799–3805.

[18] S.-H. Lee, J.-S. Yim, J.-H. Lee, and S.-K. Sul, "Design of speed control loop of a variable speed diesel engine generator by electric governor," *in Proc. IEEE IAS Annu. Meet.*, 2008, pp. 1–5.

[19] S. Di Cairano, D. Yanakiev, A. Bemporad, I. V. Kolmanovsky, and D. Hrovat, "Model predictive idle speed control: Design analysis and experimental evaluation," *IEEE Trans. Control Syst. Technol.*, vol. 20, no. 1, pp. 84–97, 2012.

[20] S. K. Yee, J. V. Milanovic, and F. M. Hughes, "Overview and comparative analysis of gas turbine models for system stability studies," *IEEE Trans. Power Syst.*, vol. 23, no. 1, pp. 108–118, 2008.

[21] M. Farasat, A. S. Arabali, and A. M. Trzynadlowski, "A novel control principle for all-electric ship power systems," *in Proc. IEEE Electric Ship Technologies Symposium (ESTS)*, 2013, pp. 178–184.

[22] S. D. Umans and D. J. Driscoll, "Excitation system for rotating synchronous machines," *U.S. Patent 6 362 588*, Mar. 26, 2002.

[23] O. Palizban, K. Kauhaniemi, and J. M. Guerrero, "Microgrids in active network management—Part I: Hierarchical control energy storage virtual power plants and market participation," *Renewable Sustainable Energy Rev.*, vol. 36, pp. 428–439, 2014.

[24] Z. Jin, G. Sulligoi, R. Cuzner, L. Meng, J. C. Vasquez, and J. M. Guerrero, "Next-generation shipboard DC power system: Introduction smart grid and

DC microgrid technologies into maritime electrical networks," *IEEE Electrific. Mag.*, vol. 4, no. 2, pp. 45–57, 2016. Available from: https://ieeexplore.ieee.org/document/7480937/.

[25] T. Dragičević, J. M. Guerrero, J. C. Vasquez, and D. Škrlec, "Supervisory control of an adaptive-droop regulated DC microgrid with battery management capability," *IEEE Trans. Power Electron.*, vol. 29, no. 2, pp. 695–706, 2014.

[26] J. M. Guerrero, J. C. Vasquez, J. Matas, L. G. de Vicuna, and M. Castilla, "Hierarchical control of droop-controlled AC and DC microgrids: A general approach toward standardization," *IEEE Trans. Ind. Electron.*, vol. 58, no. 1, pp. 158–172, 2011.

[27] A. Riccobono, M. Cupelli, A. Monti, E. Santi, T. Roinila, H. Abdollahi, S. Arrua, and R. A. Dougal, "Stability of shipboard DC power distribution: Online impedance-based systems methods," *IEEE Electrific. Mag.*, vol. 5, no. 3, pp. 55–67, 2017.

[28] J. Shi, R. Amgai, and S. Abdelwahed, "Modelling of shipboard medium-voltage direct current system for system level dynamic analysis," *IET Electr. Syst. Transp.*, vol. 5, no. 4, pp. 156–165, 2015.

[29] W. Tang and R. H. Lasseter, "An LVDC industrial power supply system without central control unit," *in Proc. IEEE Power Electron. Spec. Conf.*, 2000, vol. 2, pp. 979–984.

[30] A. Riccobono and E. Santi, "Stability analysis of an all-electric ship MVDC power distribution system using a novel passivity-based stability criterion," *in Proc. IEEE ESTS*, 2013, pp. 411–419.

[31] U. Javaid, F. D. Freijedo, D. Dujic, and W. van der Merwe, "Dynamic assessment of source-load interactions in marine MVDC distribution," *IEEE Trans. Ind. Electron.*, vol. 64, no. 6, pp. 4372–4381, 2017.

[32] R. M. Cuzner and V. Singh, "Future shipboard MVDC system protection requirements and solid-state protective device topological tradeoffs," *IEEE J. Emerg. Sel. Top. Power Electron.*, vol. 5, no. 1, pp. 244–259, 2017.

[33] E. Christopher, M. Sumner, D. Thomas, and F. De Wildt, "Fault location for a DC zonal electrical distribution systems using active impedance estimation," *in Proc. IEEE Electric Ship Technologies Symposium* (*ESTS*), 2011, pp. 310–314.

[34] P. Cairoli and R. A. Dougal, "New horizons in DC shipboard power systems," *IEEE Electrific. Mag.*, vol. 1, no. 2, pp. 38–45, 2013.

[35] Q. Deng, X. Liu, R. Soman, M. Steurer, and R. A. Dougal, "Primary and backup protection for fault current limited MVDC shipboard power systems," *in Proc. IEEE Electric Ship Technologies Symposium* (*ESTS*), 2015, pp. 40–47.

[36] R. M. Cuzner, T. Sielicki, A. E. Archibald, and D. A. McFarlin, "Management of ground faults in an ungrounded multi-terminal zonal DC distribution system with auctioneered loads," *in Proc. IEEE Electric Ship Technologies Symposium* (*ESTS*), 2011, pp. 300–305.

Chapter 13
DC-based EVs and hybrid EVs

Ruoyu Hou[1], Jing Guo[1], and Ali Emadi[1]

13.1 Introduction

The recent combination of environmental, geopolitical, economic, and health concerns related to internal combustion engine powered transportation has created fertile ground for renewed interest and investment in electric drive vehicles and electrified transportation [1]. The paradigm shift has already begun and is gaining momentum. Figure 13.1 shows the U.S. plug-in electric vehicle sales by model from 2011 to 2015 [2]. Note that the listed vehicles include plug-in hybrid electric vehicles (PHEV) and range extended electric vehicles (REEV). Only full-sized vehicles sold in the United States and capable of 60 mph are listed. It is clear that, in recent years, more car manufacturers have started to enter the competitive market of PHEV and REEV. In addition, sales for most vehicle models keep increasing.

13.1.1 Power electronic system in electrified vehicles

The power electronics system in the electrified vehicle is DC-based. Its possible architecture is shown in Figure 13.2. It has a high-voltage (HV) battery and a low-voltage (LV) battery. The HV battery is mainly for the traction system where the voltage is usually much higher than the vehicular electrical system's voltage. Based on its hybridization factor and the type of its powertrain, the HV battery voltage range varies accordingly. Typically, the LV mild hybrid obtains an HV battery voltage lower than 60 V. This is mainly because the precautions required for HV systems do not need to be applied here, as HV standards are applied over 60 V [3]. For example, in belt-driven starter generator applications, traction power is usually supplied by lead-acid batteries and the voltage is set around 48 V. Currently, for HV mild hybrids, the HV battery voltage could be around 110 V. For full hybrid, PHEV, or REEV, the HV battery voltage is rated from 200 to 650 V, or even higher.

The HV battery charging method can be classified as inductive charging and conductive charging. The inductive charging is also known as wireless charging

[1]Electrical and Computer Engineering Department, McMaster Automotive Resource Centre (MARC), McMaster University, Canada

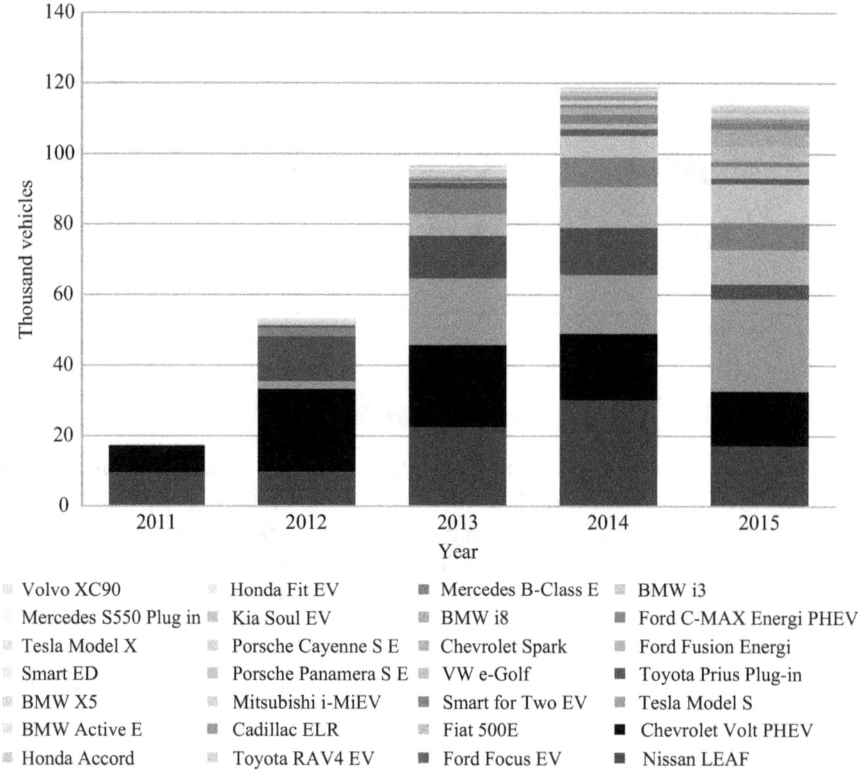

Figure 13.1 2011–2015 U.S. plug-in electric vehicle sales by model [2]

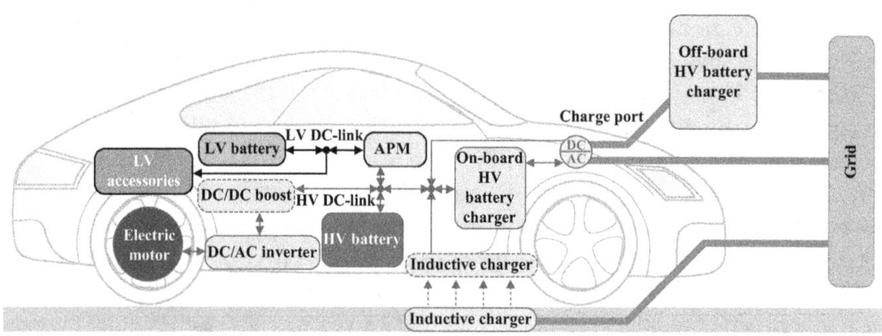

Figure 13.2 Power electronic system architecture of an electrified vehicle powertrain

and transfers power magnetically. The AC/DC power factor correction (PFC) converter, high-frequency DC/AC primary side, and the primary transducer are all located away from the vehicle. The secondary transducer and AC/DC rectifier are installed inside the vehicle. The conductive charging systems use direct contact between the connector and the charge inlet [4]. These can be categorized as onboard and off-board battery chargers. The onboard charger is a must for electrified vehicles needing external charging. It has power limitations because of cost, weight, and space constraints. The onboard charger connects to the grid through the AC charge port from the vehicle and creates the AC/DC power conversion inside the vehicle. The off-board charger has higher power rating and thus reduces the charging time. It has the AC/DC power conversion away from the vehicle and charges the HV battery directly through the DC charge port of the vehicle.

Typically, the propulsion system is an electric motor with a three-phase DC/AC inverter as the power processing unit between the HV battery and the motor. Some electrified powertrains apply an additional DC/DC boost converter in between the HV battery and the inverter to improve the system's efficiency by increasing the DC-link voltage and providing an adjustable DC-link voltage for the traction inverter.

The LV vehicular system voltage is typically 12 V; therefore, the HV battery cannot be utilized to supply power directly to the vehicular loads. For this reason, power converters are required, which convert HV from the traction battery to a lower voltage in order to supply power to the LV vehicular loads, and to charge the LV battery, which is usually a lead-acid battery. This power converter is called an auxiliary power module (APM).

In this chapter, several converter topologies of PFC converter, isolated DC/DC converter, APM, and active power filter (APF) in electrified vehicle applications are presented. The practical design considerations, including power switch, DC-link capacitor, and DC bus bar, are given. Finally, the state-of-the-art topological system reconfigurations in electrified vehicles are presented.

13.1.2 DC auxiliary loads in electrified vehicles

The loads that are not directly related to the powertrain of the vehicle are called auxiliary loads. The auxiliary loads in vehicle can be mainly categorized as air conditioning, heating, ventilation, lighting, wiper and window system, electronic accessory, and others. Figure 13.3 shows the auxiliary load distribution in the electrified vehicles. Among them, air conditioning, heating, and ventilation are the largest auxiliary loads, which are also highly dependent on the ambient temperature. At extreme temperature, these loads can have up to 32% impact on the range of the vehicle. Therefore, these loads are, most likely, directly connected to the HV battery due to the potential high-power demand. The rest loads are all connected to the 12-V DC-link and are all powered by either the 12-V battery or the APM converter.

Most of the auxiliary loads in the LV system are resistive DC loads. The resistance that is seen from the supply side changes with the current drawn by the load. The total power of a typical LV auxiliary load is around 2 kW. A typical LV auxiliary load power distribution is shown in Figure 13.4. Lighting loads consume around 32% of the total power. They are composed of headlights, fog lamps, park lamps, flashers, turn signals, etc. Among these, back lights, headlamps, and fog

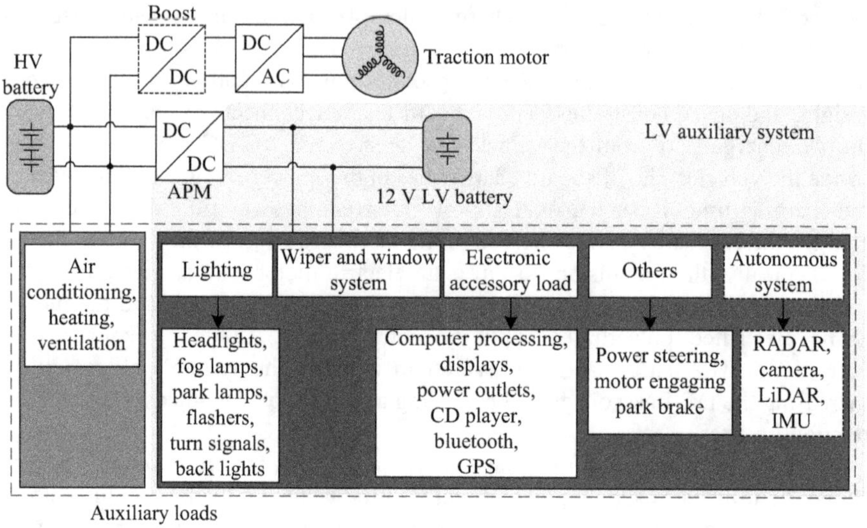

Figure 13.3 Possible DC auxiliary load distribution in the electrified vehicle

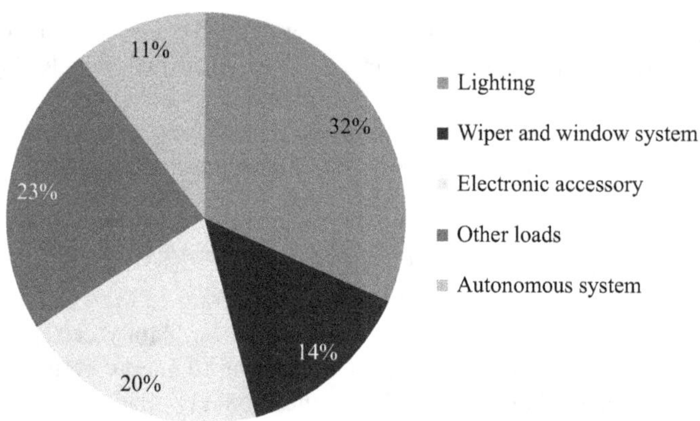

Figure 13.4 Typical LV auxiliary load power distribution in the electrified vehicle

lamps draw most of the power. Wiper and window system related loads draw around 14% of the total power. Electronic accessory loads include the computer, displays, power outlets, CD player, Bluetooth, and Global Positioning System. Electric power steering and motor engaging park brake are some other loads. Other than that, in vehicles where additional luxury loads are requested, such as power sunroof, active suspension system, or entertainment systems, the total power level could be higher.

Besides these conventional auxiliary loads, with the recent development on the autonomous vehicle, the needed power from autonomous system in the vehicle application has also been evaluated. The sensors could be used in the autonomous system are camera, radio detection and ranging (RADAR), light detection and ranging (LiDAR), and inertial measurement unit. LiDAR is a sensor used to detect the presence of objects round the vehicle, so the computer can interpret the vehicle's relation to its surroundings. Besides these sensors, extra computer processing is required for the autonomous system. The total amount of power occupied by this system is around 200 W.

13.2 Converter topologies in electrified vehicles

13.2.1 Conductive HV battery chargers

AC conductive charging topologies can be categorized into two types: single-stage and two-stage chargers. The single-stage charger is compact and combines both AC/DC and DC/DC into one stage. It can achieve smaller size, lower cost, and higher efficiency. The drawbacks are that it might not provide galvanic isolation and usually contains large low-frequency current ripple in the converter. The two-stage charger is an AC/DC PFC converter followed by either a nonisolated or an isolated DC/DC converter. The role of the front-end AC/DC converter is to convert the supplied AC voltage into DC voltage and to also act as an input current shaper for PFC and harmonic reduction. Power factor and total harmonic distortion are the major factors that influence the selection of specific PFC rectifier topology [5].

Boost type topologies and their variants are widely used for PFC rectification purposes. A conventional PFC circuit is shown in Figure 13.5(a). It consists of a full bridge rectifier and a boost preregulator. The boost stage can be continuous conduction mode (CCM), discontinuous conduction mode (DCM), or critical

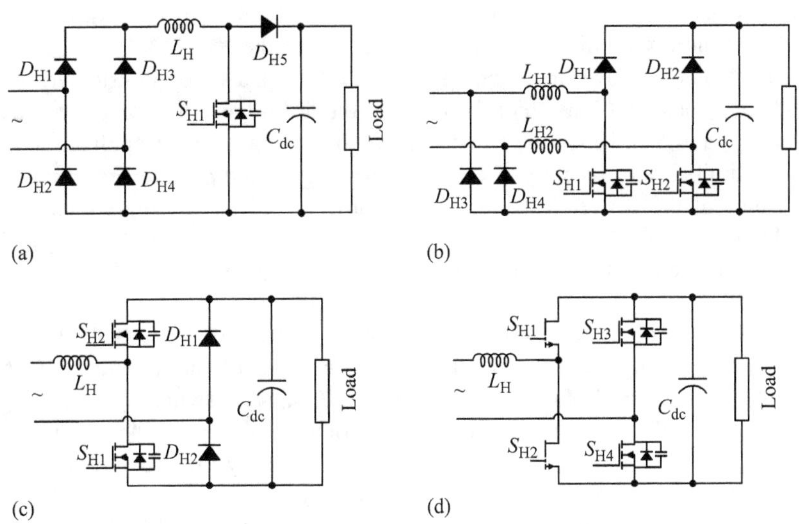

Figure 13.5 Single-phase PFC converter: (a) boost PFC converter. (b) 2-Phase bridgeless PFC converter. (c) Bridgeless totem-pole PFC converter. (d) GaN-based bridgeless totem-pole PFC converter with SR

conduction mode (CRCM) to improve the efficiency. However, large portions of system loss are in the diode bridge and cannot be avoided even with zero-voltage switching (ZVS) at the boost stage. This inherently limits the peak efficiency of a conventional PFC. The typical forward voltage of a rectifier diode is 1 V, and there are two diodes in the current path, which could reduce the total efficiency by 2%. The efficiency of a well-designed PFC can probably achieve about 97% to even 98%, but efficiency higher than 98% becomes relatively challenging for a standard PFC due to the fixed diode bridge loss.

In a bridgeless PFC, the diode losses can be eliminated, so efficiencies of 99% or higher are made possible to meet highest efficiency standards. Various bridgeless PFC topologies have been proposed to overcome the high diode bridge losses [6]. Among all the bridgeless PFC topologies, the two-phase bridgeless PFC and the bridgeless totem-pole PFC (BTPPFC) are the most popular topologies.

The topology of the two-phase bridgeless PFC is shown in Figure 13.5(b). This topology is essentially two boost legs with each one taking control during half of the AC cycle. S_{H1}/S_{H2} are typically MOSFETs and D_{H1}/D_{H2} can be diodes, or for higher efficiency, silicon carbide (SiC) diodes. In past years, this has been the most popular bridgeless PFC topology on the market because it is easy to implement using conventional Si MOSFETs with the control algorithm's being similar to a standard PFC circuit, and the efficiency is improved as it eliminates one diode from the current path. However, it has several drawbacks. (1) It has low-power density and component utilization, as it doubles the part counts, and each boost stage only works during half a cycle. (2) It needs additional return diodes D_{H3}/D_{H4} to provide a return current path and a reference DC-link ground to AC line to reduce the common mode noise. (3) The diode D_{H1}/D_{H2} need to be fast SiC diodes whose cost is relatively higher than AC rectifier diodes. (4) This topology is not bidirectional, which limits the HV battery charger to be unidirectional.

Figure 13.5(c) shows the topology of BTPPFC. It can be considered as a conventional boost PFC in which half of the diode bridge is replaced by active switches S_{H1} and S_{H2}, hence the name "totem pole." However, its application has been very limited until recently. The major challenge is the poor reverse recovery performance of the conventional silicon MOSFETs in the half-bridge configuration, which makes CCM operation impractical due to excess reverse recovery Q_{rr} loss during turn-on transient. To avoid body diode conduction, BTPPFC with silicon MOSFETs must work in DCM or CRCM modes, which is only suitable for lower power and with more complicated control. The absence of a body diode (zero reverse recovery charge Q_{rr}) and the fast switching nature of gallium nitride (GaN) make a GaN-based BTPPFC a good fit for a CCM hard switching half-bridge power stage. In addition, to further improve the converter efficiency, the diode D_{H1}/D_{H2} in the 50/60-Hz line-frequency leg can be replaced by low-Rds(on) synchronous MOSFETs as shown in Figure 13.5(d). The GaN-based CCM BTPPFC overcomes many issues which exist in the two-phase bridgeless PFC and has the following advantages: (1) Efficiency improvement. The main current only flows through two switches at a time. S_{H1}/S_{H2} is driven with complimentary pulse width modulation (PWM) signals and the S_{H3}/S_{H4} on the line-frequency leg is synchronous

MOSFETs which further reduce the conduction loss. (2) Lower part counts, higher power density, and lower bill of material cost. It uses fewer parts and has a simpler circuit. It needs only one inductor and neither SiC diodes nor AC return diodes are required. (3) Bidirectional power flow. BTPPFC is inherently capable of bidirectional operation, which is ideal as the application requires power flow in both directions. Therefore, the BTPPFC is a promising bidirectional PFC candidate for achieving bidirectional power flow from grid to vehicle and from vehicle to grid.

The main function of the second-stage DC/DC converter is to offer regulated voltage and current to better charge the HV battery. Two-stage topology provides high-power factor, wide line regulation performance and clean charge current. However, cost also needs to be taken into account, especially for the onboard charger. The most widely used unidirectional isolated topologies are the phase-shift full bridge and LLC resonant converter as shown in Figure 13.6. As a frequency-modulated resonant converter, the LLC converter has a resonant tank between the primary and secondary sides. The capacitor C_s and the inductors L_s and L_m form the resonant tank. The L_m is the transformer magnetizing inductor and L_s can be the transformer leakage inductor or an additional auxiliary inductor. The phase shift full bridge is a PWM-modulated converter. It utilizes the transformer leakage inductor and the parasitic capacitors from the primary switches to form a resonant circuit and creates a switching pattern that utilizes the lagging current in the inductor to fully discharge the parasitic capacitor before turning on the/a switch.

As shown in Figure 13.6, two contactors T_1 and T_2 need to be installed on the DC terminals of a HV battery. They provide the isolation between the battery and

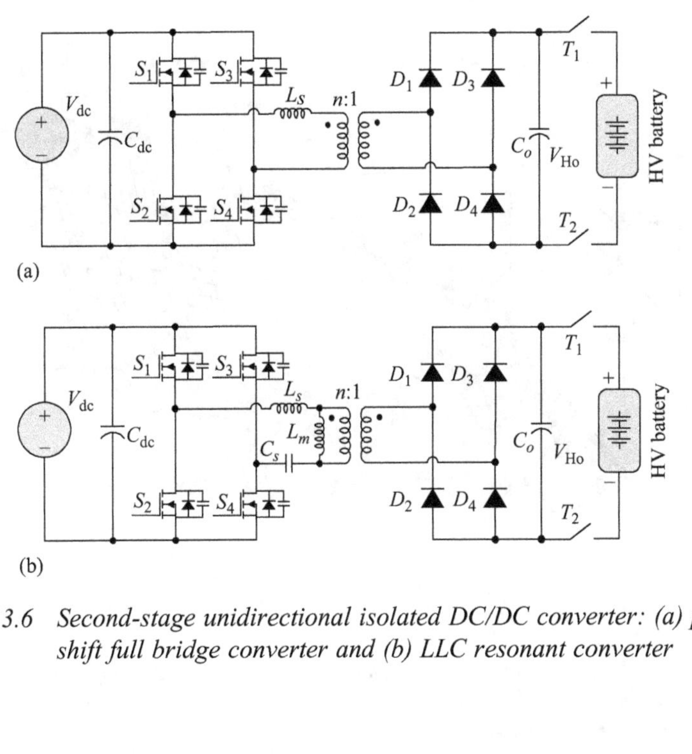

Figure 13.6 Second-stage unidirectional isolated DC/DC converter: (a) phase-shift full bridge converter and (b) LLC resonant converter

remaining vehicle systems. Among them, T_1 is the main contactor and T_2 is the backup contactor from a safety consideration.

13.2.2 Active power filters in HV battery chargers

In a typical 50/60 Hz single-phase AC HV battery charger for vehicle applications, as the input current is forced to vary sinusoidally in phase with the input voltage, the pulsating power at twice the line frequency will be seen on the DC-link after the single-phase AC/DC PFC converter. In the meantime, by applying a wide band-gap power device, such as GaN or SiC, higher switching frequency and higher efficiency can be achieved. Therefore, the size of the most passive components in the power electronic system can be shrunk.

However, large capacitance C_{DC} is still required to balance out the instantaneous current difference between the pulsating input and the constant output current. In this case, the corresponding required capacitance is dependent on the harmonic energy rather than the switching ripple. In other words, the bulky capacitor is mainly used to store the low-frequency second-order harmonic energy, whose frequency is 100/120 Hz. If the average output voltage of the HV DC-link is 400 V and the input AC line frequency is 60 Hz, the relation between the power level, capacitance, and its voltage ripple is shown in Figure 13.7. For example, if the requirement of the output peak-to-peak voltage ripple is limited to 10%, the required output capacitance will be about 4 mF for a 6.6-kW HV battery charger. Hence, the output capacitor bank becomes a major barrier in terms of power density.

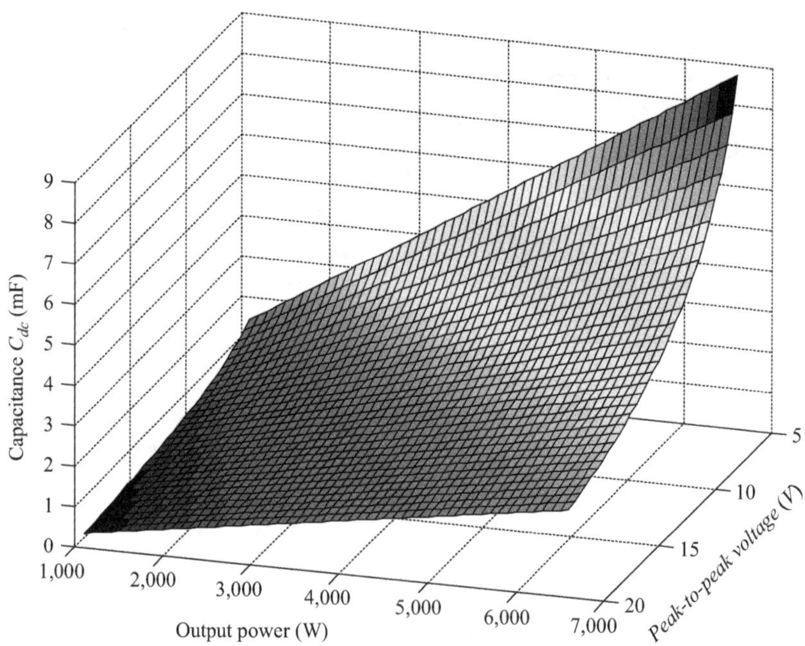

Figure 13.7 Relation among the power level, capacitance, and its voltage ripple

To increase the power density and reliability of the power electronic system in vehicle applications, capacitor-less designs have been proposed. APF can be a potential solution to reduce the required capacitance. The HV battery charger structure with additional APF is shown in Figure 13.8.

Four conventional single-phase APF converter topologies are shown in Figure 13.9. They are a voltage source inverter, a current source inverter, a

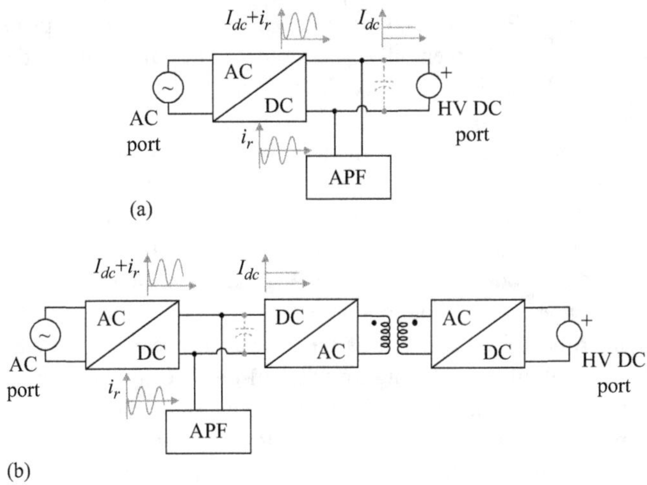

Figure 13.8 HV battery charger structures with additional APF: (a) single-stage charger, (b) two-stage charger with isolated DC/DC converter

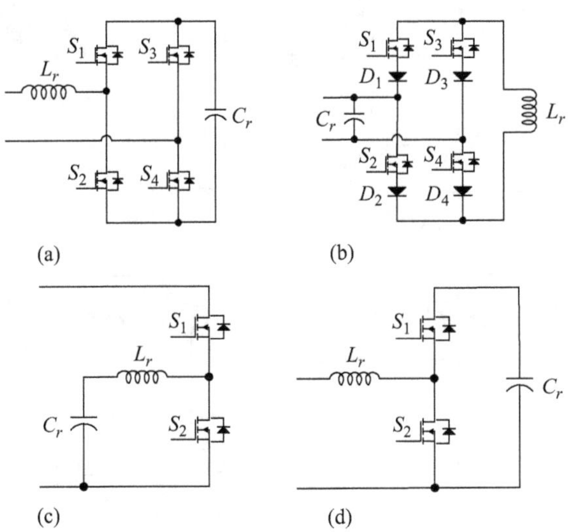

Figure 13.9 Four typical conventional APFs. (a) Voltage source inverter. (b) Current source inverter. (c) Bidirectional buck converter. (d) Bidirectional boost converter

bidirectional buck converter, and a bidirectional boost converter. With these additional APF circuits, the required capacitance can be reduced significantly. However, by adding an extra APF circuit, additional power electronic components are needed, and thus the system's complexity is also increased.

13.2.3 LV auxiliary power modules

In the vehicle, the LV battery is in parallel with all the auxiliary loads. The topology of the LV APM must incorporate galvanic isolation to protect the LV electronic system from the potentially hazardous HV [7]. In addition, there is a HV ratio between the HV and LV battery. These requirements restrict the available topologies to those containing a transformer. Two popular topologies are shown in Figure 13.10. Figure 13.10(a) shows the topology which combines the phase shift full bridge with a synchronous rectification current doubler. Synchronous rectification is critical for this application as the conduction loss becomes a significant percentage of overall loss. The synchronous MOSFET rectifier operates in the third quadrant. The body diode of the MOSFET conducts prior to the turn-on of the switch. In other words, the conduction loss of the body diode is generated just before the synchronous MOSFET turns on. However, it can be turned on in ZVS, which results in negligible switching loss at turn-on. During the turn-off transient, the MOSFET stops conducting prior to the body diode, which means that the synchronous rectifier still has reverse recovery loss from its body diode. If the APM requires bidirectional operation, the dual-active-bridge (DAB) converter is a

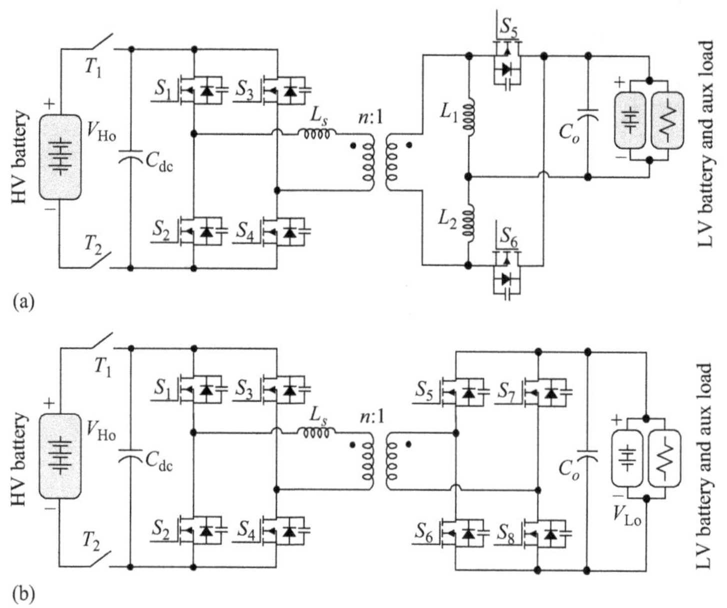

Figure 13.10 LV APM isolated DC/DC converter candidates (a) phase-shift full bridge with current doubler, (b) DAB converter

promising candidate. The DAB converter has active full bridges on both primary and secondary sides. DAB can provide ZVS on both primary and secondary sides by utilizing the leakage inductor. Its symmetry and fixed switching frequency make it a robust converter.

13.3 Practical design considerations

13.3.1 Selections of switching devices

In the converter design, switching devices are selected according to the DC-link voltage and the peak operating current. The voltage and current ratings of a device cannot be determined to be the same as the system requirements. Certain margins need to be allowed since current and voltage spikes exist in the device during turn-on and turn-off transients, as shown in Figure 13.11. The figure shows the current and voltage waveforms of an insulated-gate bipolar transistor (IGBT). The current spike in Figure 13.11(a) is caused by the diode reverse recovery and the parasitic capacitor charge during the switch turn-on period. The voltage spike in

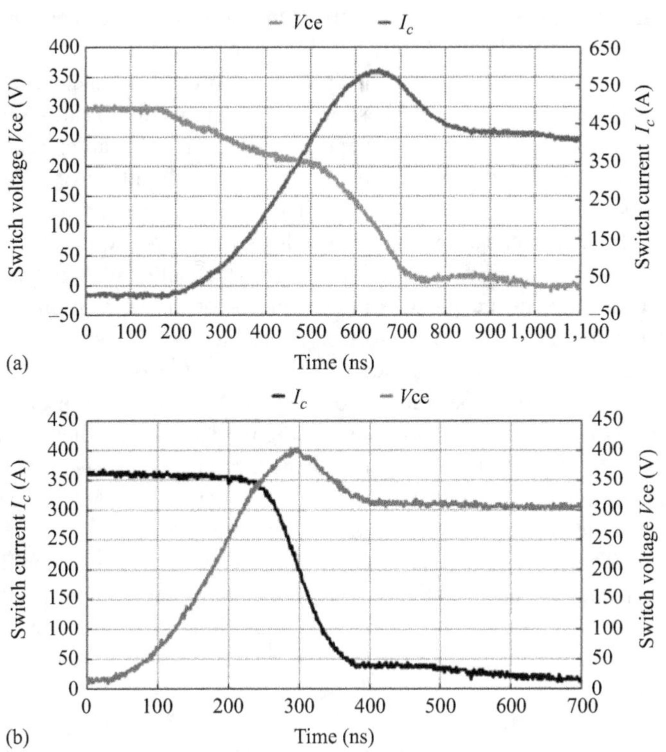

Figure 13.11 Current through and voltage across an IGBT during turn-on and turn-off transients: (a) during turn-on transient and (b) during turn-off transient

Figure 13.11(b) is generated by the voltage across the circuit stray inductance during the switch turn-off period. Generally speaking, the voltage and current ratings of the device are chosen to be about 1.5–2 times the values required by the system.

In addition, other parameters of the switching devices must also be considered, such as equivalent turn-on resistance, series equivalent inductance, switching time, and switching energy loss. The large equivalent resistance, switching time, and energy loss result in high conduction and switching power loss, while the large series equivalent inductance contributes to the total circuit stray inductance and leads to a HV spike that may damage the switching device [8].

13.3.2 Selections of DC-link capacitors

It has been proven that the power switch devices and capacitors are the most vulnerable components in power electronic systems [9]. The reliability of the system can be improved by minimizing the number of power switching devices and capacitors. In power electronics converters, the DC-link capacitor can be electrolytic, film, or ceramic capacitors. Ceramic capacitors can be operated at a relatively high temperature; however, they are brittle. One typical failure mechanism of the ceramic capacitor is cracking when the PCB is bent.

Meanwhile, the HV electrolytic capacitor is not promising in electrical systems for automotive applications due to its short lifetime and safety issues. One issue is the electrolyte leakage which poses several potential safety hazards: human contact with electrolyte and electrolyte residue, and the short-circuiting of adjacent electronic systems. This could become a disaster for both the automakers and users.

The merits of film capacitors are low ESR and ESL, no liquid electrolyte, and strong overvoltage capability. In addition, a film capacitor also has self-healing capability. Therefore, film capacitors are preferred for installation on HV DC-link in electrified vehicles rather than electrolytic capacitors. However, compared to electrolytic capacitors, the capacitance to volume ratio of film capacitors is much smaller. A size comparison between a 500-μF/450-V film capacitor and a 330-μF/450-V electrolytic capacitor can be seen in Figure 13.12, regardless of the ripple current rating. By adding the bulky film capacitor, it results in larger converter volume and lower power density. This issue becomes more stringent for the onboard power electronics system, which is preferred to be light weight, small size, and low cost. Therefore, the type of capacitor in the vehicle is dependent on the required capacitance. For a relatively small capacitance requirement, a film capacitor should be applied, especially on the HV DC-link. However, for a relatively large capacitance requirement, an electrolytic capacitor or other alternative solution (APF, etc.) might be applied in the vehicle.

The DC-link capacitors are selected based on the converter system requirements. The capacitance, rated maximum continuous current, and rated voltage are three important parameters in DC-link capacitor selection. First, the capacitance can be calculated according to the DC-link voltage ripple. The voltage ripple required by most converter systems is 5%–10% of the DC-link voltage. Second,

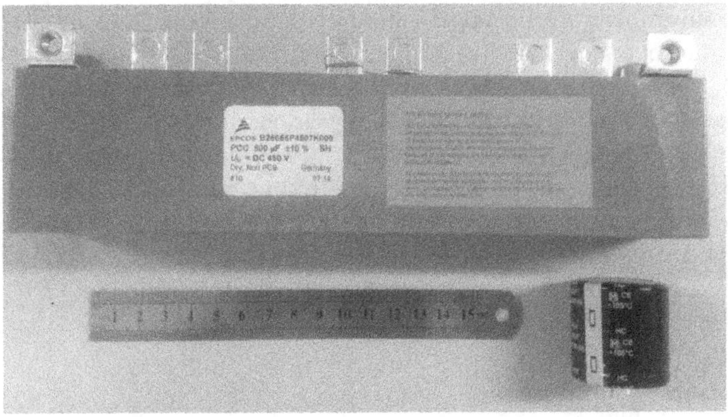

Figure 13.12 A comparison between a 500-μF/450-V film capacitor and a 330-μF/450-V electrolytic capacitor

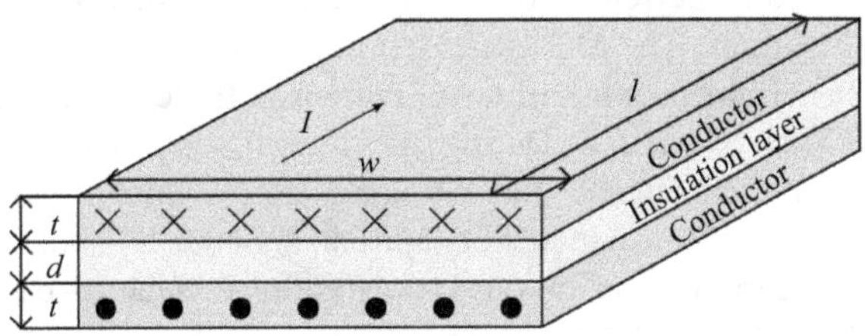

Figure 13.13 A laminated bus bar example where t, w, l, and d are conductor thickness, width, length along the current direction, and the distance between two conductors

in order to prevent overheating and the lifetime degradation of the capacitor, the rated current of the selected capacitor must be higher than the root mean square (RMS) value of the DC-link ripple current. Finally, the rated voltage of the film capacitor could be selected as the same as the DC-link voltage, since its over-voltage capability is high and can reach up to twice the rated voltage.

13.3.3 Design considerations of DC bus bars

In the high-power converter systems, bus bars are usually utilized instead of cables to simplify the assembly. The shape of a bus bar must be designed based on the conducting current requirements, the switching device packages and the DC-link capacitor numbers and packages. An example of a bus bar structure is shown in Figure 13.13, in which the positive and negative conductor plates of the DC-link are laminated. The width and length of the conductor are usually determined by the packages and locations of the converter's main components, such as switching devices, DC-link capacitors, and inductors, while the thicknesses of the bus bar conductors and insulation layer are designed according to the conducting current and voltage.

Generally, there are no upper limits of the conductor and insulation layer thickness; however, there are lower limits. For the thickness of the insulation layer, in order to sustain the operating DC-link voltage, the value should be larger than a minimum value. The dielectric strength of various insulation materials with different thicknesses is not a constant value, thus the insulation layer thickness should be defined depending on the materials and the practical dielectric strength. For conductor thickness, to avoid overheat on the entire bus bar or at some hot spots, the current density in each conductor must meet the requirement of 5 A/mm^2 [10]. The cross-section area A in the current direction is defined by (13.1), and the current density J is the amount of current I flowing through the unit area of this cross section, (13.2).

$$A = w \cdot t \left(\text{mm}^2\right) \tag{13.1}$$

$$J = \frac{I}{A} \left(\text{A/mm}^2\right) \tag{13.2}$$

There are two types of currents flowing in the DC-link bus bar conductors, DC current and AC current. The DC current flows from the input terminals to the switching device terminals, while the AC current is the current ripple that flows through the DC-link capacitor. During the design, both DC and AC current densities need to be considered, and the conductor thickness and width can be determined by the required conducting current and the DC-link current ripple. Finally, the total current density must satisfy the conductor current's density requirement [11].

DC current on the bus bar spreads out evenly in the conductor. As a result, DC current flows through the entire cross section area of the conductor. Here, the thickness of the current layer is equal to the thickness of the conductor. However, the AC current tends to be concentrated near to the conductor's surface due to the skin effect that is shown in Figure 13.14. Thus, the thickness of the current layer,

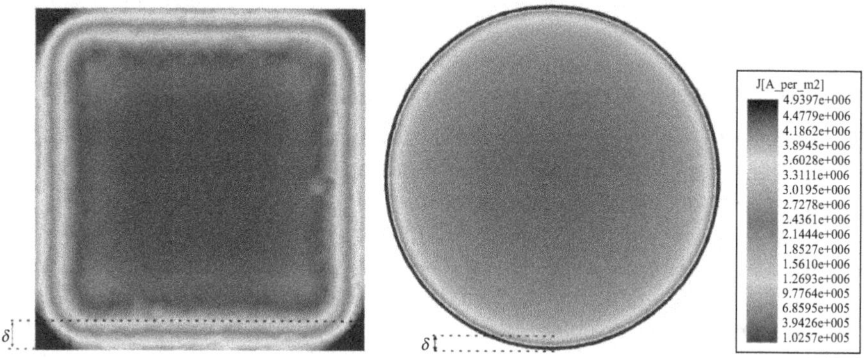

Figure 13.14 AC current on a square and a round cross section in conductors, where δ is the skin depth

which is called skin depth δ, is smaller than the thickness of the conductor. Hence, adjusting the thickness of the conductor only might not be an effective solution to improve the AC current density, which can be calculated by replacing t by δ in (13.1). To satisfy the current density requirement, the conductor width also needs to be considered under AC current conditions.

In addition, when more than one DC-link capacitor is utilized, the current distribution on the bus bar needs to be balanced to avoid imbalanced current sharing among the capacitors. Unbalanced current sharing between DC-link capacitors may result in higher current stress and a shorter lifetime for some capacitors than others. This will also impact the reliability of the system. To achieve identical impedance from different current paths on the DC bus bar and balanced current sharing between capacitors, a symmetric structure of the bus bar is usually preferred.

In the bus bar design, besides the requirements of current density and current distribution, the bus bar parasitic parameters should also be considered.

The parasitic parameters include bus bar resistance, stray inductance, and stray capacitance. When current flows through a bus bar, power loss is mainly generated by the resistance. Voltage spikes during switching device turn-off transients are caused by stray inductance. Besides, high-frequency noise can be filtered by stray capacitance. As a result, large stray capacitance, small resistance, and small stray inductance are preferred. On the one hand, the bus bar resistance should be designed to be as small as possible and is usually under 1 mΩ. Then, the amount of power loss caused by this resistance could be several watts. On the other hand, the thickness of dielectric material layer should be designed to be as thin as possible to achieve relatively large stray capacitance. Nevertheless, the dielectric layer thickness also affects the insulation ability of the bus bar; to sustain the required DC-link voltage, the thickness of the dielectric layer must be above a minimum value. In fact, other than the optimization of resistance and stray capacitance, most efforts are usually focused on minimizing the bus bar's stray inductance during the design; this is because the ohmic losses generated by the bus bar resistance is negligible compared to the total converter power losses, and there are limited improvements that could be achieved in stray capacitance maximization.

Total bus bar stray inductance L_{total} can be estimated by the summation of conductor self-inductance L_{self} and mutual-inductance L_M. Self-inductance is determined by the shape of the conductor, while mutual-inductance is influenced by both the shape and the overlapping area of the two conductors. When the current directions are opposite to each other in two DC-link bus bar conductors, mutual-inductance is negative. Thus, the total stray inductance of the bus bar can be obtained as (13.3). In order to minimize the bus bar's stray inductance, the overlapping area of the two conductors should be maximized. By laminating two conductors of the bus bar, the stray inductance can be reduced significantly. That is the reason why the laminated bus bar structure is popular in various bus bar designs, seen in Figure 13.13.

$$L_{total} = 2 \cdot \left(L_{self} - L_M\right) \tag{13.3}$$

The designed bus bar can be analyzed by FEA simulations, which is a straightforward method for estimating the parasitic parameters. Then, the voltage spike could be evaluated with the estimated bus bar stray inductance. An equivalent circuit of a converter phase-leg during the switching transient is shown by Figure 13.15, where the upper switch S_1 is being turned off, while its corresponding lower diode is turning on. The direction of current i does not change during this time period, and the current in the high-side switch is transferring to the low-side diode. During this turn-off period, voltage spike occurs as it is plotted by Figure 13.11. It is obvious that the voltage spike across switch S_1 is caused by the released energy from the inductances in the circuit, including series equivalent inductance L_c of the DC-link capacitor, series equivalent inductance L_1 of the switching device, and the bus bar stray inductance L_{busbar}. Thus, the voltage spike and the peak voltage across switch S_1 can be calculated by (13.4)–(13.6).

$$V_{ce,peak} = U_{DC} - L_c \cdot \frac{di_{a1}}{dt} - L_1 \cdot \frac{di_{a1}}{dt} + L_2 \cdot \frac{di_{a2}}{dt} - L_{busbar} \cdot \frac{di_{a1}}{dt} \tag{13.4}$$

$$V_{spike} = \left| L_c \cdot \frac{di_{a1}}{dt} \right| + \left| L_1 \cdot \frac{di_{a1}}{dt} \right| + \left| L_2 \cdot \frac{di_{a2}}{dt} \right| + \left| L_{busbar} \cdot \frac{di_{a1}}{dt} \right| \tag{13.5}$$

$$V_{ce,peak} = U_{DC} + V_{spike} \tag{13.6}$$

During the switching period shown by Figure 13.15, i_{a1} is reducing and i_{a2} is increasing; therefore, the derivative of i_{a1} with respect to time t is negative and that of i_{a2} is positive. Finally, the voltage spike is represented by (13.5), and the peak voltage across S_1 is obtained as (13.6). To prevent the device being damaged, the peak voltage across the switch should be lower than its rated voltage. Otherwise, the bus bar design must be revised.

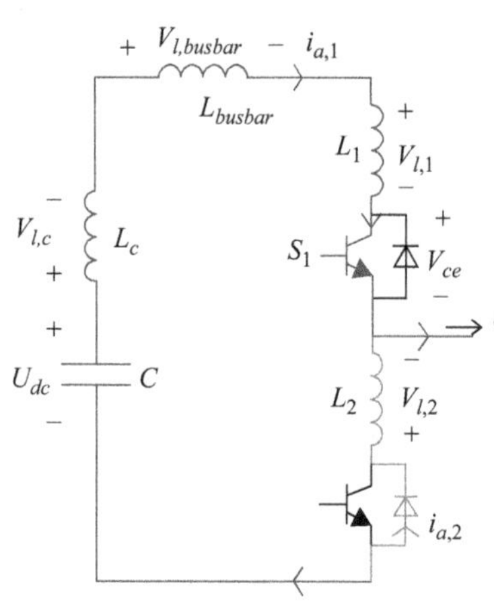

Figure 13.15 The equivalent circuit of a converter phase-leg during the turn-off transient of switch S_1

13.4 Topological reconfigurations of DC systems in electrified vehicles

The integration of power electronics is critical, as it can reduce the size, weight, and cost of the overall system. A simple type of system-level integration combines several parts to better utilize the space in the three dimensions. For example, the APM converter and traction inverter in the vehicle can be placed together in one enclosure to share the DC-link capacitors, controller, heat sink, and housing cover as presented in [12]. More advanced system-level integrated converters also allow sharing of the power switches and gate drivers. However, these converters usually either employ a relatively complex control strategy or apply a complicated topology to regulate the power flows.

Figure 13.16 shows an integrated converter topology that achieves the bidirectional power flows among four DC terminals [13]. The converter has the functions of one DAB and two two-phase bidirectional buck converters but only need eight switches. Two coupled inductors are needed so that different inductances can be obtained according to different current flow directions for different power conversions. The duty cycle is used to control the buck converter, and the phase shift angle between the primary and secondary side is applied to control the DAB. Therefore, the integrated converter can work independently and simultaneously with multiple functions. However, there is a trade-off between the duty cycle and phase shift angle which might limit the performance of both power conversions, especially for the conversion with wide voltage and wide power range, such as a battery charger. In addition, those integrated multifunctional switches need to carry the overall power for both power conversions, leading to a higher current rating on the power switches. This might become an obstacle for the power switches and cooling system to reach higher power. Eventually, the reduced part might be the gate-driver circuit only, as the required die size in the power switch devices and the amount of heat-dissipation are still the same compared to the nonintegrated method.

The operation of electrified vehicle has two modes. These are HV battery charging and vehicle running mode. This makes electrified vehicle a unique application. In other words, the role of operating time can also be considered as the

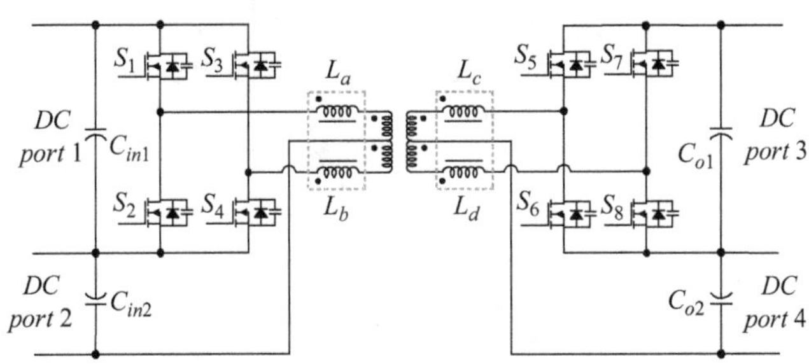

Figure 13.16 An integrated multiport converter

fourth dimension for the system-level integration. Therefore, in the recent years, the more integrated power electronic systems in vehicle applications have been evaluated and proposed with an attempt to further reduce the size, weight, and cost. These systems are dual-mode, which are application-oriented and fully utilize the characteristics of the vehicle application.

13.4.1 Topological reconfigurations in HV charging and propulsion systems

The first type of dual-mode system integration is between the HV battery charging system and the traction drive system as shown in Figure 13.17 [14,15]. The system shown in Figure 13.17(a) utilizes the traction motor and its inverter as the PFC for the HV battery charging. If a second-stage DC/DC is required in the HV battery charging system, a DC/DC converter can be added and the two contactors beside the HV battery pack are utilized to provide a path to bypass this second-stage DC/DC converter in the vehicle's running mode. Therefore, the efficiency of the traction system will not be sacrificed by this system integration.

Depending on the powertrain requirements, the design parameters of the PFC and traction drive system might not be matched. Another system integration approach is shown in Figure 13.17(b). In this integrated system, the secondary-stage isolated DC/DC converter is shared between the traction drive system and HV battery charging system. Its DC/AC stage and the magnetizing inductor of the transformer create a two-phase interleaved boost converter for the traction drive system. Therefore, no additional power switch devices and inductors are needed to

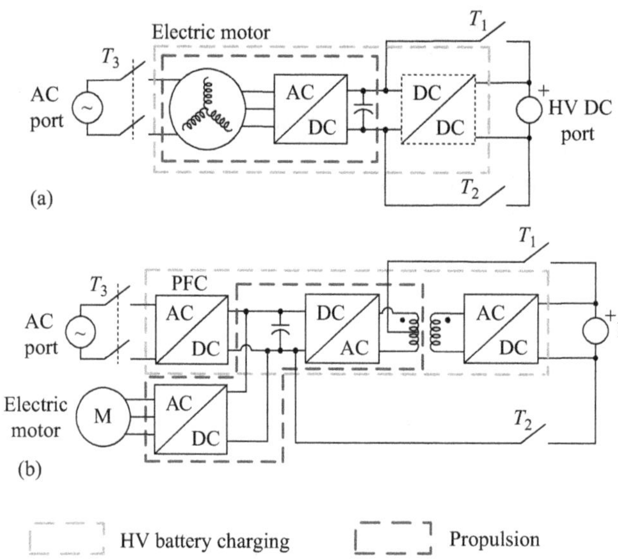

Figure 13.17 Topological reconfigurations between HV charging and propulsion systems

build the boost converter. As with the previous case, contactors T_1 and T_2 are employed for reconfiguration between the two functions and isolation between the battery and remaining vehicle systems.

13.4.2 Topological reconfigurations in dual-voltage charging systems

Other than the integration between the HV battery charging system and propulsion system, dual-mode integration can also be achieved between the HV battery charging system and LV APM charging system. Several integrated multifunctional dual-mode dual-voltage charging configurations have been proposed in [16–20]. Basically, they can be summarized and mainly classified into three types as shown in Figure 13.18.

As shown in Figure 13.18(a), the first system integrates the primary DC/AC side of the APM with the secondary AC/DC side of the second-stage DC/DC converter of the HV battery charger. The transformer is also shared between the APM and second-stage DC/DC converter.

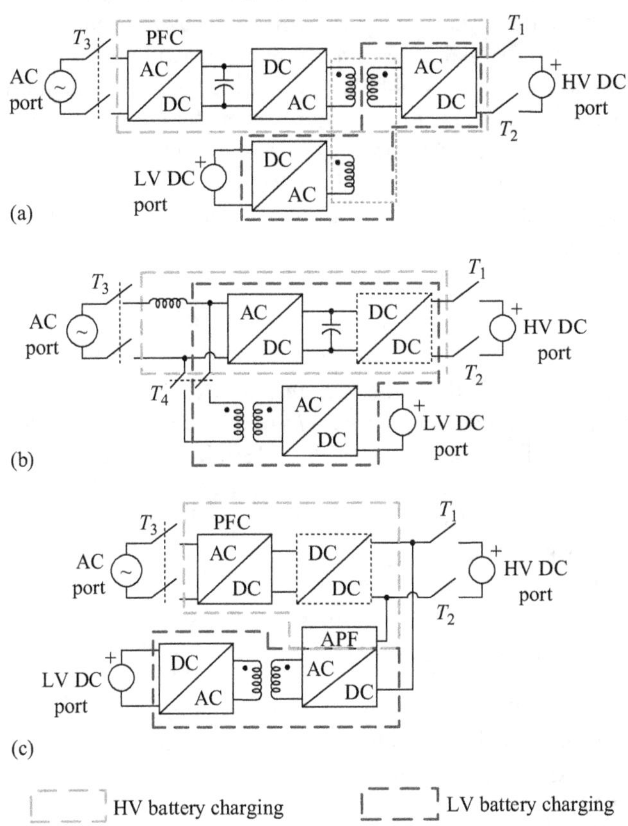

Figure 13.18 Topological reconfigurations in dual-voltage DC systems

Figure 13.18(b) shows the second system which integrates the primary DC/AC side of the APM with the first-stage AC/DC PFC converter of the HV battery charger. An extra inductor is still needed for the PFC, and the extra relay T_4 is required to achieve the reconfiguration.

The third type is shown in Figure 13.18(c). It integrates the primary DC/AC side of the APM with the APF to alleviate the second-order harmonic current from the single-phase grid line for the HV battery charger. As a result, the additional power switches and their gate drivers, heat sink, and extra inductors can be eliminated. However, the extra second-order harmonic energy storage capacitor is still required for the APF.

13.5 Conclusions

In this chapter, the power electronics converters and inverter in vehicle application are introduced. Several possible topologies for PFC, isolated DC/DC converter, APM, and APF are presented. The practical design of DC bus bars and the selection of power switch devices and DC-link capacitors are given. Finally, the state-of-the-art topological reconfigurations of DC systems in electrified vehicles are presented.

One of the biggest challenges for the next generation of power electronic systems in vehicle application will be the cost reduction to provide more affordable solutions. This has become one of the major barriers for electrified vehicle for mass commercialization. These dual-mode integration approaches are dedicated to vehicle application and can reduce the cost, size, and weight of the system. Therefore, they will also help promoting and accelerating the paradigm shift to the transportation 2.0.

References

[1] A. Emadi and J. G. Petrunic, *Chapter 1: Automotive Industry and Electrification, in Advanced Electric Drive Vehicles, Edited by A. Emadi*, pp. 1–13. Boca Raton, FL: CRC Press, ISBN: 978-1-4665-9769-3, 2014.

[2] U.S. Department of Energy Alternative Fuels Data Center, U.S. Plug-in Electric Vehicle Sales by Model, 2016. [Online]. Available: http://www.afdc.energy.gov/data/.

[3] S. G. Wirasingha, M. Khan, and O. Gross, *Chapter 10: 48-V Electrification Belt-Driven Starter Generator Systems, in Advanced Electric Drive Vehicles, Edited by A. Emadi*, pp. 331–367. Boca Raton, FL: CRC Press, ISBN: 978-1-4665-9769-3, 2014.

[4] M. Yilmaz and P. T. Krein, "Review of battery charger topologies, charging power levels, and infrastructure for plug-in electric and hybrid vehicles," *IEEE Transactions on Power Electronics*, vol. 28, no. 5, pp. 2151–2169, May 2013.

[5] A. Khaligh and S. Dusmez, "Comprehensive topological analysis of conductive and inductive charging solutions for plug-in electric vehicles," *IEEE*

Transactions on Vehicular Technology, vol. 61, no. 8, pp. 3475–3489, Oct. 2012.

[6] L. Huber, Y. Jang, and M. M. Jovanovic, "Performance evaluation of bridgeless PFC boost rectifiers," *IEEE Transactions on Power Electronics*, vol. 23, no. 3, pp. 1381–1390, May 2008.

[7] R. Hou, P. Magne, and B. Bilgin, *Chapter 9: Low-Voltage Electrical Systems for Nonpropulsion Loads, in Advanced Electric Drive Vehicles*, Edited by A. Emadi, pp. 317–328. Boca Raton, FL: CRC Press, ISBN: 978-1-4665-9769-3, 2014.

[8] J. Guo, *Modeling and Design of Inverters using Novel Power Loss Calculation and DC-Link Current/Voltage Ripple Estimation Method and Bus Bar Analysis*. Ph.D. Thesis, McMaster University, Hamilton, Canada, 2017.

[9] Y. Song and B. Wang, "Survey on reliability of power electronic systems," *IEEE Transactions on Power Electronics*, vol. 28, no. 1, pp. 591–604, Jan. 2013.

[10] J. M. Allocco, "Laminated bus bars for power system interconnects," in *Proc. 1997 IEEE Applied Power Electronics Conference and Exposition*, Atlanta, GA, Feb. 1997, pp. 585–589.

[11] A. D. Callegaro, J. Guo, M. Eull, *et al.*, "Bus bar design for high-power inverters," *IEEE Transactions on Power Electronics*, vol. 33, no. 3, pp. 2354–2367, Mar. 2018.

[12] Delphi, Delphi Integrated Inverter, Converter and Controller, 2013. [Online]. Available: http://www.delphi.com/manufacturers/auto/hevevproducts/integrated-solutions/integrated-i-c-c.

[13] K. Itoh, M. Ishigaki, N. Yanagizawa, S. Tomura, and T. Umeno, "Analysis and design of a multiport converter using a magnetic coupling inductor technique," *IEEE Transactions on Industry Applications*, vol. 51, no. 2, pp. 1713–1721, Mar. 2015.

[14] G. J. Su and L. Tang, "An integrated onboard charger and accessory power converter using WBG devices," in *Proc. 2015 IEEE Energy Conversion Congress and Exposition*, Montreal, QC, Sep. 2015, pp. 6306–6313.

[15] S. Anwar, W. Zhang, F. Wang, and D. J. Costinett, "Integrated DC–DC converter design for electric vehicle powertrains," in *Proc. 2016 IEEE Applied Power Electronics Conference and Exposition*, Long Beach, CA, Mar. 2016, pp. 424–431.

[16] A. Khaligh and Y. Tang, "An integrated dual-output grid-to-vehicle (G2V) and vehicle-to-grid (V2G) onboard charger for plug-in electric vehicles," US patent US20160016479A1, Jun. 15, 2015.

[17] J. G. Pinto, V. Monteiro, H. Goncalves, and J. L. Afonso, "Onboard reconfigurable battery charger for electric vehicles with traction-to-auxiliary mode," *IEEE Transactions on Vehicular Technology*, vol. 63, no. 3, pp. 1104–1116, Mar. 2014.

[18] R. Hou and A. Emadi, "Applied integrated active filter auxiliary power modules for electrified vehicles with single-phase onboard chargers," *IEEE Transactions on Power Electronics*, vol. 32, no. 3, pp. 1860–1871, Mar. 2017.

[19] R. Hou and A. Emadi, "A primary full-integrated active filter auxiliary power module in electrified vehicle applications with single-phase onboard chargers," *IEEE Transactions on Power Electronics*, vol. 32, no. 11, pp. 8393–8405, Nov. 2017.

[20] Y. Kim, C. Oh, W. Sung, and B. K. Lee, "Topology and control scheme of OBC–LDC integrated power unit for electric vehicles," *IEEE Transactions on Power Electronics*, vol. 32, no. 3, pp. 1731–1743, Mar. 2017.

Chapter 14
DC data centers

Enver Candan[1] *and Robert C.N. Pilawa-Podgurski*[2]

As high-performance computing and data storage transition toward becoming Internet-based services, the world has witnessed an ever-increasing demand for both size and capacity of data centers. The growth of cloud-based services and applications shows no sign of slowing down, with additional custom-hardware for machine learning algorithms beginning to be deployed at scale in dedicated data centers. Today's data centers accommodate many pieces of information technology (IT) equipment such as data-processing units, data storage units, and communication devices. A recent report estimated the energy usage of data centers in the United States (US) alone at 70 billion kW h in 2014, corresponding to 1.8% of the total electric energy consumed in the country [1]. Since the IT equipment requires low DC voltage (typically ranging from a few volts to a few dozen volts) to operate, various power delivery architectures are established to provide low DC voltage from utility and renewable resources. In this case, the power delivery infrastructure in data centers can be considered as a microgrid due to the high installed power capacity and dynamic loads. However, data centers are also quite different than typical DC microgrids in many regards, both in the characteristics of the loads (extraordinarily rapid transients, but all controlled/managed from a central load scheduling interface) and the extreme up-time requirements. This chapter addresses major aspects of power-delivery architectures in data centers such as efficiency, reliability, integration with renewable resources, and protection. A critical evaluation of the technical and commercial barriers to widespread DC power distribution in data centers will be performed, along with a few examples of existing DC data centers.

14.1 Introduction

In data centers, since the major energy supply is the AC utility and the primary power consumers (i.e., IT equipment) require low voltage DC, both AC and DC power infrastructures are concurrent. As the power is delivered from the utility to the low voltage DC loads, rectification (power conversion from AC to DC form)

[1]Department of Electrical and Computer Engineering, University of Illinois at Urbana-Champaign, USA
[2]Department of Electrical Engineering and Computer Sciences, University of California, USA

can be performed at various points, resulting in different power architecture configurations [2].

Traditionally, the utility power is distributed in AC form inside the data center, and then, rectification and voltage step-down conversion are performed at the load end of the power architecture by a dedicated power supply unit (PSU) for each piece of IT equipment as shown in Figure 14.1(a). Conversely, the utility power can be rectified at the source end of the power architecture (i.e., data center input), and

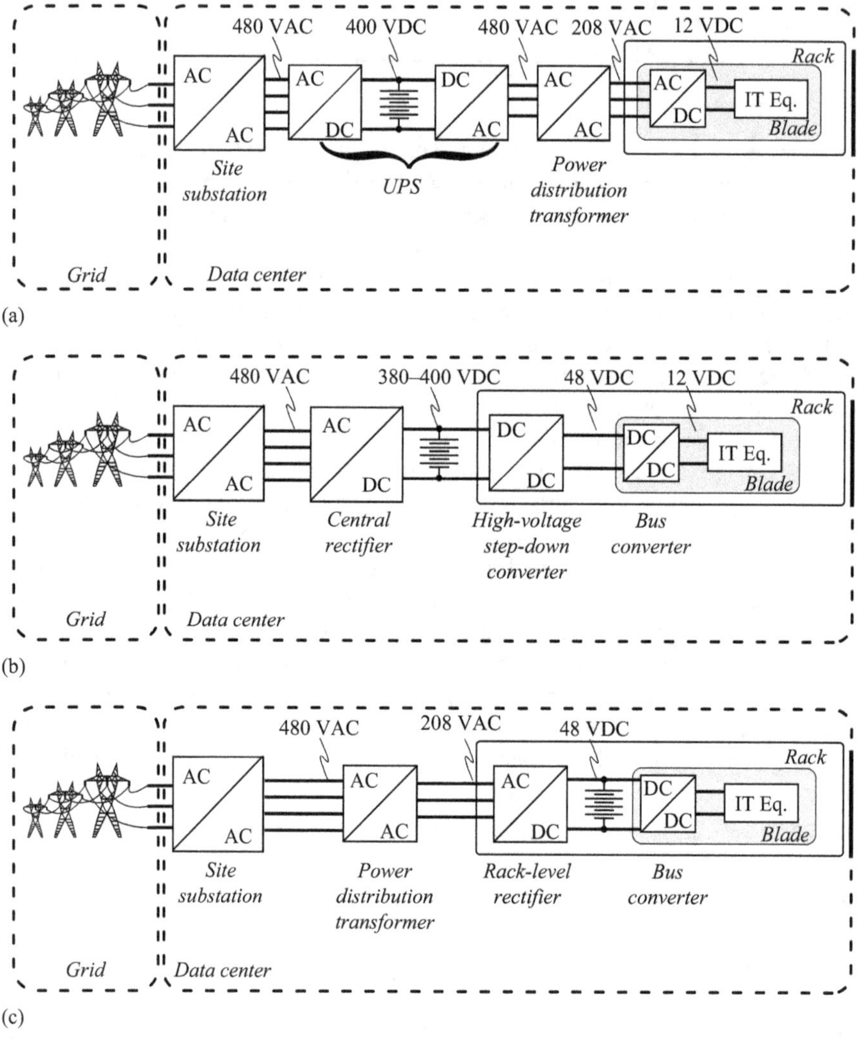

Figure 14.1 Simplified drawings of common power delivery architectures in data centers: (a) AC power distribution in data centers, (b) DC power distribution in data center, and (c) AC power distribution in data center, DC power distribution in the rack or within a few racks

distributed in high DC voltage form within the data center. Then, as shown in Figure 14.1(b), a dedicated DC–DC converter per IT equipment steps down the high DC voltage at the IT equipment input. Alternatively, the AC power can be distributed to racks that host the IT equipment, and rectifiers that are in the same rack as the IT equipment (or in another rack that is in close proximity) provide DC power to the nearby IT equipment as shown in Figure 14.1(c). Simplified diagrams of these common data center power distribution architectures are depicted in Figure 14.1. For prioritizing transition from AC to DC power distribution in Figure 14.1, isolation transformers, protection equipment, and cooling devices are not depicted in these figures but of course exist in practical designs and may introduce additional power conversion stages.

Because of the extensive background, acceptability and well-established standardization of the AC distribution in many other applications, a high percentage of the existing data centers use derivations of the power delivery architectures depicted in Figure 14.1(a) or (c), which are fundamentally inherited from telecom applications. Recently, DC power delivery architecture, as depicted in Figure 14.1 (b), has gained attention, mainly because it involves fewer conversion stages and could potentially simplify the integration of ancillary distributed energy resources, such as solar PV and fuel cells (FC).

14.2 Development of DC power distribution in data centers

The technological developments in the early 2000s led to rapid expansion of data centers, and as a result, their increasing energy consumption has been noted in a several reports [3,4]. A well-cited 2008 report has underlined the benefits of high voltage DC power distribution in data centers to date [5], although the idea of using a higher voltage level than 48 V for data center power distribution appears in the literature as early as 1999 [6].

Over the years, the literature has addressed various high DC voltage levels such as 270 [6], 300 [7], 325 [8], 380 [9,10], and 400 V [9,11]; however, the consensus for high voltage DC distribution in data centers appears to eventually have become 380 and 400 V. The protracted discussion on voltage levels over 15 years, combined with relatively slow development of standards for the IT equipment and power distribution such as ETSI EN 300 132-3-1 [12] for 400 V DC bus voltage, ETSI EN 301 605 [13] for grounding, and IEC 61643-21 [14] for protection, has arguably discouraged short-term adoption of DC distribution in data centers.

Following the potential of DC distribution for data centers noted by [5], AC and DC power distribution architectures for data centers have been both quantitatively and qualitatively compared by both academia and industry [15–19]. In addition, [20] has reviewed some of the highly cited comparison reports and stated that the results of the studies significantly vary and overstate the benefits of the DC power architectures. In order to fairly evaluate the comparison studies, it should be noted that over the years, significant barriers such as lack of standardization, market share, and compatibility with the IT equipment have prevented the widespread adoption of high voltage DC in data centers, while already well-established AC distribution architectures have kept developing to meet expectations. This chapter assesses the state of high voltage DC power delivery architecture for data centers and points out its unique advantages and challenges but does not attempt to provide a quantitative comparison between the AC and DC distributions.

14.3 Efficiency

By far, the largest operating cost of data centers is the cost of electricity. For this reason, improving the energy efficiency of the computing and power delivery is critical. As shown in Figure 14.1, regardless of the preferred distribution architecture, the utility power has to go through several cascaded power conversion stages before it reaches the IT equipment. Since the power consumed by the IT equipment must be processed by each power converter, the overall power infrastructure efficiency is limited by the stage with the lowest efficiency. Power delivery studies must thus consider the entire power conversion stage, from the high voltage AC input to the building, all the way down to the CPU and memory voltages, around 1 V.

While at a high-level, it may appear that simply reducing the number of conversion stages would yield increased efficiency. However, one must be careful to consider that for a constant power converter volume, a high-step down power converter generally has lower efficiency than a converter with a modest voltage step-down. This is particularly apparent in the case of the recently proposed 48 V to the point-of-load concept, where the conventional architecture shown in Figure 14.2 involves first a 48 to 9–12 V conversion, followed by (typically several) 9–12 V to the point-of-load (e.g., 1–2 V) converters. While it may be tempting to simply eliminate the two-stage conversion and design a single 48–1 V converter, such a converter is significantly more difficult to design to be highly efficient and highly power dense. For example, consider the reference design of [21], which represents a single-stage buck converter, achieving a peak efficiency of 84%.

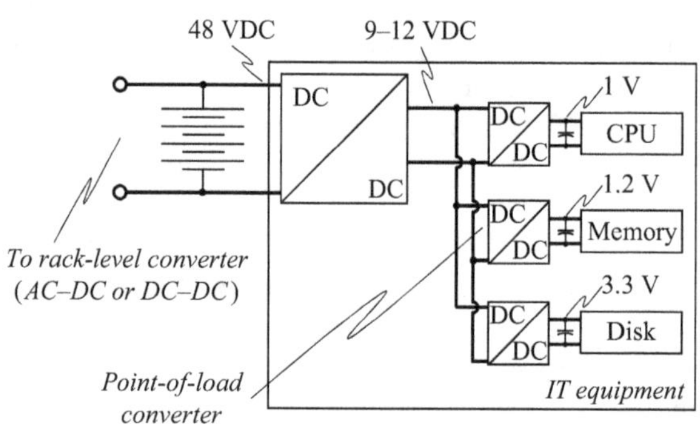

Figure 14.2 48 V to point-of-load architecture

Other, more complex topologies, likewise achieve limited performance in both efficiency and power density. In comparison, recent hybrid switched-capacitor power converters have been shown to achieve near 99% efficiency with extraordinarily high power density (106.5 kW/L) for 48–12 V conversion [22], and separate 12–1 V converters can similarly achieve high overall efficiency and power density [23–25]. While there may be other considerations (such as reliability, cost, reduced complexity) to prefer fewer numbers of stages, the above discussion highlights that increased efficiency is not necessarily a given outcome of this approach. Regardless of the power distribution architecture, power electronics plays a critical role in achieving high power conversion efficiency in data centers. An in-depth review of power electronics, ranging from utility-scale to chip-level converters in data centers, can be found in [26].

However, one key approach to increasing the system-level conversion efficiency is to eliminate, to the greatest extent possible, any double conversion, where power is stepped up and down (or converted from AC to DC or DC to AC) more times than the absolute minimum. An example of such a double conversion is the back-to-back AC–DC and DC–AC conversion of the centralized uninterrupted power supplies (UPS) approach shown in Figure 14.1(a), which is one reason why it is no longer a preferred approach. The DC power distribution for data centers offers some opportunities to reduce the number of double conversion power stages.

- A facility-level battery system for power backup in a DC system does not require an inverter stage while providing backup power to both the IT equipment and auxiliary loads such as lighting and cooling. Since electrochemical energy storage is inherently in DC form, a battery bank for a utility outage or failure scenario can be connected to the high voltage DC bus. Of course, additional circuitry may be needed to connect the battery banks to the high voltage DC bus for regulatory or other operating reasons; however, such circuitry does not process the requisite power similar to an inverter that outputs a tightly regulated AC waveform and synchronizes with multiple converters across the bus.
- Twice-line frequency energy buffering is a well-known issue in single phase rectifiers [27]. The recent Google Little Box Challenge [28] has accelerated research efforts in the area of twice-line energy buffering, and as a result, extremely high efficiencies for twice-line frequency energy buffering have been reported in the literature [29]. Nevertheless, moving from single phase rectification at the server input to a centralized three-phase rectification at the high voltage DC bus eliminates the need for twice-line frequency energy buffering and the associated circuitry from the cascaded power chain. It should be noted, however, that similar benefits can be realized if three-phase rectifiers are employed in an AC distribution approach, generally at the rack level (i.e., rack-level three-phase rectifiers). Generally, however, it is likely less costly and more practical to achieve very high efficiency when designing fewer large rectifiers at the data center input (DC distribution), than at each rack or server (AC distribution).

- Similar to the twice-line energy buffering, power factor correction (PFC) circuitry is an essential requirement in both single and multiphase high power rectifiers. The most common PFC architecture is the boost-type converter, meaning the output of the PFC circuit is at a higher voltage than its input. Since the IT equipment requires low voltage DC, employing PFC closer to the low voltage load requires a back-end, high conversion ratio, voltage step-down converter. Although the centralized three phase rectification does not eliminate the PFC circuitry, it can remove a cascaded voltage step up and down conversion at the IT equipment input, which is fundamentally counterintuitive since the load is at low DC voltage. Here, recent developments on step-down (e.g., buck-derived) PFC rectifiers [30] show promise to achieve increased system-level efficiencies.

Another potential advantage of the reduced number of cascaded stages in DC data centers is the increase in available volume inside the IT equipment blade or rack for more data processing and/or storage units. For example, the AC to DC conversion stage in the PSU of the IT equipment requires twice-line frequency energy buffering and PFC circuitries (as explained above), which consist of bulky passive components since they need to be designed for grid frequency. Removal of the AC to DC stage from the PSU thus allows more space in the IT equipment blade or rack.

Since all the power consumed by the IT equipment must be processed through the cascaded power stages, the system-level efficiency is still limited by the lowest efficient power stage. Thus, research efforts have focused on the efficiency improvements of each major power electronics converter type for high voltage DC data center applications. Development and commercial availability of wide bandgap devices have been leveraged in converter designs. Consequently, high-90% efficiencies were achieved in three-phase rectifiers with 400 V DC output by eliminating the boost PFC stage [30,31], and mid-90% efficiencies in 400–12 V or 48 V DC to DC converters by using resonant topologies [32–34].

Entirely new power delivery architectures also hold great promise for dramatic increase in power-delivery efficiency. Instead of focusing on the efficiency improvements of the individual power converters in the DC to DC conversion stage, a series-stacked architecture (Figure 14.3) that focuses on reducing the amount of

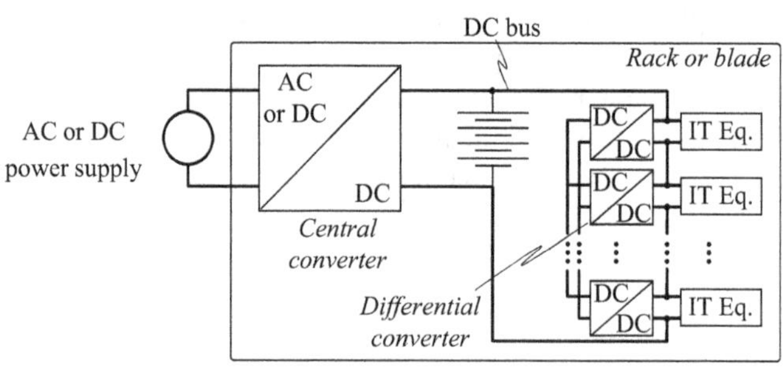

Figure 14.3 Series-stacked power architecture

power processed between the high voltage DC bus and multiple IT equipment has been proposed in [35, 36]. By electrically connecting IT equipment of the same type (servers in these works), the series-stacked architecture greatly reduces the requisite power processed while achieving inherent high voltage step-down. The bulk power to the servers is delivered by the DC bus current that passes through all the servers without being processed. Since the series connected loads (servers) must conduct the same amount of current, their input voltages drift in the case of a power consumption mismatch. Bidirectional differential converters are employed at intermediate nodes to maintain voltage balance under all circumstances. These converters are able to exchange the *difference* between the server currents in the case of a power mismatch between the servers. Through series stacking and active voltage balancing using differential power processing, only the *difference* power between the servers needs to be processed. Therefore, the amount of processed power is decoupled from the total power delivered, yielding greatly reduced power loss during the power conversion in comparison to the conventional power-delivery systems where the *full* server power must be processed by the converters. Experimental verification of the series-stacked architecture has demonstrated higher than 99.8% peak efficiency for the steady-state operation of the servers, corresponding to an up to 40× reduction in average power loss in comparison to a state-of-the-art PSU to date [36]. In addition, practical challenges of a series-stacked architecture such as load balancing in software [37], hot-swapping [38], and unregulated bus operation [39] have been analyzed and experimentally addressed without sacrificing high power conversion efficiency. Although experimental verification of this architecture is limited to a stack of four servers to a 48 V, its superior performance has been analyzed in case studies for 32 series-stacked servers connected to a high voltage (380 V) DC bus [35,36,38].

14.4 Reliability

Maintaining high reliability in data centers is crucial because of our society's dependence on the uninterrupted IT services. Nowadays, any outage of the IT services can have a large impact, both financial and in terms of societal impact. A typical desired target reliability for a data center is 99.99% uptime (often called as "four nines"), which corresponds to 52.5 min downtime per year [40]. This requirement, combined with maintaining high efficiency, makes data center power delivery architecture design challenging. Fortunately, similar to improving the power delivery architecture efficiency, reducing the number of cascaded power stages typically reduces the overall risk of system failure and the mean time between failures. Therefore, the opportunities to reduce the number of power conversion stages in DC data centers have been also considered to facilitate higher reliability and uptime. Elimination of the conversion stages such as power distribution transformers and the inverter at the UPS output in high voltage DC distribution has been considered as the main advantage for the DC data centers [11,41,42]. However, the analysis in these works is qualitative, and the details are unclear. A 2010 study quantitatively compared the reliability of AC and DC power distribution for data centers with emphasis on UPS and concluded that the DC distribution would be more reliable than the AC distribution without supplementary effect of redundancy [16]. To the best of our knowledge, a more recent quantitative reliability analysis for the DC data centers, which considers UPS, power converter failure mechanisms and inclusion of different redundancy options, is missing from the literature on this topic. Nevertheless, reducing the number of conversion stages alone is not sufficient to assure reliable power distribution; back-up power and fault tolerance must be incorporated in the data center power architecture design to achieve the desired reliability level.

14.4.1 Fault tolerance

Fault tolerance is the ability of a system to maintain its operation in the case of component failures. A fault tolerant system must be able to detect faults, protect the system from the failed component, and provide redundancy. Fault detection and redundancy for DC data centers is explained here, while protection methods are explained later in Section 14.7.

There are two main types of faults to be detected in DC power delivery architectures: over voltage and over current. Over voltage detection can be relatively easily implemented in DC systems using voltage dividers and comparators. However, over current detection is nontrivial in DC architectures because of in-rush current phenomena and limited converter power ratings. Moreover, in data centers, the rapidly changing loads may appear as an over current situation if load-scheduling causes many servers to initiate computing at the same time. In-rush, current can be thought of as an instantaneous and unusual rise at the terminal current of a DC source or load and is present in all DC systems. Fundamentally, the in-rush current can occur when a DC source or load connects to another DC source or load due to the initial voltage difference between their filter capacitances. The magnitude of the in-rush current depends on the voltage difference, filter capacitance size, series inductive elements, and the contact resistance. If not managed, in-rush currents may lead to equipment damage and voltage droops, which can interfere with the converter control algorithm. Leveraging the fact that in-rush currents occur during physical connection of different DC sources or loads, in-rush current limitation techniques such as precharge circuits and soft-start routines have been developed. However, from a fault detection perspective, differentiating between in-rush currents and an over current due to a component failure is the main challenge in over current detection in DC data centers. Similarly, the limited converter power rating complicates over current detection due to the faults. Because of the finite power rating of the converters in a DC bus, a fault current due to a component failure is limited by the rated converter power. A fault current may also result in reduced voltage in the DC bus. Especially in data centers where the instantaneous load current changes dynamically, differentiating between the full load current and fault current requires careful analysis and tuning.

Redundancy is typically achieved through the incorporation of additional and separate power conversion stages, UPS and power distribution paths in the data center power infrastructure. The redundant components may be operated at all

Table 14.1 Tier certificate requirements summary

	Tier I	Tier II	Tier III	Tier IV
Redundancy level	N	$N+1$	$N+1$	N After any failure
Distribution path	1	1	1 Active, 1 Alternate	2 Active

times (e.g., each running at partial load to increase peak efficiency) but, strictly speaking, only needed to meet the power demand of the load in case of failures. Typical redundancy levels for data centers are $N+1$, $2N$, and $2N+1$, where N represents the exact number of power converters or UPS systems in parallel to meet the load demand. Uptime Institute defines Tier Classification levels for data centers depending on the redundancy level of the data center [43]. Tier I represents basic data center infrastructure without any redundancy. Tier II certification requires redundant power stages and UPS; however, the power distribution path is not redundant. Tier III certification requires the data center to have both redundant power stages and multiple independent power distribution paths, although only one distribution path is actively used at any time, while the other is for maintenance purposes. Tier IV certification requires both redundant power stages and multiple active power distribution paths configured to serve the entire data center under any infrastructure failure. Tier Certificate requirements are summarized in Table 14.1, and the details can be found in [43].

The DC power distribution in a data center simplifies redundant power architecture design. Since there is no synchronization and frequency control needed, replacing redundant power stages is straightforward, given in-rush currents are managed. In addition, if modular and redundant design is preferred, DC–DC converters, UPS, and even the centralized rectifiers can be replaced without disrupting the IT equipment operation. In addition to contributing to the overall reliability through improving redundancy level, such a feature intuitively leads to reduced equipment repair time in data centers, which further improves availability.

14.4.2 Back-up power

Utility power loss is considered as an expected scenario in data centers instead of a failure. Data centers employ UPS, backup generators, and recently FCs at specific points in the power delivery architecture to be able to compensate for both utility loss and power stage failure. Various possible configurations are depicted in Figure 14.4.

In the case of a utility loss, the facility level UPS must be able to provide enough back-up power to the critical loads until the back-up generators can initialize and output sufficient energy to the facility. On the other hand, UPS placement close to the load (IT equipment) can compensate for any component failure between the utility and the IT equipment. Uptime Tier Certification requires data centers to have on-site fuel storage for at least 12 h of utility loss [43].

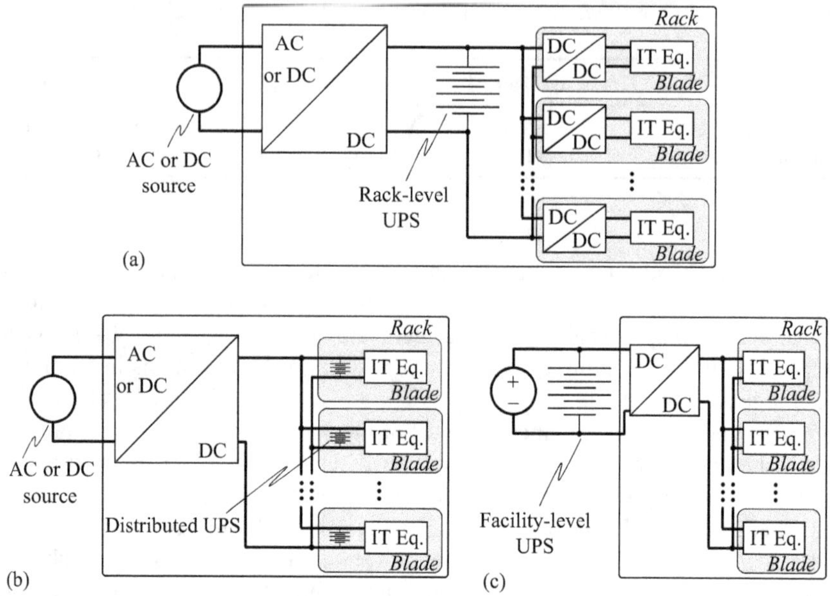

Figure 14.4 Various UPS configurations for data centers. (a) Rack-level UPS. UPS is located inside the racks and provides back-up power for multiple IT equipment. (b) Distributed UPS. Each IT equipment has a dedicated UPS. This configuration offers uninterruptible power in the case of any converter failure. (c) Facility-level UPS. UPS is located outside of the rack and provides back-up power for multiple racks

The high voltage DC bus is the prime location for UPS and backup generators to interface. Due to the DC voltage output of UPS, its connection to the high voltage DC distribution architecture is straightforward. However, any backup generator requires a rectifier stage to be able to supply power to the high voltage DC bus. Although this rectifier is an additional power conversion stage, since it only operates during a utility level power loss, the system-level efficiency and reliability are not considerably penalized. In addition, connection of backup generators to high voltage DC bus does not require synchronization and frequency control which are essential in AC architectures. Control methodologies are studied for the DC data centers that employ diesel generators and battery banks for utility power loss scenarios as early as 2008 [44].

14.5 Integration with other DC sources and loads

DC data centers have a clear advantage over their AC counterparts in interfacing renewable and/or distributed energy sources and modern auxiliary electric loads such as cooling and lighting to the power infrastructure.

14.5.1 Renewable and distributed energy sources

Photovoltaics (PVs) [45] and FC [46] are two energy sources for data centers commonly preferred by the industry. Since both PV and FC are inherently DC energy sources, their integration into the DC power infrastructure is simplified. While it is theoretically possible to design a PV array to directly interface with the high voltage bus, typically it is preferred to do so through a dedicated DC–DC converter that ensures that the PV system operates at its maximum power point, which varies with irradiation and temperature. Alternatively, multiple panel-embedded power converters [47,48] can be employed to provide this voltage matching, in addition to improved PV array performance.

Importantly, the integration ease reduces the power infrastructure control complexity since frequency control and phase synchronization are not required when DC sources and loads are connected. Power infrastructure control objectives thus can focus purely on maximizing available energy from renewable resources and minimizing energy drawn from the utility. A recent study investigated energy savings and carbon dioxide emission reductions when a solar power generation system is connected to a 380 V DC data center power distribution architecture [49].

While from a conceptual point, interconnection of a PV array (DC source) to a high voltage DC bus is simpler than having to go through intermediate AC voltages, one must carefully weigh the availability and cost of commercial power converters. Here, the rapidly growing solar PV industry has driven down the cost of high voltage DC–AC inverters due to economies of scale, whereas high-voltage DC–DC converters for solar PV are less prevalent. Consequently, the apparent advantage of DC data centers for renewable integration may not be so clear when considering the low-cost availability of inverters driven by the much larger solar PV industry.

14.5.2 Cooling

In addition to the computational data storage and networking load, cooling requires substantial electrical energy in data centers. Since the IT equipment mainly consists of digital circuits, combined with the losses in the power distribution architecture, almost all energy consumed in data centers results in heat that must be dissipated. In order to maintain safe operation of the IT equipment, temperature control is critical at all times. Therefore, the reliability discussion above also applies to the cooling system, and back-up power should be designed to support the cooling load in the case of a utility outage.

In order to provide sustainable heat removal from the IT equipment, a typical data center cooling infrastructure includes air conditioning equipment and chilled water system, which require pumps and fans. Such equipment involves electric motors which are typically controlled by adjustable speed drives (ASD) for maximum efficiency and performance. Similar to UPS, DC data centers require only an inverter stage for ASD, compared to AC-driven ASDs, which first perform rectification, followed by the adjustable frequency and voltage inverter. Overall, one would expect increased higher efficiency and reliability [50] owing to the reduced number of stages (and conversions) in the direct DC approach.

14.5.3 Lighting

Although the lighting load in data centers is small in comparison to the computational and cooling load, DC data centers present unique advantages if modern and efficient lighting technologies such as LEDs are used. Because of the DC nature of the modern lighting loads, power conversion stages like rectification, twice-line energy buffering, and PFC are not required when DC power distribution architecture is employed. As discussed in the efficiency section, reducing the number of power conversion stages reduces the power conversion losses for the lighting loads as well. Recent literature has demonstrated LED drivers with 380 V input for DC microgrids, featuring dimming and short circuit protection [51].

14.6 Installation

Early papers promoting high voltage DC distribution in data centers noted familiarity of the IT equipment with high voltage DC since the PFC stage of the PSU typically involves a 400 V DC intermediate voltage. However, the installation of high voltage DC distribution in data centers requires more than familiarity with 400 V DC. Major installation considerations include isolation, grounding, wiring, connectors, and total cost of ownership (TCO).

14.6.1 Isolation

Electrical (galvanic) isolation has been an essential part of data center power-delivery architectures. Provided by transformer usage, the electrical isolation offers to filter grid disturbances, harmonic currents, and electrical noise. Also, the electrical isolation limits ground loops and circulation of DC currents between the IT equipment and racks [2]. Because of these benefits, the electrical isolation through isolation transformers is a recommended practice by IEEE STD 1100 in AC power distribution architectures [52].

Inherited from the AC power delivery, the DC power delivery architecture in data centers also employs electrical isolation between the high voltage DC bus and the IT equipment input [11]. The electrical isolation in DC data centers can be achieved through incorporation of transformers in the DC to DC conversion stage as the high DC voltage is stepped down to 12 or 48 V for the IT equipment input. Recent developments in transformer design and wide bandgap devices [33,34] and optimization approaches [53] enabled high efficiency and compact converters for this DC to DC conversion stage. Also, in order to provide only electrical isolation (without voltage conversion), unity transformation ratio is demonstrated in DC to DC converters for data center applications at 400 [54] and 48 V [55].

Although electrical noise, ground loops and circulation of DC currents may still be present in DC data centers, enforcing the electrical isolation may not be the only and must-have practice to overcome such issues. For example, modern communication links typically provide inherent isolation, either through the medium itself (fiber optics), signal isolation transformers (Ethernet), or AC coupling capacitors (high speed serial links). Provided adequate system grounding, high safety may be obtainable without requiring inherent electrical isolation. An example of this transition is the case of grid-tied PV inverters, which until recently were required to have galvanic isolation in the United States, while transformer-less inverters were adopted earlier in Europe since they offered higher power efficiency and density at a lower cost [56]. Similar efforts may lead to the elimination of electrical isolation from data center power distribution architectures in the future.

14.6.2 Grounding

Proper grounding is an essential requirement in power infrastructure both for safety and signal integrity. Grounding contributes to the safety from electrical hazards by routing damaging currents away from the IT equipment and personnel. Although safety is the primary driver for choosing a grounding configuration, proper grounding also enables a common voltage reference for the overall electrical system in the data center including the power infrastructure and communication equipment.

A review of grounding concepts for DC data centers can be found in [57], and ETSI EN 301605 [13] defines the grounding standard for high voltage DC distribution in data centers. Both recommend high resistive mid-point ground as depicted in Figure 14.5 for DC data centers. This configuration keeps the fault current flow at a level that prevents both electric shock of the human body and fire hazard to the IT equipment.

14.6.3 Wiring

Wiring is an important consideration when transferring energy from sources to loads in data centers. Cable size for wiring in data centers is determined by considering maximum expected current, allowed voltage drop between the source and load, and target worst case power transfer efficiency. Moreover, electrical

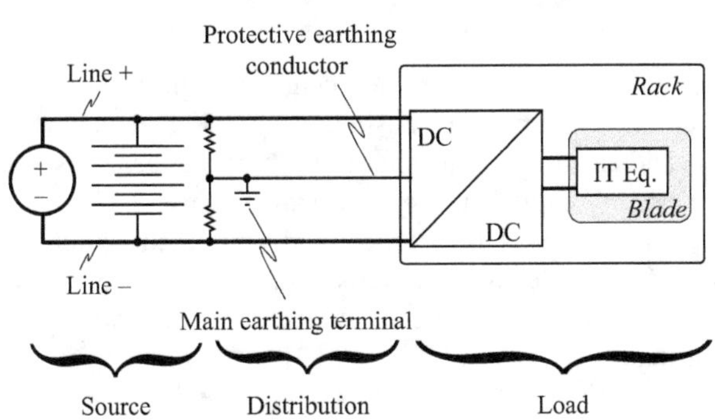

Figure 14.5 Mid-point high resistive grounding in DC data centers

insulation to protect from short circuit faults must be considered, with some differences between AC and DC rating. Because of the difference in the peak and idle power of the IT equipment, the maximum expected current in data centers results in oversizing the cables for the average operation. In addition, allowed voltage drop is restricted by standards and also depends on line regulation capabilities of the front-end power electronics converter of the load. On the other hand, although the wiring power efficiency has little effect on overall power infrastructure efficiency, the target worst-case power transfer efficiency may result in increasing the cable size. Here, the differences between AC and DC data center wiring are not viewed as significant, and DC wiring does not appear to a significant hindrance to DC data center adoption. A study of wiring practices for high voltage DC data centers can be found in [58].

14.6.4 Connectors

Connectors are another key element for transferring energy from the cables to the IT equipment or power conversion stages efficiently and reliably, which are the two most important aspects of data center power infrastructure. The efficient energy transfer through connectors requires minimal contact resistance; otherwise, substantial energy loss may occur which conflicts with the high-efficiency target of data centers as mentioned in Section 14.3. In addition, the connectors must also be able to sustain possible arcs and in-rush currents as the IT equipment or power converters are connected and disconnected during faults or maintenance. The reliable connectors are thus important to meet high uptime expectations in data centers as discussed in Section 14.4.

In [59], the basic characteristics of a 400 V 10 A connector are provided. Ongoing research on connectors for high voltage DC data centers focuses on arc and in-rush current limitation in plugs and sockets by employing permanent magnets [60] and embedded solid state electronics [61,62] while providing mechanical isolation. Here, it is likely that research and product development in other DC power transmission applications, such as solar PV and electric vehicle charging will help drive down costs and provide market acceptance.

14.6.5 Total cost of ownership

TCO is an important consideration in data center design because building a data center is a substantial investment for businesses. TCO extends beyond the cost of power equipment and electricity, but as far as power architecture is a concern, there are two main expenses: capital and operational expenses. The capital expenses include up-front costs such as installation and equipment purchases, while the operational expenses include the electricity and maintenance cost and therefore involve efficiency and reliability aspects. A detailed explanation of TCO beyond the power infrastructure can be found in [40,63].

Unfortunately, TCO is rather overlooked in the literature, as the primary motivation for most work is efficiency and reliability. A 2011 white paper quantifies the oversizing of the power infrastructure as the primary cost driver of data center TCO and suggests measuring the TCO on a per-rack basis [64]. A three phase rectifier for DC data centers is optimized for TCO in [30]. In [42], the cost advantage of high voltage DC data centers is reported as 15% less in capital cost and 36% less in lifetime cost, but the detailed analysis is missing. It should be noted that improvements in efficiency and reliability do not translate to business decisions unless the TCO analysis is incorporated into the benefits.

Critically, as TCO analysis requires accurate field data regarding reliability, it is likely that estimates of hypothetical designs without empirical results vary widely in their estimates. Hence, the TOC aspects of DC and AC data centers are difficult to assess at this time. Moreover, the major corporation that carefully track these metrics (e.g., Facebook, Amazon, Google, etc.) generally view these numbers as key competitive features of their respective designs and are unlikely to share them with researchers for dissemination.

14.7 Protection

The IT equipment and power infrastructure in data centers represent large investments; therefore, any damage due to the power system faults must be prevented by protection equipment such as fuses, relays, and circuit breakers.

Following the fault detection as mentioned in Section 14.4.1, the protection is activated to isolate the failed IT equipment or section of the power infrastructure from the rest of the system. Therefore, the location of the protection equipment in the power infrastructure is vital to enable isolation of any IT equipment and power stage whenever needed. The lack of a periodic zero-crossing of DC voltage and current complicates the protection equipment design for DC data centers. In addition, although sometimes overlooked, abrupt interruption of a DC current results in high current slew rates (di/dt), which induces high voltages in any parasitic inductive loop along the power path. This may endanger semiconductors and capacitor voltage ratings if not considered.

The periodic zero-crossing is inherent in AC distribution architectures as the polarity of voltage and current alternates 50 or 60 times per second; however, in DC distribution, the voltage and current are controlled to be constant values and do not naturally cross zero. Therefore, self-sustaining arcs are more likely to be present in DC power architectures. The basic operation principle of the protection equipment (i.e., moving electrical contacts away from each other when triggered) is similar in AC and DC systems, but with added challenges for DC protection. In order to successfully extinguish an arch in DC distribution architectures, the electrical contacts must move not only further away from each other but also faster than in AC distribution. Alternatively, protection equipment involving electronics to force the DC current toward zero can also be used. Details of the operating principles for DC protection equipment can be found in [65]. Various aspects of protection in DC microgrids, from modeling DC arcs [66] to implementation of hybrid (both electronic and mechanical) circuit breakers [67], have been investigated in the literature. We note that the availability of low-cost, standards-compliant DC protection circuitry is typically viewed as a major impediment to practical implementation to DC data centers.

14.8 Power quality

Power quality issues such as phase imbalance, frequency drifting, and harmonic currents at the multipliers of fundamental grid frequency are native to ac distribution architectures but are not present in the DC distribution architectures. However, harmonics due to specifically ASD presence for cooling loads can be severe in DC data centers. In addition, excessive use of filter capacitance to overcome electromagnetic interference in power electronics converters may result in increased in-rush currents, which reduces the power quality in DC data centers. A detailed discussion of such power quality issues in DC microgrids (with an example of DC data centers) can be found in [68].

14.9 Stability

The DC power distribution architectures in data centers can also present stability challenges due to rapid variation of the IT equipment workload. The IT equipment in data centers interfaces with the power distribution system through point-of-load converters and thus acts as constant power loads. Dynamic utilization of the IT equipment at different workloads results in rapid transitions between idle and peak power consumption, resulting in challenging the power electronics converter filter and controller design for stable operation. In addition, since high efficiency is the primary objective in DC data centers, every power electronics converter is designed to be highly efficient which results in little damping in the power flow path. Therefore, the stability of DC distribution architectures is an ongoing research topic in the literature. A comprehensive examination of stability issues in DC microgrids with a focus on constant power loads can be found in [69]. Recently, DC electric springs have been proposed as an active suspension system to stabilize the DC bus in microgrids. Theoretical analysis and an experimental proof-of-concept in a 48 V DC electric spring implementation can be found in [70]. In addition, another stability analysis with a focus on droop-controlled DC microgrids can be found in [71].

14.10 Existing high voltage DC data centers

A survey of the key data centers or telecom facilities that use high voltage DC power distribution in 2014 is provided in [72]. In this section, different types of the high voltage DC installations ranging from laboratory prototypes to extensions of existing data centers are given as examples. Note that these DC data centers are not completely configured as DC, and some of them are even limited to a few rows of racks. Although this means the high voltage DC distribution is not throughout the entire data center facility, such installations represent a significant step toward

Table 14.2 Summary of example DC data centers

Name	Location	Power	Purpose	Date
Duke Energy	Charlotte, NC, USA	100 kW	Demonstration	2011
NTT Cloud	Tokyo, Japan	500 kW	Cloud	02/2014
Green	Zurich, Switzerland	1 MW	Expansion	05/2012
Hikari	Austin, TX, USA	208 kW	Supercomputer	08/2016
Level 3	Broomfield, CO, USA	120 kW	Lab evaluation	06/2016

completely DC powered data centers. Table 14.2 summarizes the examples given in this section.

- *Duke Energy Data Center, Charlotte, NC, USA*
  Duke Energy Data Center in Charlotte, NC is one of the earliest demonstrations of distributing 380 V DC to the IT equipment racks in the United States. In [17], 15% reduction in energy consumption compared to the conventional AC power distribution system is reported. The system includes AC to DC and DC to DC converters from Delta Products Corporation for the power infrastructure, and DL 385 G7 servers from Hewlett-Packard (HP), Power795 servers from IBM, and Clarion CX4-960 data storage units from EMC as the IT equipment. Total compute power was 100 kW, and the power delivery architecture achieved 72.04% efficiency in 2011 [17].
- *NTT Cloud, Tokyo, Japan*
  NTT facilities has installed a 500 kW system with 380 V high voltage DC distribution for 4-MW scale cloud facility in February 2014. In this system, $N+1$ redundant rectifier modules (each rated at 100 kW) provide the high voltage DC bus to 80 server racks. The system spans 250 m^2, has 5 min of backup power provided by 380 V batteries, and uses mid-point high impedance grounding. Further details of the implementation including safety and protection practices can be found in [72].
- *Green Datacenter, Zurich, Switzerland*
  Collaborating with ABB, Green Datacenter installed a 1 MW 380 V DC power distribution architecture as an extension to an existing data center in May 2012 in Zurich, Switzerland [73]. They reported 15% investment cost reduction in comparison to an AC architecture. The overall architecture, including both power distribution and cooling, is estimated to provide up to 20% energy savings. The IT equipment used is compatible with high voltage DC distribution and includes HP X1800 G2 Network Storage System, HP DL385 servers, and the HP BladeSystem c3000.
- *Texas Advanced Computing Center, Austin, TX, USA*
  A supercomputer called "Hikari", which consists of 432 HP Enterprise Apollo 8000 XL730f servers, has been installed by NTT at Texas Advanced Computing Center (TACC) in August 2016. The Hikari supercomputer is energized through to a 380 V DC bus which can be supplied by the solar panels

at the TACC parking lot up to 208 kW. The 380 V DC bus can also be connected to the AC utility using rectifiers to be functional at night [74].

- *Level 3, Broomfield, CO, USA*
  Level 3, a US-based service provider, has recently published their transition from 48 to 380 V DC distribution [75]. The laboratory implementation is considered as a proof-of-concept for 380 V DC distribution. The installed capacity is 60 kW, although extension up to 288 kW is mentioned. Currently, two 60 kW rectifiers supply the high voltage DC bus at 380 V, followed by two 30 kW 380 V to 48 V DC to DC converters. In addition, the current implementation has 15–30 min of backup power provided by 380 V batteries, and preferred +/−190 V with mid-point high impedance grounding as recommended by the ETSI standard. The system was put in service in June 2016.

14.11 Key obstacles to widespread adoption of DC data centers

Despite some key benefits associated with DC distribution over AC distribution in data centers, commercial adaptation of this technology has been slow. Below, we provide our view of some of the key obstacles for market acceptance.

14.11.1 Overly optimistic claims

As with most emerging technologies, the proponents of DC power distribution in data centers overstated the advantages compared to the incumbent (AC) solution. In particular, early studies that highlighted the benefit of no double conversion (AC–DC to UPS to DC–AC) compared a hypothetical best case DC distribution system to a below average AC system, especially in terms of efficiencies. Moreover, early studies also failed to account for the fact that AC–DC and DC–AC converter efficiencies (and power densities) have continued to improve, as is evident by recent commercial and research prototypes [76].

14.11.2 Emergence of rack-level UPS

A key selling feature of the DC distribution data center in the early days was the elimination of the double conversion penalty associated with first AC to DC then DC to AC conversion for the central UPS. While this remains a significant advantage (both efficiency and reliability) compared to central UPS AC distribution systems, today's design are moving away from such system toward a single-conversion AC distribution system, with rack-level UPS [e.g., Figure 14.1(c)] that does not suffer from double-conversion in the UPS stage. While there are some benefits in terms of maintenance and costs associated with a central UPS system, the rack-level UPS can also help mitigate power distribution faults in the data centers, as each rack can operate directly from its own UPS.

14.11.3 Protection and DC circuit breakers

One area where the DC data center has a distinct disadvantage is in the protection, circuit breakers, and fuses. Owing to the increased difficulty of extinguishing any arcing in DC systems, the circuit breakers and other protective equipment will necessarily be bulkier and more expensive. In addition to the increased cost associated with more advanced circuit breaker technology, the cost for DC breakers is also higher due to the smaller market size than that for AC breakers.

14.11.4 Incumbent cost and familiarity advantages

Arguably the biggest obstacle to overcome for DC data centers is the fact that they represent a new entry, with many unknowns. Generally, in order to replace any incumbent technology, the new technology must provide sufficient advantages compared to conventional techniques to warrant the increased risk with an untested design. This is particularly difficult in data center applications, where stringent up-time requirements call for a conservative approach to introducing new components. Here, one must compare the DC power distribution to an AC power distribution with rack-level UPS, which represents arguably the best AC power distribution solution. In this comparison, the DC solution must then promise sufficiently large improvements (in efficiency, cost, up-time, etc.) to warrant the transition.

Moreover, in this case, the AC power distribution can leverage significant cost and scaling advantages associated with being widely adopted in other markets (commercial building power distribution, factories, etc.). Solutions for DC data centers that require custom hardware (e.g., protection circuitry, power converters, etc.) that do not have widespread use in other markets are unlikely to realize the low cost of AC distribution systems.

14.12 Conclusion

Data centers consume a significant amount of electric energy; therefore, their power delivery architectures require critical consideration for their sustainable growth. In this chapter, we discussed the efficiency, reliability, integration with renewable resources, installation, and protection in the DC power delivery architecture in data centers. We also provided a short list of some existing DC data center installations and our thoughts on the critical obstacles for widespread acceptance of this architecture. Although high voltage DC distribution throughout the entire data center facility is still not widely utilized, we expect that the performance outcome of the early adopters of this architecture will promote the use of entirely DC power distributed data centers.

References

[1] Shehabi A, Smith SJ, Sartor DA, *et al.* United states data center energy usage report. Ernest Orlando Lawrence Berkeley National Laboratory; 2016. LBNL-1005775.

[2] Rasmussen N. AC vs. DC power distribution for data centers [White Paper]; 2011. Available from: http://www.apc.com/salestools/SADE-5TNRLG/SADE-5TNRLG_R6_EN.pdf.

[3] Koomey JG. Estimating total power consumption by servers in the US and the world; 2007. Available from: http://www-sop.inria.fr/mascotte/Contrats/DIMAGREEN/wiki/uploads/Main/svrpwrusecompletefinal.pdf.

[4] Brown R, Masanet E, Nordman B, *et al*. Report to congress on server and data center energy efficiency: public law 109-431; 2007. Available from: http://eta-publications.lbl.gov/sites/default/files/pdf_3.pdf.

[5] Ton M, Fortenbery B, Tschudi W. DC power for improved data center efficiency; 2008. Available from: http://www.chip2grid.com/docs/DCDemo FinalReport.pdf.

[6] Yamashita T, Muroyama S, Furubo S, Ohtsu S. 270 V dc system – a highly efficient and reliable power supply system for both telecom and datacom systems. In: Proc. IEEE Int. Telecommun. Energy Conf.; 1999. p. 1–6.

[7] Carlsson U, Flodin M, Akerlund J, Ericsson A. Powering the internet – broadband equipment in all facilities – the need for a 300 V dc powering and universal current option. In: Proc. IEEE Int. Telecommun. Energy Conf.; 2003. p. 164–169.

[8] Sannino A, Postiglione G, Bollen MHJ. Feasibility of a dc network for commercial facilities. IEEE Trans Ind Appl. 2003 Sep;39(5):1499–1507.

[9] Bodi F, Lim EH. 380/400 V dc powering option. In: Proc. IEEE Int. Telecommun. Energy Conf.; 2011. p. 1–8.

[10] Salato M, Zolj A, Becker DJ, Sonnenberg BJ. Power system architectures for 380V dc distribution in telecom datacenters. In: Proc. IEEE Int. Telecommun. Energy Conf.; 2012. p. 1–7.

[11] Pratt A, Kumar P, Aldridge TV. Evaluation of 400V dc distribution in telco and data centers to improve energy efficiency. In: Proc. IEEE Int. Telecommun. Energy Conf.; 2007. p. 32–39.

[12] Final draft ETSI EN 300 132-3-1 environmental engineering (EE); power supply interface at the input to telecommunications and datacom (ICT) equipment; part 3: operated by rectified current source, alternating current source or direct current source up to 400 V; sub-part 1: direct current source up to 400 V; 2011. Available from: https://www.etsi.org/deliver/etsi_en/300100_300199/3001320301/02.01.01_40/en_3001320301v020101o.pdf.

[13] Draft ETSI EN 301 605 environmental engineering (EE); earthing and bonding of 400 VDC data and telecom (ICT) equipment; 2013. Available from: https://www.etsi.org/deliver/etsi_en/301600_301699/301605/01.01.01_20/en_301605v010101a.pdf.

[14] IEC 61643-21 low voltage surge protective devices – part 21: surge protective devices connected to telecommunications and signalling networks – performance requirements and testing methods; 2012. Available from: https://webstore.iec.ch/preview/info_iec61643-21%7Bed1.2%7Db.pdf.

[15] Aldridge T, Pratt A, Kumar P, Dupy D, AlLee G. Evaluating 400 V direct-current for data centers a case study comparing 400 Vdc with 480–208 Vac power distribution for energy efficiency and other benefits [White Paper]. Available from: https://blogs.intel.com/wp-content/mt-content/com/research/Direct%20400Vdc%20White%20Paper.pdf.

[16] Sithimolada V, Sauer PW. Facility-level dc vs. typical ac distribution for data centers: a comparative reliability study. In: TENCON - IEEE Region 10 Conf.; 2010. p. 2102–2107.

[17] Kintner D. Duke energy – EPRI dc powered data center demonstration executive summary; 2011. Available from: http://portal.emergealliance.org/DesktopModules/Inventure_Document/FileDownload.aspx?ContentID=20674.

[18] Rasmussen N, Spitaels J. A quantitative comparison of high efficiency ac vs. dc power distribution for data centers [White Paper]; 2012. Available from: http://www.apc.com/salestools/NRAN-76TTJY/NRAN-76TTJY_R4_EN.pdf.

[19] Szpek M, Sonnenberg BJ, Lisy SM. 400 VDC distribution architectures for central offices and data centers. In: Proc. IEEE Int. Telecommun. Energy Conf.; 2014. p. 1–6.

[20] Rasmussen N. Review of four studies comparing efficiency of ac and dc distribution for data centers [White Paper]; 2012. Available from: http://www.apc.com/salestools/VAVR-8Q7K7N/VAVR-8Q7K7N_R0_EN.pdf.

[21] Reusch D, Glaser J. Dc–dc converter handbook – a supplement to GaN transistors for efficient power conversion. Power Conversion Publications; 2015.

[22] Ye Z, Lei Y, Pilawa-Podgurski RCN. A resonant switched capacitor based 4-to-1 bus converter achieving 2180 W/in3 power density and 98.9% peak efficiency. In: Proc. 33rd Annu. IEEE Appl. Power Electron. Conf. Expo.; 2018. p. 1–6.

[23] Reusch D, Lee FC, Gilham D, Su Y. Optimization of a high density gallium nitride based non-isolated point of load module. In: Proc. IEEE Energy Conversion Congr. Expo.; 2012. p. 2914–2920.

[24] Rhea BK, Jenkins LL, Abell WE, *et al.* A 12 to 1 V five phase interleaving GaN POL converter for high current low voltage applications. In: Proc. IEEE Workshop Wide Bandgap Power Devices and Appl.; 2014. p. 155–158.

[25] Shenoy PS, Amaro M, Morroni J, Freeman D. Comparison of a buck converter and a series capacitor buck converter for high-frequency, high-conversion-ratio voltage regulators. IEEE Trans Power Electron. 2016 Oct;31(10):7006–7015.

[26] Krein PT. Data center challenges and their power electronics. CPSS Trans Power Electron Appl. 2017;2(1):39–46.

[27] Krein PT, Balog RS, Mirjafari M. Minimum energy and capacitance requirements for single-phase inverters and rectifiers using a ripple port. IEEE Trans Power Electron. 2012 Nov;27(11):4690–4698.

[28] Detailed inverter specifications, testing procedure, and technical approach and testing application requirements for the little box challenge, 2015. Available from: https://littleboxchallenge.com/pdf/LBC-InverterRequirements-20150717.pdf.

[29] Qin S, Lei Y, Barth C, Liu WC, Pilawa-Podgurski RCN. A high power density series-stacked energy buffer for power pulsation decoupling in single-phase converters. IEEE Trans Power Electron. 2017 Jun;32(6):4905–4924.

[30] Schrittwieser L, Kolar JW, Soeiro TB. 99% Efficient three-phase buck-type SiC MOSFET PFC rectifier minimizing life cycle cost in DC data centers. CPSS Trans Power Electron Appl. 2017;2(1):47–58.

[31] Xu F, Guo B, Tolbert LM, Wang F, Blalock BJ. An all-SiC three-phase buck rectifier for high-efficiency data center power supplies. IEEE Trans Ind Appl. 2013 Nov;49(6):2662–2673.

[32] Biela J, Badstuebner U, Kolar JW. Design of a 5-kW, 1-U, 10-kW/dm3 resonant dc–dc converter for telecom applications. IEEE Trans Power Electron. 2009 Jul;24(7):1701–1710.

[33] Huang D, Ji S, Lee FC. LLC resonant converter with matrix transformer. IEEE Trans Power Electron. 2014 Aug;29(8):4339–4347.

[34] Seeman MD, Bahl SR, Anderson DI, Shah GA. Advantages of GaN in a high-voltage resonant LLC converter. In: Proc. 29th Annu. IEEE Appl. Power Electron. Conf. Expo.; 2014. p. 476–483.

[35] McClurg J, Pilawa-Podgurski RCN, Shenoy PS. A series-stacked architecture for high-efficiency data center power delivery. In: Proc. IEEE Energy Conversion Congr. Expo.; 2014. p. 170–177.

[36] Candan E, Shenoy PS, Pilawa-Podgurski RCN. A series-stacked power delivery architecture with isolated differential power conversion for data centers. IEEE Trans Power Electron. 2016 May;31(5):3690–3703.

[37] McClurg J, Zhang Y, Wheeler J, Pilawa-Podgurski R. Re-thinking data center power delivery: regulating series-connected voltage domains in software. In: Proc. IEEE Power Energy Conf. Illinois; 2013. p. 147–154.

[38] Candan E, Heeger D, Shenoy P, Pilawa-Podgurski R. Hot-swapping analysis and implementation of series-stacked server power delivery architectures. IEEE Trans Power Electron. 2017;32(10):8071–8088.

[39] Candan E, Shenoy PS, Pilawa-Podgurski RCN. Unregulated bus operation of server-to-virtual bus differential power processing for data centers. In: Proc. 32nd Annu. IEEE Appl. Power Electron. Conf. Expo.; 2017.

[40] Barroso LA, Clidaras J, Hölzle U. The datacenter as a computer: an introduction to the design of warehouse-scale machines. 2nd ed.; San Rafael, CA: Morgan & Claypool Publishers, 2013.

[41] Maiquet D, Kervarrec G. New flexible powering architecture for integrated service operators. In: Proc. IEEE Int. Telecommun. Energy Conf.; 2005. p. 575–580.

[42] AlLee G, Tschudi W. Edison redux: 380 Vdc brings reliability and efficiency to sustainable data centers. IEEE Power Energy Mag. 2012 Nov;10(6):50–59.

[43] Tier standard: topology; 2012. Available from: https://uptimeinstitute.com/publications/asset/tier-standard-topology.

[44] Salomonsson D, Soder L, Sannino A. An adaptive control system for a dc microgrid for data centers. IEEE Trans Ind Appl. 2008 Nov;44(6):1910–1917.

[45] Sverdlik Y. Apple reaches 100% renewable energy across all data centers; 2013. Available from: http://www.datacenterdynamics.com/content-tracks/design-build/apple-reaches-100-renewable-energy-across-all-data-centers/74708.fullarticle.

[46] Sverdlik Y. eBay's utah data center offers a glimpse into the future; 2013. Available from: http://www.datacenterdynamics.com/news/ebays-utah-data-center-offers-a-glimpse-into-the-future/83495.fullarticle.

[47] Pilawa-Podgurski RCN, Perreault DJ. Submodule integrated distributed maximum power point tracking for solar photovoltaic applications. IEEE Trans Power Electron. 2013 Jun;28(6):2957–2967.

[48] Qin S, Cady ST, Domínguez-García AD, Pilawa-Podgurski RCN. A distributed approach to maximum power point tracking for photovoltaic submodule differential power processing. IEEE Trans Power Electron. 2015 Apr;30(4):2024–2040.

[49] Yajima H, Usui K, Hayashi T, *et al.* Energy-saving effects of super computers by using on-site solar power and direct HVDC feeding systems. In: Proc. IEEE Int. Telecommun. Energy Conf.; 2016. p. 1–4.

[50] Kwasinski A. Analysis of electric power architectures to improve availability and efficiency of air conditioning systems. In: Proc. IEEE Int. Telecommun. Energy Conf.; 2008. p. 1–8.

[51] Alonso JM, Perdigão M, Costa MAD, Martínez G, Osorio R. Analysis and experiments on a single-inductor half-bridge LED driver with magnetic control. IEEE Trans Power Electron. 2017 Dec;32(12):9179–9190.

[52] 1100 IEEE recommended practice for powering and grounding electronic equipment; 2015. Available from: https://ieeexplore.ieee.org/stamp/stamp.jsp?tp=&arnumber=1638205.

[53] Badstuebner U, Biela J, Kolar JW. Design of an 99%-efficient, 5 kW, phase-shift PWM dc–dc converter for telecom applications. In: Proc. 25th Annu. IEEE Appl. Power Electron. Conf. Expo.; 2010. p. 773–780.

[54] Simanjorang R, Yamaguchi H, Ohashi H, *et al.* High-efficiency high-power dc-dc converter for energy and space saving of power-supply system in a data center. In: Proc. 26th Annu. IEEE Appl. Power Electron. Conf. Expo.; 2011. p. 600–605.

[55] Vicor BCM bus converter BCM48Bx480y300A00 isolated fixed ratio dc-dc converter; 2016. Available from: http://cdn.vicorpower.com/documents/datasheets/BCM48B_480_300A00.pdf.

[56] Araujo SV, Zacharias P, Sahan B. Novel grid-connected non-isolated converters for photovoltaic systems with grounded generator. In: Proc. IEEE Power Electron. Specialists Conf.; 2008. p. 58–65.

[57] Hirose K, Tanaka T, Babasaki T, *et al.* Grounding concept considerations and recommendations for 400 VDC distribution system. In: Proc. IEEE Int. Telecommun. Energy Conf.; 2011. p. 1–8.

[58] Tanaka T, Hirose K, Marquet D, Sonnenberg B, Szpek M. Analysis of wiring design for 380-VDC power distribution system at telecommunication sites. In: Proc. IEEE Int. Telecommun. Energy Conf.; 2012. p. 1–5.

[59] Iino T, Hirose K, Noritake M, Nakamura A, Kiryu K, Sekikawa J. Characteristics of 400 V dc plug and socket-outlet for dc distribution systems. In: Proc. Int. Conf. Renewable Energy Res. Appl.; 2012. p. 1–6.

[60] Iino T, Hirose K, Noritake M, Yuba T, Miyazawa H, Sekikawa J. Development of 400 Vdc plug and socket-outlet for dc distribution system. In: Proc. IEEE Int. Telecommun. Energy Conf.; 2012. p. 1–6.

[61] Tan K, Huang AQ, Martin A. Development of solid state arc-free socket for dc distribution system. In: Proc. 29th Annu. IEEE Appl. Power Electron. Conf. Expo.; 2014. p. 2300–2305.

[62] Tan K, Peng C, Liu P, Song X, Huang AQ. Zero standby power high efficiency hot plugging outlet for 380 VDC power delivery system. In: Proc. 31st Annu. IEEE Appl. Power Electron. Conf. Expo.; 2016. p. 132–137.

[63] Koomey J. A simple model for determining true total cost of ownership for data centers [White Paper]; 2007. Available from: https://www.missioncriticalmagazine.com/ext/resources/MC/Home/Files/PDFs/(TUI3011B)SimpleModelDetermingTrueTCO.pdf.

[64] Rasmussen N. Determining total cost of ownership for data center and network room infrastructure [White paper]; 2011. Available from: http://www.apc.com/salestools/CMRP-5T9PQG/CMRP-5T9PQG_R4_EN.pdf.

[65] Salomonsson D, Soder L, Sannino A. Protection of low-voltage dc microgrids. IEEE Trans Power Delivery. 2009 Jul;24(3):1045–1053.

[66] Uriarte FM, Gattozzi AL, Herbst JD, *et al.* A dc arc model for series faults in low voltage microgrids. IEEE Trans Smart Grid. 2012 Dec;3(4):2063–2070.

[67] Meyer JM, Rufer A. A dc hybrid circuit breaker with ultra-fast contact opening and integrated gate-commutated thyristors (IGCTs). IEEE Trans Power Delivery. 2006 Apr;21(2):646–651.

[68] Whaite S, Grainger B, Kwasinski A. Power quality in dc power distribution systems and microgrids. Energies. 2015 May;8(5):4378–4399. Available from: http://dx.doi.org/10.3390/en8054378.

[69] Kwasinski A, Onwuchekwa CN. Dynamic behavior and stabilization of dc microgrids with instantaneous constant-power loads. IEEE Trans Power Electron. 2011 Mar;26(3):822–834.

[70] Mok KT, Wang MH, Tan SC, Hui SYR. Dc electric springs – a technology for stabilizing dc power distribution systems. IEEE Trans Power Electron. 2017 Feb;32(2):1088–1105.

[71] Anand S, Fernandes BG. Reduced-order model and stability analysis of low-voltage dc microgrid. IEEE Trans Ind Electron. 2013 Nov;60(11):5040–5049.

[72] Inamori J, Hoshi H, Tanaka T, Babasaki T, Hirose K. 380-VDC power distribution system for 4-MW-scale cloud facility. In: Proc. IEEE Int. Telecommun. Energy Conf.; 2014. p. 1–8.

[73] Schmidt T, Ligi A, Zangger E, Monachesi A. World's most powerful dc data center online [Press Release]. Available from: http://www04.abb.com/global/CNABB/CNABB050.NSF!OpenDatabase&db=/global/cnabb/cnabb054.nsf&v=BF6&e=us&url=/global/seitp/seitp202.nsf/0/C30D32AC816B2D5BC1257A140026A31F!OpenDocument.

[74] Salazar J. New Hikari supercomputer starts solar HVDC; 2016. Online. Available from: https://www.tacc.utexas.edu/-/new-hikari-supercomputer-starts-solar-hvdc.

[75] Ambriz R, Kania M. A service provider's decision to move from 48 V to 380 V powering: the problem statement, technical assessment, financial analysis and practical implementation plan. In: Proc. IEEE Int. Telecommun. Energy Conf.; 2016. p. 1–7.

[76] Lei Y, Barth C, Qin S, *et al.* A 2-kW single-phase seven-level flying capacitor multilevel inverter with an active energy buffer. IEEE Trans Power Electron. 2017 Nov;32(11):8570–8581.

Chapter 15
DC microgrid in residential buildings

Rajeev Kumar Chauhan[1]*, Francisco Gonzalez-Longatt*[2]*, Bharat Singh Rajpurohit*[1]*, and Sri Nivas Singh*[3]

15.1 Introduction

A microgrid is an innovative control and management architecture at the distribution level, which makes it easy to implement smart grid techniques at power distribution level [1]. Depleting of fossil fuels, reducing the emission of greenhouse gases and increasing energy demand are main factors responsible for the growing penetration of photovoltaic (PV) generators to power distribution grids [2]. The distribution generation (DG) has the potential to provide ancillary services under these circumstances. For example, they may have many functions such as instantaneous power reserve, emergency supply and peak power saving. The PV system with DG is a good solution for electrification to poorly grid connected or isolated area. This concept easily works in both the AC and DC microgrid system [3]. However, the implementation of the DC microgrid simplifies and provides the opportunity to integrate the renewable energy resource (RES) such as PV system, and fuel cells (FC) without converting into AC [4]. These are intrinsically DC power source at higher efficiency. Additionally, the loads used in buildings may be DC or AC load which can be converted into DC. The rapid developments in the DC technologies, the DC loads, are more energy efficient than the AC loads [5]. The DC output power of PV systems can be used directly without conversion into the low voltage direct current (LVDC) distribution system [6].

There are no inductive, capacitive and skin effects in the DC system. The DC system has a better voltage regulation than the AC system because there is no inductive effect on steady state [3]. Additionally, the power loss due to charging and discharging of the capacitor is eliminated in the DC system. So overall, there is less power loss in the DC system [7]. Due to the absence of the skin effect, the entire cross-section area of the line conductor is used in the DC system. It means a thinner conductor could be used for DC systems, and it reduces the line conductor weight. Additionally, as the cross-sectional area is inversely proportional to the

[1]School of Computing and Electrical Engineering, Indian Institute of Technology Mandi, India
[2]Department of Electrical Engineering, University of Loughborough, United Kingdom
[3]Department of Electrical Engineering, Indian Institute of Technology Kanpur, India

conductor resistance, the DC system has less line resistance than the AC system [7]. In this way, the direct current distribution system improves the system efficiency. Additionally, the local utilisation of the energy reduces the transmission and distribution losses and may reduce the shortage of electricity. In this chapter, analysis and comparison between AC and DC microgrid in residential buildings have been done based on appliances, converters and their power losses in both systems. The layouts for LVDC distribution network have been discussed. Both unipolar and bipolar layouts of LVDC system have been discussed. Two microgrid system configurations have been discussed: AC residential building [i.e. AC distribution system (ACDS) with DC appliances] and DC residential building (i.e. distribution network with DC appliances).

15.2 Conceptualisation: DC microgrids in buildings

A microgrid refers to a power distribution system integrated with distributed energy resources (DERs) and controllable loads, which can either operate with the main grid (i.e. grid-connected mode) or use the DERs to supply the loads without the main grid (i.e. islanded mode) [6]. It can be viewed as a small-scale grid, and it is widely considered as one of the key components to integrate renewable DERs and save transmission losses for efficiency.

In a conventional power system, the power flow is a single directional from bulk generation, e.g. a power plant, to load via a transmission and distribution network. As more DERs (such as roof-top PV system, small wind turbines – WTs, battery system, etc.) are becoming available on the distribution side, the bi-directional power flow is now possible, and it makes the conventional power distribution system into a microgrid which can use only the DERs as the power supply without the main grid (i.e. islanded mode) or even sell back its surplus power to the main grid. The size of a microgrid may vary depending on the application. For example, the concept of microgrid can be implemented as smart home, smart building, even a smart campus. Figure 15.1 shows a microgrid example consisting of PVs, WTs, electrical vehicle, batteries and loads. Two-way communications are available between the local controllers and the microgrid central controller.

15.3 Classification of microgrids

The microgrids can be classified into three categories based on the type of supply system used by the distribution grid.

15.3.1 AC microgrid system

A typical structure of an AC microgrid system interconnected with medium voltage (MW) system at the point of common coupling (PCC) is shown in Figure 15.2. The main system might be an AC or DC bulk system. The DG units and energy storage system (ESS) are connected at some points within the distribution networks. Part of

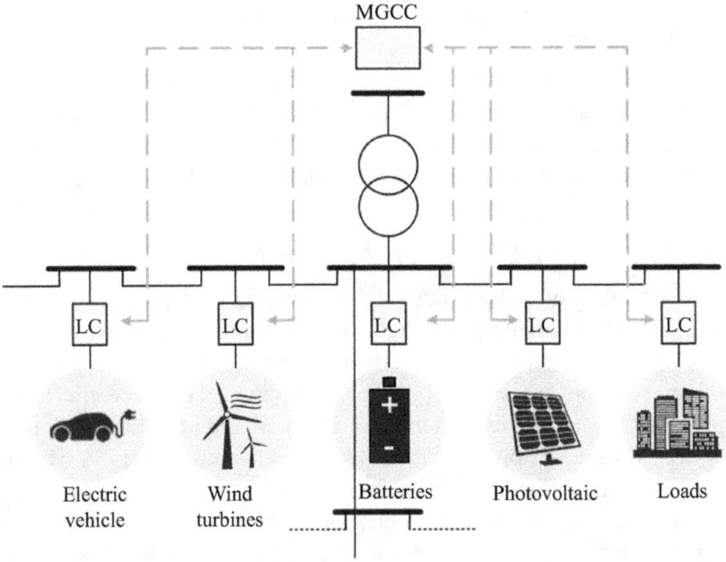

Figure 15.1 Conceptual layout of a microgrid including microgrid central controller (MGCC) and local controllers (LCs)

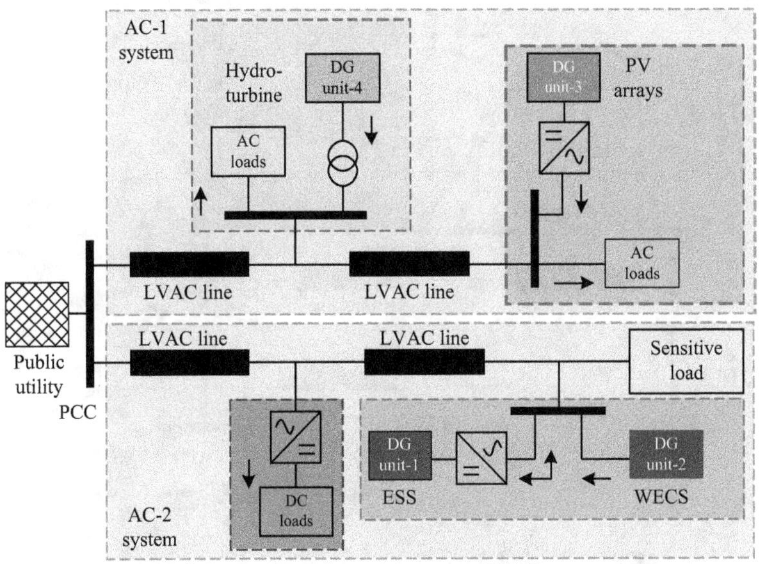

Figure 15.2 The general structure of AC microgrid with DG units and mixed types of loads

the network consisting of the DG units and load circuits can form a small isolated AC electric power system, i.e. an 'AC microgrid'. During normal operating conditions, the two networks are interconnected at the PCC, while the loads are supplied from the local sources (e.g. the RES-based DG units) and if necessary from the utility. If the load demand power is less than the power produced by DG units, excess power can be exported to the public utility (PU).

15.3.2 Hybrid AC–DC microgrid system

The general structure of a hybrid AC–DC microgrid is depicted in Figure 15.3. After staying on AC technology in the area of electric power supply, the DC power systems join it due to technological advancement in DC technology for power conversion, generation, transmission, and consumption. However, there are many challenges in DC technologies. Therefore, DC technologies should be integrated into the power system by applying the algorithms at some places [8]. The scope to explore the DC technologies with its specific advantages is the microgrid. The hybrid AC–DC microgrids facilitate benefits of both the technologies, and these are having the integration of AC technology with the DC technology. In a leading hybrid AC–DC microgrid system, the AC and DC buses are connected through interlinking, bi-directional converters. However, this interlinking creates a stability issue and requires control algorithms to maintain the power quality. Microgrids,

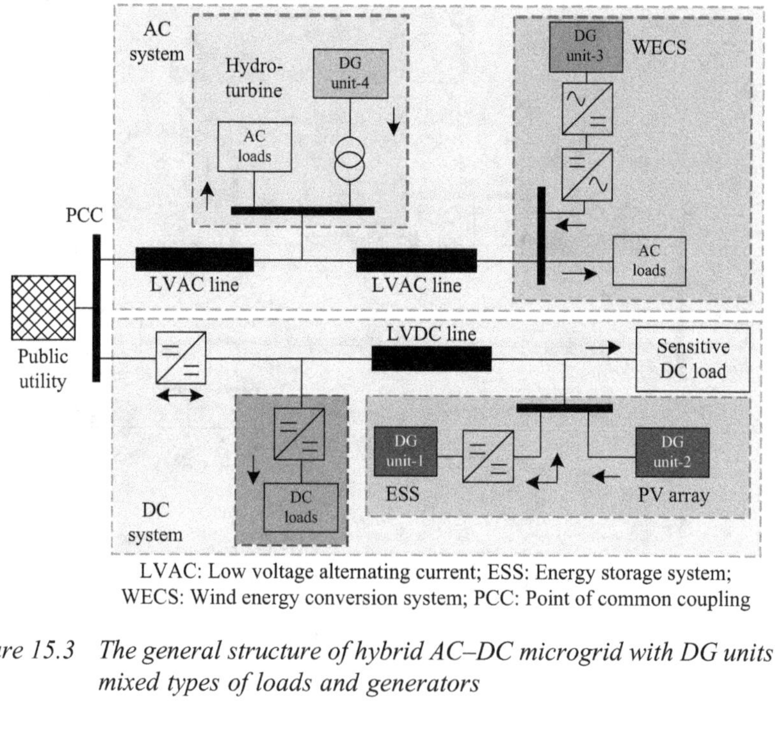

LVAC: Low voltage alternating current; ESS: Energy storage system; WECS: Wind energy conversion system; PCC: Point of common coupling

Figure 15.3 The general structure of hybrid AC–DC microgrid with DG units and mixed types of loads and generators

which are having different types of sources, and loads are the type of AC–DC systems [9].

15.3.3 DC microgrid system

The traditional electric power system was designed to move the central station AC power, via high-voltage AC transmission lines and lower voltage distribution lines to households and businesses that use the power in incandescent lights, AC motors and other AC equipment for heating and cooling. Meanwhile, the DC power system has been used in industrial power distribution systems, telecommunication infrastructures and point-to-point transmissions over long distances or via sea cables and for interconnecting AC grids with different frequencies. Power electronics devices dominate today's consumer equipment and tomorrow's DG units [10]. These devices (such as computers, fluorescent lights, variable speed drives, households, businesses, industrial appliances and equipment) need DC power for their operation. However, all these DC devices require conversion of the available AC power into DC for use, and the majority of these conversion stages typically use inefficient rectifiers. Moreover, the power from DC-based DG units must be converted into AC to tie with the existing AC electric network, only later to be converted to DC for many end users. These DC–AC–DC power conversion stages result in substantial energy losses. Using the positive experiences in the high voltage direct current (HVDC) operation and the advance in power electronics technology, interests to use effective solutions have increased. Figure 15.4 shows typical DC

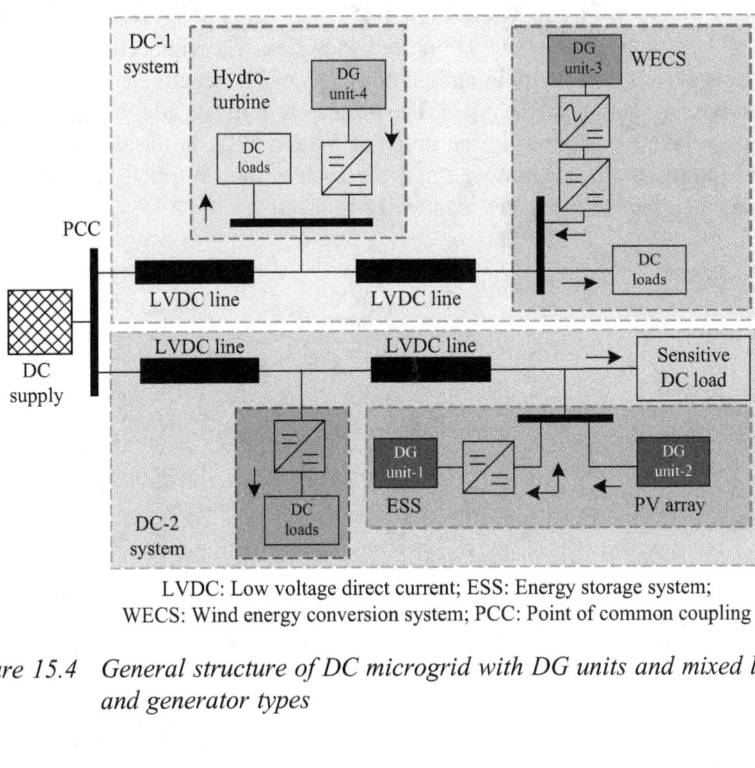

LVDC: Low voltage direct current; ESS: Energy storage system;
WECS: Wind energy conversion system; PCC: Point of common coupling

Figure 15.4 General structure of DC microgrid with DG units and mixed load and generator types

microgrid system interconnected with the main systems at PCC which can be a medium voltage AC network from the conventional power plants or an HVDC transmission line connecting an offshore wind farm.

15.4 Topologies for DC microgrid for residential applications

There are several topologies used by the DC microgrid. However, the unipolar or bipolar type structure is typically used to configure the DC microgrid.

15.4.1 Unipolar LVDC system

The unipolar system consists of a two-winding transformer and a line converter, connected as shown in Figure 15.5. In a unipolar DC system, the line and loads are connected via two conductors, a neutral and other a positive polarity DC voltage. The unipolar system has a one voltage level for energy transmitted. All the customers are connected to this one voltage level [10]. The unipolar system has the small power transfer capacity compared to bipolar type system with the same voltage level (i.e. 12 and ±12 V) and does not have a broad range of choices of DC voltage level.

15.4.2 Bipolar LVDC system

The bipolar system is a combination of a three-winding transformer, and two line converters, connected as shown in Figure 15.6, i.e. the bipolar system is a combination of two unipolar systems connected in series. The connection alternatives may be between a positive pole and neutral, between a negative pole and neutral, between positive and negative poles. The bipolar system provides a broad range of DC voltage levels compared to the unipolar system [11]. In the bipolar type, the system requires more components, and the system may result in an unbalanced situation when the loads are not identically connected.

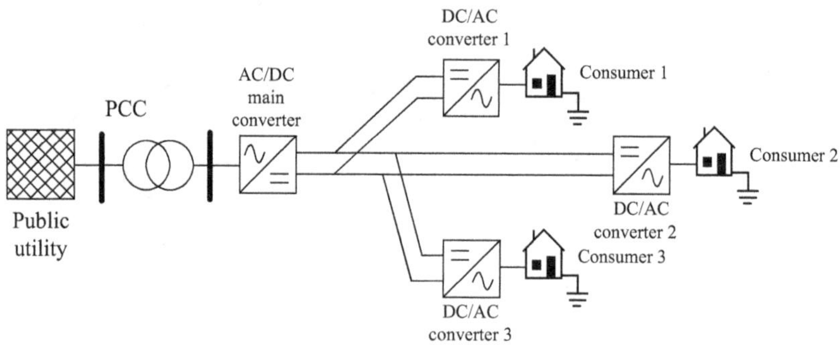

Figure 15.5 Unipolar LVDC distribution system used in a cluster of residential buildings

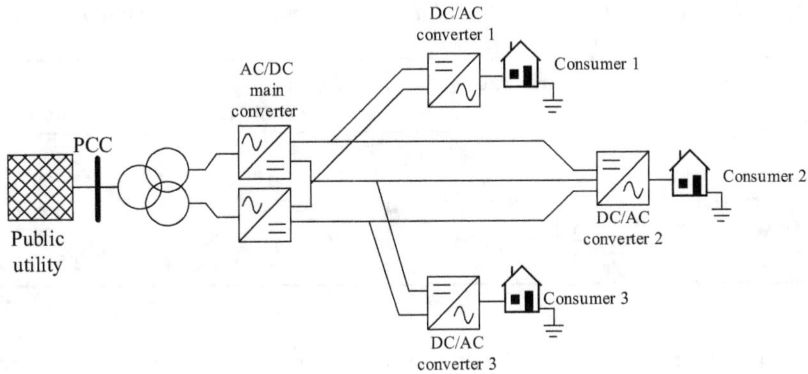

Figure 15.6 Bipolar LVDC distribution system used in a cluster of residential buildings

15.5 Mathematical analysis of AC vs DC microgrid system

In this section, mathematical descriptions have been derived to calculate voltage, current and power losses during an AC as well as DC distribution system supply.

15.5.1 Total daily load

The total daily load (TDL) has been calculated with respect to the different DC voltage level supply as well as the different AC voltage levels and inverter efficiencies. When the DC load is connected to the system, the DC load (γ) in amperes is given as

$$\gamma = \frac{k}{V_{DC}} \tag{15.1}$$

where k is the load rating in kilowatt; V_{DC} is the DC system voltage; and TDL, L_{DC}, can be obtained by multiplication of the DC load with the number of operating hours (λ) of the load per day in ampere hours, and it can be expressed as:

$$L_{DC} = \gamma \times \chi \tag{15.2}$$

If there is a variable DC load connected to the system, the calculation of total daily load is as follows:

$$L_{DC} = \sum_{j=1}^{\Omega} \gamma_j \chi_j \tag{15.3}$$

where, $\gamma_1, \gamma_2, \ldots, \gamma_\Omega$ and $\chi_1, \chi_2, \ldots, \chi_\Omega$ are variable DC loads in amperes and different time instants at which DC loads are switched 'ON' in a day.

Table 15.1 Illustrative values of total daily load on DC and AC systems

System voltage	Power rating (kW)	Inverter efficiency (%)	Total daily load (A h)
24 V DC	2.4	–	2,400.0
48 V DC	2.4	–	1,200.0
120 V AC	2.4	92	2,608.7
220 V AC	2.4	92	2,608.7
220 V AC	2.4	95	2,526.3

If the AC load is connected to the system, then the AC voltage must be converted to DC voltage and the inverter efficiency is also considered. The DC load of the AC system can be expressed as:

$$\gamma = \frac{k}{V_{DC}\eta_{inv}} \tag{15.4}$$

Table 15.1 shows the characteristic values of a DC system based on the type of load (AC or DC load). If the DC load 2.4 kW is supplied at 24 and 48 V, the TDL is 2,400 and 1,200 A h, respectively. In the case of 2.4 kW AC load at 120 and 220 V supplied by the same efficiency inverter (92%), then TDL is 2,609 A h remains the same but higher than the TDL as obtained in the case of 2.4 kW DC load at 24 V. If a 2.4 kW AC load supplied at 220 V by the 95% efficiency inverter, the obtained is TDL 2,526 A h. The above calculation shows that TDL depends upon the DC system voltage and inverter efficiency. If the inverter efficiency is high, the losses in DC to AC conversion are less, and the system TDL will be reduced. As the system voltage is high, the system TDL will also decrease, i.e. system losses may decrease. On the other hand, if there is a lower inverter efficiency or DC supply voltage, the TDL will be higher. It means the battery bank (BB) will discharge at the fastest rate, which may decrease the battery lifetime and efficiency.

15.5.2 Voltage, current and power losses in DC supply

In India, the ACDS for residential buildings is single phase 230 V (RMS). The equivalent DC voltage (325 V) applied to the same load can reduce the current ratings. For example, 230 and 110 V AC supply the current rating can be reduced to 30% and 70%, respectively [6,7]. The DC system voltage stress equivalent to the single-phase AC system can be calculated as

$$V_{DC} = \sqrt{2}V_{AC} \tag{15.5}$$

The DC system has less potential stress compared to the AC system for the same voltage. For example, if a system is designed for 230 V AC, it can bear 325 V DC without any rapture in insulation. The voltage level helps to reduce the gap between two conductors of the distribution line. The less potential stress and weight of conductor reduce the size of the tower and insulator. This decreases the cost of the system and makes the system more economical. The power transfer in a DC system can be expressed as

$$P = V \times I \tag{15.6}$$

The current in a DC system can be expressed as

$$I = \frac{P}{V} \tag{15.7}$$

where P is the transferred power, V and I are the system voltage and current, respectively. As shown in (15.7), if DC voltage is applied instead of AC voltage then the insulation can bear higher voltage stress, which allows applying higher DC voltage for the same system. The current is inversely proportional to the voltage. It means as the system voltage increase, their current will decrease. Therefore, the system copper losses (p) will also decrease.

$$p = I^2 R \tag{15.8}$$

where p represents power losses in the system and R is the resistance of the feeder cable.

The AC and DC system comparison in terms of current and power losses can be found in Table 15.2. The current and power losses in the 325 V DC system (equivalent to 230 V AC system) are approximately one-third and a half respectively as compared to the 230 V AC system (single-phase AC voltage level in India). Additionally, it has around 2/3 times and 7/8 times less current flow and power losses, respectively, as compared to the 110-V AC system (single-phase AC voltage level in the United States).

Table 15.3 shows the energy savings, which can be obtained by switching from AC technologies to the most energy efficient DC-internal technologies. It is seen from Table 15.3 that the total energy saving using DC technology in the residential loads is varying between 30% and 71%. Some rectifier losses have already existed in a few cases like a refrigerator, cloth washer and fans, in AC system.

The integration of DC sources to conventional AC system necessitates the introduction of DC–AC converter at the generation end, thereby adding conversion losses and complexity [11]. In last two decays, the continued development of DC technologies to produce an energy efficient DC appliance is a cause of significant decrement in the building load but insists on introducing AC–DC converter and

Table 15.2 Comparative analysis of AC and DC system based on current rating and power losses

DC voltage	AC voltage	Reduction in current rating (%)	Reduction power losses (%)
325	230	29.2	50.1
325	110	66.2	88.5

Table 15.3 Possible energy saving using most efficient DC internal technology-based appliances

Appliances	Efficient DC compatible replacement technology	Total energy savings (%)
Fan	Run by brushless DC motor in place of single-phase AC induction motors	47
Room air conditioners	Variable-speed compressor	35
Lighting-incandescent	14 LPW incandescent goes to CFL (electronic ballast) at 52 LPW	73
Lighting-reflector	15 LPW goes to CFL (electronic ballast) at 52 LPW	71
Lighting-touchier	Assuming 80%, incandescent at 14 LPW goes to CFL at 52 LPW, and 20% CFL stays the same	69
Electric water heaters	Heat pump	50
Refrigerators	Assuming 85% standard-size at 587 kW h AEU with savings of 51% and 15% compact at 331 kW h AEU with savings of 75%	53
Clothes washers	Brushless DC permanent magnet (BDCPM) variable speed	30
Ceiling fans	BDCPM variable speed system	30

Table 15.4 Description of DC appliances and their AC–DC converter in India

Appliance name	Voltage rating (V)	Current rating (A)	Power rating (W)	AC–DC converter efficiency (%)
LED bulb	12	0.6	7	79
CFL bulb	12	1.0	12	79
Electric geyser	96	10.5	1,000	89
Sandwich maker	24	23.0	550	87
Water purifier	24	0.5	11	79
Refrigerator	24	3.0	72	87
Coffee maker	12	11.0	135	87
Washing machine	24	3.0	70	86
Water pump	24	14.9	350	87
Vacuum cleaner	12	8.0	95	87
Air conditioner	24	33.3	800	88
Hybrid car	96	32.0	3,000	90
Cell phone	12	0.3	4	78
Ceiling fan	12	1.7	20	83
Hair drier	24	15.0	425	87
TV	12	2.5	30	83
Computer	12	14	170	87

increase the conversion loss and complexity of the system [12]. The details of the voltage and power ratings of the DC-appliance and their AC–DC converters efficiency to connect these appliances to the conventional AC system can be found in Table 15.4. The AC–DC converter efficiencies vary from 78% to 90% according to

Table 15.4. It can be noted that higher the converter power rating is, the higher is the AC–DC efficiency, as the highest efficiency 90%, which is in the case of a hybrid car with a converter power of rating 3,000 W. The cell phone converter of 4-W power rating has the lowest efficiency of 78%, as mentioned in Table 15.4.

15.6 Comparison between AC and DC residential buildings

A grid-connected residential building consists of DC appliances for AC and DC distribution system is shown in Figures 15.7 and 15.8, respectively. The building is supplied by the PU and PV in both cases. The energy storage (BB and EV) is used to store the energy to supply the future load in case the PV and PU are not acting. Additionally, the ES is also responsible for storing the surplus power produced by the PV and supply the surplus load to balance the power in the system in case of PU acts as outage source. The residential building consists of six rooms. The electrical specifications of the loads are mentioned in Table 15.4. The voltage ratings of the appliances are 12, 24 and 96 V for the low, medium and high power rating loads. As the building appliances have a wide range of voltage levels, the bipolar type of distribution system is used in order to optimise the conversation stage in the building.

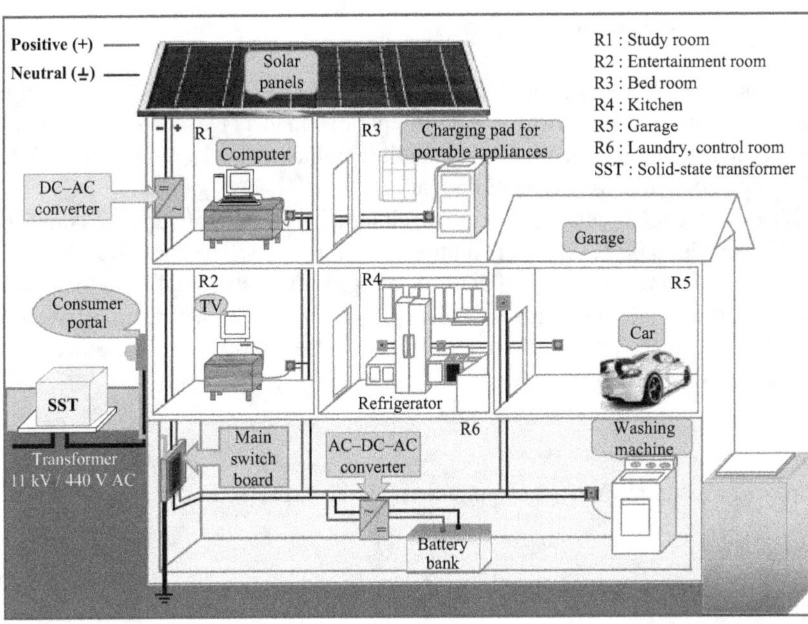

Figure 15.7 Conceptual layout of AC residential building

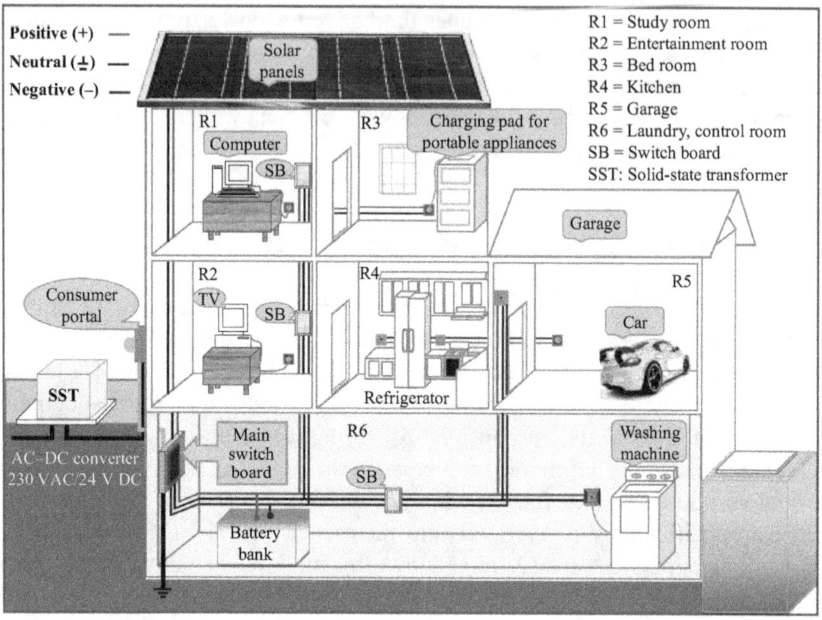

Figure 15.8 Conceptual layout of DC residential building

15.7 AC residential buildings

In this case, each line has a single voltage level of 230 V AC. The DC compatible loads are more efficient than the AC compatible load [11]. Therefore, it is also assumed that each appliance is DC compatible, which helps to reduce the building load compared to the AC compatible load.

Moreover, each appliance has its own internal AC–DC converter to connect to the AC line, which is the cause of additional power losses. The specifications of the appliances and their converter efficiencies are given in Table 15.4. The total power consumption in buildings with ACDS (P_{ACB}) is the combination of power consumed by appliances (P_A) and power loss (p_C) in the converters. The power equation can be expressed as

$$P_{ACB} = P_A + p_C \tag{15.9}$$

The power consumption by the appliances can be expressed as

$$P_A = \sum_{j=1}^{j=n_a} P_{aj} \tag{15.10}$$

where P_{aj} is the power consumed by the jth appliance.

The total power loss (p_C) in the converters is the addition of power loss in the internal converters of the appliances (p_{ac}) and the power loss in the source converter (p_{sc}) and can be expressed as

$$p_C = \sum_{j=1}^{j=n_a} p_{acj} + \sum_{j=1}^{j=n_s} p_{scj} \tag{15.11}$$

where n_a is the number of appliances, and n_s number of source converters.

15.8 DC residential buildings

In this case, it is also assumed that each appliance is DC compatible which helps to reduce the building load compared to the AC compatible load. The main DC bus has 24 V voltage level. Moreover, one boost DC–DC converter to increase the voltage level from 24 to 96 V and supply EV and electric geyser. A DC–DC buck converter is used to tie 24–12 V DC bus. The appliances of 12 V ratings, such as CFL, LED, computer, TV, etc. are directly connected to the 12 V DC bus, while for the remaining 24 V rating, appliances are also connected to the 12 V DC bus between the conductors of positive and negative polarity as shown in Figure 15.8.

The total power consumption in the buildings for DC distribution system (DCDS) (P_{DCB}) is the addition of power consumed by appliances (P_A) and power losses in DC–DC and AC–DC converters (p_c). The power expression is as given below [8]:

$$P_{DCB} = \sum_{j=1}^{j=n_a} p_{aj} + \sum_{j=1}^{j=n_c} p_{cj} \tag{15.12}$$

where $n_c = 3$, the number of converters in buildings with DCDS.

15.9 Automation architecture for smart DC residential buildings

The building automation system is where centralised control and monitoring of the building services are done. The purpose of the automation system is to maintain the building environment more efficiently to reduce the building's environmental impact and energy costs. There may be different types of architecture for the building automation. A general architecture having all type of complex activities and facilities has been shown in Figure 15.9, and it can be divided into the following levels [13].

15.9.1 Field level

The control and detection of the devices consist of this level. These devices may be sensors, actuators, light or smoke detectors, valves, switches and other intelligent sensors.

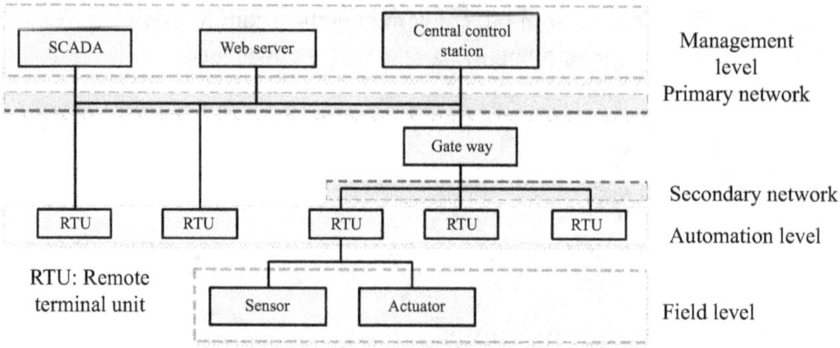

Figure 15.9 The basic network architecture of a building automation system

15.9.2 Field network

It is the connection network between the field level and automation level. The main purpose of this network is to connect the field level devices to an RTU (remote terminal unit), in the automation level. These connections may be of four types:

1. Hardwired
2. Bus system
3. Powerline
4. Wireless

15.9.3 Automation level

This level has different advanced controllers to control and regulate the sensors, actuators and other types of field-level devices. Usually, the digital-based microprocessor is used to freely programme them with different control and control logic like proportional control, integral control, differential control and any other logic control as well as a combination of logic controls, etc. [14]. The controllers include the RTU.

15.9.4 Primary and secondary network

Most building automation networks consist of a primary and secondary bus which connects high-level controllers with lower level controllers, input/output devices and user-interface devices. The ASHRAE's open protocol BACnet or the open protocols LonTalk specify how most such devices interoperate. Modern systems use the simple network management protocol (SNMP) to track events, building on decades of history with SNMP-based protocols in the computer networking world.

The primary network is the management network, and it is the backbone of the system. The primary network connects the automation level and the management level in the building automation system. A primary network can either be separated or shared with the conventional local area network in a building.

Secondary networks connect the automation level and the primary network as a sub-network to the primary network. The purpose of this level is to connect the device to the automated level, but these devices are working with a different protocol compared to the devices connected to primary network devices directly.

15.9.5 Management level

It is the level that has the capability to manage activities and monitor the building automation system. The interaction method may be personal or with the internet. Some examples of these devices are databases that log activity, web servers, operator's panels, central control station and servers that translate messages into different protocols.

15.10 Advantages, challenges and barriers of smart DC residential buildings

Some factors will always be there to make the candidature of DC systems more strong then the AC system such as [15] (i) DC systems have renewable energy sources, like PV panels and FC, and ESSs, as batteries, (ii) around 50% less energy consumption of the complete load that appears in the AC system in some operating point, (iii) the scope of the integration of new sources like electric vehicle will increase the use of DC devices, (iv) zero skin effect makes the DC systems more efficient then the AC systems, (v) use of DC devices, sources and storage to interconnect and distribute the energy between them avoid the DC–AC and AC–DC conversions stages and reduces the losses.

A detailed discussion and analysis of LVDC are required to see the potential of DC technology. However, some advantages of it already are being discussed. There are various advantages including the system efficiency improvement and energy saving in the residential building. In [16,17], a study of the US-based region was carried out with different topologies and at various locations of the country [16]. Different topologies have been considered with or without some ESS. The results of this study show that the efficiency is significantly improved with the storage device. Annual 5% energy savings are estimated without storage system, and 14% energy saving is observed with ESS. This difference in energy saving is because of the consequence of PV generation and consumption time according to the residential load. The consumption peaks in the afternoon and evening while the peak PV generation is in the afternoon. Because of this reason, storage system is required to store the excess energy that is available during the peak PV generation and utilised during high peak demand. There are some more studies that are showing to achieve little more energy saving up to 25% energy saving [18,19]. The building loads, which are affected by environmental conditions such as cooling or heating, need to be considered for such type of studies. Another condition of comparison is that the AC or DC comparable loads have to consider for the different AC or DC distribution systems that have to be compared [20]. One important consideration is also that energy saving should not be considered which is achieved with extremely efficient DC loads in the place of AC loads.

As several advantages and features of DC distribution system are discussed, it also faces critical challenges and some barriers when going to implement it into residential buildings. There are some challenges and obstacles to the DC distribution network:

- There is a lack of proper standardisation and codes for the DC system. However, continuous work is going on in this area. The leads, in these directions, are taken by some organisations like IEEE, the International Electrotechnical Commission, Emerge Alliance, the European Telecommunications Standards Institute, and others, which are already actively developing the necessary regulation and standards. However, much work is required in this direction.
- Protection schemes for the use of DC systems have to be developed. New protection devices have to be designed for the safe use of DC systems [4,21,22].
- Currently, there are fewer industries and devices available those will work well for the DC distribution systems. There are some rare products available when going to analyse the DC systems for residential buildings; there is a lack of products which can be used on DC voltage directly. However, there are many DC devices that can be used directly on the DC voltage as having internal conversion unit [4,22].

In respect to energy efficiency and the fulfilment of demand, DC systems taking ahead with ACDSs. However, for residential use, more time is needed in order to use it in residential buildings safely with proper standards. Lack of standards, regulation methods and protection schemes are the main challenges that need to be discussed.

15.11 Comparison AC and DC residential buildings: an illustrative example

To investigate the performance of electrical distribution system, an LVDC distribution network for a residential building supplied by PV including a BB and the PU has been simulated in MATLAB®. The 24 V BB is configured with the series, and parallel combination of 16 cells of 150 A h. The 1.44 and 2.28 kW with a 24 V rated voltage PV is considered for the DCDS with DC compatible appliances (DC residential building), and ACDS with DC compatible appliances (AC residential building), respectively. The PU is tied to the consumer portal via AC–DC converter and a step-down transformer (solid-state transformer) to 24 V DC bus and 230 V AC for AC and DC residential building, respectively. The distribution lines of the power system are considered as lossless. The power consumption in the AC and DC residential building is shown in Figure 15.10.

In the case of DC residential building, the building demand curve consists of power consumption in the DC compatible appliances of the building and power loss in PU inverter and it is represented by the blue line in Figure 15.10. In the case of AC residential building, the building demand curve consists of power consumption

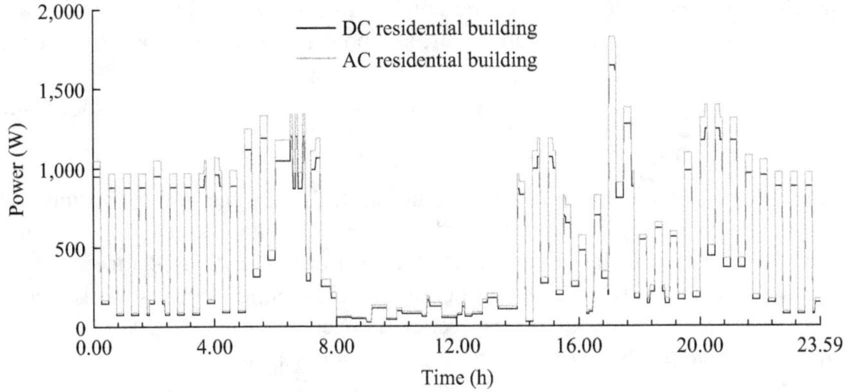

Figure 15.10 Demand for AC and DC residential building

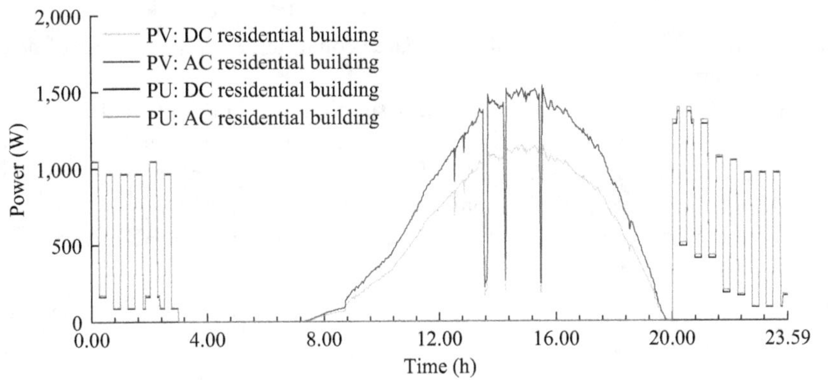

Figure 15.11 Power consumed from PV and PU in AC and DC residential building

in the DC compatible appliances of the building, power loss in the internal converter of appliances, power loss in the BB and PV converter as represented in Figure 15.10 by the red line. The building demand always remains higher for AC residential building compared to the DC residential building.

The PU and PV power curves for AC and DC residential building are shown in Figure 15.11. The PU and PV power curve remains always lower for the DC residential building compared to the AC residential building. Different operation modes are discussed (Mode I–Mode III).

Mode I (PU as a source): The PU supplies the building load during the 0.00–03.00 and 21.30–23.59 h time interval. In the case of AC residential building, all the building loads are DC compatible and connected via AC–DC internal converters. The converters loss is the combination of power loss in the internal converters of the switch 'ON' appliances, and power equation can be expressed as

$$P_{pu}(t) = \sum_{j=1}^{j=n_{ao}} \{p_{aj}(t) + p_{acj}(t)\} \tag{15.13}$$

where $P_{pu}(t)$ is the PU power at the time instant t, $P_{aj}(t)$ is the power consumption in the jth appliance, n_{ao} is the number of switched 'ON' appliances in the building and $p_{acj}(t)$ is the power loss in the internal converter of the jth appliance.

Moreover, in the DC residential building, all the building loads are DC compatible and directly (without any converter) connected to the ±12 V bipolar DC distribution system. The load is connected to the PU via AC–DC converter. So the power loss in the converter is the power loss in the PU converter. The power equation can be expressed as

$$P_{pu}(t) = \sum_{j=1}^{j=n_{ao}} p_{aj}(t) + \sum_{j=1}^{j=n_s} p_{scj}(t) \tag{15.14}$$

where $p_{scj}(t)$ is the power loss in the jth source converter, n_s is the number of power source converters.

Mode II (BB as a source): The BB supplies the building load during the 03:01–07:21 and 19:51–20.00 h time interval as represented in Figure 15.12. In the AC residential building, all the building loads are DC compatible and connected via AC–DC internal converters. The converters losses are a combination of power loss in the internal converters of a switch 'ON' appliances and the BB inverter. The power equation can be expressed as

$$P_{bb}(t) = \sum_{j=1}^{j=n_a} \left[P_{adj}(t) + p_{adcj}(t)\right] + \sum_{j=1}^{j=n_s} p_{scj}(t) \tag{15.15}$$

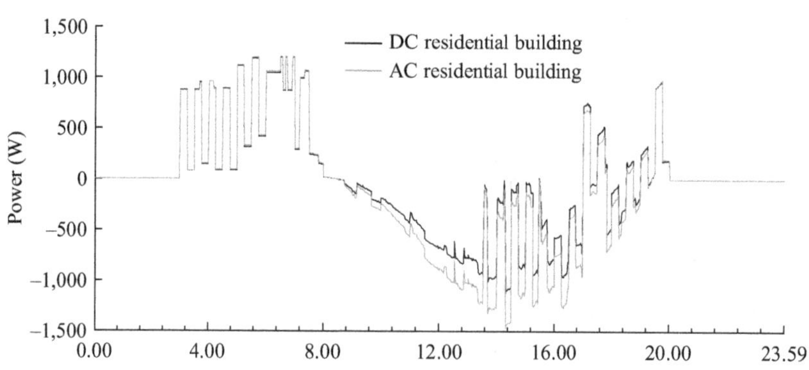

Figure 15.12 Battery bank power in AC and DC residential building

In the case of DC residential building, all the building loads are DC compatible. The appliances and DC power source (i.e. BB) are directly (without any converter) connected to the DC bus. Therefore, the power loss in the sources converters and internal converter of the appliances remain zero in this mode as represented by a green line in Figures 15.13 and 15.14, respectively. The power equation can be expressed as:

$$P_{bb}(t) = \sum_{j=1}^{j=n_a} P_{aj}(t) \tag{15.16}$$

Mode III (BB and PV): The PV generates the power during the 07.22–19.50 h time interval. The BB balances the power by supplying the surplus load and absorbing the surplus PV power. In the case of AC residential building, all the

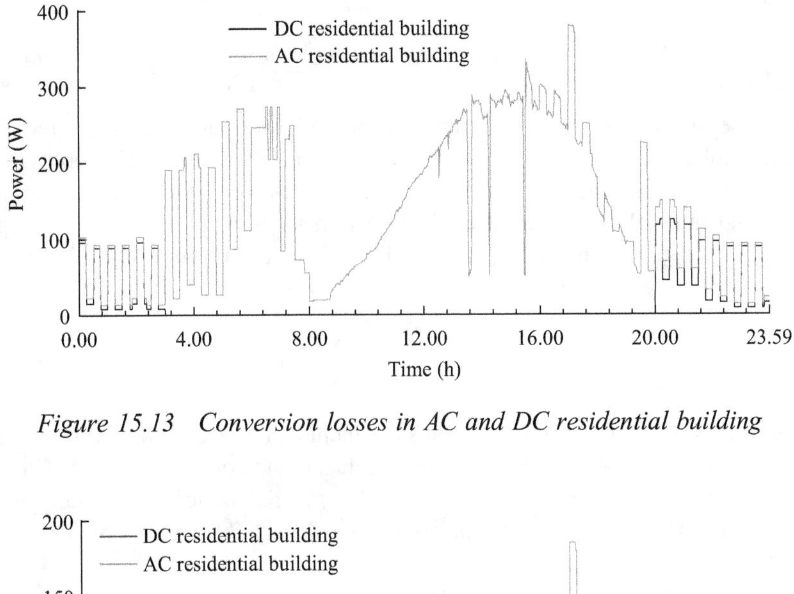

Figure 15.13 Conversion losses in AC and DC residential building

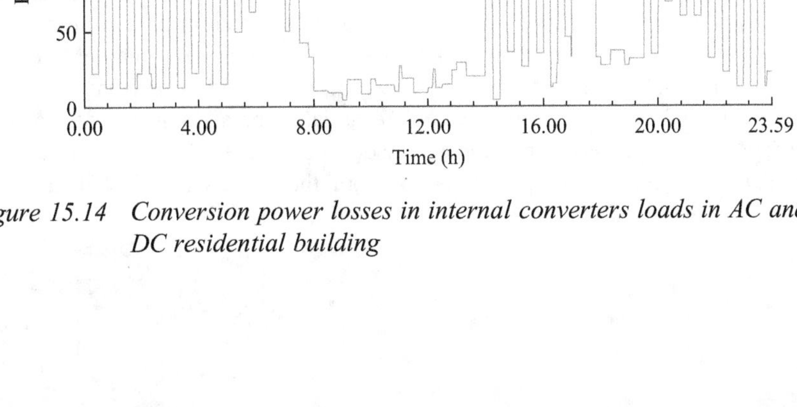

Figure 15.14 Conversion power losses in internal converters loads in AC and DC residential building

Table 15.5 Description of energy consumption in the building

System	Demand (kW h)	Conversion loss (kW h)	Total energy consumption (kW h)	PV energy (kW h)	PU energy (kW h)
AC residential building	12.47	3.34	15.81	11.38	4.43
DC residential building	11.12	0.44	11.56	7.2	4.36

building loads are DC compatible and connected via AC–DC internal converters. The converter losses are the combination of power losses in the internal converters of a switch 'ON' appliances including with PV and BB inverter. The power equation can be expressed as:

$$P_{pv}(t) \pm P_{bb}(t) = \sum_{j=1}^{j=n_a} \left[P_{adj}(t) + p_{adcj}(t)\right] + \sum_{j=1}^{j=n_s} p_{scj}(t) \tag{15.17}$$

In the case of DC residential building, all the building loads are DC compatible. The DC appliances and DC power source (i.e. BB and PV) are directly connected to the DC bus without any converter. Therefore, the power loss in the source converters and internal converter of appliances remains zero in this mode as represented by the green line in Figures 15.13 and 15.14, respectively. The power equation can be expressed as:

$$P_{pv}(t) \pm P_{bb}(t) = \sum_{j=1}^{j=n_a} P_{aj}(t) \tag{15.18}$$

The energy consumption, conversion losses, including energy supplied by PV and PU can be found in Table 15.5. The energy demand and conversion loss for the AC residential building are higher than the DC residential building. The conversion losses in the AC residential building is approximately 7.5 times higher than the conversion losses in the DC residential building.

15.12 Conclusions

In this chapter, the concept of DC microgrid for the residential buildings has been discussed. Comparison of AC vs DC system and DC microgrids architecture has been discussed. The distribution topologies discussed in this chapter are very helpful to understand the most efficient way of interconnection. The data of energy dissipated in DC appliances, and the cable cost data representing are used to show the correlation of different parameters associated with the losses. This chapter demonstrates different configurations for both the ACDS and DC distribution system. A power system control strategy based approach is used for the voltage standardisation. This approach enables the development of energy efficient economy and flexible LVDC system and as well as medium-voltage standardisation. A comparative analysis of AC and DC residential building shows the superiority DC residential building, in respective energy saving. Simulation results also show that the power consumed in the DC residential building is less than the power consumed in the AC residential building. Because the converter stages and conversion losses are much less in the DC residential building compared to the AC residential building.

References

[1] F. Gonzalez-Longatt, B. S. Rajpurohit, and S. N. Singh, "Optimal structure of a smart DC micro-grid for a cluster of zero net energy buildings," in *Proc. IEEE International Energy Conference (ENERGYCON)*, Leuven, Belgium, Apr. 4–6, 2016.

[2] F. Gonzalez-Longatt, B. S. Rajpurohit, and S. N. Singh, "Smart multi-terminal DC micro-grids for autonomous zero-net energy buildings: implicit concepts," in *Proc. IEEE Innovative Smart Grid Technologies-Asia (ISGT ASIA),* Bangkok, Thailand, Nov. 04–06, 2015, pp. 1–6.

[3] R. K. Chauhan, B. S. Rajpurohit, S. N. Singh, and F. M. Gonzalez-Longatt, "DC grid interconnection for conversion losses and cost optimization," in *Renewable Energy Integration: Challenges and Solutions*, J. Hossain and A. Mahmud, Eds. Singapore: Springer Singapore, 2014, pp. 327–345.

[4] L. Mackay, L. Ramirez-Elizondo, and P. Bauer, "DC ready devices – is redimensioning of the rectification components necessary?," in *Proc. 16th International Conference on Mechatronics – Mechatronika*, Dec. 03–05, 2014, pp. 1–5.

[5] F. Katiraei, R. Iravani, N. Hatziargyriou, and A. Dimeas, "Microgrids management," *IEEE Power and Energy Magazine*, vol. 6, no. 3, pp. 54–65, May 2008.

[6] X. Fang, S. Misra, G. Xue, and D. Yang, "Smart grid – the new and improved power grid: a survey," *IEEE Communications Surveys & Tutorials*, vol. 14, no. 4, pp. 944–980, Dec. 2011.

[7] J. Stamp, "The SPIDERS project – smart power infrastructure demonstration for energy reliability and security at US military facilities," in *Proc. 2012 IEEE PES Innovative Smart Grid Technologies (ISGT)*, Jan. 16–20, 2012.

[8] R. K. Chauhan, and B. S. Rajpurohit, "DC distribution system for building," in *Proc. IEEE, 18th National Power System Conference*, IIT Guwahati, India, Dec. 18–20, 2014, pp. 1–6.

[9] D. Nisbet and H. Thiesen, "Review of the initial phases of the LHC power converter commissioning," in *Proc. 11th European Particle Accelerator Conference*, Jun. 2008.

[10] R. K. Chauhan, B. S. Rajpurohit and N. M. Pindoriya, "DC power distribution system for rural applications," in *Presented at the 8th National Conference on Indian Energy Sector*, Ahmedabad, India, Oct. 11–12, 2012, pp. 108–112.

[11] R. K. Chauhan, B. S. Rajpurohit, R. E. Hebner, S. N. Singh, and F. Gonzalez-Longatt, "Voltage standardization of DC distribution system for residential buildings," *Journal of Clean Energy Technologies*, vol. 4, no. 3, pp. 167–172, 2016.

[12] I. Z. D. M. Larruskain, O. Abarrategui, and Z. Aginako, "Conversion of ac distribution lines into dc lines to upgrade transmission capacity," *Electric Power Systems Research*, vol. 81, no. 7, pp. 1341–1348, Jul. 2011.

[13] R. O. J. Kensby, "Building automation systems design," Master Thesis, Division of Building Services Engineering, Chalmers University of Technology, Göteborg, Sweden, 2012.

[14] R. Kumar, K. Saini, and M. L. Dewal, "Intelligent SCADA system," *International Journal on Power System Optimization and Control*, vol. 2, no. 2, pp. 143–149, 2010.

[15] M. Saeedifard, M. Graovac, R. F. Dias, and R. Iravani, "DC power systems: challenges and opportunities," in *Proc. of IEEE PES General Meeting*, Providence, RI, USA, Jul. 25–29, 2010, pp. 1–7.

[16] K. G. V. Vossos, and H. Shen, "Energy savings from direct-DC in U.S. residential buildings," *Energy and Buildings*, vol. 68, no. A, pp. 223–231, Jan. 2014.

[17] E. Vossos, "Optimizing energy savings from direct-DC in U.S. residential buildings," Master's Thesis, The Faculty of the Department of Environmental Studies San Jose State University, San Josė, USA, 2011.

[18] R. R. N. P. Savage, and S. P. Jamieson, "DC microgrids: benefits and barriers," in *From Silos to Systems: Issues in Clean Energy and Climate Change*, pp. 51–66, 2010.

[19] V. V. Karina Garbesi, and H. Shen, "Catalog of DC Appliances and Power Systems," Lawrence Berkeley National Laboratory (Berkeley Lab) 2011, vol. LBNL-5364E.

[20] E. Rodriguez-Diaz, M. Savaghebi, J. C. Vasquez, and J. M. Guerrero, "An overview of low voltage DC distribution systems for residential applications," in *Proc. IEEE 5th International Conference on Consumer Electronics-Berlin (ICCE-Berlin)*, 2015, pp. 318–322.

[21] D. Salomonsson, L. Soder, and A. Sannino, "Protection of low-voltage DC microgrids," *IEEE Transactions on Power Delivery*, vol. 24, no. 3, pp. 1045–1053, Apr. 2009.

[22] G. Makarabbi, V. Gavade, R. Panguloori, and P. Mishra, "Compatibility and performance study of home appliances in a DC home distribution system," in *Proc. of IEEE International Conference on Power Electronics, Drives and Energy Systems (PEDES)*, Dec. 16–19, 2014, pp. 1–6.

Chapter 16

DC microgrids for photovoltaic powered systems

Rasool Heydari[1], Saeed Peyghami[1], Yongheng Yang[1], Tomislav Dragičević[1], and Frede Blaabjerg[1]

16.1 Introduction

Climate changes induced by the use of fossil fuels and also the shortage of such energy have been the major driving forces to develop environmental-friendly renewable energy systems [1,2]. Many attempts have been made to advance renewable energies, two of which are wind power and solar photovoltaic (PV) power that are still the most favorable, as it is seen in Figure 16.1. Depending on the power ratings, PV systems are mostly installed at a residential-scale, while wind power systems are connected to medium-voltage (MV) or high-voltage (HV) power grids. Hence, for microgrid applications, PV energy is more feasible also due to its high scalability. The still declining price of PV panels is another reason for a massive deployment of PV microgrid systems. Notably, in those applications, power electronic converters (e.g., dc–dc converters and dc–ac inverters) are the key, which converts the dc power from PV panels to another type of source (e.g., dc at a different level or ac power with a fixed amplitude and a constant frequency).

Additionally, the high penetration rate of dc loads including energy storage (ES) systems (e.g., batteries and electric vehicles) makes dc microgrids a promising solution for energy integration [3,4]. For one thing, in the context of a large-scale utilization of dc loads, the dc microgrid architecture eliminates at least one conversion stage (i.e., ac–dc rectification), and thus, the entire system efficiency maybe is improved. For another thing, it simplifies the system control, and the frequency stability is addressed to some extent, which, however, is a bigger challenge in ac power systems. Yet, despite the above compelling reasons to deploy PV microgrids, issues are also encountered [4]. For instance, the PV power is fluctuating, possibly leading to dc bus voltage variations, being harmful to sensitive equipment. Nevertheless, employing dc microgrids has been increasingly acknowledged in recent years, and a great deal of the focus has been put on reliability, efficiency, and cost-effectiveness, i.e., reducing the cost of PV energy. In practice, dc microgrids are utilized in various sectors, ranging from small-scale

[1]Department of Energy Technology, Aalborg University, Denmark

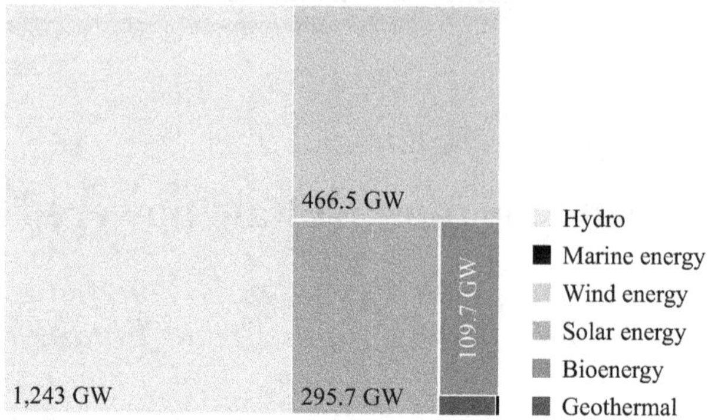

Figure 16.1 Global renewable power capacity composition in 2016 (more than 2,000 GW in total) [2], where hydro power also includes pumped storage and mixed plants; solar energy includes solar photovoltaic (290.8 GW) and concentrated solar power (4.7 GW)

residential to medium-scale industrial applications including power generation and/or distribution systems, space stations, maritime application, data centers, oil drilling platforms, remote regions (the energy accessibility is poor), and residential smart homes.

Many aspects should be considered in the design phase of dc microgrids. Efficiency, reliability, and stability are the major concerns as aforementioned. In respect to the system efficiency, the system architecture or layout and the power converter technologies should be considered. This chapter first overviews the dc microgrid architecture, where the primary energy sources are PV panels. Then, the focus is on power electronic converter technologies for PV dc microgrid systems, where an industrial case is also given. Regarding the control dc microgrids for PV collection plants, it is discussed in this chapter as well. A design example is also provided.

16.2 Architecture of dc microgrids

In the following, some outstanding applications and architecture schemes of dc microgrids are introduced.

16.2.1 MVDC and LVDC microgrid for dc distribution systems

Recently, due to the fast pace of the commercialization of distributed energy resources, a wide range of power distributed generation (DG) and ESs have been employed in power systems. Since it is known that a microgrid is dynamically separated from the main grid, the future of power systems relies on isolated ac and dc microgrids. Compared to the ac grids, dc microgrids bring various advantages, as mentioned in the previous section. For instance, it is easy to interface dc loads and renewable energy resources, in particular, PV panels. And there are no issues

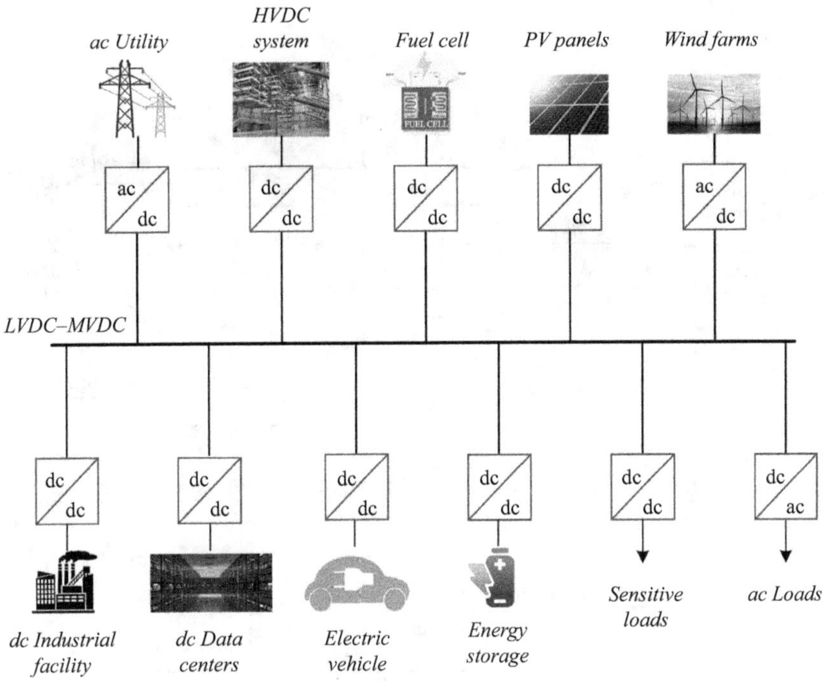

Figure 16.2 General LVDC–MVDC unipolar architecture layout

like frequency stability, reactive power compensation, and harmonics. More importantly, high efficiency, power quality, and reliability make the dc microgrid more and more popular [5–7].

A majority of renewable energy resources such as PV arrays and fuel cell modules as well as battery and ESs are naturally dc. Furthermore, by utilizing advanced power electronic devices, domestic and industrial dc loads are increasing in recent years. Therefore, utilizing dc microgrids to integrate dc loads and dc-renewable energy resources increases the efficiency and reliability of the distribution systems. Figure 16.2 shows general layout of an LVDC–MVDC distribution system, in which renewable energy resources (e.g., wind turbines, solar panels, and fuel cells) and dc loads are connected to a common dc-bus through dc–dc or ac–dc converters. The dc microgrid is employed in the distribution system to increase the controllability of DGs, which are typically installed in the vicinity of loads. It thus leads to increased reliability, reduced transmission losses, and consequently decreased operation and planning costs.

A typical bipolar LVDC microgrid structure is shown in Figure 16.3 [8]. This topology introduces better reliability in comparison to the unipolar one. In this topology, loads, sources, and storage can be connected to the different lines. A typical voltage for unipolar system can be ± 170, ± 380, and ± 750 V based on the required application.

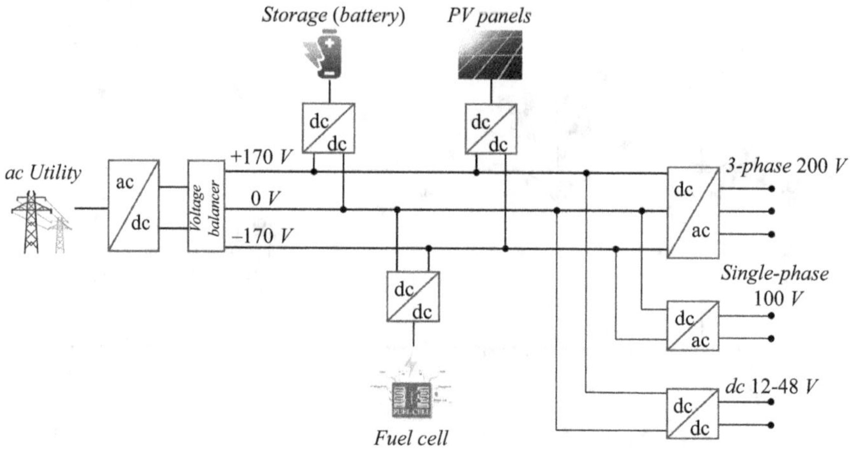

Figure 16.3 A typical bipolar LVDC distribution systems [8]

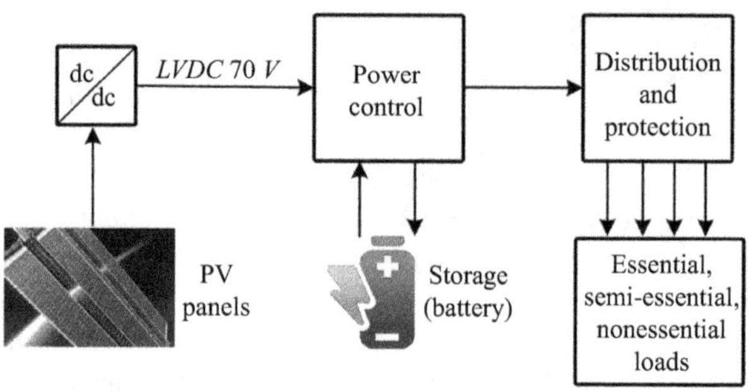

Figure 16.4 Power system structure in a space station

16.2.2 LVDC microgrid for space applications

Generally, a space station comprises various devices designed to address air craft missions. In such applications, the required power in early satellites is in the order of several watts, while, in today's space stations, it is rated at several kilowatts. Furthermore, the estimated power demand in some strategic defense space crafts is expected to be at the order of 100-MW level and it is still increasing. The basic structure of a space station power system is demonstrated in Figure 16.4, where the primary energy resources are mainly solar panels, nuclear reactors, and radioisotopes.

Solar PV panels, batteries, and dc–dc converters have been the major infrastructure in space power systems since the beginning. PV arrays supply the loads and charge the batteries. If the generated power by the PV arrays is not sufficient, batteries are then enabled to power the loads. Furthermore, dc–dc converters are employed to protect the space station grid and regulate the voltage. These components have formed an on-board dc microgrid in the space station. Since high reliability and flexibility are necessary for missions in space stations, reliable dc microgrid with high robustness in both hardware and software is required and a promising solution [9,10].

The loads in space crafts can be categorized into three groups: essential, semi-essential, and nonessential loads. dc Microgrids in space applications should solely supply the on-board loads. Therefore, nonessential loads are typically adopted to be sheded during critical situations to regulate the voltage. It should be noted that the dc microgrid employs special protection systems to maintain the stability in space stations.

16.2.3 MVDC microgrid for marine applications

Shipboard microgrids are another type of reliable microgrids, which have no access to the main grid. The shipboard power system and an islanded microgrid have a reasonable analogy in terms of supplying loads with local generation units [11]. The control structure, volume, and weight of shipboard are some major concerns in the design of shipboard microgrids, and there are important issues to deal with vitally imposing pulse loads. Some of the loads are intermittent, operating at a time-scales that can be milliseconds or even less, and the load power can be various from a few kWs to MWs, which may reach 90% of the installed power capacity in a short period [12]. Furthermore, the load priority can vary, depending on the missions of the ship or operating conditions [13]. For instance, a load may be crucial in a cruising mission, while negligible in another case. Thus, the prioritization of loads and dc microgrid control structures, at runtime, are important in shipboard microgrids. Figure 16.5 illustrates an industrial shipboard dc microgrid structure for

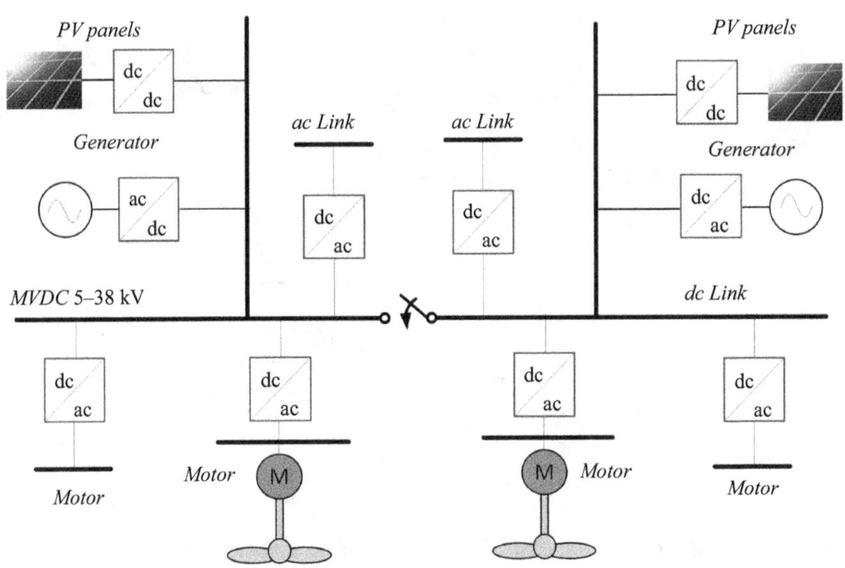

Figure 16.5 dc Microgrid for marine power applications

marine applications [14]. With such structures, not only the operation cost is reduced but also the maintenance of engines is lowered due to more efficient operation. Additionally, dynamic response enhancement and maneuverability are advantages of dc microgrids for marine applications.

16.2.4 LVDC microgrid for data centers

Figure 16.6 shows a typical data center structure consisting of a dc bus, loads, batteries, and power electronic converters. In order to supply sensitive loads in data centers, a local generation unit is implemented. The best solution for data centers to integrate local generation units and reliable power is to employ dc microgrid technologies. In such cases, dc microgrids can operate in the islanded mode and be connected to the grid simultaneously. The main challenge of data centers is to supply the loads with a high power quality and availability. Practically, data center loads are sensitive to voltage and current changes, and hence, a fast, accurate, and reliable control structure is the first requirement for data centers [15]. Furthermore, ESs (batteries) and spinning reserve are also important for reliable operation of dc microgrids in data centers. On the other hand, the required cooling system leads to high power consumption in data centers. Nevertheless, when utilizing LVDC

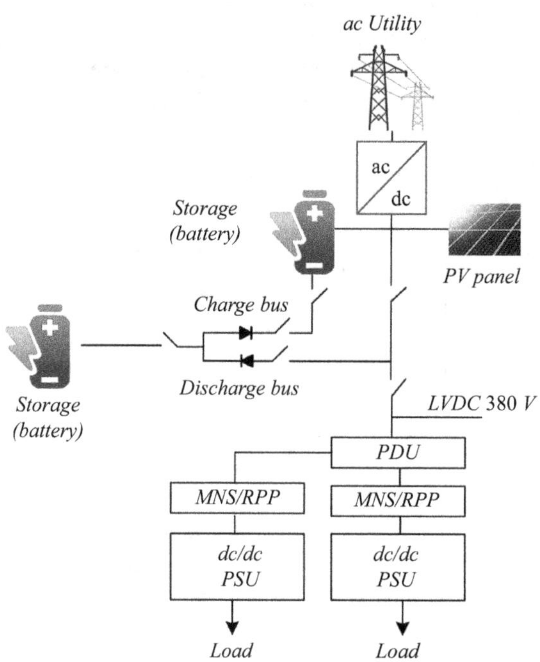

Figure 16.6 LVDC microgrid architecture for data center. PDU, power distribution unit; RPP, remote power panel; PSU, power supply unit; MNS, switchgear

microgrids, ES devices, renewable energy resources, and dc loads can be integrated to form a relatively high reliable power system [16].

16.2.5 LVDC microgrid for homes

Residential dc microgrids enable a distributed power system in a flexible manner for a smart home. In future smart homes, dc loads and electrical vehicles will be massively adopted and coordinated to achieve an optimal residential operation. As it can be seen in Figure 16.7, dc loads are the main load of future smart homes, which are connected to the dc bus via power electronic devices. In the past decade, approximately 80% of the residential power demand—from cell phones and LED lightening to air conditioning systems—are processed through power electronic converters and thereby dc-powered. Such a dc microgrid increases the overall reliability and decreases power losses in green homes by eliminating a conversion stage from dc to ac [16,17]. Furthermore, the integration of renewable energy resources and ESs into LVDC-microgrids makes it possible for zero-net-energy buildings.

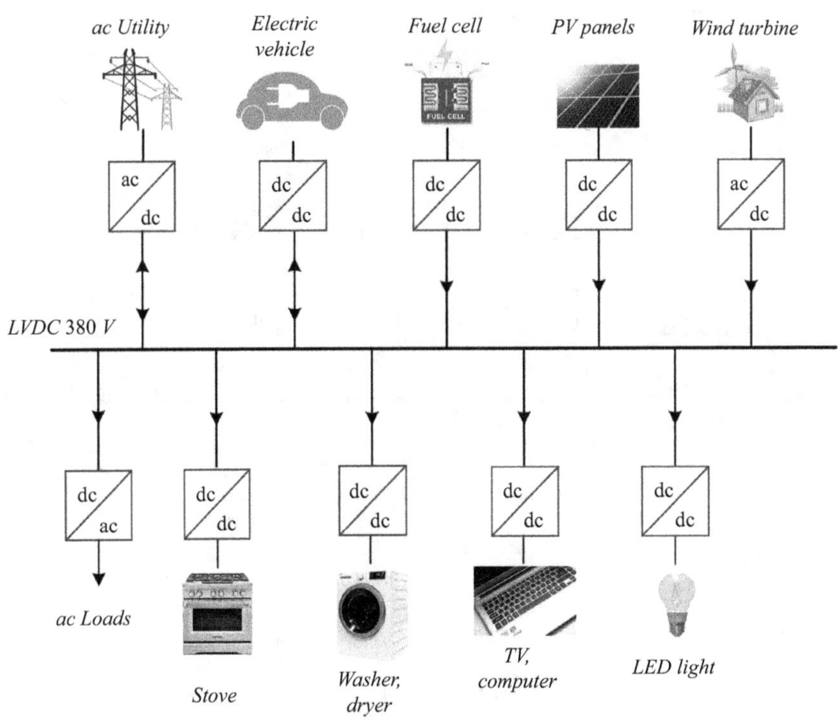

Figure 16.7 A residential dc microgrid in a future smart home, where the arrow indicates the power flow direction

A residential dc microgrid typically has two voltage levels, i.e., low voltage at 24 and 380 V. The 24-V system powers LED lighting and small appliances, while high-power heating equipment is supplied by the 380 V. In such a residential dc microgrid, the power is fed from devices and controllable converters, which can provide a wide range of voltages as well as current limitations (protections). Therefore, electromechanical protection devices can be eliminated and the system safety can be maintained.

16.2.6 MVDC microgrid for oil-drilling applications

Off-shore oil and gas drilling systems are heavy-industrial applications. Drilling, station keeping, propulsion, and pumping products are the main applications of drilling rigs. In such a heavy energy-demand, the power generation and conversion are major barriers to the optimal operation of oil drilling stations. On the other hand, prior-art investigations envision the potential deployment of large wind farms along the shore, where drilling stations are installed. That is, the MV-dc-microgrid architecture with a high penetration of wind turbines can serve as a link to harvest energy for off-shore oil-drilling rigs. Figure 16.8 shows a dc microgrid for oil-drilling applications with collocated wind farms and PV panels [18]. It is noticed in Figure 16.8 that the main grid can be connected, and in that sense, it can be considered as an onshore system [18,19].

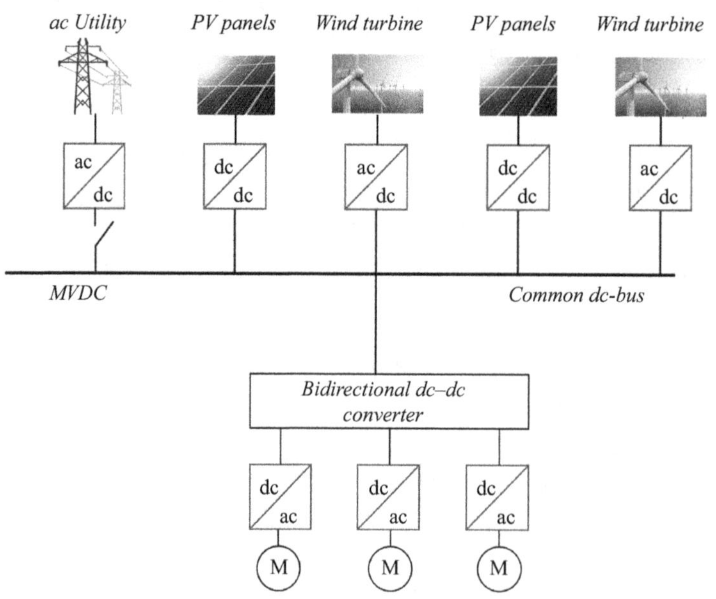

Figure 16.8 MVDC structure utilized in an offshore oil-drilling application integrated with wind farms

16.3 dc Microgrids for photovoltaic power plants

The output power of PV panels is dc, and hence, the operation of PV power plants in dc microgrids may introduce high efficiency and reliability. As the output power of PV panels is intermittent and varies according to weather changes, utilizing a battery storage will increase the system availability. Depending on the combination of PV panels and batteries, two types of dc microgrids are developed: (1) hybrid PV-battery system and (2) distributed PV panel and batteries. In the first structure, as shown in Figure 16.9, a battery storage unit is locally connected to the PV source, and then both are connected to the microgrid. In this case, the entire unit including PV panels and batteries is a unified dispatchable unit. If the maximum power of PV panels (typically achieved by a maximum power point tracking algorithm) is higher than the load demand, the excessive power will charge the battery. Furthermore, once the load power exceeds the PV maximum power, the battery is deployed to supply the load. Notably, a power management system is required for properly sharing power among the clusters of the hybrid PV-battery unit.

On the other hand, in the distributed PV and battery, every PV unit and battery storage are connected to the microgrid through a power converter, as shown in Figure 16.10. In this case, the maximum power from PV plants is used to charge the battery and supply the load. Hence, they can be modeled as current sources. Here, the dc-link voltage will be formed by the batteries. Furthermore, a supervisory control is required to monitor the state-of-charge (SoC) level of batteries and also to manage the output power of PV panels. Once the microgrid demand goes below the maximum power of PV units, the maximum power tracking control of PV panels should be adjusted (e.g., reduced power control) to ensure the demand–supply balance in the microgrids. The control system of PV panels in the

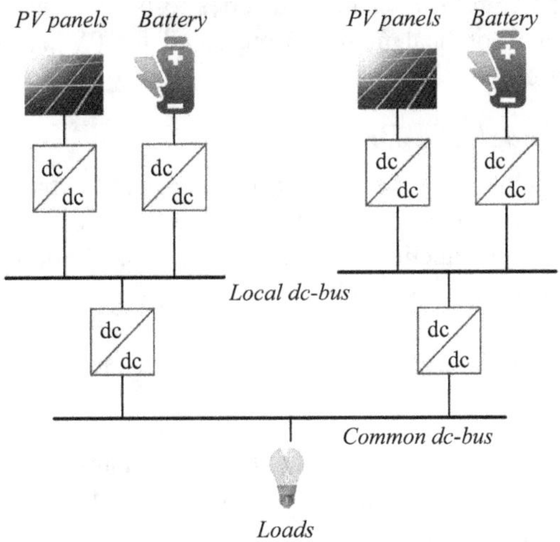

Figure 16.9 dc Microgrid structure for hybrid PV-battery units

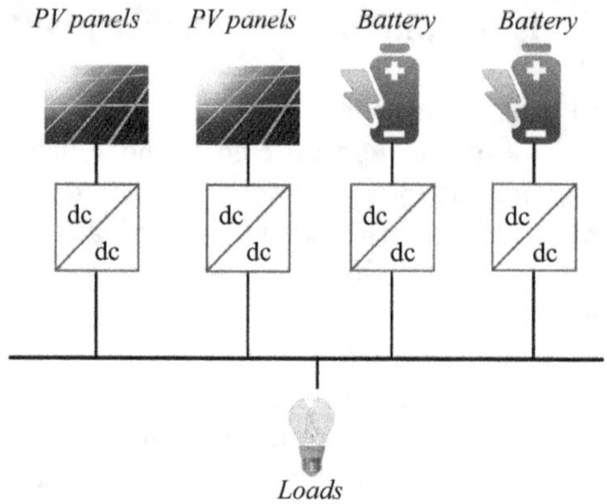

Figure 16.10 dc Microgrid structure for distributed PV and battery units

two PV-integrated microgrids will be explained after the modeling of PV systems in this chapter.

16.4 PV system modeling

With the high penetration of renewable energy resources, academic and industrial interest in utilizing PV and solar energy has been increased. Due to the intermittency, low efficiency, and high manufacturing cost of PV panels, there are challenges when integrating PV energy generation units into dc microgrids. Detailed modeling, analysis, and simulation of PV systems are the prerequisite to design an efficient PV microgrid and properly address the aforementioned challenges. In the practical model of PV modules, different types of partial shading, and the effects of temperature, irradiance, and series resistance have to be scrutinized. In the following, a theoretical model is presented for PV panels to be used for modeling and control.

16.4.1 Ideal model of a PV cell

The PV cells modeling involves *P–V* (output power–output voltage) and *I–V* (output current–output voltage) characteristic curves. Most ideal PV cells have been modeled as a p–n junction with a Shockley-based equivalent circuit. Figure 16.11 shows the basic PV model, which comprises of a diode and an ideal current source.

The mathematical description of the *I–V* characteristic of the PV model is

$$I = I_{Ph} - I_D, \tag{16.1}$$

where I is the output current of the PV cell, I_{ph} is the photo-generated current by the sunlight that is proportional to the solar irradiance level, and I_D is the Shockley current modeling the diode behavior as

$$\begin{aligned} I_D &= I_0\left[e^{qV_D/akT} - 1\right] \\ I &= I_{Ph} - I_{I_0}\left[e^{q(V_D+I)/akT} - 1\right] \end{aligned}, \tag{16.2}$$

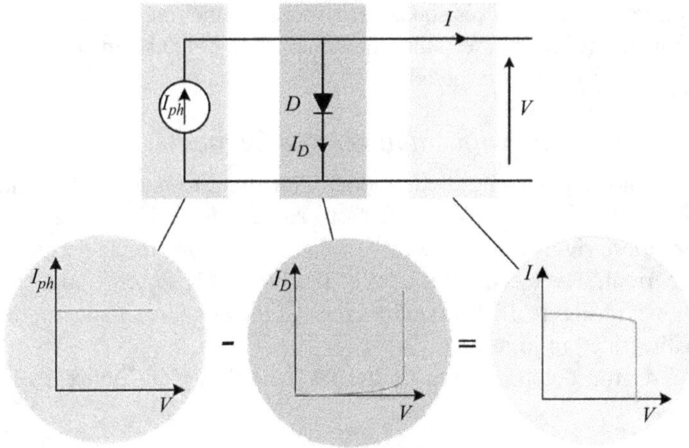

Figure 16.11 *Ideal model of a PV cell. I_{ph}, photo-generated current; I, PV cell current; I_D, diode current; V, PV voltage*

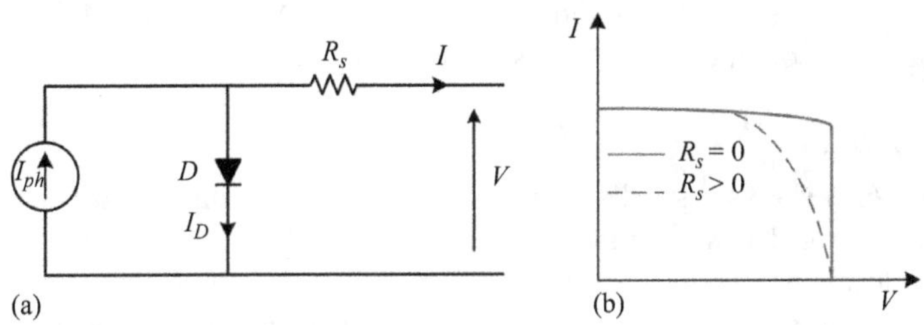

Figure 16.12 *Single-diode PV model with series resistance (R_s represents the series resistance): (a) electrical circuit and (b) V–I characteristic*

I is a reverse bias current of the diode, k is the Boltzmann's constant, q is the charge of an electron ($= 1.602 \times 10^{-19} C$), and T refers to the absolute p–n junction temperature in Kelvin, and α is the diode ideality factor typically within 1–2 [20]. Nevertheless, this simplified model cannot demonstrate the real behavior of a PV cell. Thus, a single-diode model with series resistance has to be introduced.

16.4.2 Single diode with series resistance model

By utilizing series resistance as shown in Figure 16.12(a), the current through base and emitter of solar cell is modeled. Furthermore, the series resistance can model the interconnection losses between silicon and metal contacts, as well as top and rear metal contacts [21,22]. The effect of series R_s on the output I–V characteristic is shown in Figure 16.12(b).

Therefore, the I–V characteristic in this model can be described as:

$$I = I_{PV} - I_0 \left[e^{q(V_D + IR_s)/\alpha kT} - 1 \right], \tag{16.4}$$

where R_s refers to the series resistance. In order to model the behavior of a PV cell in high-temperature conditions and manufacturing defects, an additional shunt resistor is also applied to the model [23].

16.4.3 Single-diode with shunt resistance model

Figure 16.13 shows the single-diode model with shunt resistance. The shunt resistance explains the manufacturing defects to some extent. The effect of the shunt resistance is more significant at a weak irradiance level due to the decreased photo-generated current. The electrical model of the PV cell with series and shunt resistance is shown in Figure 16.13(a). The effect of the shunt resistance on the output current is illustrated in Figure 16.13(b).

Therefore, the output current of the PV cell shown in Figure 16.13 can be obtained as

$$I = I_{PV} - I_{01}\left[e^{q(V_D+IR_s)/akT} - 1\right] - \frac{V+IR_s}{R_{sh}}, \tag{16.5}$$

where R_{sh} refers to the shunt resistance. However, this model cannot show the accurate behavior of PV cells at low solar irradiance. Therefore, the double-diode model was presented in the literature to model the PV cell characteristics at different operation conditions, e.g., weak solar irradiance.

16.4.4 Double-diode model

Equation (16.5) shows a single-diode model with the ideality factor α. However, the ideality factor depends on the voltage across the device. Thus, at a HV, $\alpha = 1$ and at a low voltage, $\alpha = 2$. By adding a parallel diode to the model in Figure 16.13, a relatively accurate PV cell model can be achieved, as shown in Figure 16.14. In order to include the effect of diode voltages in the model, two diodes with $\alpha = 1$ and $\alpha = 2$ are considered.

The comprehensive model shown in Figure 16.14 can be described by

$$I = I_{PV} - I_{01}\left[e^{q(V_D+IR_s)/\alpha_1 kT} - 1\right] - I_{02}\left[e^{q(V_D+IR_s)/\alpha_2 kT} - 1\right] - \frac{V+IR_s}{R_{sh}}, \tag{16.6}$$

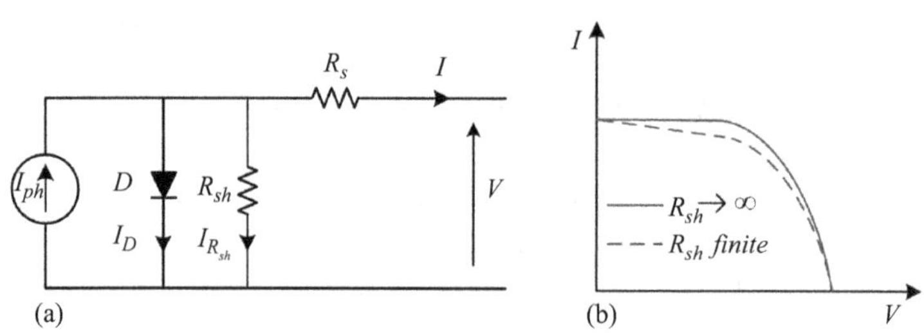

Figure 16.13 Single-diode model with shunt resistance (R_{sh} represents the shunt resistance): (a) electrical circuit and (b) V–I characteristic

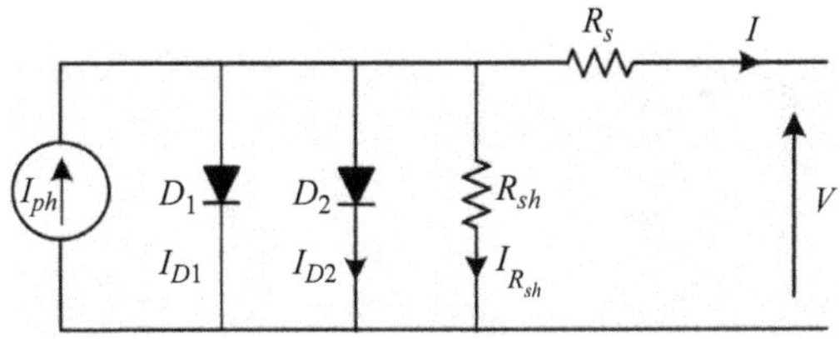

Figure 16.14 Double-diode model for PV cells

where $\alpha = 1$ and $\alpha = 2$. Hence, the diode (D_1) in Figure 16.14 models the effect of the HV and the diode (D_2) models the effect of the low voltage. A multiple-diode model is also presented in the literature which increases the model complexity.

As the power of a single PV cell is very low, a great deal of PV cells should be connected in series and parallel to form a PV module/panel up to several hundred watts. In a similar way, the PV modules/panels can be connected in series and parallel to further achieve high-power and HV PV arrays. The number of series and parallel panels in each PV array is dependent on the *P–V* characteristics of each panel, the input voltage, and the rated current of the interfacing converter.

Furthermore, as the output power of PV panels varies following the solar irradiance and ambient temperature, an interfacing converter is required to control the output current of the PV panels to harvest the maximum power. This is called a maximum power pointing tracking (MPPT) control. In the next section, different dc–dc converters for PV applications are presented.

16.5 Power converter technologies

Generally, dc–dc converters can be implemented between PV panels and the dc-bus or a dc–ac converter in order to control the output voltage/current of the PV panel for power optimization. Thus, different power converter topologies for isolated and nonisolated converters and various control strategies have been presented [24–26]. In the following, transformerless and transformer-based (typically, high-frequency transformers) topologies are explained. Since this chapter focuses on dc microgrids, some dc–dc converters are introduced in this section. Furthermore, some commercial inverter topologies will be introduced, where the first conversion stage can be used in dc microgrids as the interlinking between PV panels and the dc-bus.

16.5.1 Transformerless topologies

Transformerless topologies offer a low-cost and high-efficient solution for different PV structures. In order to perceive the foundation of transformerless technologies, a simple single-phase boost converter will first be presented. The simple but the most common dc–dc converter for PV applications is the conventional boost converter as shown in Figure 16.15. It can be seen that there are a parasitic capacitance (C_{pv}), a filter inductor (L_b), and a switch (S) representing a common mode behavior of the conventional boost converter. In this topology, the output voltage is higher than the

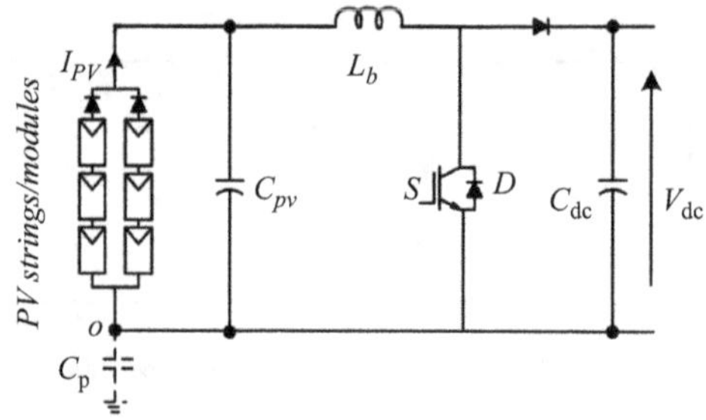

Figure 16.15 Conventional boost converter for PV applications

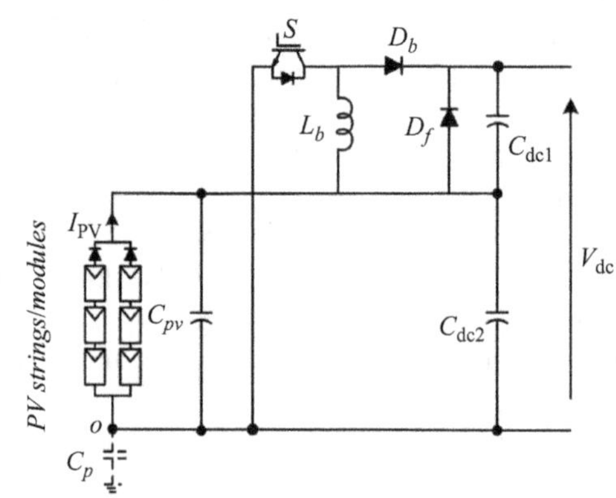

Figure 16.16 Parallel input, series output bipolar dc–dc converter for PV applications

input voltage. The input voltage depends on the PV characteristics and can be determined by an MPPT control system to obtain the maximum power. Furthermore, the output voltage is almost a constant value, which can be controlled by the microgrid.

In order to increase the efficiency of the boost converter, a dc–dc converter with a series output and a parallel input can be used. This structure can be seen in Figure 16.16. It is worth to note that the voltage step-up of this boost converter structure is also enhanced [27].

Boost converters can be used as an interface between PV panels and dc microgrids and/or between PV panels and inverters. Furthermore, a boost converter is employed in industrial PV inverters as the first stage in the entire energy conversion process to control the PV panels to operate at MPPT.

For example, Figure 16.17 shows a commercial topology of a boost converter, introduced in SMA Sunny Boy 5000 TL.

In order to increase the voltage level and standardization of the output PV voltage, safety, transformer-based topologies, which isolate the PV side-circuit from the grid side, have been presented.

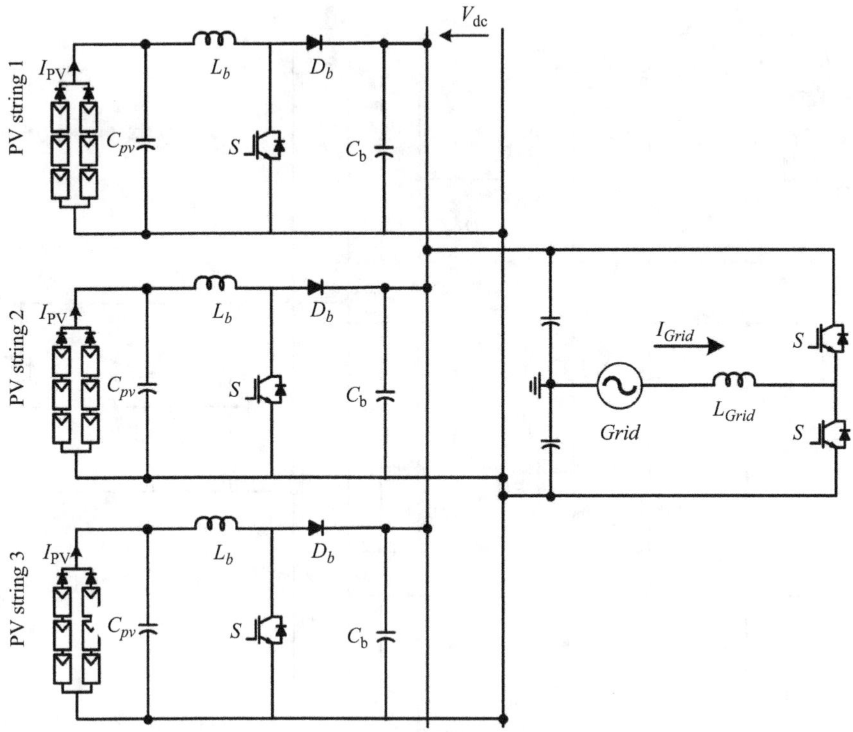

Figure 16.17 SMA Sunny Boy 5000 TL dc–dc converters

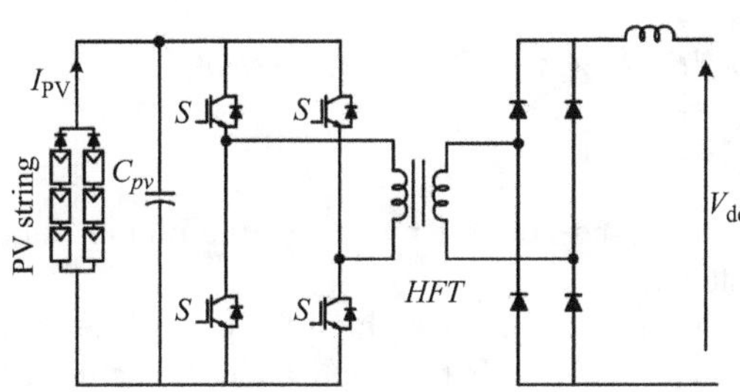

Figure 16.18 Typical transformer-based for dc converters for PV applications. HFT, high frequency transformer

16.5.2 dc Converters with high-frequency transformers

Figure 16.18 shows a basic isolation transformer-based converter for a PV string. By utilizing a high-frequency transformer, the galvanic isolation from the dc-bus is provided. The active full-bridge converter controls the PV string to operate at MPPT, and the voltage level can be increased by the transformer.

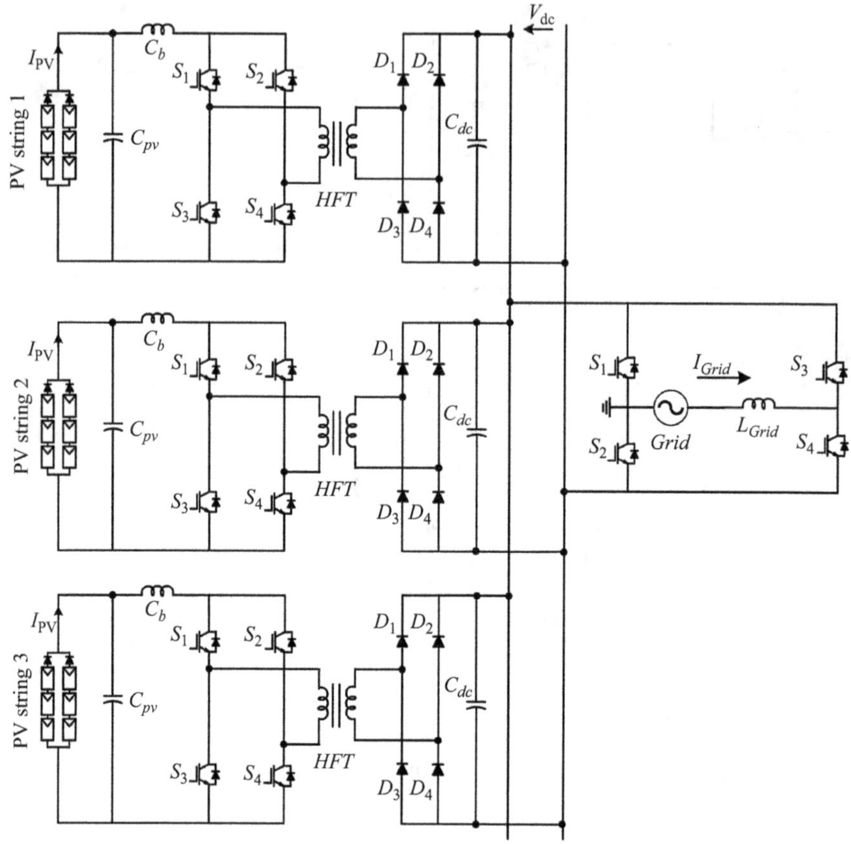

Figure 16.19 A 4.5-kW inverter system from Powerlynx

The full-bridge diode-based converter rectifies the output voltage of the transformer to a dc voltage.

Figure 16.19 shows a commercial high-frequency transformer-based dc–dc converter designed by Powerlynx. In this application, dc–dc full-bridge transformer-based converters are employed as an interface converter between PV modules/strings and the dc–ac inverter.

Another transformer-based topology, i.e., the push–pull converter, is shown in Figure 16.20, which is one of the most economic dc–dc converter configurations. Minimum number of switching devices, the galvanic isolation, and the simple design are the main reasons making this topology more popular.

16.5.3 Next generation PV converters

Due to the electric codes, material isolation, and safety issues, the maximum dc input voltage of PV converters is limited to 600 V in North America and 1,000 V in

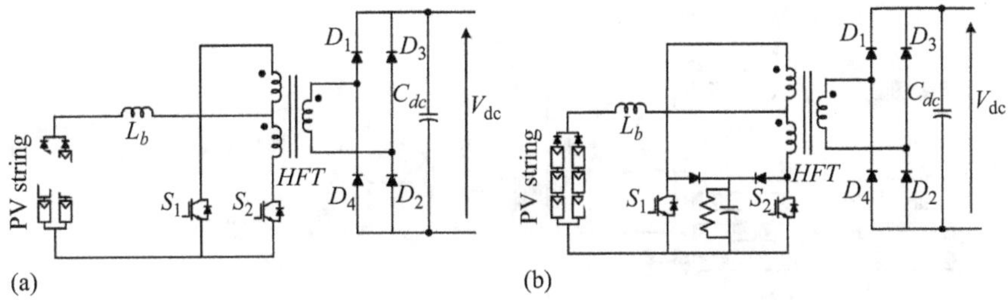

Figure 16.20 Single-phase, push–pull topology converter: (a) with hard switching operation and (b) with clamped snubber circuit

Europe [28–30]. This voltage determines the maximum number of PV modules in each string. Compared to 600-V PV systems, a PV system with the maximum dc voltage of 1,000 V will decrease the required components of the balance-of-system (BOS) (e.g., combiner boxes, dc wiring, and converters) and installation efforts, which thus contributes to reduced overall system cost and increased efficiency.

In today's market, commercial PV converters for utility-scale applications are mostly based on 600 and 1,000-V technologies. However, in order to further reduce the cost of PV power plants and increase the system efficiency, the industry is moving toward a higher voltage level of 1,500 V, where the BOS cost can be reduced [30,31]. As aforementioned, employing the 1,500-V design lowers wiring power losses in ac and dc sides, transformers, and inverters by 1.5%–2%. Another advantage of these systems is the design cost reduction due to less BOS components for 1,500-V systems. For example, in the case of 155-V PV modules, the number of modules in a string will be expanded from around 22 in 1,000-V systems to 32 in 1,500-V systems. Therefore, for the same power rating, the number of strings, combiner boxes, and dc-side cables will be decreased. Consequently, the less volume of equipment due to a higher power density will reduce the costs of installation, transportation, and maintenance.

Besides the above advantages of 1,500-V systems, there are many new challenges to be addressed including PV modules and related components (market availability at the 1,500-V level), raw materials, certifications, regulations, and safety. In the case of 1,500-V dc voltage, the main challenges for PV modules are electric safety and potential-induced degradation, which may weight down the entire benefits of 1,500-V. Furthermore, increasing the dc voltage from 1,000 to 1,500 V brings new challenges from the insulation point of view. Hence, connectors, circuit breakers, protection devices, and switching power supplies require withstanding higher voltage. Finally, a comprehensive electric code update or revision is needed to certificate all the components employed in 1,500-V dc systems. Nevertheless, many attempts by the industry have been made to dissolve the above barriers so that HV (e.g., 1,500-V) PV systems are expected to improve the entire system economics [28]. Figure 16.21 exemplifies the general configuration of 1,500-V PV systems.

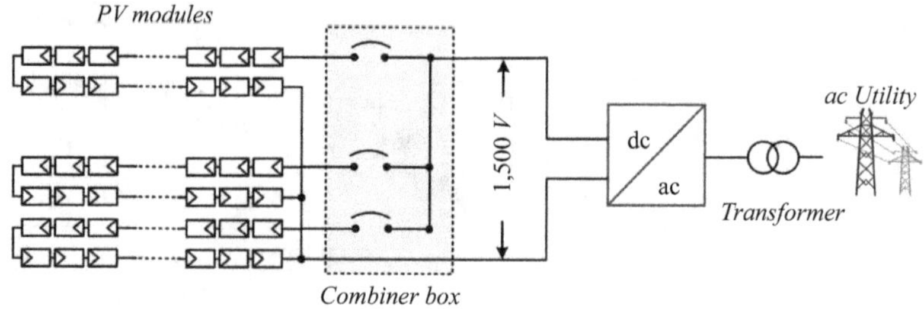

Figure 16.21 PV power systems for utility-scale applications with a high dc voltage (i.e., 1,500 V), where the grounding of PV modules is not shown. In this case, each PV string can reach to a voltage of 1,500 V and multiple strings can be adopted

16.6 Control of dc microgrids for PV collection plants

Microgrid brings an infrastructure to integrate renewable resources into the grid in order to increase the power system reliability and availability [32]. Hence, both grid-connected and islanded operation should be viable to reach the aforementioned aims [7]. As previously mentioned, PV units can be merged with battery units and then both can be connected to the dc grid through power converters. In this case, the entire unit can be seen as a dispatchable unit. Therefore, droop-based controllers can be employed for each unit. Here, the PV unit operates at MPPT and the demand can be supplied by the PV units and batteries. If the demand is lower than the PV power, the excessive power will charge the batteries; however, for the condition, the demand is higher than the PV power, and the battery will support the remaining load demand.

Furthermore, in most cases, PV units and batteries are distributed through the dc grid and individually connected to the grid by its own converters. In this condition, the PV units are not dispatchable, and hence, they cannot fully support the grid. It is obvious that in this case, the PV units cannot operate in the MPPT mode once the generation power is higher than the microgrid demand. Therefore, the microgrid power management system has to control the output power of PV units. There are some supervisory controls to monitor and control the power balance in the microgrids with MPPT-based units like PV panels in order to change the operation mode from MPPT to a constant power mode [33]. Furthermore, droop-based autonomous approaches have been proposed for MPPT-based units. Some power sharing approaches for PV-based dc microgrids have been presented [32,34,35].

For a typical dc microgrid with PV and battery units, a typical droop characteristics can be considered as shown in Figure 16.22. Following Figure 16.22(a), PV units can operate in two modes, i.e., the MPPT mode in which all the PV modules generate an MPPT power, and the droop-controlled mode in which the PV

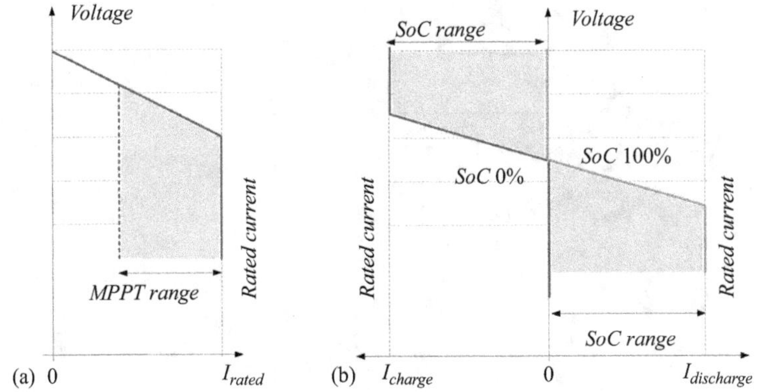

Figure 16.22 Droop characteristics for dc microgrid: (a) nondispatchable units and (b) storage converters

units are operating under the MPPT power and share the load based on their droop gains [33]. Furthermore, the maximum supplied power by the PV units are limited by the MPPT power as shown in Figure 16.22(a). By decreasing the microgrid demand, the dc voltage will be increased, and hence, the PV controller enters into the droop-controlled mode.

The battery units also should be controlled in terms of power sharing and energy level, i.e., SoC level. An autonomous droop characteristic for distributed batteries in the microgrid can be shown as Figure 16.22(b), where the maximum charging and discharging power can be set taking into account the SoC level of batteries. Hence, by adapting the droop characteristics in the local controller, the energy level of each battery can be controlled. Furthermore, the power sharing, both charging and discharging powers, among the batteries can be controlled by properly setting the droop gains shown in Figure 16.22(b).

Since the scope of this chapter focuses on the PV units in microgrids, possible control structures of PV converters are shown in Figure 16.23. For simplicity, a dc–dc boost converter is considered as shown in Figure 16.23(a). For the control structures shown in Figure 16.23(b) and (c), a supervisory control changes the PV operation mode following the system generation-load balance constraints. However, in the control structures shown in Figure 16.23(d) and (e), the operation mode of the PV units will be determined by the local parameters, i.e., the MPPT power and the droop-induced power. These approaches are suitable for autonomous droop techniques.

16.7 Simulation of a droop-controlled PV microgrid

In this section, a simulation is performed in order to demonstrate the droop-based autonomous control of dc sources in a dc microgrid with two PV units and two

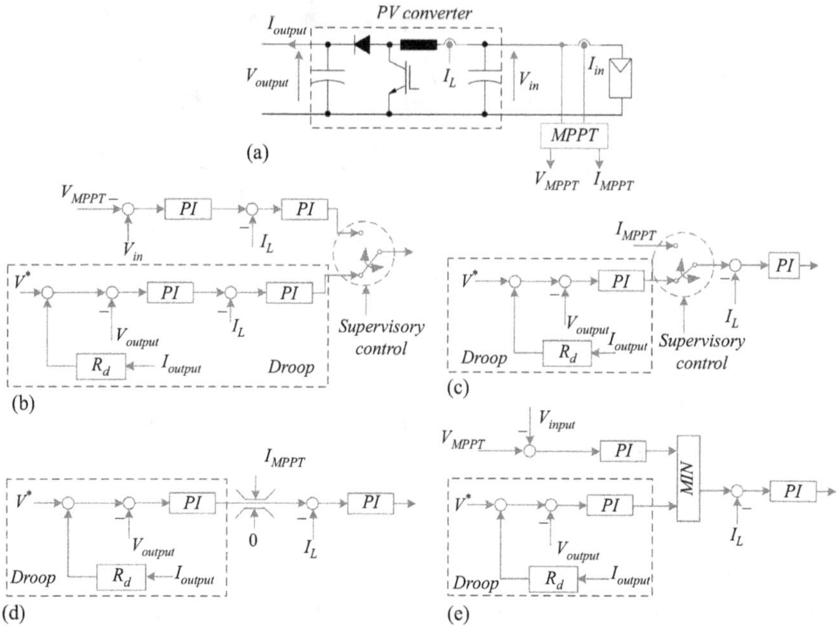

Figure 16.23 Adjustable droop control for MPPT-based units (e.g., PV): (a) dc–dc boost converter for PV, (b), (c) mode transition strategy, and (d), (e) seamless strategy

batteries as shown in Figure 16.24. The parameters of the simulated dc grid are given in Table 16.1. The droop characteristics of the units are set as shown in Figure 16.23(a) and (b), where the droop gains of batteries are 2 and 1 Ω. Furthermore, the MPPT currents of PV units are set to 2.25 and 2.75 A, and the droop gains of PV units are 10 Ω. The batteries are considered to be fully charged in the beginning. Simulation results are shown in Figure 16.25. The microgrid load is 3.2 kW and the PV units are operating in MPPT mode with $(2.25 + 2.75) \times 400 = 2$ kW. The remaining load power is supplied by batteries, where the output current of the first battery is two times that of the other one following their droop gains as shown in Figure 16.25(b). Therefore, the batteries are operating in the droop mode and forming the dc-link voltage, and the maximum power of the PV units is injected into the microgrid. Once the load is decreased to 1.6 kW, the dc-link voltage is increased as shown in Figure 16.25. Following Figure 16.22(b), by increasing the dc voltage, the PV units should operate in the droop-controlled mode and decrease the output current to balance the demand of the system. The load sharing among the dc sources is shown in Figure 16.25(b). The PV units are working in the droop-controlled mode and supplying the equal current following their droop gains. Furthermore, the battery currents are zero, since they are considered as fully charged. In this mode, the dc voltage is formed by the PV units.

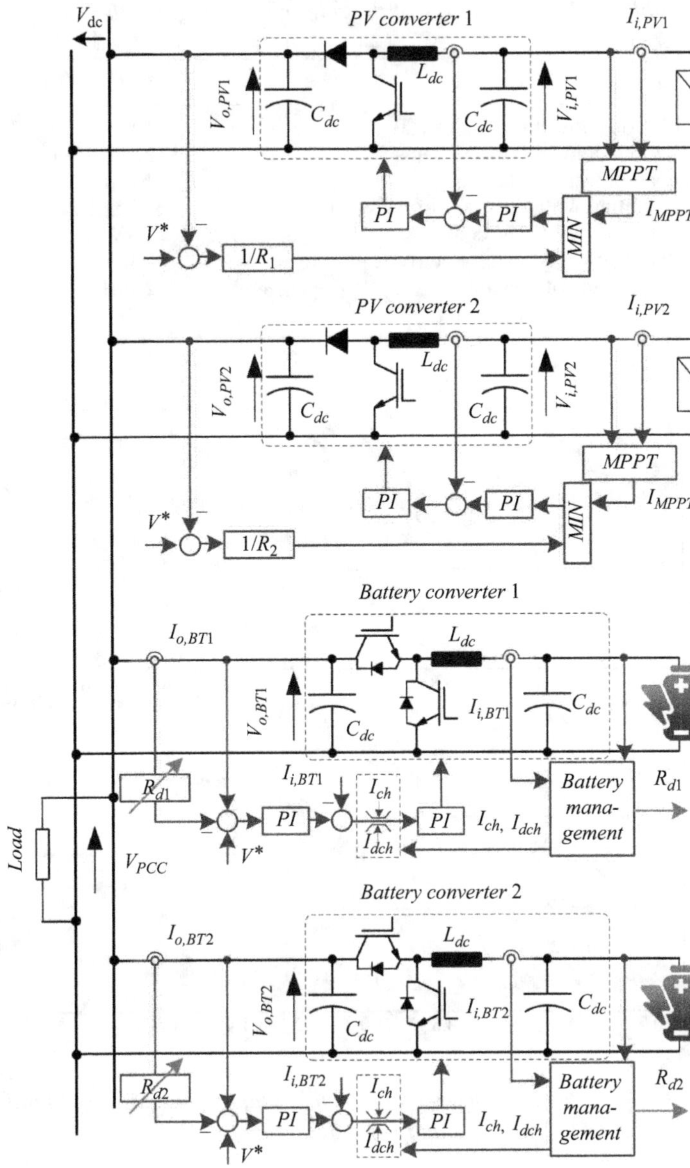

Figure 16.24 A typical dc microgrid with two dispatchable units, i.e., battery and two nondispatchable PV units

Table 16.1 Parameters of the power system controllers

Parameter	Value
PV converter droop gains (Ω)	10, 10
Battery converter droop gains (Ω)	2, 1
Load power (kW)	1.6, 1.6
Operating MPPT current of PVs (A)	2.25, 2.75
Rated power of PV converter	2 kW
Rated power of battery converter	2 kW
dc Inductor of converters	2 mH
dc Capacitor of converters	500 μF

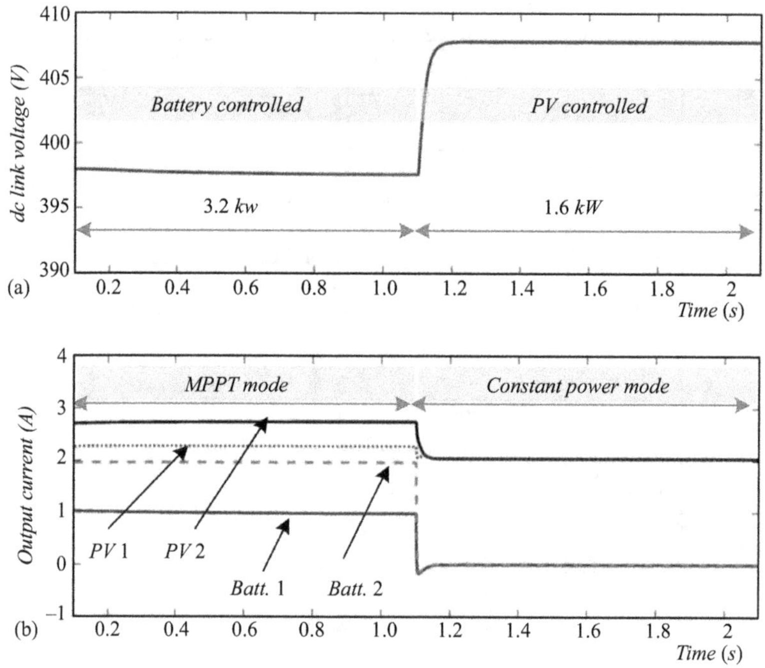

Figure 16.25 Load sharing among dispatchable and nondispatchable units: (a) dc link voltage and (b) output currents

16.8 Summary

dc Microgrids introduce an infrastructure for integrating dc-based energy resources, storages and loads with high reliability, availability, and efficiency. Meanwhile, PV systems are one of the most popular renewable resources due to its modularity as well as lower costs. This chapter discussed different applications and structures of dc microgrids, where PV units can be utilized as an energy resource. These applications can be divided into dc distribution systems, marine applications, data centers, oil-drilling platforms, off-grid applications, dc homes, etc. Different PV integration possibilities into dc microgrids were presented as well. Following the application of dc microgrids, hybrid PV-battery units or distributed PV and battery units can be employed. For example, the first structure is suitable for clusters of dc homes, where each home has its own PV and battery units. For an industrial application, the second structure, i.e., distributed PV and battery units, is more viable.

Furthermore, in this chapter, the electrical model of PV systems was explained and different dc–dc converter topologies for PV applications were presented. Moreover, various control systems of PV plants in dc microgrid were introduced as well as power sharing concepts among paralleled PV units and battery storages was provided by simulations on a typical dc microgrid employing the autonomous droop control approach.

Optimal design and control of PV power plants in dc microgrids taking into account the reliability and availability constraints are required. Since the microgrids may operate in the islanded mode, the critical loads need to be supplied with an acceptable level of availability. Therefore, this concept should be taken into consideration in future research of PV-based dc microgrids.

References

[1] Blaabjerg, F., Yang, Y., Yang, D., Wang, X.: "Distributed power-generation systems and protection" *Proc. IEEE*, 2017, **105**, (7), pp. 1311–1331.

[2] International Renewable Energy Agency: "Renewable Energy Capacity Statistics 2017," 2017.

[3] Youichi, I., Yang, Z., Hirofmi, A.: "DC Microgrid Based Distribution Power Generation System", in "Power Electronics and Motion Control Conference, 2004" (IPEMC, 2004), the 4th International, 2004, **3**, pp. 1740–1745. IEEE.

[4] Kumar, D., Zare, F., Ghosh, A.: "DC microgrid technology: system architectures, AC grid interfaces, grounding schemes, power quality, communication networks, applications, and standardizations aspects" *IEEE Access*, 2017, **5**, pp. 12230–12256.

[5] Diaz, N.L., Dragicevic, T., Vasquez, J.C., Guerrero, J.M.: "Intelligent distributed generation and storage units for DC microgrids: a new concept on cooperative control without communications beyond droop control" *IEEE Trans. Smart Grid*, 2014, **5**, (5), pp. 2476–2485.

[6] Elsayed, A.T., Mohamed, A.A., Mohammed, O.A.: "DC microgrids and distribution systems: an overview" *Electr. Power Syst. Res.*, 2015, **119**, pp. 407–417.

[7] Dragicevic, T., Lu, X., Vasquez, J.C., Guerrero, J.M.: "DC microgrids—part II: a review of power architectures, applications, and standardization issues" *IEEE Trans. Power Electron.*, 2016, **31**, (5), pp. 3528–3549.

[8] Kakigano, H., Miura, Y., Ise, T.: "Low-voltage bipolar-type DC microgrid for super high quality distribution" *IEEE Trans. Power Electron.*, 2010, **25**, (12), pp. 3066–3075.

[9] McNelis, A.M., Beach, R., Soeder, J.F., *et al.*: "Simulation and Control Lab Development for Power and Energy Management for NASA Manned Deep Space Missions," in "12th International Energy Conversion Engineering Conference" (American Institute of Aeronautics and Astronautics, 2014).

[10] Dever, T., Trase, L., Soeder, J.F., *et al.*: "Application of Autonomous Spacecraft Power Control Technology to Terrestrial Microgrids," in "12th International Energy Conversion Engineering Conference" (American Institute of Aeronautics and Astronautics, 2014).

[11] Khooban, M.-H., Dragicevic, T., Blaabjerg, F., Delimar, M.: "Shipboard microgrids: a novel approach to load frequency control" *IEEE Trans. Sustain. Energy*, 2017, **9**, (2), pp. 843–852.

[12] Hebner, R.E., Uriarte, F.M., Kwaskinski, A., Gattozzi, A.L., Estes, H.B., Anwar, A.: "Technical cross-fertilization between terrestrial microgrids and ship power systems" *J. Mod. Power Syst. Clean Energy*, 2016, **4**, (2), pp. 161–179.

[13] Im, W.-S., Wang, C., Tan, L., Liu, W., Liu, L.: "Cooperative controls for pulsed power load accommodation in a shipboard power system" *IEEE Trans. Power Syst.*, 2016, **31**, (6), pp. 5181–5189.

[14] Hansen, J.F., Lindtjørn, J.O., Vanska, K.: "Onboard DC Grid for enhanced DP operation in ships" *MTS Dyn. Position. Conf.*, Houston, 2011.

[15] Jiang, T., Yu, L., Cao, Y.: "Green Energy and Technology Energy Management of Internet Data Centers in Smart Grid." Berlin Heidelberg: Springer, 2016.

[16] Yu, L., Jiang, T., Zou, Y.: "Real-time energy management for cloud data centers in smart microgrids" *IEEE Access*, 2016, **4**, pp. 941–950.

[17] Rodriguez-Diaz, E., Vasquez, J.C., Guerrero, J.M.: "Intelligent DC homes in future sustainable energy systems: when efficiency and intelligence work together" *IEEE Consum. Electron. Mag.*, 2016, **5**, (1), pp. 74–80.

[18] Reed, G.F., Grainger, B.M., Sparacino, A.R., Zhi-Hong M.: "Ship to grid: medium-voltage DC concepts in theory and practice" *IEEE Power Energy Mag.*, 2012, **10**, (6), pp. 70–79.

[19] Kounev, V., Tipper, D., Grainger, B.M., Reed, G.: "Analysis of an Offshore Medium Voltage DC Microgrid Environment; Part II: Communication Network Architecture," in "2014 IEEE PES T&D Conference and Exposition" (IEEE, 2014), pp. 1–5.

[20] Patel, H., Agarwal, V.: "MATLAB-based modeling to study the effects of partial shading on PV array characteristics" *IEEE Trans. Energy Convers.*, 2008, **23**, (1), pp. 302–310.

[21] Mette, A., Pysch, D., Emanuel, G., Erath, D., Preu, R., Glunz, S.W.: "Series resistance characterization of industrial silicon solar cells with screen-printed contacts using hotmelt paste" *Prog. Photovoltaics Res. Appl.*, 2007, **15**, (6), pp. 493–505.

[22] Pysch, D., Mette, A., Glunz, S.W.: "A review and comparison of different methods to determine the series resistance of solar cells" *Sol. Energy Mater. Sol. Cells*, 2007, **91**, (18), pp. 1698–1706.

[23] Salam, Z., Ishaque, K., Taheri, H.: "An Improved Two-Diode Photovoltaic (PV) Model for PV System," in "2010 Joint International Conference on Power Electronics, Drives and Energy Systems & 2010 Power India" (IEEE, 2010), pp. 1–5.

[24] Liu, K.-H., Lee, F.C.Y.: "Zero-voltage switching technique in DC/DC converters" *IEEE Trans. Power Electron.*, 1990, **5**, (3), pp. 293–304.

[25] Lee, K.-H., Chung, E., Han, Y., Ha, J.-I.: "A family of high-frequency single-switch DC–DC converters with low switch voltage stress based on impedance networks" *IEEE Trans. Power Electron.*, 2017, **32**, (4), pp. 2913–2924.

[26] Rehman, M.M.U., Zhang, F., Zane, R., Maksimovic, D.: "Control of Bidirectional DC/DC Converters in Reconfigurable, Modular Battery Systems," in "2017 IEEE Applied Power Electronics Conference and Exposition (APEC)" (IEEE, 2017), pp. 1277–1283.

[27] Yang, Y., Blaabjerg, F.: "Overview of single-phase grid-connected photovoltaic systems" *Electr. Power Components Syst.*, 2015, **43**, (12), pp. 1352–1363.

[28] S. Moskowitz: "The Next Opportunity for Utility PV Cost Reductions: 1,500 Volts DC, Greentech Media," 2018. https://www.greentechmedia.com/articles/read/the-next-opportunity-for-utility-pv-cost-reductions-1500-volts-dc, accessed February 2018.

[29] Serban, E., Ordonez, M., Pondiche, C.: "DC-bus voltage range extension in 1500 V photovoltaic inverters" *IEEE J. Emerg. Sel. Top. Power Electron.*, 2015, **3**, (4), pp. 901–917.

[30] Inzunza, R., Okuyama, R., Tanaka, T., Kinoshita, M.: "Development of a 1500 Vdc Photovoltaic Inverter for Utility-Scale PV Power Plants," in "2015 IEEE 2nd International Future Energy Electronics Conference (IFEEC)" (IEEE, 2015), 2015, pp. 1–4.

[31] C. Crowell: "1,500-Volt Systems to Trend in 2017: Here's What You Need to Know – Yaskawa – Solectria Solar," 2018. https://solectria.com/company/in-the-news/1-500-volt-systems-to-trend-in-2017-heres-what-you-need-to-know/, accessed February 2018.

[32] Boroyevich, D., Cvetkovic, I., Burgos, R., Dong, D.: "Intergrid: a future electronic energy network?" *IEEE J. Emerg. Sel. Top. Power Electron.*, 2013, **1**, (3), pp. 127–138.

[33] Dragicevic, T., Guerrero, J.M., Vasquez, J.C., Skrlec, D.: "Supervisory control of an adaptive-droop regulated DC microgrid with battery management capability" *IEEE Trans. Power Electron.*, 2014, **29**, (2), pp. 695–706.

[34] Gu, Y., Xiang, X., Li, W., He, X.: "Mode-adaptive decentralized control for renewable DC microgrid with enhanced reliability and flexibility" *IEEE Trans. Power Electron.*, 2014, **29**, (9), pp. 5072–5080.

[35] Peyghami, S., Mokhtari, H., Blaabjerg, F.: "Decentralized load sharing in an LVDC microgrid with an adaptive droop approach based on a superimposed frequency" *IEEE J. Emerg. Sel. Top. Power Electron.*, 2017, **5**, (3), pp. 1205–1215.

Chapter 17
Demonstration sites of dc microgrids

Aditya Shekhar[1], *Laura Ramírez-Elizondo*[1], *Seyedmahdi Izadkhast*[1] *and Pavol Bauer*[1]

17.1 Introduction

In recent years, several demonstration projects using dc microgrids have been implemented across the world due to some distinct advantages of dc over ac systems. The purpose of such initiatives is to validate the theoretically predicted benefits of dc distribution in practical scenarios. This chapter presents a non-comprehensive overview of existing demonstrations and pilot projects for a wide range of applications such as off-grid microgrids, transportation electrification, datacenters, residential and industrial purposes. For each application, key aspects such as architecture, components, control, protection and socioeconomic impacts are highlighted. A short discussion is offered on trends in voltage levels, capacity and topology, progressing toward possible standardization approaches based on recognized best practice.

17.2 Off-grid dc microgrids

Off-grid applications can refer to remote locations with limited access to the conventional ac grid. The grid may be required to be operated in stand-alone mode for long hours during the day, e.g., from 12 to 24 h. For instance, in many states of India, 70%–90% households fall under this category [1]. In such applications, lack of market inertia of the ac grid may aid to the proliferation of dc nanogrids. The energy needs of such areas are scarcely met due to economic constraints that make load shedding a preferred option [1]. Consequently, off-grid solar dc homes similar to one shown in Figure 17.1 are the emerging solution for rural electrification [2].

17.2.1 Architecture

A typical architecture for off-grid dc solar home systems (SHS) is shown in Figure 17.2. The structure is radial with a 48-V common dc bus interconnecting the

[1]Department of Electrical Sustainable Energy, DC Systems, Energy Conversion and Storage (DCE&S) Group, Delft University of Technology, The Netherlands.

Figure 17.1 Illustration of dc powered offgrid Tribal homes at Kundithal village, The Nilgiris, Tamil Nadu, India (courtesy: [2])

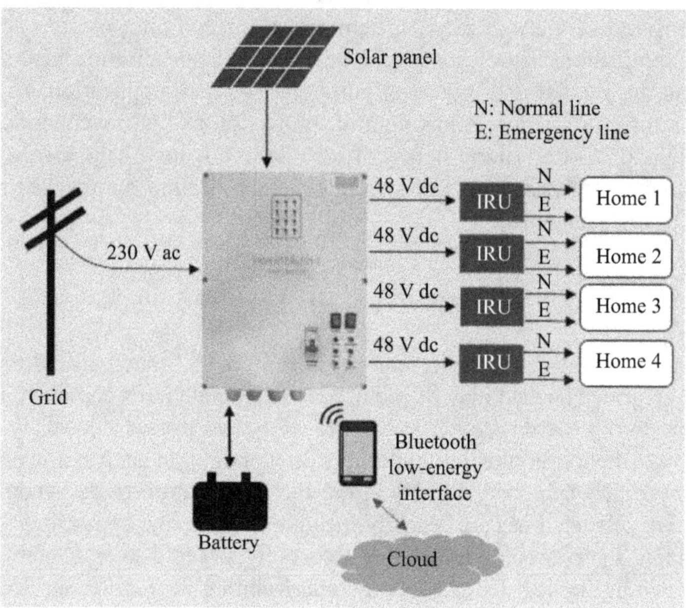

Figure 17.2 Inverterless-500 dc powered solar home system architecture [1]

distributed sources, storage and loads. To guarantee power supply to critical loads, connections are bifurcated into normal line (N) and the emergency line (E). Each home has an inverter-less remote unit (IRU). The option of remote monitoring using, for example, a Bluetooth interface is also available.

17.2.2 Components

The minimum needs for a low-income household may at least consist of loads such as a few fans, tube-lights, bulbs and mobile phone chargers. With available opportunities, some households showed preference for entertainment loads such as television and commercial loads such as tailoring [2–4], amounting to a demand of 100–500 W. With use of dc, 2–2.5 times reduction in battery and solar module size at half the cost and efficiency enhancement at upward of 15% have been reported [1,2]. The use of dc-ready appliances is encouraged, claiming efficiency gains of almost 50% with such practice [2], such as 1×18 W LED tube-light, 1×32 W brush-less dc fan, 1×5 W LED bulb and a mobile phone charger. The 125-Wp solar photovoltaic (PV) generation has a trade-off between cost and availability. Storage capacity caters for load demands when neither PV nor grid is available, designed to supply power for a downtime of 1–2 days [2]. Some ac loads can be directly supplied from the grid or through the dc bus, even though, the use of an additional inverter is discouraged.

17.2.3 Control and operational safety

Considering the small scale of application, centralized control is employed in the main control board for one or several households. Heuristic algorithms implemented by the IRU minimize grid usage and maximize PV utilization. They have to also ensure adequate electricity supply to critical loads during significant downtimes. The state of charge (SoC) of the batteries is monitored and if no power is generated, the storage supplies power to all loads until a certain threshold. If the threshold is reached, then the non-critical loads are shed. dc Voltage of 48 V is usually chosen for low capacity nanogrids, as occurrence of sustained series arcing is limited [5].

17.2.4 Socioeconomic impact

Case-studies of solar house systems for rural electrification across the world shared positive experiences from the end users, showing enhancement in the standard of living with fulfillment of basic energy needs [1–4]. Of the six cases of SHS studied in [4], only four had payback below 3 years and an internal rate of return above 39%. Demonstrating financial viability with larger capacity and/or encouragement of rural businesses, it was highlighted that sustainability of dc based SHS hinges on greater support from various government and nongovernment organizations, local sources of funds and a premium on social benefits of rural electrification.

17.2.5 Discussion

A summary of various dc offgrid applications worldwide with respect to voltage levels, capacity, topology, and control is listed in Table 17.1. In most cases, the demonstrations were dominated by low capacity (below 500 W) and low voltage levels (48 V or below, in line with future IEEE P2030.10 Standard for rural and remote dc), signifying that this technology is developed to meet the basic energy

Table 17.1 Demonstration sites for offgrid or near offgrid dc microgrids

	Demo-site description	Voltage level	Capacity	Topology	Control
1	SHS[a], Bangladesh [3,4]	12 V	≤ 100 W		
2	SHS[a], Africa [6]		20–100 W		
3	SHS[a], Kundithal, Tamil Nadu, India [2]	48 V	20 homes, 100–500 W each		Central
4	SHS[a], Irrukam Island [2] Andhra Pradesh, India	48 V	26 homes 100–500 W each	All radial	Central
5	SHS[a], Bhomji ka Gaon [1]	48 V	125 Wp, 1 kW h		Central
6	School, Tirmal, Orissa, India [7]	48 V	3 kW, 50 kW h		
7	PEC[b], Portugal [8]	24 V	1.8 kWp		Central

[a]SHS: solar house systems.
[b]PEC: polygeneration energy container.

needs of socioeconomically backward areas. The topology choice is usually radial with PV panels, small storage and dc-ready loads attached to a common dc bus, generally with no interconnection between individual nanogrids. Thus, end-user need based system design with entity level centralized control scheme was preferred.

17.3 Transportation electrification

Electrification of the transportation sector is driven by the desire to shift toward clean energy sources and reduced emissions. Potential size and weight reduction play a pivotal role in dc-based transportation electrification. To give a numerical example on the size, The University of Trieste in Italy conceived a 2 MVA, 22,500 rpm, 12 phase permanent magnet prototype generator [9], connected to 3 kV dc shipboard power system with a controlled active ac–dc converter. The weight of generator was reduced to less than 50% of its predecessor. The adoption of dc technologies in transportation can be rapid, since these systems are stand-alone. This section provides an overview of shipboard, traction and electric vehicle (EV) charging dc power systems.

17.3.1 Shipboard dc microgrids

Electric propulsion has been considered in naval destroyers (e.g., UK's Type 45 destroyer) as it may provide opportunities to improve the power density using advancements such as emerging permanent magnet and super-conducting technologies [10]. The interest in all electric ships (AES), fueled by the need for efficient space utilization, has given rise to the concept of a dc-based integrated power system (IPS) [11]. Typically, the shipboard power systems have a high capacity in the order of a few megawatts. The voltage levels are between 1 and 35 kV. The recommended practices for reliable integration of shipboard power components in a medium voltage dc (MVDC) system are discussed by the IEEE Standards Association [12].

Figure 17.3 An illustration of a dc-based all electric ship (adapted from [14])

17.3.1.1 Architecture

Keeping in mind the redundancy requirements, two buses, the port and starboard bus run through the length of the ship. An illustration of such a dc microgrid island on ship is shown in Figure 17.3. Different shipboard dc electrical distribution system architectures with trade-offs between reliability, complexity and efficiency are discussed in [11]. An interesting compact design concept of "Power Corridor" toward the IPS keeping in mind the space reservations for AES is proposed in [13].

17.3.1.2 Components

As part of The Electric Ship Research and Development consortium, the University of Texas at Austin center for electromechanics (UT-CEM) assembled a 2-kV MVDC microgrid testbed as shown in Figure 17.4 [15].

The isolated naval microgrid is envisioned to operate at 80–100 MW with different steady state and pulsed power loads. The following components are highlighted: (a) gas turbine driven 3 MW high speed generator, (b) full bridge passive diode rectifier, (c) bidirectional, active, IGBT controlled converter, (d) 1.25 MW Toshiba model variable frequency drive capable of operating up to 300 Hz modified from its commercial version to operate as a "dc-ready" device able to directly interface the load with the common dc bus, (e) Kahn model hydraulic dynamometer capable of absorbing 4.77 MW at 18,000 rpm, that can be controlled to mimic transient conditions in dc distribution systems.

17.3.1.3 Power quality, control and protection

System level studies on transients associated with microgrid islanding and reconnection were planned in [15], focusing on ancillary services with different control strategies for better power quality. In [9], the bus connected choppers integrating the prototype high speed permanent magnet generator are responsible for the output

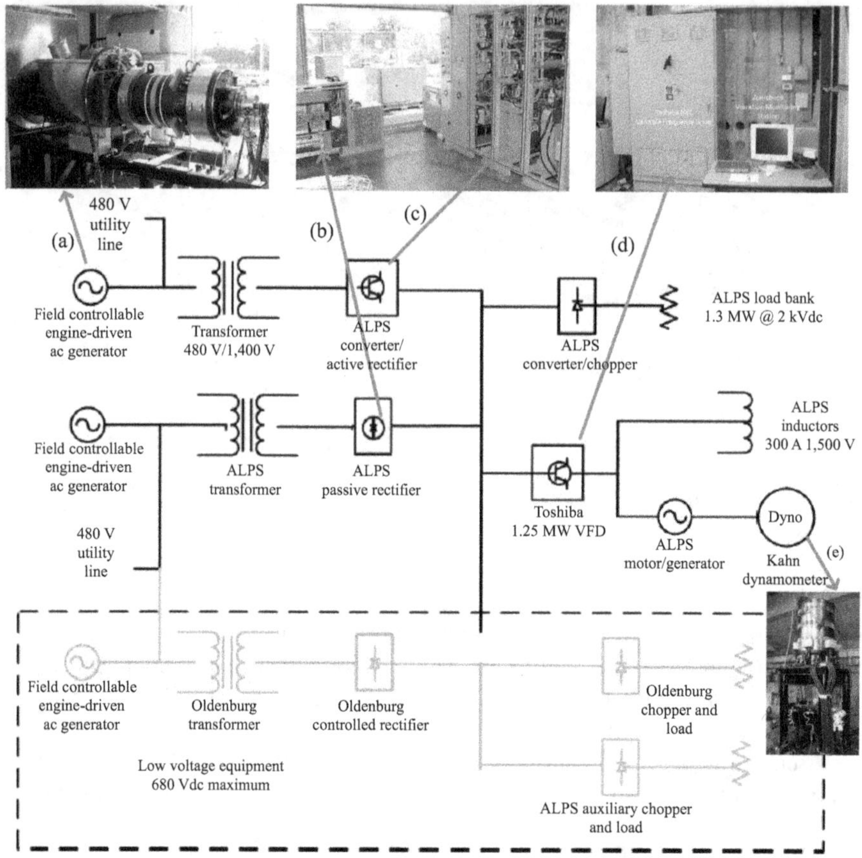

Figure 17.4 Testbed at UT-CEM for MVDC shipboard microgrid (courtesy: [15])

voltage control. The machine reacted to the power electronics and resulted in distortion in phase current, leading to flux pulsations and eddy current losses, an important challenge to consider during design.

For stable ship operation, it is important to maintain power availability during contingencies. It is identified that due to interconnectivity, the reconfiguration capability maybe a potential advantage of MVDC IPS in AES [11]. While acknowledging significant efficiency gains and lesser problems with harmonic distortion, The "ABB" group of industries noted that the challenge in the design of the marine on-board dc system was related to protection and selective localization of faults. Their approach was a coordinated use of fuses and isolation switches along with protective power electronic devices. Further, the controlled thyristor rectifier connected to the generator played an added role of protection device. It was claimed that dc fault currents can be controlled within 10–20 ms, providing a significant reduction in fault energy as compared to the conventional ac system [16].

17.3.1.4 Economical operation

The interest in onboard dc led to the commercial design of Dina Star by ABB. This eliminated the ac switchboards and need for bulky transformers, as well as reduced the power conversion stages. It was claimed that the use of 1 kV dc with up to 20 MW demand improves the vessel's energy efficiency by 20% and the weight of electrical system by 30% [16]. Based on the operational experience with Dina Star, ABB claimed that 27% efficiency gains could be achieved for 10% loading. It was concluded that between 9% and 14% efficiency gain is achievable with variable load.

17.3.2 Railway traction with dc supply

High starting torque, ability to share torque, good natural response to load variations and rapid speed control have historically made series dc motors an attractive choice for short distance traction applications [17]. For example, the share of 1.5/3 kV dc was about 37% in France in a survey conducted as recently as 2003 [18], with about 66% of railway lines running at 1.5 kV dc in Japan [19], almost 10,000 km including 1,650 dc substations with 2,700 rectifiers span the length of the country as of 2013 [20]. Similarly, most of the 20,000-km Poland's railway tracks are electrified by 3 kV dc traction supply [21]. The split between preference for short distance dc traction and long distance ac traction should be kept in mind. This section describes several dc traction projects and discusses the typical characteristics of such systems.

17.3.2.1 Architecture

A typical dc traction system with rail return is shown in Figure 17.5. Considering that catenaries of more than one conductors are usually not preferred [18], monopolar dc maybe advantageous as it simplifies the overhead catenary system for collecting traction current and reduces costs.

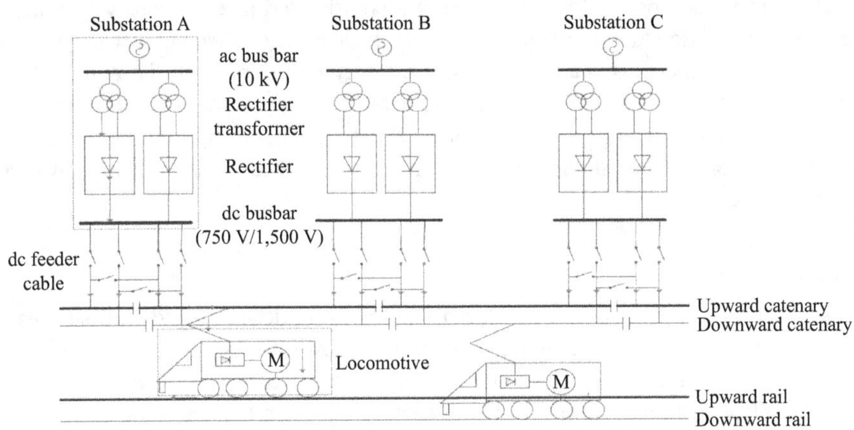

Figure 17.5 dc Traction system architecture (courtesy: [22]

With standard voltage levels of 750 V for urban metros and trams and 1.5–3 kV for interurban and regional system [17], spacing between 1 and 10 MW substations is selected between 3 and 20 km depending on the operating voltages (higher the voltage, greater the allowable spacing). The system is sectionalized using tie breakers to ensure isolation of faults to a limited zone of influence [23]. In general, the zone of influence is about 8 km for a 1.5-kV dc substation with 4 kA current [24].

17.3.2.2 Components

The key components of each dc substation are rectifiers, filters, transformers and dc switchgear. With 4 × 12-car trains nearby (2 leaving and 2 approaching), each drawing 4 kA, the demand could be as high as 25 MVA [25]. However, the nature of the load is such that it is not expected to burden the substation for a long time continuously. In order to minimize the related system costs, such substations maybe designed to operate with overload capacity for a short while, typically 150%, 300% and 450% for 1 h, 1 min and 10 s, respectively, for a 3 MW, 750 V supply [24]. The use of energy storage and tie-lines is employed in dc traction on the Yamatoji line in Japan [19]. Apart from providing ancillary services such as regenerative energy absorption, supplying power deficit and compensating catenary voltage drop, at least 3% decrease in energy from dc substation feeders was reported in the project.

17.3.2.3 Protection

Protection of 3 kV dc traction systems from lightening surges, and particularly, the vulnerability of overhead catenary conductor to lightning strike is discussed in [21,26]. For rail-return traction supply, issues associated with rail to earth potential have been cautioned in [27]. Generally, a combination of ac and dc switchgear is needed to isolate faults in traction supply systems [28]. High-speed dc circuit breakers are employed to clear the faults within 15–30 ms. An intertripping mechanism is incorporated while ensuring a coordinated discriminatory operation between ac and dc side circuit breakers [29]. Further, a reverse current tripping is incorporated in the dc switchgear to isolate faulty rectifiers from the common dc bus. In [23], it is highlighted how dynamic nature of traction loads could mimic the sharp rise in fault currents when train is accelerating. Therefore, the time setting of the *di*/*dt* tripping relay should be carefully chosen based on the traction system characteristics.

17.3.3 Discussion

Some examples of transportation electrification using dc microgrids are summarized in Table 17.2. It is observed that the trend is toward high installed capacity of operation and correspondingly medium voltage levels are chosen. The dc shipboard IPS is controlled in a decentralized way to crucially increase the reliability, while using radial or distributed topologies. Traction systems typically use hierarchical control scheme in a sectioned, monopolar with rail return topology.

Table 17.2 Demonstration sites for transportation electrification using dc

	Description	Voltage level	Capacity	Topology	Control	Protection
			Shipboard dc microgrids			
1	Dina Star (ABB)	1 kV	20 MW	Distributed, multidrive		SSCB with fuse[a]
2	Austin, USA [15]	2 kV	2–10 MW	rdl.[b] dc bus	Dctr.[c]	Series arc
3	Italy [9]	3 kV	2 MVA	rdl.[b] dc bus		SSCB[a]
			Traction systems [17–30]			
4	BE, ES, IT	3 kV	1.5–10 MW, ≈ 20 km	Sectioned, rail return		ac, dc switchgear
5	FR, NL	1.5 kV	1.5–10 MW, ≈ 10 km	Sectioned, rail return		ac, dc switchgear
6	United Kingdom	750 V	3 MW, 3 km	Sectioned, rail return		ac, dc switchgear
7	Atlanta, USA	750 V	3 MW	Sectioned, rail return		ac, dc s witchgear
8	Singapore	1.5 kV	3 MW, 5 km	Sectioned, rail return		ac, dc switchgear
9	CN	750 V, 1.5 kV	Few MW ≈ 2.4 km	Sectioned, rail return	Hrr.[d]	ac, dc switchgear
11	Yamanote, JP	1.5 kV	Few MW			
			Electric vehicles [31–33]			
12	Hawaii, USA	240 V	1–5 MW	Meshed	Dctr.[c]	
13	Italy	790 V	20 kV A	Bipolar	Hrr.[d]	

[a]Solid state circuit breaker.
[b]Radial.
[c]Decentralized.
[d]Hierarchical.

Innovator Elon Musk envisioned a synergistic, large-scale evolution of rooftop PVs, storage and EVs [34]. Signs of potential in these technologies can be observed from the launch of Tesla power wall 2.0 and the acquisition of SolarCity by Tesla [35]. The role of dc in interconnecting PV generation, EV charging and storage can be significant [36].

17.4 Datacenters

Currently, many datacenters around the globe are tending to employ dc technology due to its high reliability and efficiency. To highlight the importance of efficiency, in 2006 1.5% of total US electricity consumption was from datacenters [37]. This section provides an overview of dc datacenters at Lawrence Berkeley National Laboratory (LBNL) and then reviews various dc datacenters around the world.

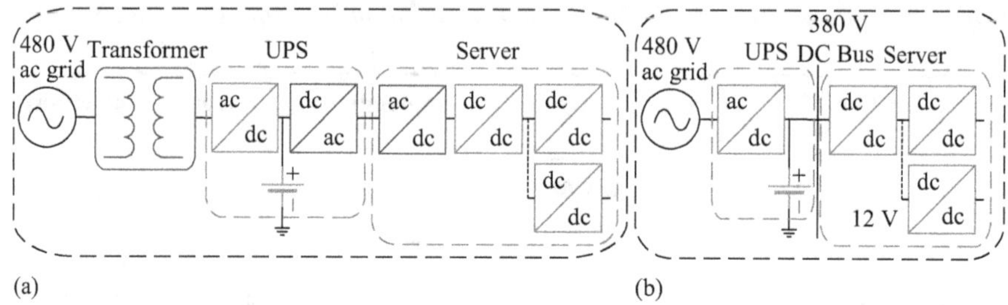

Figure 17.6 Datacenter distribution system: (a) ac based and (b) dc based [38]

17.4.1 Architecture

The schematics of ac and dc datacenters studied by LBNL are shown in Figure 17.6. The dc datacenter servers are connected to a 380 V dc bus, where from the main ac grid is routed via uninterruptible power supply (UPS). It is indicated that one dc/ac conversion step in UPS and one ac/dc conversion step in server power supply is reduced for a dc based datacenter.

17.4.2 Components

The need for standardization of various dc-ready commercially available components such as servers, connectors, power cords and in-rack distribution strips was identified in [38]. The project compared the performance of ac and dc datacenters with the same level of achievable functionality and claimed that the results demonstrated an efficiency improvement by 7%–8% with dc. Out of this, it was reported that the share of efficiency gain was 4% for UPS, 2% for transformer and 2% for server power supply. It was anticipated that with component optimization, further improvements are possible.

17.4.3 Protection and grounding system

The LBNL datacenter grounding system was designed based on the IEEE Standard 1100-2005, describing the recommended practices for powering and grounding electronic equipment [39]. According to ANSI T1.311, the highest dc voltage level is 160 V, while LBNL uses 380 V dc. Highlighting that suitable dc breakers are required for proper coordinated protection, it is deemed necessary to further investigate the over-current protection devices and disconnect devices for the dc system.

17.4.4 Discussion

Table 17.3 lists the information of some dc-based datacenters from around the world. The system level operating voltage of datacenters is standardizing at about 380 Vdc. Even the technical report of INTEL labs describing a 400 V datacenter affirms its support to the standardization around 380 V [40]. Regarding the size, there is a clear trend toward high power dc datacenters, as Intel Labs have reached

Table 17.3 Demonstration sites for data centers using dc distribution

	Description	Voltage level	Capacity	Topology	Control	Protection
1	LBNL, USA [38]	380 V	15–40 kW			
2	Intel Labs USA [40,41]	400 V	5.5 MW	Bipolar		dc-Rated IEC 309 Connectors
3	Nedo, Japan [42]	380 V	500 kW		Adaptive	
4	San Diego, USA [37]	380 V	2.8 MW	Bipolar		
5	Switzerland [37]	380 V	1 MW	Bipolar		
6	Rokkasho, Japan [43]	380 V	13.6 kW	Bipolar		

up to 5.5 MW. The key challenges highlighted by the LBNL study include (a) economics of scale may shift in favor of dc if knowledge on its benefits are properly disseminated. (b) Industry leadership in moving toward development of standardized practices is desirable for widespread market adoption on dc-based datacenters. Demonstration sites for datacenters using dc distribution

17.5 Residential and commercial dc buildings

The importance of dc microgrids in realizing a flexible, smart and efficient energy network, termed "Enernet" is envisioned in [44]. As shown in Figure 17.7, the need for large number of ac/dc conversions can be eliminated resulting in lower costs and higher efficiency in dc buildings.

17.5.1 dc House

17.5.1.1 Architecture

Many household appliances such as televisions, refrigerators, computers, LED lighting, EVs as well as sources like solar PV and storage elements can be interconnected to form a dc nanogrid. A typical grid connected dc residential house is shown in Figure 17.8.

17.5.1.2 Components

An example of a demo house from EPARC, Taiwan [45], is shown in Figure 17.9. The focus of this demonstration was to adapt commonly used ac residential appliances indicated in the figure to operate with dc input. Various components were connected radially to a common dc bus of 380 $\pm$20 V with about 5–10 kW capacity.

The products were converted from ac source with power factor correction by rerouting the protection and circuitry. It was highlighted that several appliance manufacturers such SAMPO, TECO, Eulife, Jamicon and Fego supported this modification of their products. It is encouraged to mark such products as "dc-ready" to reinforce end-user confidence in adopting dc technologies [46]. Operational data collected showed a efficiency enhancement of 8.5% compared to ac.

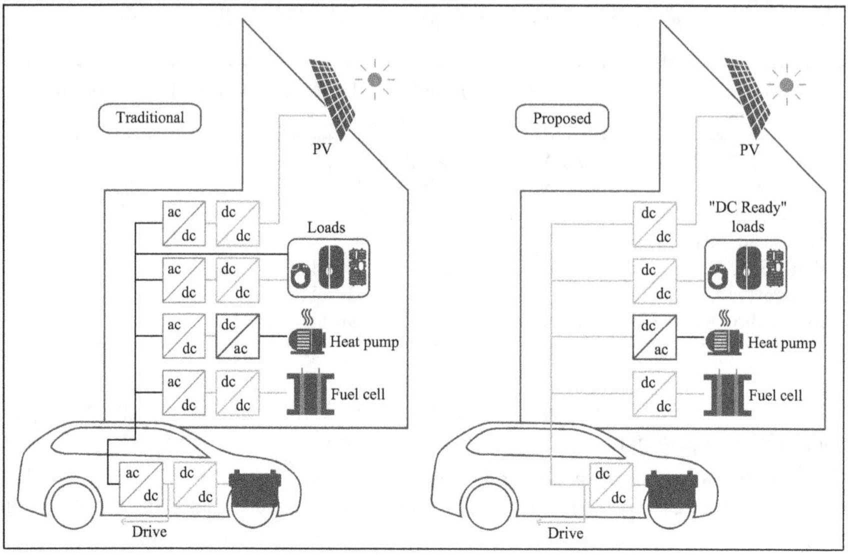

Figure 17.7 dc Buildings compared to traditional ones

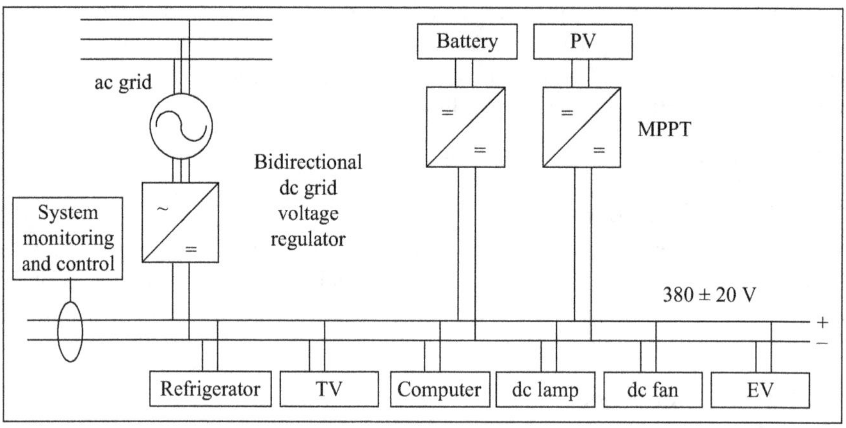

Figure 17.8 Grid connected residential dc microgrid (based on [45])

17.5.1.3 Control

Heuristic controls try to minimize the energy exchange with the main grid using storage elements while maximizing the use of distributed green energy resources in the system [45]. The grid connected bidirectional inverter performs the energy balance and voltage regulation of the dc bus. The role of this inverter can be diversified, for example, to draw or store energy in the flywheel or operating it as a high power rating charger/discharger for dc UPS [45].

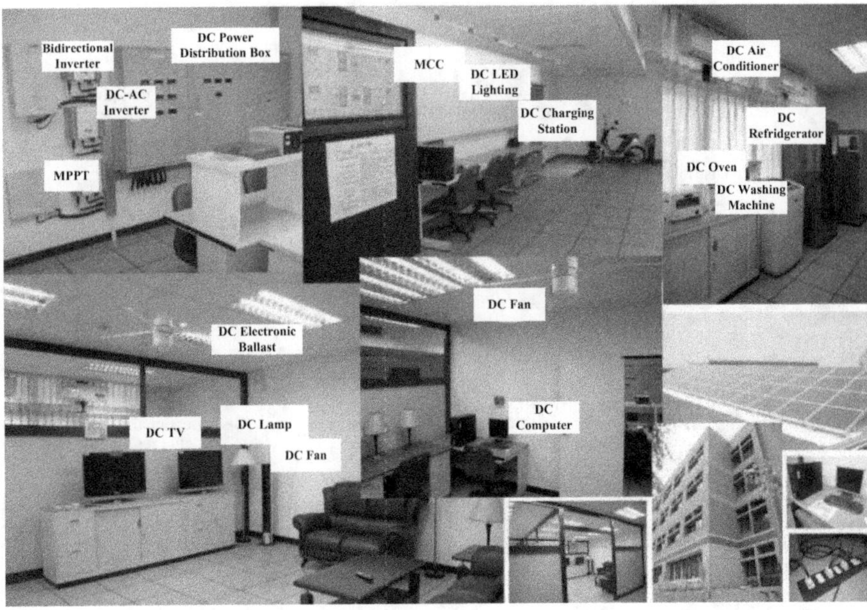

Figure 17.9 Grid connected residential microgrid (courtesy: [45])

17.5.1.4 Protection

It was reported that a special dc plug was developed by Fujitsu to eliminate arcing. Issues with series arcing are more relevant in residential dc nanogrids due to a relatively high operating voltage where sustained arcs can be expected [5]. Considering the wear and tear issues as well as added costs of such components, a novel selective series arc extinction scheme is developed in [47] with experimental proof of concept. Challenges such as localized short circuit protection and standardization of safety codes pertaining to different system components are important considerations.

17.5.2 Building interiors

17.5.2.1 Architecture

The layout of the dc office building at the Fraunhofer Institute in Erlangen, Germany is shown in Figure 17.10 [48]. This system has two bipolar dc buses. The main bus is controlled to 380±20 V and interconnects various office loads, the ac grid and the PV electrical source in a radial structure at 15 kW capacity. An alternate PV to grid bus maintains a higher dc voltage at 800 V. The chosen voltage levels are consistent with the ETSI EN 300 132-3-1 standard [49] for dc voltage range 260–400 V, existing ac infrastructure and semiconductor technology at adequate efficiency.

17.5.2.2 Components

The front-end ac/dc conversion step in individual devices can be removed, comprising filter, rectifier, power factor correction and link capacitor. A central rectifier

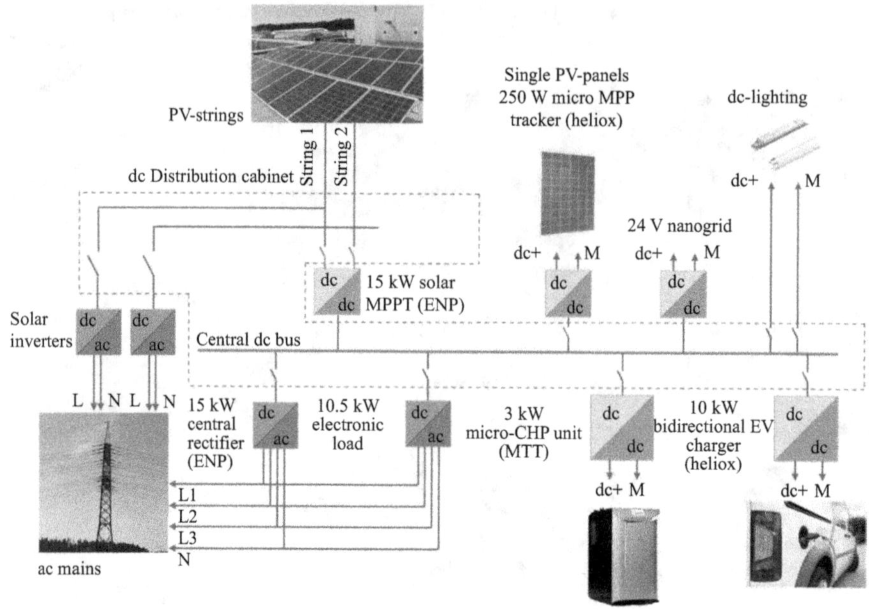

Figure 17.10 Layout of dc office building demonstrator (courtesy: [48])

with common dc bus shared among dc-ready devices can offer size, cost and energy savings [48]. This design concept can further benefit from emerging fast switching power electronic devices based on materials such as silicon carbide (SiC) and gallium nitride (GaN). The microgrid components illustrated in Figure 17.10 were developed by partners Emerson Network Power, Philips, Heliox, MTT and/or were retrofitted for 380 V, dc operation, from commercially available ac devices [48].

17.5.2.3 Control

Central energy management system (EMS) can optimize the operation of dc office building. The central rectifier, for example in [48], can regulate the output dc bus voltage. Therein, a star point programmable logic controller is employed to monitor the system parameters and govern the power flow in each subsystem. Information such as weather forecast and user behavior can be utilize by the EMS to fulfill higher level control objectives such as building CO_2 reduction. Specifically in [50], it was shown that about 10% reduction in CO_2 emissions is achieved just by converting the system from ac to dc, while above 16% are achieved if dc EMS is used. Heuristic control was used to minimize grid energy exchange, maximize the available green energy resources and optimize the use of energy storage.

17.5.2.4 Protection

In [48], dc switch technology from ABB is employed to interrupt fault currents. Fast responding power electronic based solutions may also be considered. TN-S grounding scheme is used in the test bed, consistent with the ETSI EN 301

605 standard for grounding systems in 400 V dc datacenters [51]. Special attention must be paid to arc detection, for instance, based on the methods presented in [47,52].

17.5.2.5 Economics

The study at Xiamen University showed that the installation cost of 150 kW dc microgrid to power a building is about $2.2/W [53], having a payback of 9 years and 5.5 years with and without incentives, respectively.

17.5.3 dc Distribution grids

The impediment to the grid level realization of dc distribution is the market inertia of ac system and the lack of standardization. This subsection looks into projects specifically aiming to make a bottom-up approach to transition from ac to dc [54].

17.5.3.1 Architecture

Three nanogrids, each with 2–5 kW capacity and 51.2 V internal dc bus, were interconnected by 380 V external dc bus as shown in Figure 17.11 [54]. The open energy system (OES) was a multilayered, recursively scalable, bottom-up approach as shown in Figure 17.12 and was realized for a cluster of 19 dc nanogrids [55].

17.5.3.2 Components

A full scale prototype of the idea was realized at the Okinawa Institute of Science and Technology, Japan as shown in Figure 17.13.

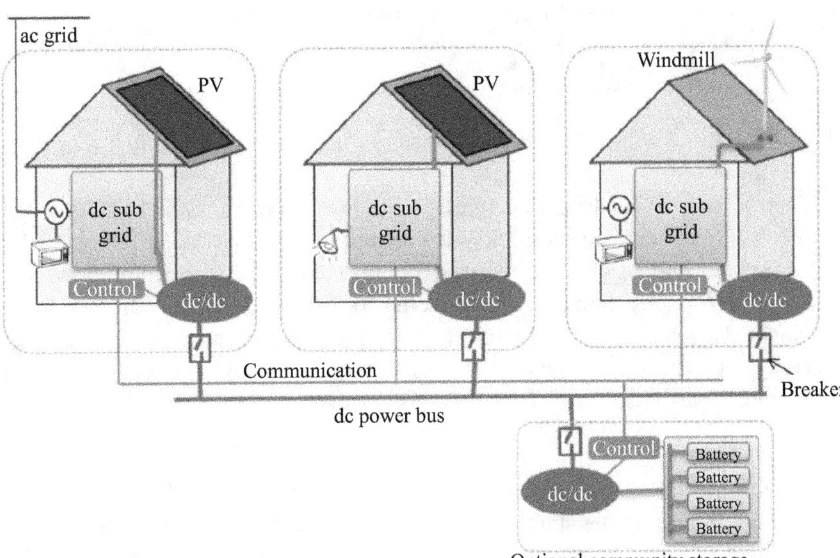

Figure 17.11 Schematic of the open energy architecture (courtesy: [54])

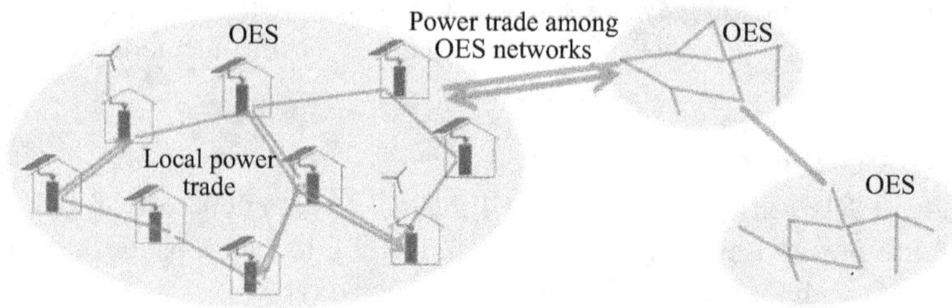

Figure 17.12 Scalability of the OES concept (courtesy: [54])

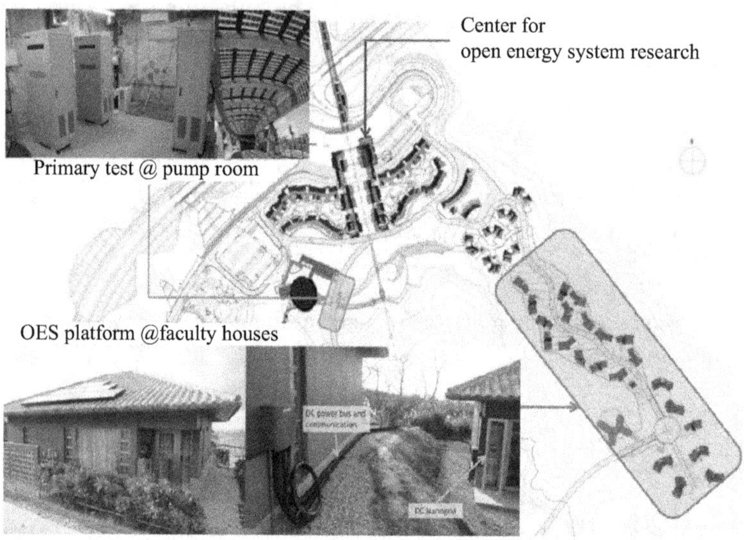

Figure 17.13 Full scale demo prototype of the OES concept (courtesy: [54])

Each nanogrid was based on the labscale prototype, consisting of 3.6 kW h, 45–57.6 V lithium-ion batteries, 2 kW PV charger, 2.5 kW ac/dc converter, 3.5 kW dc/ac converter and an internal controller, connected to the external dc power bus using a 2.5-kW bidirectional dc/dc converter. dc Breakers of 10 A are employed. Additional PV panels of 5, 2.5 and 2.5 kWp are installed at the entrance of the university. Power exchange was remotely controlled and visualized, showcasing the technical feasibility and safety of the OES-based setup.

17.5.3.3 Control

The peer-to-peer (P2P) multiagent control for OES dc microgrids is described in [56]. Highlighting that the standard hierarchical control in IEEE 1547 was limited to large generation, top-down approach, it was argued that a grid independent fully decentralized design was necessary for resilient, scalable and self-sufficient operation. It was anticipated that the role of Information Communication

Technologies for active, multilayered and collaborative P2P control in OES would be significant.

Ensuring that the failure of any single entity would not result in complete loss of network services, the 19 autonomous dc nanogrids at Okinawa, Japan were operated as community-wide, interconnected P2P controlled microgrid [56]. It was shown that the self-sufficiency ratio, representing minimization of external non-renewable energy, needs improved by more than 4% when decentralized P2P dc exchange was allowed as compared to a stand-alone nanogrid. The solar operation ratio, signifying reduction in maximizing solar generation usage and minimizing curtailment, improved by almost 10% with dc interconnections as compared to stand-alone mode. A centralized control approach provided a further improvement of 4% and 2%, respectively. While it was claimed that an improvement in resilience could not be shown experimentally, its conceptual discussion was presented [56].

17.5.4 Discussion

A selection of demonstration projects of residential and commercial buildings, transitioning toward campus dc and universal dc grids is presented in Table 17.4. Demonstration sites for dc microgrids in residential, commercial buildings and campus applications

The dc operating voltage was usually selected at 380 V with capacity as high as 150 kW. Most small-scale projects used radial topology with centralized control. A scalable, multilayered approach with multiagent control was suggested in [54]. There is potential for a universal dc distribution with meshed, multilayered and bipolar topology employing distributed control. One important issue discussed in many initiatives was the protection aspect. Specifically, the localization and extinction of arcing was considered vital. Impact and reliability of dc switch based protection is another important domain. Perhaps, a clear demonstration of the interplay between reliability and availability, supported by a selective and reconfigurable protection system, will strengthen the acceptability of a meshed universal dc grid.

Table 17.4 Demonstration sites for dc microgrids in residential, commercial buildings and campus applications

	Description	Voltage level (V)	Capacity (kW)	Topology	Control	Protection
1	Taiwan [45]	380±20	5–10	Monopolar, radial	Centralized, Heuristic	Special plug
2	Japan [57]	420	20	Hybrid ac–dc		
3	Bosch [58]	380	15	Radial	Centralized	
4	Japan [59]	380±20	15–100		Centralized	
5	Germany [48]	380±20 ±380	15	Bipolar	Centralized	dc Switches
6	China [53]	380±20	150	Radial	Centralized	
7	Japan [54]	350–380	2–5 per house	Multilayered	Multiagent	

References

[1] Jhunjhunwala A, Lolla A, Kaur P. Solar-dc microgrid for Indian homes: a transforming power scenario. IEEE Electrification Magazine. 2016 Jun;4(2):10–19.

[2] Kaur P, Jain S, Jhunjhunwala A. Solar-DC deployment experience in off-grid and near off-grid homes: Economics, technology and policy analysis. In: IEEE First International Conference on DC Microgrids; 2015. p. 26–31.

[3] Chakrabarty S, Islam T. Financial viability and eco-efficiency of the solar home systems (SHS) in Bangladesh. Energy. 2011; 36(8): 4821–4827. {PRES} 2010.

[4] Mondal MAH. Economic viability of solar home systems: case study of Bangladesh. Renewable Energy. 2010;35(6):1125–1129.

[5] Liu Z, Shekhar A, Ramírez-Elizondo L, *et al.* Characterization of series arcs in LVdc microgrids. In: IEEE Second International Conference on DC Microgrids (ICDCM); 2017.

[6] IRENA. Solar PV in Africa: Costs and Markets. International Renewable Energy Agency (IRENA). Available at: https://www.irena.org/-/media/Files/IRENA/Agency/Publication/2016/IRENA_Solar_PV_Costs_Africa_2016.pdf

[7] Shenai K, Jhunjhunwala A, Kaur P. Electrifying India: using solar dc microgrids. IEEE Power Electronics Magazine. 2016 Dec;3(4):42–48.

[8] Paleta R, Pina A, Santos Silva CA. Polygeneration energy container: designing and testing energy services for remote developing communities. IEEE Transactions on Sustainable Energy. 2014 Oct;5(4):1348–1355.

[9] Sulligoi G, Tessarolo A, Benucci V, *et al.* Shipboard power generation: design and development of a medium-voltage dc generation system. IEEE Industry Applications Magazine. 2013 Jul;19(4):47–55.

[10] Jin Z, Sulligoi G, Cuzner R, *et al.* Design and control of hybrid power and propulsion systems for smart ships: a review of developments. Applied Energy. 2017;194:30–54.

[11] Shekhar A, Ramírez-Elizondo L, Bauer P. Next-generation shipboard DC power system: introduction smart grid and dc microgrid technologies into maritime electrical networks. IEEE Electrification Magazine. 2016 Jun; 4(2):45–57.

[12] IEEE Recommended Practice for 1 kV to 35 kV Medium-Voltage DC Power Systems on Ships. IEEE Std 1709-2010. 2010 Nov; p. 1–54. doi: 10.1109/IEEESTD.2010.5623440.

[13] Shekhar A, Ramírez-Elizondo L, Bauer P, *et al.* DC microgrid islands on ships. In: IEEE Second International Conference on DC Microgrids (ICDCM); 2017.

[14] Herbst JD, Gattozzi AL, Ouroua A, *et al.* Flexible test bed for MVDC and HFAC electric ship power system architectures for Navy ships. In: 2011 IEEE Electric Ship Technologies Symposium; 2011. p. 66–71.

[15] Chryssostomidis C, Cooke CM. Space reservation for shipboard electric power distribution systems. In: 2015 IEEE Electric Ship Technologies Symposium (ESTS); 2015. p. 187–192.

[16] Hansen JF, Lindtjorn JO, Myklebust TA, *et al.* Onboard dc grid for enhanced DP operation in ships. In: Marine Technology Society Dynamic Positioning Conference; 2011.

[17] Hill RJ. Electric railway traction. I. Electric traction and DC traction motor drives. Power Engineering Journal. 1994 Feb;8(1):47–56.

[18] Steimel A. Under Europe's incompatible catenary voltages a review of multi-system traction technology. In: 2012 Electrical Systems for Aircraft, Railway and Ship Propulsion; 2012. p. 1–8.

[19] Suzuki T, Hayashiya H, Yamanoi T, *et al.* Application examples of energy saving measures in Japanese DC feeding system. In: 2014 International Power Electronics Conference (IPEC-Hiroshima 2014 – ECCE ASIA); 2014. p. 1062–1067.

[20] Uzuka T. Faster than a speeding bullet: an overview of Japanese high-speed rail technology and electrification. IEEE Electrification Magazine. 2013 Sep;1(1):11–20.

[21] Zielenkiewicz M, Maksimowicz T, Burak-Romanowski R. Coordination of surge arresters in DC 3 kV railway traction system-field tests. In: 2016 33rd International Conference on Lightning Protection (ICLP); 2016. p. 1–10.

[22] Song XM, He JH, Yip T, *et al.* Research into fault location methods for DC railway traction system. In: 12th IET International Conference on Developments in Power System Protection (DPSP 2014); 2014. p. 1–6.

[23] Thong M, Cheong A, Wijaya H. Overview of DC traction protection scheme for Singapore rapid transit system. In: 2005 International Power Engineering Conference; 2005. p. 1–587.

[24] Van Eck RA. The separately excited DC traction motor applied to DC and single-phase AC rapid transit systems and electrified railroads, Part II—Application. IEEE Transactions on Industry and General Applications. 1971 Sep;IGA-7(5):650–657.

[25] Sane S, Sharma S, Prasad SK. Harmonic analysis for AC and DC supply in traction sub station of Mumbai. In: IEEE International Conference on Electrical, Computer and Communication Technologies; 2015. p. 1–5.

[26] Hermoso B, Montanya J, Garnacho F, *et al.* Catalonia (Spain) railway traction DC system project to limit lightning transient overvoltage. In: 2010 30th International Conference on Lightning Protection (ICLP); 2010. p. 1–5.

[27] Ku BY, Hsu T. Computation and validation of rail-to-earth potential for diode-grounded DC traction system at Taipei Rapid Transit System. In: ASME/IEEE Joint Rail Conference. Proceedings of the; 2004. p. 41–46.

[28] Nikouee S, Ledbetter T. DC traction power supply and distribution system for MARTA's Armour Yard Rail Services Facility. In: Proceedings of the 2006 IEEE/ASME Joint Rail Conference; 2006. p. 241–252.

[29] Goh EJ, Chu KN, Ng NK. 1500 V DC traction system for the North East Line. In: 2004 International Conference on Power System Technology, 2004. PowerCon 2004. vol. 2; 2004. p. 1904–1909.

[30] Kumagai K, Fujita T, Nakahira M, *et al.* Study on train operation energy between commuter train and traction substations in a Japanese urban Railway. In: IEEE Electrical Power and Energy Conference (EPEC); 2015. p. 50–55.

[31] Black, Veatch. Analysis of Smart Grid Technologies. Hawai'i Natural Energy Institute; 2013.

[32] Roose LR. Hawaii Smart Technology Demonstration for a 100% RE Future [Presentation]. www.nedo.go.jp: Smart Community Summit 2016; Jun 2016 [cited July, 2017].

[33] Capasso C, Veneri O. Experimental study of a DC charging station for full electric and plug in hybrid vehicles. Applied Energy. 2015;152:131 – 142.

[34] Powerwall 2 and Solar Roof Launch [Video]. www.tesla.com; Oct 2016 [cited July, 2017].

[35] Tesla and SolarCity [Web News]. www.tesla.com; Nov 2016 [cited July, 2017].

[36] Exploring Future Power Management [Technical News]. www.delta.tudelft.nl; Apr 2016 [cited July, 2017].

[37] AlLee G, Tschudi W. Edison Redux: 380 Vdc brings reliability and efficiency to sustainable data centers. IEEE Power and Energy Magazine. 2012 Nov;10(6):50–59.

[38] Ton M, Fortenbery B, Tschudi W. DC Power for Improved Data Center Efficiency. Berkeley, CA: Lawrence Berkeley National Laboratory; 2008

[39] IEEE Recommended Practice for Powering and Grounding Electronic Equipment - Redline. IEEE Std 1100-2005 (Revision of IEEE Std 1100-1999) – Redline. 2006 May; p. 1–703.

[40] Aldridge T, Pratt A, Kumar P, *et al.* Evaluating 400V direct-current for data centers: a case study comparing 400 Vdc with 480–208 Vac power distribution for energy efficiency and other benefits. Intel Labs.

[41] Pratt A, Kumar P, Aldridge TV. Evaluation of 400V DC distribution in telco and data centers to improve energy efficiency. In: INTELEC 07 – 29th International Telecommunications Energy Conference; 2007. p. 32–39.

[42] Yajima H, Babasaki T, Usui K, *et al.* Energy-saving and efficient use of renewable energy by introducing an 380 VDC power-supply system in data centers. In: IEEE International Telecommunications Energy Conference; 2015. p. 1–4.

[43] Kaga M, Noritake M, Hirose K, *et al.* Verification of container data center using 380 V dc power distribution system. In: 2012 International Conference on Renewable Energy Research and Applications (ICRERA); 2012. p. 1–5.

[44] Patterson BT. DC, come home: DC microgrids and the birth of the "Enernet". IEEE Power and Energy Magazine. 2012 Nov;10(6):60–69.

[45] Wu TF, Chen YK, Yu GR, *et al.* Design and development of dc-distributed system with grid connection for residential applications. In: 8th International Conference on Power Electronics – ECCE Asia; 2011. p. 235–241.

[46] Mackay L, van der Blij NH, Ramirez-Elizondo L, *et al.* Toward the universal DC distribution system. Electric Power Components and Systems. 2017;45 (10):1032–1042.
[47] Shekhar A, Ramirez-Elizondo L, Bandyopadhyay S, *et al.* Detection of series arcs using load side voltage drop for protection of low voltage DC systems. IEEE Transactions on Smart Grid. doi: 10.1109/TSG.2017.2707438.
[48] Wunder B, Ott L, Szpek M, *et al.* Energy efficient DC-grids for commercial buildings. In: 2014 IEEE 36th International Telecommunications Energy Conference (INTELEC); 2014. p. 1–8.
[49] Environmental Engineering (EE); Power supply interface at the input to telecommunications and datacom (ICT) equipment; Part 3: Operated by rectified current source, alternating current source or direct current source up to 400 V; [ETSI EN 300 132-3-1]. European Standards; 2012.
[50] Noritake M, Yuasa K, Takeda T, *et al.* The demonstration of the CO2 reduction effect of 400V class DC micro grid for offices. In: 2015 IEEE International Telecommunications Energy Conference (INTELEC); 2015. p. 1–6.
[51] Environmental Engineering (EE); Earthing and bonding of 400 VDC data and telecom (ICT) equipment [ETSI EN 301 605]. European Standards; 2013.
[52] Strobl C. Arc fault detection in DC microgrids. In: 2015 IEEE First International Conference on DC Microgrids (ICDCM); 2015. p. 181–186.
[53] Zhang F, Meng C, Yang Y, *et al.* Advantages and challenges of DC microgrid for commercial building a case study from Xiamen university DC microgrid. In: IEEE First International Conference on DC Microgrids (ICDCM); 2015. p. 355–358.
[54] Werth A, Kitamura N, Tanaka K. Conceptual study for open energy systems: distributed energy network using interconnected DC nanogrids. IEEE Transactions on Smart Grid. 2015 Jul;6(4):1621–1630.
[55] Werth A, Tokoro M, Tanaka K. Bottom-up and recursive interconnection for multi-layer DC microgrids. In: 2016 IEEE 16th International Conference on Environment and Electrical Engineering (EEEIC); 2016. p. 1–6.
[56] Werth A, André A, Kawamoto A, *et al.* Peer-to-peer control system for DC microgrids. IEEE Transactions on Smart Grid. 2018 Jul;9(4):3667–3675.
[57] Hirose K, Reilly JT, Irie H. The Sendai microgrid operational experience in the aftermath of the Tohoku earthquake: a case study. New Energy and Industrial Technology Development Organization (NEDO); 2013.
[58] Fregosi D, Ravula S, Brhlik D, *et al.* A comparative study of DC and AC microgrids in commercial buildings across different climates and operating profiles. In: IEEE First International Conference on DC Microgrids; 2015. p. 159–164.
[59] Noritake M, Yuasa K, Takeda T, *et al.* The demonstration of the CO2 reduction effect of 400V class DC micro grid for offices. In: 2015 IEEE International Telecommunications Energy Conference (INTELEC); 2015. p. 1–6.

Index